ETHNOBOTANY OF INDIA

Volume 5

The Indo-Gangetic Region and Central India

ETHNOBOTANY OF INDIA

Volume 5

The Indo-Gangetic Region and Central India

Edited by
T. Pullaiah, PhD
K. V. Krishnamurthy, PhD
Bir Bahadur, PhD

AAP APPLE ACADEMIC PRESS

Apple Academic Press Inc.
3333 Mistwell Crescent
Oakville, ON L6L 0A2 Canada

Apple Academic Press Inc.
9 Spinnaker Way
Waretown, NJ 08758 USA

© 2018 by Apple Academic Press, Inc.
First issued in paperback 2021
Exclusive worldwide distribution by CRC Press, a member of Taylor & Francis Group
No claim to original U.S. Government works
Need to add somewhere, preferably on the copyright page.

Cover photo 1 by Wouter Hagens. Used with permission via public domain. https://commons.wikimedia.org/wiki/
File:Houttuynia_cordata_A.jpg
Cover photo 2 by Karsten Heinrich. Used with permission via the Creative Commons Attribution license. https://commons.
wikimedia.org/wiki/Category:Mahonia_napaulensis#/media/File:Mahonia_napaulensis_Nepal.JPG
Cover photo 3 by Kristian Peters. Used with permission under the terms of the GNU Free Documentation License. https://
commons.wikimedia.org/wiki/Mentha_arvensis#/media/File:Mentha_arvensis.jpeg
Cover photo 4 by JeremiahsCPs. Used with permission by via public domain. https://commons.wikimedia.org/wiki/
File:Nepenthes_khasiana.jpg

ISBN-13: 978-1-77463-123-2 (pbk)
ISBN-13: 978-1-77188-599-7 (hbk)

Ethnobotany of India, 5-volume set
ISBN-13: 978-1-77188-600-0 (hbk)

Library and Archives Canada Cataloguing in Publication

Ethnobotany of India / edited by T. Pullaiah, PhD, K. V. Krishnamurthy, PhD, Bir Bahadur, PhD.
Includes bibliographical references and indexes.
Contents: Volume 4. Western and central Himalayas -- Volume 5. The Indo-Gangetic Region and Central India.
Issued in print and electronic formats.
ISBN 978-1-77188-550-8 (v. 4 : hardcover).--ISBN 978-1-77188-599-7 (v. 5 : hardcover).-- ISBN 978-1-315-20739-1 (v. 4 : PDF).--ISBN 978-1-315-18784-6 (v. 5 : PDF)
1. Ethnobotany--India. I. Pullaiah, T author, editor II. Bahadur, Bir, author, editor III. Krishnamurthy, K. V., author, editor
GN635.I4E85 2016   581.6'30954   C2016-902513-6   C2016-902514-4

Library of Congress Cataloging-in-Publication Data

Names: Pullaiah, T., editor. | Krishnamurthy, K. V., editor. | Bahadur, Bir., editor.
Title: Ethnobotany of India. Volume 5, The Indo-Gangetic Region and Central India / editors: T. Pullaiah, K.V. Krishnamurthy, Bir Bahadur.
Other titles: Indo-Gangetic Region and Central India
Description: Waretown, NJ : Apple Academic Press, 2017. | Includes bibliographical references and index.
Identifiers: LCCN 2017022227 (print) | LCCN 2017022998 (ebook) | ISBN 9781315187846 (ebook) | ISBN 9781771885997 (hardcover : alk. paper)
Subjects: LCSH: Ethnobotany--India.
Classification: LCC GN476.73 (ebook) | LCC GN476.73 .E824 2017 (print) | DDC 581.6/30954--dc23
LC record available at https://lccn.loc.gov/2017022227

Apple Academic Press also publishes its books in a variety of electronic formats. Some content that appears in print may not be available in electronic format. For information about Apple Academic Press products, visit our website at **www.appleacademic-press.com** and the CRC Press website at **www.crcpress.com**

CONTENTS

LIST OF CONTRIBUTORS

S. John Adams
R&D – Phytochemistry and Pharmacognosy, Sami labs Ltd, Peenya Industrial Area, Bangalore, India,
E-mail: s.johnadams13@gmail.com

Bir Bahadur
Department of Botany, Kakatiya University, Warangal–506009, India,
E-mail: birbahadur5april@gmail.com

S. Noorunnisa Begum
Centre of Repository of Medicinal Resources, School of Conservation of Natural Resources,
Foundation for Revitalization of Local Healthand Traditions, 74/2, Jarakabande Kaval, Attur P.O.,
Via Yelahanka, Bangalore – 560106, India, E-mail: noorunnisa.begum@frlht.org

E. Chamundeswari
Department of Botany, Kakatiya University, Warangal – 560009, Telangana, India

Baljot Kaur
Stri Roga & Prasuti Tantra (Gyne & Obs), SKSS Ayurvedic Medical College, Sarabha, Ludhiana,
Punjab, India, E-mail: bhbharaj@gmail.com

K. V. Krishnamurthy
R&D – Phytochemistry and Pharmacognosy, Sami Labs Ltd, Peenya Industrial Area,
Bangalore–560058, India, E-mail: kvkbdu@yahoo.co.in

K. Ravi Kumar
Centre of Repository of Medicinal Resources, School of Conservation of Natural Resources,
Foundation for Revitalization of Local Healthand Traditions, 74/2, Jarakabande Kaval, Attur P.O.,
Via Yelahanka, Bangalore – 560106, India, E-mail: k.ravikumar@frlht.org

Suman K. Mandal
Department of Botany, Visva-Bharati University, Santiniketan–731235, West Bengal

R. Ratna Manjula
Department of Botany, Andhra University, Visakhapatnam – 530003, India

K. Sri Rama Murthy
R&D Center for Conservation Biology and Plant Biotechnology, Shivashakti Biotechnologies Limited,
S. R. Nagar, Hyderabad – 500038, Telangana, India, E-mail: drmurthy@gmail.com

Gorti Bala Pratyusha
Department of Genetics, Shadan P. G. Institute of Biosciences for Women, Osmania University,
Hyderabad – 500004, Telangana, India, E-mail: bala.pratyusha@yahoo.com

T. Pullaiah
Department of Botany, Sri Krishnadevaraya University, Anantapur–515003, Andhra Pradesh, India,
E-mail: pullaiah.thammineni@gmail.com

Chowdhury Habibur Rahaman
Department of Botany, Visva-Bharati University, Santiniketan – 731235, West Bengal, India,
E-mail: habibur_cr@rediffmail.com or habibur_cr@yahoo.co.in

Maddi Ramaiah
Department of Pharmacognosy, Hindu College of Pharmacy, Guntur – 522002, A.P., India,
E-mail: rampharma83@gmail.com

J. Koteswara Rao
Department of Botany, Andhra University, Visakhapatnam – 530003, India

Sudip Ray
Department of Botany, PMB Gujarati Science College, Indore – 452001, Madhya Pradesh, India,
E-mail: sudbot@yahoo.com, raysudip8@gmail.com

T. V. V. Seetharami Reddi
Department of Botany, Andhra University, Visakhapatnam – 530003, India,
E-mail: reddytvvs@rediffmail.com

Manickam Tamil Selvi
Value Added Corporate Services Pvt. Ltd, Chennai – 600090, Tamil Nadu, India

R. L. S. Sikarwar
Arogyadham (J.R.D. Tata Foundation for Research in Ayurveda and Yoga Sciences),
Deendayal Research Institute, Chitrakoot, Dist. Satna (M.P.) – 485334, India,
E-mail: rlssikarwar@rediffmail.com, sikarwarrls@gmail.com

Ankanagari Srinivas
Department of Genetics, Osmania University, Hyderabad – 500007, Telangana, India,
E-mail: srinivasmessage@gmail.com

J. Suneetha
Department of Botany, Andhra University, Visakhapatnam – 530003, India

D. K. Ved
Centre of Repository of Medicinal Resources, School of Conservation of Natural Resources,
Foundation for Revitalization of Local Healthand Traditions, 74/2, Jarakabande Kaval, Attur P.O.,
Via Yelahanka, Bangalore – 560106, India, E-mail: dk.ved@frlht.org

Vijay V. Wagh
Plant Diversity, Systematics and Herbarium Division, CSIR – National Botanical Research Institute,
Rana Pratap Marg, Lucknow – 226001, Uttar Pradesh, India, E-mail: vijaywagh65@gmail.com

LIST OF ABBREVIATIONS

AAAS	American Association for the Advancement of Science
AICRPE	All India Co-ordinated Research Project on Ethnobiology
AIT	Automated Identification Technology
ASU	Ayurveda, Siddha, or Unani
AZ	Azadirachtin
CBD	Convention on Biological Diversity
CDER	Centre for Drug Evaluation and Research
CDRI	Central Drug Research Institute
CDSCO	Central Drugs Standards Control Organization
CER	comparative effectiveness research
CIMAP	Central Institute of Medicinal and Aromatic Plants
CPI	conservation priority index
CSIR	Council of Scientific and Industrial Research
DL	Digital Library
DNTI	drugs from nature targeting inflammation
EBKDL	Ethnobotanical Knowledge Digital Library
EMA	European Medicine Agency
EOLLS	Encyclopedia of Life Support Systems
ESTs	expressed sequence tags
Fic	informant consensus factor
FL	fidelity level
FSH	follicle stimulating hormone
GAPs	good agricultural practices
GC	gas chromatography
GCP	good clinical practice
GLP	good laboratory practices
GMP	good manufacturing practice
HTS	high throughput screening
IPR	Intellectual Property Rights
ISMs	Indian Systems of Medicine
IUCN	International Union for Conservation of Nature
LC	liquid chromatography
LCPI	local conservation priority index

LH	luteinizing hormone
MS	mass spectrometry
NBA	National Biodiversity Authority
NBRI	National Botanical Research Institute
NCL	National Chemical Laboratory
NGS	next generation sequencing
NMR	nuclear magnetic resonance
NPM	non-pesticidal management
PIC	prior informed consent
PTGs	particularly vulnerable tribal groups
RCT	randomized clinical trials
RFC	relative frequency of citation
RRL	Regional Research Laboratories
SK	scientific knowledge
ST	scheduled tribals
TGA	Therapeutic Goods Administration
TK	traditional knowledge
TKDL	Traditional Knowledge Digital Library
TKS	traditional knowledge systems
UTs	Union Territories
UV	use value
WHO	World Health Organization

PREFACE

Humans are dependent on plants for their food, medicines, clothes, fuel and several other needs. Although the bond between plants and humans is very intense in several 'primitive' cultures throughout the world, one should not come to the sudden and wrong conclusion that post-industrial modern societies have broken this intimate bond and interrelationship between plants and people. Rather than plants being dominant as in the 'primitive' societies, man has become more and more dominant over plants after the industrial revolution, leading to over-exploitation of plants, and resulting in a maladapted ecological relationship between the two. Hence a study of the relationships between plants and people—ethnobotany—and, thus, between plant sciences and social sciences, is central to correctly place humanity in the earth's environment. Because ethnobotany rightly bridges both of these perspectives, it is always held as a synthetic scientific discipline that bridges science and humanity.

Most people tend to think that ethnobotany, a word introduced by Harshberger in 1896, is a study of plants used by 'primitive' cultures in 'exotic' locations of the world, far removed from the mainstream. People also think wrongly that ethnobotany deals only with non-industrialized, non-urbanized and 'non-cultured' societies of the world. Ethnobotany, in fact, studies plant-human interrelationships among all peoples. However, since indigenous non-Westernized societies form the vast majority of people now as well as in the past a study of their interrelationships with people becomes important. Tens of thousands of human cultures have existed in the past and a number of them persist even today. They contain the knowledge system and wisdom about the adaptations with nature, particularly with plants, for their successful sustenance. Thus, ethnobotanical information is vital for the successful continuance of human life on this planet.

Ethnobotany is of instant use in two very important respects: (i) providing vital ecological knowledge, and (ii) acting as a source for economically useful plants. The first will help us to find solutions to the increasing environmental degradation and the consequent threat to our biodiversity. In indigenous societies, biodiversity is related to cultural diversity and, hence, any threat to biodiversity would lead to erosion of cultural diversity. Indigenous cultures are not only repositories of past experiences and knowledge but also form the frameworks for future adaptations. Ethnic knowledge on economically useful plants has resulted in detailed studies on bioprospection for newer sources of food, nutraceuticals, medicines and other novel materials of human use. Bioprospecting has resulted in intense research on reverse pharmacology and pharmacognosy. This has given rise to attendant problems relating to intellectual property rights, patenting and the sharing of the benefits with the traditional societies who owned the knowledge. This has also resulted in efforts to seriously document all types of traditional knowledge of the different cultures of the world and to formalize the methods and terms of sharing this traditional knowledge. It has also made us to know not only *what* plants people in different cultures use and *how* they use them, but also *why* they use them. In addition it helps us to know the biological, sociological and cultural roles of plants important in human adaptations to particular environmental conditions that prevailed in the past, and may prevail in future.

This series of the five edited volumes on ethnobotany of different regions of India tries to bring together all the available ethnobotanical knowledge in one place. India is one of the most important regions of the Old World which has some of the very ancient and culturally rich diverse knowledge systems in the world. Competent authors have been selected to summarize information on the various aspects of ethnobotany of India, such as ethnoecology, traditional agriculture, cognitive ethnobotany, material sources, traditional pharmacognosy, ethnoconservation strategies, bioprospection of ethnodirected knowledge, and documentation and protection of ethnobotanical knowledge.

The first volume was on Eastern Ghats and Adjacent Deccan Region of Peninsular India, while the second one is on Western Ghats and Western Peninsular India. Third volume is on North-East India and Andaman and Nicobar Islands, one of the hot spots of biodiversity. The fourth volume is on *Western and Central Himalayas*. And the fifth volume is on *The Indo-Gangetic Region and Central India*. Published information is summarized on different aspects. We have added seven general chapters on ethnobotany of neem, skin diseases, hepatoprotective plants, plant contraceptives, trade in medicinal plants, ethnogenomics, ethnobotany of post-genomic horizons. Our intention is that the information contained in this volume may lead in the future to discovery many new drugs, nutraceuticals, novel molecules, and other useful products for the benefit of mankind.

Since it is a voluminous subject we might have not covered the entire gamut but we have tried to put together as much information as possible. Readers are requested to give their suggestions for improvement of future volumes in this series.

ACKNOWLEDGEMENTS

We wish to express our grateful thanks to all the authors who have contributed their chapters. We thank them for their cooperation and erudition. We also thank several colleagues for their help in many ways and for their suggestions from time to time during the evolution of this volume.

We wish to express our appreciation and help rendered by Ms. Sandra Jones Sickels and her staff at Apple Academic Press. Above all, their professionalism that has made this book a reality is greatly appreciated.

We thank Mr. John Adams, Senior Research Fellow of Prof. K. V. Krishnamurthy for his help in many ways.

We wish to express our grateful thanks to our respective family members for their cooperation.

We hope that this book will help our fellow teachers and researchers who enter the world of the fascinating subject of ethnobotany in India with confidence.

—Editors

Ethnobotany of India 5-volume Series

Editors: T. Pullaiah, PhD, K. V. Krishnamurthy, PhD, and Bir Bahadur, PhD

ABOUT THE EDITORS

T. Pullaiah, PhD
Former Professor, Department of Botany, Sri Krishnadevaraya University, Andhra Pradesh, India

T. Pullaiah, PhD, is a former Professor at the Department of Botany at Sri Krishnadevaraya University in Andhra Pradesh, India, where he has taught for more than 35 years. He has held several positions at the university, including Dean, Faculty of Biosciences, Head of the Department of Botany, Head of the Department of Biotechnology, and Member, Academic Senate. He was President of the Indian Botanical Society (2014), President of the Indian Association for Angiosperm Taxonomy (2013), and Fellow of the Andhra Pradesh Akademi of Sciences. He was awarded the Panchanan Maheswari Gold Medal, the Dr. G. Panigrahi Memorial Lecture Award of the Indian Botanical Society, the Prof. Y. D. Tyagi Gold Medal of the Indian Association for Angiosperm Taxonomy, and a Best Teacher Award from Government of Andhra Pradesh. He has authored 45 books, edited 15 books, and published over 300 research papers, including reviews and book chapters. His books include *Flora of Eastern Ghats* (4 volumes), *Flora of Andhra Pradesh* (5 volumes), *Flora of Telangana* (3 volumes), *Encyclopedia of World Medicinal Plants* (5 volumes), and *Encyclopedia of Herbal Antioxidants* (3 volumes). He was also a member of the Species Survival Commission of the International Union for Conservation of Nature (IUCN). Professor Pullaiah received his PhD from Andhra University, India, attended Moscow State University, Russia, and worked as postdoctoral fellow during 1976–78.

K. V. Krishnamurthy, PhD
Former Professor, Department of Plant Sciences, Bharathidasan University, Tiruchirapalli, India

K. V. Krishnamurthy, PhD, is a former Professor and Head of Department, Plant Sciences at Bharathidasan University in Tiruchirappalli, India, and is at present a consultant at Sami Labs Ltd., Bangalore. He obtained his PhD degree from Madras University, India, and has taught many undergraduate, postgraduate, MPhil, and PhD students. He has over 48 years of teaching and research experience, and his major research areas include plant morphology and morphogenesis, biodiversity, floristic and reproductive ecology, and cytochemistry. He has published more than 170 research papers and 21 books, operated 16 major research projects funded by various agencies, and guided 32 PhD and more than 50 MPhil scholars. His important books include *Methods in Cell Wall Cytochemistry*, *Textbook of Biodiversity*, and *From Flower to Fruit*. One of his important research projects pertains to a detailed study of the Shervaroy Hills, which form a major hill region in the southern Eastern Ghats, and seven of his PhD scholars have done research work on various aspects of Eastern Ghats. He has won several awards and honors that include the Hira Lal Chakravarthy Award (1984) from the Indian Science Congress; Fulbright Visiting Professorship at the University of Colorado, USA (1993); Best Environmental Scientist Award of Tamil Nadu state (1998); the V. V. Sivarajan Award of the Indian Association for Angiosperm Taxonomy (1998); and the Prof. V. Puri Award from the Indian Botanical Society (2006). He is a fellow of the Linnaean Society, London; National Academy of Sciences, India; and Indian Association of Angiosperm Taxonomy.

Bir Bahadur, PhD
Former Professor, Department of Botany, Kakatiya University, Warangal, Telangana, India

Bir Bahadur, PhD, was Chairman and Head of the Department, and Dean of the Faculty of Science at Kakatiya University in Warangal, India, and has also taught at Osmania University in Hyderabad, India. During his long academic career, he was honored with the Best Teacher Award by Andhra Pradesh State Government for mentoring thousands of graduates and postgraduate students, including 30 PhDs, most of whom went onto occupy high positions at various universities and research organizations in India and abroad. Dr. Bahadur has been the recipient of many awards and honors, including the Vishwambhar Puri Medal from the Indian Botanical Society for his research contributions in various aspects of plant Sciences. He has published over 200 research papers and reviews and has authored or edited dozen books, including *Plant Biology and Biotechnology* and *Jatropha*,

Challenges for New Energy Crop, both published in two volumes each by Springer Publishers. Dr. Bahadur is listed as an Eminent Botanist of India, the Bharath Jyoti Award, New Delhi, for his sustained academic and research career at New Delhi and elsewhere. Long active in his field, he is a member of over dozen professional bodies in India and abroad, including Fellow of the Linnean Society (London); Chartered Biologist Fellow of the Institute of Biology (London); Member of the New York Academy of Sciences; and a Royal Society Bursar. He was also honored with an Honorary Fellowship of Birmingham University (UK). Presently he is an Independent Director of Sri Biotech Laboratories India Ltd, Hyderabad, India.

CHAPTER 1

INTRODUCTION

K. V. KRISHNAMURTHY,[1] T. PULLAIAH,[2] BIR BAHADUR,[3]
and S. JOHN ADAMS[1]

*[1]R&D – Phytochemistry and Pharmacognosy, Sami Labs Ltd,
Peenya Industrial Area, Bangalore, India,
E-mail: kvkbdu@yahoo.co.in*

*[2]Department of Botany, Sri Krishnadevaraya University, Anantapur –
515003, India, E-mail: pullaiah.thammineni@gmail.com*

*[3]Department of Botany, Kakatiya University, Warangal – 505009,
India, E-mail: birbahadur5april@gmail.com*

CONTENTS

ABSTRACT

The present chapter introduces the articles that form the content of volume 5 of Ethnobotany of India. This volume relates to the Indo-Gangetic plains which form a vast area in northern India. This study region and its physical features, land use patterns, vegetations and floristics are introduced first. This is followed by an introduction to ethnic diversity of the region, Indus valley civilization, one of the oldest civilizations of the world that saw the emergence of great traditional knowledge on plants and utilitarian traditional knowledge (on food, medicine, veterinary medicine and other useful items). These are followed by introductions to chapters on plant contraceptives, Ethnomedicinal plants of skin diseases, and liver diseases as well as on the ethnobotany of neem. The importance of ethnogenomics and the developments that happened during the post –genomic period on ethnobotanical research is also introduced. Finally the future aspects that need to be taken on Indian ethnobotany are also discussed.

1.1   PHYSICAL FEATURES OF THE STUDY REGION

The Himalayas girdling the northern border, the flat Indo-Gangetic plains in the middle, the peninsular India along with its hills and plateaus in the south and the narrow coastal plains forming the seaboard are the major provinces of India (Valdiya, 2010). The Indo-Gangetic plains are separated from the Himalayas by the Siwalik or outer Himalayas and are one of the world's largest alluvial plains. These plains were built in the Holocene times by the Brahmaputra, Ganges and Indus river systems and their tributaries and extend 3,200 km from the southern limit of Brahmaputra-Ganges delta in the east to the terminus of the Indus delta and the Rann of Kutch in the west. The width of the plains varies from 90 km (in Assam) to 550 km (in Punjab).

In the Indus basin the alluvial gives way southeastwards to the Thar Desert (with several sand dunes) that extends in Rajasthan and adjoining Sindh region. The desert region is believed to represent the basin of the legendary Saraswati River that got lost in the Later Holocene period. It is now represented by the dry flood water-channels of the Ghaggar-Hakra-Nara River. The average elevation of the Indo-Gangetic plains range from 150 m to 290 m. The underground extension of the Aravallis towards Haridwar forms the boundary between Indus and Ganga plains. The Indus plain is predominantly the Khaddar expanse that is known as Chung in the Punjab. This plain is fringed in the west by Piedmont Belt, which is 16–24 km wide. Within the Indus plain the rivers and streams are confined the Luni river forms an alluvial plain in the Arid part of long and 300 km wide, on the foot of Aravalli hills, there are many saline lakes like Sambhar, Dadwana, Degana, Lunkaransar, etc. (which turn into water bodies during rainy days) (Valdiya, 2010).

The Ganges plain exhibits little variation in landscape or relief for several hundred kilometers but may be relieved in some areas by bluffs, leaves and abandoned channels, oxbow lakes and ravines, with occasional formations of Chars (Uplands) and Bhils or Jheels (Marshes and lakes). The main Ganges Domain is in Uttar Pradesh with two distinct physiographic regions: Older Alluvium comprising the Banda, Varanasi and Bhangar units of Coarse sedimentary deposits and newer Alluvium comprising the Khaddar, Bhaur and the Bhabhar units of gravelly sediments. The fans of rivers like Ganga, Sharada, Gandak and Kosi continue to grow in size and thickness (Valdiya, 2010).

The central India has some important mountain ranges. The Aravalli Range extends for more than 800 km from Palanpur in Gujarat to Delhi. This is made up of the Proterozoic metasedimentary rocks intruded extensively by granites. The general height is 400 to 600 m, although in the southwestern segment it is more than 1,000 m. The highest peaks are Gurushikhar (1,722 m) near Mount Abu, Jaga (1431 m) and Bhorat (1,225 m), north-west of Udaipur. Southwest of Udaipur the Aravalli bends in the southeast direction and appears to link with the Satpura Hill range (Valdiya, 2010).

The Satpuras have similar rock types and belongs to the same age as that of Aravallis. The average height of Satpura range is around 900 to 1,000 m. It has seven hills: Rajpipla hills, Mahadeo hills, Maikal hills, Ranchi hills, Rajmahal hills, Meghalaya hills and Malaygiri hills. The highest peaks are Amarkantak in Maikal hills, (1,127 m), Dhungarp in Mahadev hills (1,350 m) and Astambatongar (1,127 m). The rivers associated with this hill range are Narmada (flowing west), Son (flowing north) and Mahanadi (flowing southeast). The Vindhya hill in Madhya Pradesh has a several elevation of 300 to 650 m. The most important are Kaimur in the north the Bhainu in the south. It is made up of Proterozeric sedimentary rocks. The rivers associated with the hills are Chambal, Sind, Betwa and Ken (Valdiya, 2010).

The narrow seaboard on the western side of the study region comprises of the Makran coast, where the margin of the Indus deltaic plain is made up of an arcuate zone of older tidal deposits (Valdiya, 2010). The Rann of Kutch is located on the east-southwest of this deltaic plain; this is a salt-impregnated and encrusted tidal flat (350 km long and 150 km wide) that remains inundated by sea waters during the monsoon months. It was once connected to the Gulf of Khambhat through the Nalsarovara shallow, brackish water lake east of Saurastra. The estuaries in this coast cover an area of 3,76,000 hectares (Valdiya, 2010).

The seaboard on the east is characterized dominantly by Sundarban delta, a combined product of the Ganges and the Brahmaputra. This delta forms the head of the Bengal Basin. It is just about 3–20 m m.s.l. The delta is characterized by abandoned channels, lakes and swamps and tidal flats. The Arakan coast is a narrow rocky belt and is characterized by a steep rocky shore overlooking the 600 m high Arakan Yoma. Many rocky bays and small beaches are found here. Very small rivers like Mayor, Kaladam, and Lemro are found and these make a 70 km wide coastal plain in the Akyab region. The total area of estuaries in the West Bengal coast is around 4,05,000 hectares (Valdiya, 2010).

1.2   CLIMATE AND ECOLOGY

The present-day ecology of the Indo-Gangetic plain is dominated to a very large extent by the influence of the Himalayas located on its northern side, as well as by the increasing pressure of the growing human population, particularly due to rapid urbanization and industrialization (Mani, 1974). This is also the region which has been under intensive and continuous agricultural activity for the last five to six thousand years and hence deforestation has been complete, at least in the plains. Even during the Moghul period Uttar Pradesh was fairly densely forested (up to 16th Century CE) and was used for hunting wild animals in the doabs of Yamuna and Ganges rivers. Whatever secondary vegetation that occupied the cleared forested regions, it was markedly xeromorphic and scrub-savanna-like with many thorny elements.

Of the 11 biogeographical zones recognized in India, this region contains four: desert, semi-arid, plains and coasts. The plain is somewhat more humid and receives more rainfall in its eastern part than in its western part. The average rainfall ranges from >500 mm to <3000 mm depending upon the region (Sudhakar Reddy et al., 2015). However, in the last 50 years arid conditions are extending eastwards. Ecologically and biogeographically the

Indo-Gangetic region is a transition region between the south and north as well as between west and east. According to Mani (1974) these transitions are also with respect to seasons. For example, the "ecosystems oscillate between monsoon-rains and post-monsoon seasons." In the upper Gangetic plain nearly 90% of the rainfall is received from the southwest monsoon, except in the extreme northwest where the summer northwesters also bring some amount of rain. Rainfall decreases from 145 cm in the east to 100 cm in the west and also from the Himalayas southward into the Ganges River. In the delta region of Bengal, there are violent cyclonic norwesters very often accompanied by very heavy rains during March–April.

1.3 LAND USE PATTERNS, VEGETATION AND FLORA

There are 29 land use/land cover classes, as recognized by remote-sensing data, in the Indo-Gangetic region. Among these are 14 forest types, 7 scrub types and the rest 8 to others (Sudhakar Reddy et al., 2015). The main forest types recognized are tropical wet evergreen forests (353 km^2), tropical semi-evergreen forests (730 km^2), tropical moist deciduous forest in the Gangetic plains (11,080 km^2), in the coast (102 km^2), tropical dry deciduous forests in the Gangetic plains (3359 km^2) in the semi-arid regions and in the coast (5 km^2), Littoral and Swamp forests in the coast, tropical thorn forest in the Gangetic plains (248 km^2), in the semi -arid regions (3491 km^2) and in the desert region (1121 km^2), and subtropical broad leaved hill forest in the semi-arid regions (86 km^2). In addition, this part of India has grasslands, plantations, agricultural lands, barren land, various water bodies, etc. The total forest area (in km^2) in different states of this region are: Bihar (5,454), Chhattisgarh (55,116), Delhi (34), Gujarat (10,737), Haryana (866), Jharkhand (22, 766), Madhya Pradesh (77,590), Punjab (1,280), Rajasthan (15,358), Uttar Pradesh (10,415), West Bengal (8,524) and Chandigarh (6) (Sudhakar Reddy et al., 2015).

Extensive floristic studies have been done on this vast region of India. The plants are mainly listed in the following works (Aitchison, 1869; Bamber, 1916; Bhandari, 1990; Duthie, 1903–1922; Haines, 1961; Maheshwari, 1963; Mudgal et al., 1997; Nair, 1978; Naskar, 1993; Prain, 1903; Shah, 1978; Singh et al., 2001; Verma et al., 1993). A critical study of these works reveals that this region may be having about half of the total number of flowering plants of India, that is, about 11,000 to 12,000 species. Many aquatic plants of India (around 2/3) are found here. In the Punjab and Chandigarh region there are 1,320 species of angiosperms, 31 pteridophytes and gymnosperms of which 141 are threatened taxa (including 11 pteridophytes and 1 gymnosperm) (Sharma, 2012). Gujarat State has about 2,200 species of angiosperms and these include 374 tree species and 285 grass species. In the Kachchh district of Gujarat above 13 RET taxa have been reported. In the Aravallis 56 rare, 11 endemic and 6 threatened species (totally 73 RET taxa) of angiosperms and 1 rare gymnosperm has been reported. The Aravallis also have around 300 species of grasses. In the state of Delhi originally Maheswari (1963) reported 170 angiosperms, but now around 273 species of angiosperms and 15 gymnosperms are known to occur.

There are several wetlands in the Indo-Gangetic plains. The most important among them is the ox-bow wetland in West Bengal. The riparian community along the Sabarimati River in Gujarat has 542 species of flowering plants.

There are not many protected areas in the Indo-Gangetic plains. One of them is the Bethuadahai Wildlife Sanctuary located in Nadia (West Bengal). It has an area of 121 km^2 and 199 species of flowering plants. The marine and coastal protected areas located in the study region are the following: Sundarbans National Park and Sajnekhali, Lothian and Holiday Wildlife Sanctuaries in West Bengal, and the Gulf of Kachchh National Park and Wildlife Sanctuary in Gujarat coast.

1.4   ETHNIC DIVERSITY

The contemporary populations of India are known to have evolved from peopling of India by waves of migrants for the last several thousands of years (Majumder, 1998). We have almost all the primary ethnic strains in India. Proto-Australoid/Australoid (skin color – dark brown), Mediterranean (skin color – light brown), mongoloid (skin color – yellow), Negrito (skin color – black), Caucasoid (skin color – white) and a number of composite strains: but in the Indo-Gangetic region the Negrito ethnic group is totally absent. The Proto-Australoids are akin to modern aborigines of Chota Nagpur and Central India (particularly Baiga, Gond, Bhil, Santal and Oroan tribes). The Mangoloids are mainly present in West Bengal and E-Bihar and to some extent in Madhya Pradesh and Chhattisgarh, although in fairly small numbers in all these places and many people are a mix-up of mongoloid and proto-Australoid groups. These include the Munda, Gadaba and some Santals

(Munda-Speaking groups) and Maria, Muria, Gond and Oroan (Dravidian-speaking groups). The Mediterraneans also were among the ancient people of this region, particularly among the Indus Valley civilization people; this ethnic group is now seen in Gujarat, Rajasthan, Punjab and Haryana.

In India, now, people speak about 750 dialectics which can be classified under four language groups: Austro-Astem (called Nishada), Dravidian, Sino-Tibetan (called Kirata) and Indo-European (called Aryan). The main Austro-Asiatic language group is Munda which is spoken in central and Eastern India (in the states of Bihar, Jharkhand, and Madhya Pradesh). It is spoken by Santals, Kherwar, Mundari, Bhumij, Birhar, Koda/Kora, Ho, Asuvi, Korwa, Korku, Kharia, Savara, Gadaba, etc. The other Austro-Asiatic language group is Mon-Khmer, which is not spoken in the study region by any tribe. The Sino-Tibetan language is spoken by Lepcha. Toto, Gwung, Dhimal, Rabha, Deori and Lotha Tribes. The Dravidian language is spoken by Gond, Parji, Kuruch, Oroan, and Bharia tribes. The Indo-European language is spoken by Dardi, Sindhi, Gujarat, Bhil and Pahari tribes.

Of the total population of India 8% are tribal groups as per the 2001 census. Among the 461 tribal groups of India (Singh, 1993) the vast majority 402 live in the hilly terrains and dense forests. In the Indo-Gangetic plains more than 90 different important tribes are reported to live. Some tribes are common to more than one state while few tribes are exclusive to some states. In Chapter 2 of this volume Sikarwar has given a detailed account on the ethnic diversity of Indo-Gangetic plains, along with socio-cultural aspects of the various tribes.

It is generally accepted that the peopling of India by the modern-human species began around 70 to 50 thousand years back, although there were several other waves of peopling subsequent to this period by different major ethnic groups and major language groups. But it is very certain that major peopling happened in the Indo-Gangetic plains where Indus-valley civilization and Aryan occupation dominated. The birth of settled life, urban civilization, literary compositions and establishment of agricultural activities all happened in this region. In Chapter 3 Krishnamurthy and Bir Bahadur give a precise account on the Ethnobotany of Indus Valley Civilization.

1.5 TRADITIONAL KNOWLEDGE OF PLANTS

Traditional knowledge systems (TKS) on plants includes all kinds of knowledge and belief systems of ancient ethnic tribes and people relating to subsistence production systems, ecosystems, logics of subsistence, social, cultural and utilitarian aspects of plants and community decision-making relating to use of plants. Such TKS is available in India in both non-codified and codified forms. In non-codified form it is generally passed on through several generations orally as well as actually by putting into practice the knowledge so gained. In codified form it is documented in writing. The Indo-Gangetic region is one of the earlier regions or even the only earliest region of India, which excelled in both non-codified and codified TKS. Several classical literary works starting from Rig Vedas to Samhitas and Sastras were produced in this region all of which showed the great depth of the TKS of the people of this region, with respect to plants. The several ancient ethnic tribes of this region also had a vast reservoir of TKS on plants. Both categories of TKS formed the sources of information summarized in chapters 4 to 12 of this volume.

1.5.1 FOOD PLANTS

There are archaeobotanical/archaeological evidences to show that the Indo-Gangetic region of India was one of the earliest regions that were peopled when the modern human race initiated its out- of-Africa migration around 70,000 to 50,000 years ago. Several waves of migration of humans belonging to different ethnic groups and language groups happened. Such early-migrated human tribes of India were only hunter-gatherers and nomadics who met their subsistence requirements by hunting and foraging plants from the wild. Around 12,000 to 10,000 years ago, perhaps due to drastic climate change people in northern India, particularly in the Indo-Gangetic region started to have a settled life along the Indus, Ganges and other riverbanks and started intense agricultural activity and trying to grow and domesticate wild/land races of cereals, legumes, and fruits. Artifacts of such crop plants are found in various archaeological sites from the above period onwards. Thus, food was beginning to be obtained from cultivated sources, although in remote areas and forests ethnic tribes continued to live on hunted and foraged food materials. However, it is the latter sources of food that are of importance to us as we are now looking for newer and newer sources of food, apart from those that are produced through agriculture. It is also important for us to bring into practice the traditional food preparation methods not only to preserve them for posterity and to prevent them from being corrupted by more modern methods. It is to be emphasized here that traditional food/food preparation systems cover the full spectrum of life which the modern systems do not and also that traditional foods and preparations have been developed based on locally available resources and on local environmental and cultural conditions at least over the past ten thousand years. An impressive array of traditional food plants and food preparations are available and these need to be documented before they vanish. In Chapter 4, Sudip Ray gives a very detailed account on ethnic food plants and food preparations of the Indo-Gangetic region.

1.5.2 ETHNOMEDICINAL AND ETHNOVETERINARY PLANTS

India is known for its very rich traditional knowledge, with both codified and non-codified systems already very well developed, both in theory and practice (Bahadur et al., 2007); the therapeutic, physical and biological bases of these systems were very clearly established. Almost all ailments and diseases that are known to afflict modern humans were also known to ancient Indian Vaidyas (although known in different terminologies) (Kudumbia). The sources of drugs/medicines for all these ailments were also known to ancient Indian doctors from plants, animals or minerals. At least 8,000 plants were known to have been used (Murthy et al., 2008) and different kinds of medicinal preparations and administrations were in practice which did not cause any side-effects. The codified Ayurvedic system, particularly Atharva Veda and Ayurvedic texts (Charaka Samhita, Susruta Samhita, Vagbhatta text, Agnipurana and many other texts) have details on plants used as medicine. Similarly during the Moghul Rule, the codified Unani-Tibb medical system got perfected in the Indo-Gangetic region, these texts contained details on medicinal plants used in that system. However, the plants used in the non-codified folk medicinal and tribal medicinal systems have not been documented systematically for this study region. In Chapter 5 of this volume, Rahman et al., have taken great pains to list out these medicinal plants along with details on their medicinal uses.

India has a very rich stock of domesticated animals, local breeds and wild relatives with particular reference to cattle, poultry and draft animals (Aruna Kumara and Anand, 2006). Most breedings were done by the local tribal communities and pastoral people. The Ganges-Yamuna basin particularly is an important region of cattle breeding and animal husbandry, besides a region where prevention and curing of diseases of cattle and poultry got perfected using locally available plants. Thus, ethnoveterinary medicine in the Indo-Gangetic plains is as old as peopling in that region. More than 250 diseases of animals have been reported along with preventive and curative details of these diseases. However, the details on this aspect are highly scattered and Sikarwar has taken great pains to document these details in Chapter 6 of this volume

Subsequent to Alma Ata Declaration and the recognition of alternative and complementary medical systems by World Health Organization there was a great impetus on the use of medicinal plants worldwide, which was till then restricted to particular country/region depending on the type of alternative system of medicine (e.g., Ayurvedic plants in India). However, the greatest impetus to use of medicinal plant resources was given by sustained and increased interest in bioprospection for newer medicinal plants, nutraceuticals, cosmeceuticals and biomolecules of interest. This promoted trade in botanicals both within a country and across countries. The annual demand for botanicals steadily increased over the years. In 2012, the herbal dietary supplement sales were the highest ever, and this was around $5.6 billion (Lindstrom et al., 2013). Another estimate of the global herbal personal and cosmetic care products in 2005 sets a value at $12 billion (Bird, 2005). The trade scenario of Indian medicinal plants is presented by Ved et al. in Chapter 7 of this volume.

1.5.3 OTHER USEFUL PLANTS

Other than as sources of food and medicines plants are known to provide a great variety of useful products for human consumption. There are some plants which are directly exploited from the wild while the remaining ones are from cultivated sources. The plant or their parts got are used directly or are converted into useful products through various means and methods. In spite of great overall development in diverse spheres of human life, potential resources or those that are locally used by ethnic communities largely remain poorly understood, underexploited (or not at all exploited) and, above all, very poorly documented, if at all, since most of these are under the domain of ancient/primitive ethnic societies (Krishnamurthy, 2003). Many plant species are known as ornamentals, timbers, fibers, dyes, fuel/renewable energy sources and as a source of a host of other plant products. Wagh has dealt with in detail the other useful plants of the Indo-Gangetic region in Chapter 8 of this volume.

1.5.4 ETHNIC PLANTS FOR LIVER AND SKIN PROTECTION

Liver is one of the very vital organs of human body and the principal site of control of metabolism, secretion, excretion, storage and homeostasis. Hence, any injury to liver or impairment of its functions has grave implications for the affected person. Every year about 18,000 people are reported to die due to live cirrhosis and about million die of viral hepatitis infection alone. Viral infection is one of the main causes for hepatic injury and dysfunction. The herbal agents that are important in liver problems are hepatotoxy-healing agent, agents that protect liver and agents that antagonize the effects of any hepatotoxins-causing hepatitis or any other liver disease. Herbal-based therapeutics for liver disorders has been in use in India, in Ayurveda, Siddha and folk medicine, for a long time. A large number of plants and formulations have been claimed to have hepatoprotective activity. Nearly 160 phytoconstituents from around 100 plants have been claimed to possess liver-protecting activity. In India more than 87 plants are used in 33 patented and proprietary multi-ingredient plant formulations (Jannu et al., 2012). Most of these plants have been identified from ethno-directed knowledge. In Chapter 9 of this volume Ramaiah had dealt with in detail on hepatoprotective plants that are in the ethnodomain-knowledge of various tribes of Indo-Gangetic region.

Skin diseases are known to occur worldwide and constitute about 34% of all occupational diseases known so far. They affect people of all ages of both genders and form one of the five major reasons for consulting a doctor. They present a major health burden in all parts of the world since socio-economic environments play a vital role in the spread of these diseases, particularly in hot and humid conditions. Mortality rate due to skin diseases is extremely low, although the impact it has on the quality of life is enormous (Hay et al., 2006). Traditionally, plants have been used by many ethnic tribes of India in managing skin diseases and in treating of skin ailments thereby contributing significantly in the primary healthcare of the population. In Chapter 10 of this volume, Koteswara Rao et al. have given a detailed account on the ethnomedicine for skin diseases.

India is one of the most populated countries of the world with a population of about one sixth of that of the world. It has already attained a size of about 133 crores and is still growing. Overpopulation continues to be a critical contributor to environmental degradation, poverty, health and unemployment. A significant percentage of population growth is due to unintended pregnancies and about a half of these pregnancies are due to failure in using the correct contraceptive (male or female). There are also economic and cultural aspects to the use of contraceptives or to abort the conceived fetus. There are more female contraceptive methods than male. This is also a major factor in population growth. The WHO has been promoting the development of safe and cheap contraceptives that can be used in poor countries which are the main targets of the ill-effects of population growth. One among these efforts is the identification and use of herbal contraceptives, which are better accepted socially and culturally than other types of contraceptives. In Chapter 11 of this volume Baljot Kaur has given an exhaustive account on plant contraceptives.

The neem tree (*Azadirachta indica* A. Juss.) is well-known in India and neighboring countries for more than 2000 years for its beneficial properties (Schmutterer, 1995). All parts of the plant are used by Ayurvedic, Siddha, Unani and folk medicinal systems of India. The villagers/ethnic tribes use neem leaves to protect stored grains, particularly rice, and use twigs to brush their teeth, although the insect-repellent property of neem was first reported in India only in 1928, while its efficacy in treating dental diseases was proved much later. Its medicinal efficiency on small pox and related viral diseases was also known for a very long time in India. However, only in the past five decades a marked increase in interest to study neem was noticed and most of which were triggered by the vast amount of traditional knowledge on neem. A number of phytochemicals, especially Azadirachtins, were reported in neem. In Chapter 12 of this volume Sri Rama Murthy et al. have given a detailed review on the ethnobotany of neem.

1.6 GENOMIC AND POST-GENOMIC ASPECTS OF ETHNOBOTANY

Traditional biodiversity knowledge is being studied at three levels: genetic, specific and ecosystem levels. Traditional communities throughout the world have been promoting efforts to maintain all these three levels of biodiversity in their own traditional conservation methods. People interested in bio-prospecting are also now increasingly interested in studying ethnic plant resources from a genomic perspective. A new branch of ethnogenomics has emerged in the last few decades and this branch studies not only genomics of ethnic human groups but also of plants and animals used by various ethnic communities. Useful properties, particularly medicinal properties, are not randomly distributed in plants. Instead, some plant groups have more properties than others and here phylogeny-directed methods are used to identify a new medicinal plant taxon related to a known medicinal plant using phylogenetic genomics because of the possibility of phylogenetic conservation of medicinal properties (Lagoudakis et al., 2011). Plant phylogenies (based on genomics) are extremely versatile and are very valuable tools that are used to recover a variety of patterns on biogeography, ecology, development and chemistry of plants of ethnic value. In Chapter 13 of this volume Bir Bahadur et al. discuss in detail on Ethnogenomics, while in Chapter 14, Tamil Selvi and Srinivas describe the various developments that had taken place in the discipline or Ethnobotany in the post-genomic era.

1.7   PRESENT STATUS AND FUTURE PROSPECTS OF ETHNOBOTANY IN INDIA

Since this is the final volume of the series of books on the ethnobotany of India, a critical review of all articles contained in all volumes have been made in order to make an unbiased assessment of the work that has so far been done on Indian Ethnobotany with reference to its achievements and merits as well as its shortcomings. The latter are particularly important so that proper planning of future work can be done so that Indian Ethnobotany fruitions to its full potential. Although Chowdhary Habibur Rahman had dealt with a detailed review of Indian Ethnobotanical work and its future in India in the last chapter of this volume, the following discussion should also be kept in mind on what should be done in the coming years on Indian Ethnobotany.

The origin and development of Indian ethnobotanical knowledge is largely due to peopling of India in several ways from around 70,000–50,000 years ago with several ancient tribes that differed in ethnicity as well as in the language they spoke. Although detailed studies have been made on the genetic basis of the great ethnic diversity seen in India (see, Gadgil et al.,

1996; Thangaraj, 2011), still a lot needs to be done on initial location of occupation and their further spread in India, so that a proper correlation of the traditional knowledge on plants held by these tribes and its relation to the environment where they lived or got spread to. Especially important in this respect is data on South Indian and N. E. Himalayan regions.

The approaches and methods to study the perception of the various ethnic tribes of India towards plants form a very important component of ethnobotanical research. This perception of ethnic tribes was largely triggered by important motivations, which include, among others, purely survival, social and cultural and purely utilitarian/economic motivations, while the motivations of modern ethnobotanists are purely theoretical and academic besides being predominantly utilitarian largely triggered by serious bioprospection attempts. The utilitarian motive seems to have largely suppressed the socio-cultural motivations among ethnobotanists of India thanks to the influence of western science and the gradual waning of traditional knowledge. In fact, traditional knowledge has been exploited to glorify western science and its bioprospection attempts. Here, the problems of utilitarian approach of ethnobotany in reference to India will be highlighted followed by an attempt of stress the importance of future focusing of attention on socio-cultural aspects of TKS so dear to ethnic people.

Although the study of utilitarian ethnobotany has clearly changed enormously in many other parts of the world, in India it has not changed much from the mere compilation of "laundry list' of traditionally useful plants. As far as we could count there are already around 3,000 such lists published on Indian ethnobotanically important Indian plants. Even these lists have often been prepared without serious and fairly long-term interaction with the ethnic tribes who were the discoverers and owners of this knowledge; quite often these details are obtained from secondary and tertiary sources. Proper ethnobotanical inventory procedures as outlined by Martin (1995), Bellany (1993), Jain (1987) and others have also not been followed strictly. Many investigators have not even seen the efficacy of these plants while being used by the concerned tribes; neither the efficacy of most of the listed plants has been tested by scientific methods to validate the traditional knowledge.

As per articles 15 and 8(j) of CBD to which India is a signatory, the very important prior informed consent (PIC) of ethnic communities with whom the research is to be conducted (Berlin and Berlin 2004), to our knowledge, has not been obtained while listing the traditional sources. This naturally denies the source community of the benefits due to them (EWG, 2003).

When a traditional plant material is reported to have pharmaceutical, nutra-ceutical or cosmeceutical value, as per ethnic knowledge, this should be fol-lowed by detailed phytochemical analysis, identification of the chemical(s) responsible for the above value and verification of the therapeutic action. For example, out of the about 8,000 plants of medicinal value in the Indian flora, the phytochemistry has hardly been investigated even for 10% of the plants; the number investigated for therapeutic efficacy is not even 1% of these. In other words, the "laundry list" of Indian ethnobotanical plants continues to remain as a laundry list without the listed plants being subjected to further work to prove the values known as per ethnic knowledge. Detailed phyto-chemical, reverse pharmacological and reverse pharmacognostic works on these plants should be immediately undertaken in the coming future on a war footing. Otherwise the utilitarian approach of ethnobotany would miserably fail to yield the desired results.

The ethnocultural approach to ethnobotanical research involves in demystifying many socio-cultural practices which involve the use of plants/ plant products, particularly in agricultural, food and socio-cultural rituals, rites and taboos. The importance of these plants can be realized only when their relevant cultural influences are known. Despite their culture- specific nature and importance many such traditional practices may at first glance appear irrational. Unless we know how ethnic people perceive plants and how such perceptions are influenced by socio cultural factors and spiritual beliefs the actual role of plants in these rituals cannot be understood cor-rectly This approach also involves rituals/symbolic behaviors that fall in the realms of religion, spiritual, supernatural, magic, shamanism, etc. Detailed research should be carried out immediately to know the significance of plants in such acts in the line of work done by Quiroz et al. (2016) to under-stand the ethnopharmacological significance of ritual plant use in Benin and Gabon (Africa).

The first and foremost requirement towards conservation of traditional botanical knowledge is a very serious and effective documentation of this knowledge. Traditional communities have their strategies for documenta-tion which is mostly oral but in order to prevent loss of such oral knowl-edge it should be recorded and documented. Although several efforts have been made in India to document ethnic botanical knowledge (in the form of research papers, reports, data, books, banks, etc.) there are still informa-tion that are yet to be documented. Indian ethnobotanists are slow to adopt and apply tools of information revolution and to integrate research data collaboratively. The data sets collected should allow for a "triangulation of information that increases the validity (truthful and logical measures of what is intended to measure), accuracy (qualitative evaluation of the closeness of agreements with truth or fact) and reliability (reproducibility)" of results (Berlin and Berlin, 2005).

KEYWORDS

- **ethnic diversity**
- **ethno food plants**
- **ethno medicinal plants**
- **ethnogenomics**
- **future of ethnobotany in India**
- **Indo-Gangetic plains**
- **neem**
- **plant contraceptives**

REFERENCES

Aitchison, J. E. T. (1869). A catalog of the plants of the Punjab and Sindh. Taylor and Francis, London.

Aruna Kumara, V. K., & Anand, A. S. (2006). An initiative towards the conservation and development of Indian Cattle Breeds. In: A. V. Balasubramanian & T. D. Nirmala Devi, (Eds.) Traditional Knowledge Systems of India and Sri Lanka. Centre for Indian Knowledge Systems, Chennai, India. pp. 104–113.

Bahadur, B., Janardhan Reddy, K., & Rao, M. L. N. (2007). Medicinal Plants: An overview. In: Janardhan Reddy, K, Bir Bahadur & Rao, M. L. N. (Eds.): Advances in Medicinal Plants, Hyderabad: Universities Press, pp. 1–50.

Bamber, C. J. (1916). Plants of the Punjab: a descriptive key to the flora of the Punjab, North-West Frontier Province, and Kashmir. Supt. Govt. Printing, Punjab, Lahore.

Bellany, B. (1993). Ethnobiology-Expedition Field Techniques. Expedition Advisory Centre, Royal Geographical Society, London.

Berlin, E. A., & Berlin, B. (2004). Prior informed consent and bioprospecting in Chiapas. In: Riley, M. (Ed.). Indigenous intellectual Property Rights. Altamira Walnut Creak, CA, USA. pp. 341–372.

Berlin, E. A., & Berlin, B. (2005). Some field Methods in medical Ethnobiology. *Field Methods 17,* 235–268.

Bhandari, M. M. (1990). Flora of the Indian Desert. Scientific Publishers. Jodhpur, Rajasthan, India.

Bird, K. (2005). North American naturals market returns to growth. http://www.cosmeticsdesign. Com/market-trends/North American-naturals-market-returns-to-growth.

Duthie, J. F. (1903–1922). Flora of the Upper Gangetic Plain and of the Adjacent Siwalik and Sub. Himalayan Tracts. Calcutta, India.

EWG. (2003). Intellectual Imperatives in Ethnobiology. Ethnobiology working group, Missouri Botanical Garden, St. Louis, USA.

Gadgil, M., Joshi, N. V., Manoharan, S. Patil, S., & Shambu Prasad, U. V. (1996). Peopling of India. In: D. Balasubramanian & N. Appaji Rao (Eds.). The Indian Human Heritage. Universities Press, Hyderabad, India. pp. 100–129.

Haines, H. H. (1961). Botany of Bihar and Orissa. Vol. 3, BSI, Calcutta, India.

Hay, R., Bendec, S. E., Chen, S., Estrada, R., Haddix, A., & McLead, T. (2006). Matr: Skin diseases. In: Jamison, D. T., Breman, J. G., Meashan, A. R., Alleyne, G., Claeson, M., Evans, D. B., Jha, P., & Mills, A. (Eds.). Disease control Priorities in developing countries. Musgrove, Washington DC.

Jain, S. K. (1987). A manual of Ethnobotany, Scientific publishers, Jodhpur, India.

Jannu, V., Baddam, P. G., Boorgula, A. K., & Jambula, S. R. (2012). A review on Hepatoprotective plants. *Intl. J. Drug Dev. & Research 4,* 1–8.

Krishnamurthy, K. V. (2003). Textbook of Biodiversity. Science Publishers, Enfield (NH), USA.

Lagoudakis, C. H. S., Klitgaard, B. B., Forest, F., Francis, L., Savolainen, V., Williamson, E. M., & Hawkins, J. A. (2011). The use of phylogeny to intepret cross-cultural patterns in plant use and guide medicinal plant discovery: an example from *Pterocarpus* (Leguminosae). *PLoS One. 6* (7), e22275.

Lindstrom, A., Ooyen, C., Lynch, M. E., & Blumenthal, M. (2013). Herb supplement sales increase 5.5% in 2012. *Herbalgram 99,* 60–64.

Maheshwari, J. K. (1963). The Flora of Delhi, CSIR, New Delhi, India.

Majunder, P. P. (1998). People of India: Biological diversity and affinities. In: Balasubramanian, D., & Appa Rao, N. (Eds.). The Indian Human Heritage. Universities Press, Hyderabad, India. pp. 45–59.

Mani, M. S. (1974). Chapter XIII. Biogeography of the Indo-Gangetic Plain. In: Mani, M. S. (Ed.) Ecology and Biogeography in India. Dr. W. Junk Publishers, The Hague, Netherlands.

Martin, G. J. (1995). Ethnobotany—A Conservation Manual. Chapman & Hall, London.

Mudgal, V., Khanna, K. K., & Hajra, P. K. (1997). Flora of Madhya Pradesh. Vol. II. BSI, Kolkata, India.

Murthy, G. V. S., Benjamin, J. H. F., & Bir Bahadur (2008). Medicinal plants of Andhra Pradesh. In: Plant Wealth of Andhra Pradesh: Special Issue. Bahadur, B. (ed.) *Proc. A. P. Akad. Sci., 12,* 120–137.

Nair, N. C. (1978). Flora of the Punjab plains. *Rec. Bot, Surv. India 21,* 1–326.

Naskar, K. (1993). Plant Wealth of the Lower Ganga Delta. 2 Vols. Daya Publication House, Delhi, India.

Prain, D. (1903). Bengal Plants Vols. 1–2. Calcutta, India

Quiroz, D., Sosef, M., & Van Andel, T. (2016). Why ritual plant use has ethnopharmacological relevance. *J. Ethnopharmacol. 188,* 48–56.

Schmutterer, H. (1995). The Neem Tree. VCH, Weinhem, Germany.

Shah, G. L. (1978). Flora of Gujarat State-Univ. Press, Sardarpatel Univ. Vallabh Vidhyanagar, Gujarat, India.

Singh, K. S. (1993). Peoples of India (1985–92). *Curr. Sci. 64,* 1–10.

Singh, N. P., Khanna, K. K., Mudgal, V., & Dixit, R. D. (2001). Flora of Madhya Pradesh. Vol. III, BSI, Kolkata, India.

Sudhakar Reddy, C., Jha, C. S., Diwakar, P. G., & Dadhwal, V. K. (2015). Nationwide classification of forest types of India using remote sensing and GIS. *Environ. Monit. Assess. 187,* 777.

Thangaraj, K. (2011). Evolution and migration of modern human: Inference from peopling of India. In: Symposium volume on 'New Facets of Evolutionary Biology.' Madras Christian College, Tambaram, Chennai, India. pp. 19–21.

Valdiya, K. S. (2010). The Making of India: Geodynamic Evolution. MacMillan Publishers Ltd., New Delhi, India.

Verma, D. M., Balakrishnan, N. P., & Dixit, R. D. (1993). Flora of Madhya Pradesh. Vol. I. BSI, Kolkata, India.

CHAPTER 2

ETHNIC DIVERSITY OF THE INDO-GANGETIC REGION AND CENTRAL INDIA

R. L. S. SIKARWAR

Arogyadham (J. R. D. Tata Foundation for Research in Ayurveda and Yoga Sciences), Deendayal Research Institute, Chitrakoot, Dist. Satna (M.P.), 485334, India, E-mail: rlssikarwar@rediffmail.com

CONTENTS

ABSTRACT

The Indo-Gangetic region consists mainly with Rajasthan, Haryana, Punjab Plains, Delhi, Uttar Pradesh, Jharkhand, Bihar and Plains of West Bengal and Central India consists of Madhya Pradesh and Chhattisgarh of Indian states. Except Jharkhand, Rajasthan, Plains of West Bengal, Bihar and Uttar Pradesh, the Indo-Gangetic region is very poor in ethnic diversity because some states of Indo-Gangetic region like Haryana, Punjab and Delhi have no tribal communities. Central India is very rich in ethnic diversity because out of Total population (9,81,72,007) of Central India (Chhattisgarh: 25,545,198 and Madhya Pradesh: 72,626,809), 23.57% population (30.6% Chhattisgarh and 21.1% Madhya Pradesh) belongs to various tribal communities. The main tribes of Indo-Gangetic region and Central India are Abujhmaria, Baiga, Bhil, Bharia, Bhattra, Bhilala, Gond, Kol, Korku, Korwa, Oraon, Sahariya Munda, Oraon, Santhal, Meena, Garasia, Dhodia, etc. The details of the each ethnic community are given in this chapter.

2.1 INTRODUCTION

India has over 84.3 million tribal people belonging to 550 communities of 227 ethnic groups as per the classification made by anthropologists on linguistic basis. They inhabit in about 5000 forested villages. Each tribal community has a distinct social and cultural identity of its own and speaks a common dialect. There are about 116 different dialects and 227 subsidiary dialects spoken by tribals of India. With great antiquity the rich and varied culture the colorful traditions of tribals add to the texture and luster of the great civilization and the heritage of India (Pushpangadan and Pradeep, 2008).

India is the land of tribal people. The tribal people of India mostly live in forests, hills, plateaus and naturally isolated regions (Table 2.1) and are differently termed as Adivasi (original settlers), Adim niwasi (oldest ethnological sector of population), Adimjati (primitive caste), Anusuchit Janjati (scheduled tribe) and several names signifying their ecological or economic or historical or cultural characteristic. Among these the most popular is 'Adivasi,' while in India constitution name for them is 'Anusuchit Janjati' (Scheduled tribe) (Jain, 1981).

TABLE 2.1   Region Wise Important Tribes of Central India and Indo-Gangetic Region

S. No.	Region	Percentage	Important tribes
1.	Central India	Chhattisgarh	Durva, Gond, Halba, Maia, Mudia, Abhujmarhia, Agaria
		Madhya Pradesh	Baiga, Bhil, Bhillala, Bhatra, Bharia, Gond, Kol, Kanwar, Mawasi, Oraon, Sahariya, etc.
2.	Indo-Gangetic region	Rajasthan	Bhil, Mina, Sahariya, Garasia, Gaduliya Lohar, Bisnoi
		Jharkhand	Asur, Baiga, Chero, Kharia, Khairwar, Munda, Santal
		Bihar	Bathudi, Binjhia, Birjia, Chik Baraik
		Uttar Pradesh	Sahariya, Kol, Tharu
		West Bengal	Oraon, Munda, Santal, Bhumij, Kora
		Punjab	0
		Haryana	0
		Delhi	0

2.2   TRIBAL POPULATION OF INDIA

India in South East Asian subcontinent is an abode of nearly 2000 ethnic groups of people that includes some 550 tribal communities found inhabited with their language and culture in 30 different States and Union Territories (except Punjab, Haryana, Delhi NCT, Chandigarh UT and Pondicherry UT). As per the census 2011, the tribal population of India is 10, 42,81,034 (i.e., 8.2% of country population). The states and Union territories with tribal population in the descending order can be arranged as follow: Lakshadweep Islands UT (94.8%), Mizoram (94.4%), Nagaland (86.5%), Meghalaya (86.1%), Arunachal Pradesh (68.8%), Dadra Nagar Haveli UT (52.0%), Manipur (35.1%), Sikkim (33.8%), Tripura (31.8%), Chhattisgarh (30.6%), Jharkhand (26.2%), Odisha (22.8%), Madhya Pradesh (21.1%), Gujarat (14.8%), Rajasthan (13.5%), Assam (12.4%), Jammu & Kashmir (11.9%), Goa (10.2%), Maharashtra (9.4%), Andaman & Nicobar Islands UT (7.5%), Andhra Pradesh (7%), Karnataka (7%), Daman & Diu UT (6.3%), West Bengal (5.8%), Himachal Pradesh (5.7%), Uttarakhand (2.9%), Kerala (1.5%), Bihar (1.3%), Tamil Nadu (1.1%) and Uttar Pradesh (0.6%) (Sahoo et al., 2013).

The state wise tribal population of Central India and Indo-Gangetic region is given in Table 2.2.

The Central India consists mainly with states of Madhya Pradesh and Chhattisgarh which is very rich in ethnic diversity, Indo-Gangetic region except Jharkhand, Rajasthan, West Bengal and Uttar Pradesh has good ethnic diversity while states of Indo-Gangetic region like Haryana, Punjab and Delhi have no tribal communities. The forests of Indo-Gangetic region and Central India consist mainly of Tropical semi-evergreen forests, Tropical moist deciduous forests, Tropical dry deciduous forests and Tropical thorn forests. Indo-Gangetic region enjoys the widespread Indian monsoon climate with maximum rain falling between the ends of June to September. The average annual rainfall varies from 700 mm to 2000 mm. The maximum temperature recorded highest as 47°C in May and June and minimum sometimes reaching as low as 1°C in December and January.

The various tribal communities inhabit in several states of Central India and Indo-Gangetic region like Madhya Pradesh, Chhattisgarh, Jharkhand, West Bengal, Bihar, southern parts of Uttar Pradesh and Rajasthan. The dominant tribes are Abujhmaria, Baiga, Bhil, Bhilala, Gond, Kol, Korku, Korwa, Oraon, Sahariya Munda, Santhal, Meena, Garasia, Dhodia, etc. They inhabit in and around the forest areas and mainly dependent on forest resources for fulfillment of their routine requirements such as for food, medicine, fodder, fiber, hunting and fishing, household and agricultural equipments, etc. Although they mainly depend upon forest resources

TABLE 2.2   State Wise Tribal Population of Central India and Indo-Gangetic Region (Census of India, 2011)

S. No.	Name of State	Total population	Tribal population	% of Tribal Population	Region
1	Chhattisgarh	25,545,198	7,822,902	30.6	Central India
2	Madhya Pradesh	72,626,809	15,316,784	21.1	
3	Jharkhand	32,988,134	8,645,042	26.2	Indo-Gangetic Region
4	Rajasthan	68,548,437	9,238,534	13.5	
5	West Bengal	91,347,736	5,296,953	5.8	
6	Bihar	103,804,637	1,336,573	1.3	
7	Uttar Pradesh	199,581,477	1,134,273	0.6	
8	Delhi (NCT)	16,753,235	0	0.0	
9	Haryana	25,353,081	0	0.0	
10	Punjab	27,704,236	0	0.0	

for their livelihood but apart from local and rural people they also raise domestic animals such as cows, buffaloes, oxen, goats, sheep, hen, dogs, pigs, etc., for milk, agriculture and commercial purposes.

2.3   THE TRIBES OF CENTRAL INDIA

The central India, consisting of states of Madhya Pradesh and Chhattisgarh, is the largest region of India with great physical and ethnic diversity. It is also richest region in mineral and forest wealth. The famous valley of Narmada runs almost through the middle of region flanked by the two great mountains ranges: the Vindhyachal and the Satpuras. This valley has been the home of oldest aboriginal tribes of India, who hunted and fought with axes and stones implements (Russel and Hiralal, 1916; Dube and Bahadur, 1966).

The central India today has the largest tribal population in the country. According to the 2011 census, the total tribal population is about 23,139,686 constituting nearly 23.57% of the central India population. The majority of the tribals live in forest areas. The tribes of Central India belong to the two great stocks—the Kolarian or the Munda and the Dravidian. The Kolarian or the Mundas are known to be the oldest known inhabitants of this country. The main tribes are Bhil, Bhilala, Gond, Kol, Korku, Pardhan, Oraon, Korwa, Kawar, Maria, Muria, Sahariya, etc. A few primitive tribal groups such as Abujhmarhia, Bharia, Pahari Korwa and Sahariya live in the interior of forest with a primitive life-style. Some tribes have taken up settled cultivation in forest villages and countryside. A small number of tribals have migrated to suburban areas and are working in fields, factories and construction projects (Raizada, 1984).

2.3.1   *THE MAIN TRIBES MADHYA PRADESH*

There are 46 recognized Scheduled Tribes in Madhya Pradesh, India, three of which have been identified as Particularly Vulnerable Tribal Groups (PTGs) (formerly known as Special Primitive Tribal Groups). The population of Scheduled Tribals (ST) is 21.1% of the state population (15.31 million out of 72.62 million), according to the 2011 census. Bounded by the Narmada River to the north and the Godavari River to the southeast, tribal people occupy the slopes of the region's mountains.

The term *Scheduled Tribes* refers to specific indigenous peoples whose status is acknowledged to by the Constitution of India. The term Adivasi also applies to indigenous peoples of this area.

The diversity in the tribes across the state comes from differences in heredity, lifestyle, cultural traditions, social structure, economic structure, religious beliefs and language and speech. Due to the different linguistic, cultural and geographical environments, the diverse tribal world of Madhya Pradesh has been largely cut off from the mainstream of development.

Madhya Pradesh holds 1st rank among all the States/Union Territories (UTs) in terms of special tribal population and 12th rank in respect of the proportion of ST population to total population.

The main tribal groups in Madhya Pradesh are Gond, Bhil, Baiga, Korku, Bhariya, Halba, Kol, Mariya, and Sahariya. Dhar, Jhabua, and Mandla districts have more than 50% tribal population. Khargone, Chhindwara, Seoni Sidhi and Shahdol districts have 30–50% of the population of tribals. The largest population is that of Bhil tribe. The main tribes and their sub types given in Table 2.3.

According to the 2011 Census of India, Bhil is the most populous tribe with a total population of 4,618,068, constituting 37.7% of the total ST

TABLE 2.3   Tribes and Sub Tribes of Central India

Name of tribe	Sub-tribe	Districts inhabited
Gond	Pardhan, Agariya, Ojha, Nagarchi, Solhas	All districts mainly spread on both banks of Narmada river in Vindhyas and Satpura, Balaghat
Bhil	Barela, Bhilala, Patliya	Dhar, Jhabua, East Nimar
Baiga	Bijhwar, Narotia, Bharotiya, Nahar, Rai Bhaina, Kadh Bhaina	Mandla, Balaghat
Korku	Movasiruma, Nahala, Vavari, Bodoya	East Nimar Hosangabad, Betul, Chhindwara, Burhanpur
Bharia	Bhumiya, Bhuihar, Pando	Chhindwara, Jabalpur
Halba	Halbi, Bastariya	Balaghat
Kaul	Rohiya, patel, Rauthail	Rewa, Satna, Shahdol, Sidhi
Maria	Abujh Maria, Dandami Maria, Metakoitur	Jabalpur, Mandla, Panna, Shahdol, Chhindwara
Sahariya		Guna, Shivpuri, Sheopur, Gwalior, Vidisha, Rajgarh

population. Gond is the second largest tribe, with a population of 4,357,918 constituting 35.6%. The next four populous tribes are: Kol, Korku, Sahariya and Baiga. These six tribes constitute 92.2% of the total ST population of the State. Pardhan, Saur and Bharia Bhumia have a population ranging from 105,692 to 2152,472; together, they form 3.2% of state population. Four tribes, namely, Majhi, Khairwar, Mawasi and Panika have populations in the range of 47,806 to 81,335, and account for another 2.2% of the ST population. The remaining 33 tribes (out of the total of 46 tribes) along with the generic tribes constitute the residual 2.5% of total ST population. Tribes having below 1000 population are 12 in number. Bhils have the highest population in Jhabua district, followed by Dhar, Barwani, and Khargone districts. Gonds have major concentrations in Dindori, Chhindwara, Mandla, Betul, Seoni, and Shahdol districts. Other four major groups Kol, Korku, Sahariya and Baiga have registered highest population in Rewa, Khandwa, Shivpuri and Shahdol districts, respectively.

2.3.1.1 Agariya

The Agariya, a small Dravidian tribe, is considered to be an offshoot of the Gond. They by virtue of their occupational specialization as iron smelters and blacksmiths, form a separate sub-tribe. They live primarily in Uttar Pradesh and Madhya Pradesh. Those in the vicinity of Mirzapur were involved in mining and smelting iron during the British Raj. The Agariya speak the Agariya language as well as Hindi and Chhattisgarhi. There is a group known as the Agariya in Gujarat that are salt makers in the desert. It is not clear if these Agariya have any relation to the others. In the early 20th century, the Agariya in Mirzapur were divided into totemic groups. They had been heavily influenced by Hinduism. They called themselves Hindu but did not worship any of the major Hindu deities which other Hindus did. They inhabit mostly in Mandla, Raipur, Bilaspur and Shahdol districts in Central India.

2.3.1.2 Baiga

Baiga is tribe found in Madhya Pradesh, Chhattisgarh and Jharkhand states of India. The largest number of Baigas is found in Madhya Pradesh. They have sub-castes—Bijhwar, Narotia, Bharotiya, Nahar, Rai Bhaina, and Kadh

Bhaina. The Baiga tribe is concentrated mainly in Baiga-chuk in Mandla and Balaghat districts, Rewa and Shahdol districts of Madhya Pradesh and Durg and Bilaspur districts of Chhattisgarh. The Baigas are considered as a branch of the Bhuiyas of Bengal and Bihar having seven sub divisions viz. Binjhwar, Bharotia, Narotia, or Nahar, Raibhaina, Kath Bhaina, Kondwan or Kundi and Gondwaina (Elwin, 1939). Tattooing is an integral part of their lifestyle of Baiga tribe. This tribe inhabits the dense hilly forests in the eastern part of the Satpuras, in Shahdol, Bilaspur, Rajnandgaon, Mandla, and Balaghat districts. The Baigas are of Dravidian stock and are one of eight primitive tribes of Madhya Pradesh. It is believed that this tribe is an offshoot of the Bhuiya tribe of Chhota Nagpur. A distinguishing feature of the Baiga tribe is that their women are famous for sporting tattoos of various kinds on almost all parts of their body. The women who work as tattooing artists belong to the Ojha, Badni and Dewar tribes of Madhya Pradesh and are called Godharins. They are extremely knowledgeable about the different types of tattoos preferred by various tribes. Their mothers traditionally pass on this knowledge to them. Tattooing amongst the tribals commences with the approach of winter and continues until summer. The Baiga takes coarse food and shows no extravagance in this aspect. They eat coarse grain, Kodo, and Kutki, drink Pej, eat little flour and are normally content with what little that they get. One of the prime foods is 'Pej' that can be made from grounding Macca (*Zea mays*) or from the water left from boiling rice. Local people gave testimony that this food is much better and healthier than many other food that they eat. Also, beyond doubt they eat several items from the forest that includes primarily Gular leaves, Chirota, Chinch, Chakora, Sarroota, Peepal, etc. They also eat Birar Kand, Kadukand and other rhizomes. Mushroom is also a delicacy. Numerous fruits such as Mango, Char, Jamun, Tendu are also eaten. They hunt as well, primarily fish and small mammals. Baigas used to practice bewar or shifting cultivation. Inside the Baiga chuk, they are still allowed to pursue their traditional method of shifting cultivation in a restricted manner. Baigas do not use the iron plow share for tilling the land because using it is believed as lacerating "The breast of Mother Earth." However, they maintain that God made the forests to produce the necessities of life and made the Baigas, the King of Forests, giving them wisdom to discover the things provided for them. Their major implement is the iron axe. Besides primitive cultivation, the Baigas are engaged in making bamboo artifacts. They also collect honey and other forest produce. Jungle fruits and roots are plentiful. The time spent in gathering them is the happiest and most romantic of the Baigas life. The Baigas also go for hunting and fishing. Very apt in the use of bow and arrow, they have an inborn ability for hunting.

2.3.1.3   Bharia

The Bharia is another little known tribe of Madhya Pradesh, is concentrated mainly in the districts of Chhindwara and Jabalpur districts of Madhya Pradesh and Bilaspur of Chhattisgarh. Patalkot is a bowl shaped formation situated on the Satpura plateau in Chhindwara district. The shape of the area is like a horse shoe, surrounded on three sides by hill ridges, locally known as 'Kamat'. There are 12 villages in the area and almost all the inhabitants belong to Bharia tribe. Bharia is a Dravidian tribe and Bharia of Patalkot is a small group which has been living in this peculiar physiographic formation where they practice shifting cultivation. With the ban on shifting cultivation and dwindling of forests, their economy has been disturbed. They now depend on primitive agriculture, forest animal husbandry and wage labor. Due to the physical barriers, the attendance in school is very thin and the percentage of literacy is very low. As Bharia have lived for generations with the more influential and prosperous Gonds, they have many things common with them.

2.3.1.4   Bhilala

The Bhilala principally inhabits the districts of Dhar, Jhabua and Khargone. They are considered to be a mixed caste born from the alliances between immigrant Rajputs and the Bhils of Madhya Bharat region. Several Bhilala families hold estates in Malwa and Khargone and their chiefs now claim to be pure Rajputs. The Bhilala landlords usually have the title of Rao or Rawat.

2.3.1.5   Bhil

The Bhils are inhabitants of Dhar, Jhabua, Khargone and Ratlam districts of Madhya Pradesh. The Bhil tribe is one of the most important tribe of India. The name has been derived from Dravidian word 'Bil' or 'Vil' meaning a bow because they always keep bow and arrow for hunting. Bhils are primarily an Adivasi people of North India. Bhils are also settled in the Tharparkar

district of Sindh, Pakistan. They speak the Bhil languages, a subgroup of
the western zone of the Indo-Aryan languages. According to Census, 2011,
Bhils were the largest tribal group in India followed by Gond tribe. Bhils
are listed as Adivasi residents of the states of Gujarat, Madhya Pradesh,
Chhattisgarh, Maharashtra and Rajasthan – all in the western Deccan regions
and central India as well as in Tripura in far-eastern India, on the border with
Bangladesh. Bhils are divided into a number of endogamous territorial divisions, which in turn have a number of clans and lineages. Most Bhils now
speak the language of the region they reside in, such as Marathi and Gujarati.
They mostly speak a dialect of Hindi. In Gujarat and Maharashtra, the Bhil
are now mainly a community of settled farmers, with a significant minority
who are landless agricultural laborers. A significant subsidiary occupation
remains hunting and gathering. The Bhil are now largely Hindu, with Nidhi
and Tadvi Bhil following Islam, and few sub-groups in the Dangs following
Christianity. They continue to worship tribal deities such as Mogra Deo and
Sitla Matta. The Bhil are classified as a Scheduled Tribe in Andhra Pradesh,
Chhattisgarh, Gujarat, Karnataka, Madhya Pradesh, Maharashtra, Rajasthan
and Tripura under the Indian government's reservation program of positive
discrimination. The Bhil are divided into a number of endogamous territorial divisions, which in turn have a number of clans and lineages. The
main divisions in Gujarat are the Barda, Dungri Garasia and Vasava. While
in Maharashtra, the Bhil Mavchi and Kotwal are their main sub-groups.
In Rajasthan, they exist as Bhil Garasia, Dholi Bhil, Dungri Bhil, Dungri
Garasia, Mewasi Bhil, Rawal Bhil, Tadvi Bhil, Bhagalia, Bhilala, Pawra,
Vasava and Vasave.

2.3.1.6 Gond

The Gonds are a Dravidian people of central India, spread over the states
of Madhya Pradesh, eastern Maharashtra (Vidarbha), Chhattisgarh, Uttar
Pradesh, Telangana, Andhra Pradesh and Western Odisha. With eleven million people, they are the second largest tribe in Central India. The Gonds are
also known as the Raj Gond. The term was widely used in 1950s, but has
now become almost obsolete, probably because of the political eclipse of the
Gond Rajas. The Gondi language is closely related to the Telugu, belonging
to the Dravidian family of languages. About half of Gonds speak Gondi languages while the rest speak Indo-Aryan languages including Hindi. Scholars

believe that Gonds settled in Gondwana, now known as eastern Madhya Pradesh, between the 13th and 19th centuries AD. Muslim writers described a rise of Gond state after the 14th century. Gonds ruled in four kingdoms (Garha-Mandla, Deogarh, Chanda, and Kherla) in central India between the 16th and 18th centuries. They built number of forts, palaces, temples, tanks and lakes during the rule of the Gonds dynasty. The Gondwana kingdom survived till late 16th century. They also gained control over the Malwa after the decline of the Mughals followed by the Marathas in the year 1690. The Maratha power swept into Gondland in the 1740s. The Marathas overthrew Gond Rajas (princes) and seized most of their territory. Some Gond zamindaris (estates) survived until recently. During the British regime in India, Gonds challenged the Britishers in several battles.

2.3.1.7 Kol

The Kol tribe is regarded as a Kolarian or Mundari tribe. The term Kol as generic category occurs in the Sanskrit literature along with Bhil and Kirat. The Kol tribe is concentrated principally in Jabalpur, Rewa, Narsinghpur, Damoh and Satna districts of Madhya Pradesh. They are offshoot of Munda. They practice traditional system of agriculture and grow kodo, barley, maize, bagaridhan as staple food crops. Kols are very fond of Mahua liquor. The Kols are mainly laborers. There are two important divisions of the Kols known as the Raitia and the Rautele.

2.3.1.8 Korku

The Korkus fall in Munda or Kolarian tribal group. The term Korku means group of humans, Kor-Manav and Ku is used for plural. As per the Korku myth, Mahadev sent a crow to fetch Kavi clay from Maalik to the world with it. Mula-Muli was the first human couple created by him, from whom all the tribes originated in course of time. The crow informed Mahadev that the Maalik had given the Kavi maati (clay) on the condition that it would be returned. Mahadev gave an herb also to the crow – one for Mula-Muli and the other for itself. He exchanged the herbs and consumed the herb meant for the humans and gave to human the one meant for it. The herb was Sanjeevani on consuming the Sanjivani herb the crow became immortal, whereas the human and all other creatures dies or are, as required to return to the Malik

(God) again and again to return the clay. The Korku houses are built oppo-
site each other in two rows and the Otala or platform decorated with ochre,
chalk and yellow clay give a distinct identity to Korku Dhada. On Deepawali
night the women of Thathya, that is, Gwala families paint auspicious pictures
called Gudaniyan with red clay and chalk on the walls of Korku houses. The
Thathya men go door-to-door playing Bhugadu or a long flute and dancing all
the night. The Korku women are fond of tattooing. On the forehead a figure
resembling English letter M is tattooed and on its upper and lower sides two
dots are made which are called Kapar Godai. Dots are also tattooed on the
chin, cheek and nose. On the hands-chowk, Rani Godai, etc. are tattooed. In
the figures of tribal tattooing the dots often symbolize food grain or fire. Here
tattooing is ornamentation of the body and also device to stay disease-free
and strong. Tattoos are also created to ensure overflowing granaries in the
house. They are concentrated principally in East Nimar, Hosangabad, Betul
and Chhindwara districts of Mahakaushal region. The Korku tribe as a whole
is usually divided into two groups, Raj Korku and Potharia.

2.3.1.9 Nagarchi

The Nagarch is one of the lesser known tribes of Madhya Pradesh. The tribe
is found scattered over a large area in the districts of Chhindwara, Seoni,
Balaghat, Mandla and Durg. The Gagarchi is considered to be a branch of
the Gond. They are occupational group of musicians.

2.3.1.10 Panika

Panika or Panka is a Hindu caste that is found in Chhattisgarh, Madhya
Pradesh and Uttar Pradesh. They were earlier a sub-group of Kotwar which
have now got separated. As per the history, their name originated from
the word Pankha, which meant hand fan. Historically, the community was
involved in fan manufacturing and hence the origin of their name. They
were fully involved in the music, dance and party during celebrations like
the marriage, Barahon, Ramleela, etc. The Kotwar Panika worked as the
watchmen in south eastern areas of Uttar Pradesh. They are usually found in
the areas of Mirzapur and Sonbhadra and have now been incorporated into
Hinduism. In the olden times, this tribe was known amongst the other tribes
for their honesty. However, in the recent times, this characteristic seems to
have been disappeared.

2.3.1.11   Pao

The Pao tribe is also known as Pabra. The tribe is found principally in three districts viz. Chharpur, Satna and Shahdol.

2.3.1.12   Pardhan

The Pardhan tribe is considered to be an inferior branch of the Gond tribe and their traditional occupation is to act as the priest and minstrel of the Gonds.

2.3.1.13   Sahariya

The Sahariyas are dispersed in in 21 districts but their largest concentration and their native home is in Gwalior and Chambal divisions of Madhya Pradesh. The Sahariyas are found principally in Shivpuri, Gwalior, Datia and Guna districts of Gwalior division and Sheopur and Bhind district of Chambal division. They are also found in some parts of Bidisha, Raisen and Sehore districts of Madhya Pradesh, Kota and Bundi districts of Rajasthan and Jhansi and Lalitpur districts of Uttar Pradesh. The term Sahariya comprises of two words, for example, *Sa* means 'Sahachar' (Companion) and *Haria* means 'Lion'. The term Sahariya indicates that it lives in forest like a lion. The Term Sahariya probably means inhabitants of forests. Even now all adult male Sahariyas always keep axe on their shoulders. Because this tool is of multiple use in thick forest to cut tree, to cut down branches, to takcout the bark, to make wooden tools, to chop-off bamboos for making baskets or any other uses like hunting wild animals and defense, etc. Their economic condition is extremely poor with hardly any regular source of sustenance. They are by and large landless and mostly work as casual laborers. Literacy percentage is very low in this tribe. Generally Sahariyas are of medium statured and dark complexioned. They are very fond of liquor. The Sahariyas men and women along with their children drink liquor. Gudna (inscribing the body with some marks) is a must for all Sahariya women. They believe this ritual to be the highest and permanent ornament of women. These Sahariya women like to inscribe their bodies with tattoos marks through imparting permanent pigmentation on the skin with various designs (Sikarwar, 1997).

2.3.2 TRIBES OF CHHATTISGARH

Chhattisgarh, a state that is at the vanguard of Indian industries and also a repository of minerals has a diverse cultural legacy. Chhattisgarh and tribal culture are two tautological terms since a third of the state's populace is dominated by tribals. Chhattisgarh, the "rice bowl" of India is famed for its mind-blowing natural splendor, cultural extravaganzas, storehouse of minerals and power and large iron and steel plants. The populace of Chhattisgarh is mainly dominated by tribals, of which the Muria race of aborigines holds special place. The tribes of Chhattisgarh are an unique race who mainly inhabits the dense forests of Bastar. In fact more than 70% of Bastar's population is composed of tribals who account for 26.76% of Chhattisgarh's entire tribal populace. The lifestyle of the tribal people is unique and imbibed with traditional rituals and superstitions. They are a friendly and jovial lot who are industrious and diligent. Although shrouded in poverty, they live life to the hilt and love to celebrate every joyous occasion. Food, drink, music, dance, mirth and merriment add color to their otherwise simple lives. The tribal women also love to adorn themselves in ethnic jewelry. One of the Chhattisgarh's eminent tribes is the Gonds or the Kotoriya tribe. The etymological connotation of their name comes from the Telugu term "Kond" meaning hills. The Gonds dominate most of Chhattisgarh's tribal population and primarily depend upon agriculture, forestry, cottage industries, hunting and fishing for their subsistence. The talking point of the lifestyle and culture of the Gond tribals is their Ghotul marriage policy, a one of a kind arrangement for conducting the nuptial rituals. Another tribe, the Abuj Maria lives in isolation in the dense and secluded enclaves of the forests of Narayanpur Tehsil in Basir. They are a ferocious and barbaric tribe who believe in primitive customs and are hardly tempted by the material pleasures of life. The main tribes of Chhattisgarh state of Central India are:

2.3.2.1 Abujhmaria

The term 'Abujhmaria' is an anthropological make shift and is not the name by which the tribe is known or refers to itself. The Maria live in Abujhmarh region (hence known as Abujhmaria) which extends over Narayanpur, Bijapur and Dantewar districts, but most of the area and population is in Narayanpur district. The area is covered with hills, dales and dense forests.

The forest is rich in flora and fauna. Because of their isolation, the hill Marias has retained most of their religious and cultural traditions. The areas inhabited by the Abhujmaria tribals are a dense forest that sprawls across nearly 1500 miles of lush greenery. They are very much feared by mankind. They are a primitive race whose mannerisms are rather ferocious. The savage and barbaric tribesmen are hostile to strangers and sometimes directly shoot them with their arrows. This hill Maria tribals live in the forest enclosure in a world of their own, completely out of the touch with human society. Money and other material pleasures seldom tempt the people of this race. They are hardly affected by the ravages of time and their recluse lifestyle not only keeps them out of touch with modern civilization, it also helps to preserve their archetypal tradition and customs. In fact the Abhujmaria is one of the few tribes that have many to keep their quintessential culture alive and unaffected by the vestiges of time. The tribal people are scantily clad and simply cover themselves with a loincloth. However, they are very fond of traditional ethnic jewelry and adorn themselves with several iron rings strung around their neck. The women love wearing earring and sometimes pierce as many as 14 holes in their ears and hang two rings or studs from them. The Abhujmaria tribals are more bestial than human in their characteristics. They seldom clean themselves or their garments. They do slash and burn cultivation known as 'Penda', 'Korsa', 'Madia' and other millets which form their staple food are grown in this 'Penda'. They also have some fields close to the village where paddy is grown. Kitchen gardens for growing tobacco plants and vegetables are invariably attached with all houses. They rear pigs, raise poultry and collect forest produces mainly for their own consumption. Hunting and fishing are also common (Grigson, 1949).

2.3.2.2 Bhaina

The Bhainas are found only in Chhattisgarh. They are concentrated mainly in Bilaspur district and its neighboring areas. The tribe is considered to have a mixed descent from the Baigas and the Kawars. Their connection with the Baigas is shown by the fact that in Mandla. The Baigas have two sub divisions which are known as Rai or Rai-Bhaina and Kath-Bhaina. A Bhaina is also frequently found to be employed as village priest and magician. The Bhainas are also closely connected with the Kawars who still own many large estates north of Bilaspur.

2.3.2.3 Bhattra

The Bhattra tribe found principally in Bastar and south of Raipur district. They are cultivators, farm servants and also work as village watchmen.

2.3.2.4 Bison Horn Maria

Bison Horn Maria is one of the famous tribal groups of India. Mostly found in Chhattisgarh's Bastar region, they are a major sub-caste of a tribal community called Gond. Apart from the Jagdalpur Tehsil towards the south of the river Indravati in the state of Chhattisgarh, they mainly reside in the district of Garhichiroli in Maharashtra as well as some parts of Madhya Pradesh. Their introvert nature makes them live in isolation in the interiors of dense forest areas of these states. This tribal community of Chhattisgarh derived their name from their unique custom of wearing a distinctive headdress, which resembles the horns of a wild bison. They generally wear that headdress during marriage dances or other ceremonies. This main distinct language spoken by this tribe of Chhattisgarh is Dandami Maria. Some of them even speak mutually unintelligible Gondi dialects, which is an oral language of Dravidian origin. Most the members of this community of Bison Horn follow the traditions and customs of the Hindu religion. Some of them are ethnic religionist though. As per the World Evangelization Research Center's estimation, about 70% of this tribe is Hindu. Besides worshipping the earth goddess Danteshwari for their retention, Bison Horn Marias worship spirits and non-human objects. Any resource for the religion of Christianity in Dandami Maria language is not available among Bison Horn Maria tribal community. This in turn has made them unaware of this religion. Being followers of the Hindu religion that makes them believe in a super power, this tribe of India believes non-human objects to have spirits. Their religious belief is a combination of Hinduism with animistic beliefs. They worship varied gods. The outskirts of different village are enshrined for the god of that particular clan. Apart from being related to each other, these Clan Gods are supposedly territorial. These Clan Gods are housed on the border of every village so as to protect the village from any external or black magic. Besides this, they even belief in rebirth. This unifying feature makes them check for an identification mark of ancestors in the body of a new born baby to know whose soul has been reincarnated into that baby's body. The spiritual belief of this tribal group includes sorcery or black magic. Legendary and strange stories about their physical and spiritual powers are quite popular. They attribute any illness to be caused by a negative force manipulated by an enemy. Though medicine people are very powerful, still the villages of this tribe remain constantly concerned about black magic and occult forces. The Bison Horn Maria men have got a distinct hairstyle of long pony tail. Besides that, they carry a tobacco box and a special kind of comb. This comb remains attached to their loincloth. Women of Bison Horn tribal group generally dress in white skirts. They even use varied jewelry for adornment. The bison horn shaped headdress worn by them are nowadays made of cattle horns because of the scarcity of bison horns. Those headdresses are placed on a frame of bamboo and decorated with feathers of peacock or chicken and hanging cowry shell strings. Such a headdress is passed on from one generation to another.

2.3.2.5   Dhurva

Durvas (Parjaas) are the third largest tribal community of Bastar coming next to the Maria and Muria. The Dhurvas are possibly the most significant indigenous tribe that occupies the domicile of Chhattisgarh's Jagdalpur, Dantewara and Konta. In terms of social hierarchy, the Dhurvas rank second only after the elite Bhatra tribals. The Dhurvas are also recognized by the popular nom de plume Parjaas that locally mean the Public. However, the tribesmen prefer the nomenclature Dhurva that in their native dialect means a local village chieftain. The Dhurvas are a proud, courageous and highly caste conscious race who only mix with people of an equal social standing. Their society is progressive and broadminded and polygamy is a common and accepted practice. The women, who are responsible for all domestic matters are held in high esteem and thus they are very haughty. The men are generally indolent and except for the routine cultivation and hunting, they don't take much interest in domestic affairs. The Dhurvas depend upon agriculture for their economic subsistence. The tribal people are also talented craftsmen whose expertise is manifested by the exquisite handicrafts that they make out of cane and other forest products. They are highly religious and pious and worship several local cult gods and goddesses. Mirth and merrymaking are an eminent part of all celebrations and no religious celebration is complete without animal sacrifice and coconut is also offered to mollify the deities.

2.3.2.6 Dorla

The Dorlas are found in the southern parts of Bastar. Their territory starts from about 40 km south to Sukma in Konta tesil. The river Kolab or Sabri divides the districts from Orissa, forming the eastern border of the Dorla tract. The Dorla tract extends southwards up to Konta and in the west the Dorlas are distributed over the adjoining tehsil of Bijapur (Dubey, 1970).

2.3.2.7 Gond

Chhattisgarh is a nature lover's paradise. The state provides a glimpse of central India's cultural potpourri and of the prevailing lifestyle. The tribes of Chhattisgarh are mostly a primitive race who faithfully follows all traditional customs and their archetypal age-old ritual. The oldest and most populous tribes of Chhattisgarh are the Gonds. The Gonds Tribals, who are also recognized as the Koytorias are widely dispersed throughout the state. However, they mainly predominate the dense forests enclosed in southern Chhattisgarh's Bastar District that accounts for more than 20% of Chhattisgarh's population. The three principal sub castes of the aboriginal Gonds are the Dorla, Maria and Muria races. The etymological significance of the term Gond is derived from the Telugu connotation "Kond" meaning hill. The tribal economy is predominantly agrarian. But the poverty stricken people also depend upon forestry, local cottage industries, hunting and fisheries for their economic subsistence. Some of the Gond people are however employed in cushy primary sector jobs as well as other allied industries. The unique and one of a kind Ghotul marriage tradition of the Gonds is renowned all across the world. They mainly practice the traditional Hindu customs and marry within the family in order to preserve the customary completion of the nuptial vows within the family. Of course some of the romantic daredevils choose to elope with their beloved. Gond marriages however are not a bed of roses. Remarriage, widow marriage, divorce and marrying in laws as well as brothers and sisters are a common affair. Gond society is somewhat matriarchal where the groom has to pay a substantial dowry top the bride's family to pay his due respects.

2.3.2.8   Halba

The Halba tribe is a popular tribe who has happily settled in the bucolic lands of Chhattisgarh. The Halba tribals are widely dispersed all over Chhattisgarh, Maharashtra, Madhya Pradesh and Orissa. One of India's predominant tribes, the Halba tribals inhabits the districts of Durg, Bastar and Raipur in Chhattisgarh. The mannerisms and lifestyle of the Halbas who inhabit Bastar closely resemble that of their counterpart who resides in Andhra Pradesh's Warangal District. The Halba tribe owes its nomenclature to the term 'Hal' that locally means plowing or farming. This clearly implies the Halbas were primarily farmers although nowadays they are involved in a myriad of professions of their choice. Of all the tribes that occupy Chhattisgarh, the Halbas are possibly the most affluent and progressive lot. They also enjoy the privileged status of a high local caste and hence are deeply revered in the tribal society. The unique individuality of the Halbas is evinced by their apparels, dialects and traditional customs. What add to the diversity of their dialect are the pronounced traces of Oriya, Marathi and Chhattisgarhi languages.

2.3.2.9   Kamar

The Kamars are concentrated principally in the former Zamidaries of Bindranawagarh, Saurmarh and Fingeshwar and in the Wagri and Sihawa tracts in Dhamtari tehsil of Raipur district.

2.3.2.10   Kawar

The Kawar tribe is also called Kanwar and Kaur. The home of the Kawars is the hilly area of Chhattisgarh, north of the Mahanadi in the districts of Bilaspur, Raigarh and Surguja. Their sub-divisions include Cherw, Rautia, Tanwar, Kamalbansi, Paikara, Dudh Kawar, Rathia and Chanti.

2.3.2.11   Khairwar

The Khairwar tribe is also known as Kharwar, Khaira and Khairwa. The tribe is found mainly in the districts of Bilaspur, Surguja districts and parts of Vindhyan region of Madhya Pradesh.

2.3.2.12   Korwa

The Korwa tribe belongs to the Kolarian family. This tribe is concentrated principally in the districts of Bilaspur, Surguja and Raigarh of Chhattisgarh state. The principal sub-divisions of the tribe are the Diharia or Kisan Korwas and the Paharia Korwas (also called Benwaria).

2.3.2.13   Muria

The Murias are one of one of the innumerable tribes that inhabit Chhattisgarh. They are a prominent sub caste of the Gonds who dominate the populace of Chhattisgarh. The Muria tribesmen primarily reside in the dense forest zones of Narayanpur 1 and Kondagon districts of Chhattisgarh, the home of majority of the tribals. Unlike the primitive social outcasts like the Abhujmaria and Bison Maria tribes who live in isolation in the secluded corners of the jungles. The Murias are more advanced and broadminded and live in the open amidst the vast rolling plains and valleys. The Muria economy is predominantly agrarian. They cultivate rice in plenty. Some Muria tribals also depend up on collecting forest products. The forest products are not only used to make useful products, the edible parts are also consumed by the poverty ridden tribals. In case of illness and maladies they seek the remedial powers of the Mahua plant. The tribals are a highly superstitious lot who believe in worshiping the cult gods and goddesses. The Muria society is devoid of a caste system and the people also practice magic, dark arts and wizardry. Their society is quite progressive and although Ghotul marriages are the common practice, dating and also indulge in free sex (Elvin, 1947).

2.3.2.14   Nagesia

The Nagesia live principally in Surguja and Jaspur districts of Chhattisgarh. The Nagesia is considered to be an offshoot of the Mundas.

2.3.2.15   Oraon

The Oraon are an important Dravidian tribe of Chhattisgarh, is concentrated principally in Raigarh and Surguja districts. They are commonly known as 'Dhangar' which means a farm servant. The Oraon call them Kurukh. The primary occupation of the Oraon is agriculture and some of their women make mats and sell them. To most of the Oraon, the economic importance of forest has considerably been reduced due to its extensive destruction. Only those who live near the forests have some degree of dependence upon it.

2.4 THE TRIBES OF JHARKHAND

The Scheduled Tribe (ST) population of Jharkhand State is as per 2001 census 7,087,068 constituting 26.3% of the total population (26,945,829) of the State. Among all Sates and UTs, Jharkhand holds 6th and 10th ranks terms of the ST population and the percentage share of the ST population to the total population of the State, respectively. The state has a total of 30 Scheduled Tribes and all of them have been enumerated at 2001 census. The Scheduled Tribes are primarily rural as 91.7% of them reside in villages. District wise distribution of ST population shows that Gumla district has the highest proportion of STs (68.4%). The STs constitute more than half of the total population in Lohardaga and Pashchimi Singhbhum districts whereas Ranchi and Pakaur districts have 41.8–44.6% tribal population. Kodarma district (0.8%) preceded by Chatra (3.8%) has the lowest proportion of the STs Population. Jharkhand has 32 tribal groups:

The tribes of Jharkhand consist of 32 tribes inhabiting the Jharkhand state in India. The tribes in Jharkhand were originally classified on the basis of their cultural types by the Indian anthropologist, Lalita Prasad Vidyarthi. His classification was as follows:
  • Hunter-gatherer type — Birhor, Korwa, Hill Kharia
  • Shifting Agriculture — Sauria Paharia
  • Simple artisans — Mahli, Lohra, Karmali, Chik Baraik
  • Settled agriculturists — Santhal, Munda, Oraon, Ho, Kurmi, Bhumij, etc.

All 32 tribes of Jharkhand are Baiga, Asur, Banjara, Bedia, Bathaudi, Binjhia, Bhumij, Birjia, Birhor, Chick Baraik, Chero, Gorait, Gond, Karmali, Ho, Kharwar, Khond, Kisan, Kharia, Korba, Kora, Mahli, Lohar, Munda, Mal Paharia, Parhaiya, Oraon, Sauria Paharia, Santhal and Savar. Most of the people belonging to different Jharkhand tribes dwell in villages. These villages are grouped into tolas. They live in mud houses, which are devoid of any window. They often adorn the external surface of their houses with paintings. The major food for these tribal people is rice and the flesh of birds and animals.

2.4.1 ASUR

Asur in Jharkhand is one of the thirty major tribes of people who have made the state of Jharkhand their home. The people who belong to this tribe form quite a big part of the total population of the state of Jharkhand. It ranks 21st among all the 30 tribal groups of the state, in terms of population, that is, there are as many as 9 tribal groups in the state that have a smaller population than the Asur of Jharkhand. The people belonging to Asur at Jharkhand stay within houses made of clay. They live in villages that are grouped into different tolas for the convenience of the people. The houses in which the people belonging to Asur tribe live do not have any window. The people love to make their houses look even more beautiful by painting them on their external walls. They thrive mostly on the flesh of animals and birds and rice. The total population of the tribal group of Asur is 7783. The rate of literacy among the people of the state of Jharkhand is not very satisfying. The rate is only 10.62%. Though their total number is not ignorable, the percentage of the total population of the state that they cover is not a massive one. The people who belong to the Asur tribe cover only 0.13% of the total population of the state.

2.4.2 BAIGA

The Baiga is one of the most important tribes in the state of Jharkhand in India. The people who belong to the Baiga tribe of Jharkhand are reportedly least civilized of all the different tribes of the state. The people of the tribe of Baiga in Jharkhand inhabit in a particular district of the state. The name of this district of Jharkhand is the Garwa district. The people who belong to the tribe of Baiga constitute a Kolerian ethnic community. The name of this tribe of Jharkhand has quite a few meanings. One of them is 'ojha' or a person who makes medicines. Many of the people who belong to the Baiga tribe make medicines by profession, though their chief traditional occupation has been shifting cultivation

2.4.3 BANJARA

This is another group that stands threatened by rapidly dwindling numbers. Their villages are located near hills and forests. They are skilled weavers and make mats, baskets, trays, brooms, etc. from grass growing wild in the forest. They move residence often and in a group. They also go around villages to sing prayers on the birth of a child. They constitute the 'smallest' tribal population in Jharkhand. They use local language. Though smallest in number, the Banjara tribe in Jharkhand is a recognized part of the tribal community. Unlike the Banjara tribe of Rajasthan, the Banjaras of Jharkhand lead a settled life. They generally live in thatched huts with kuchcha walls. Though they remain unperturbed by the modernization around, recent years has seen far reaching changes in the relationship between the Banjaras and the large society. The literacy rate of the Banjaras is about 12.38%. The colorful lives of the Banjaras now has become the source of entertainment to the entire state. Tribal festivals like Sarhul, Tusu and Sohrai are celebrated throughout the state. Banjara music and dances like Chaw, Natua, Ghatwari and Matha now-a-days has become sources of recreation even to the tourists to Jharkhand. They now seem to plan their visit to Jharkhand in the festive seasons of the tribes in Jharkhand.

2.4.4 BIRHOR

The Birhors, though much less in number, are found in Jharkhand. The name Birhor is derived from words 'bar' meaning jungle and 'har' meaning man and thus the Birhors are forest dwellers in true sense. They are a nomadic community, though the Government tries to settle them. In Jharkhand they are distributed in districts of Ranchi, Gumla and Hazaribagh in Chhotanagpur plateau. Their language is Birhor, which is considered to be an Austro Asiatic language. They also speak Sadri and Hindi. Regarded as landless community, Birhors are mainly gatherers. They are also found engaged in rope-making.

2.4.5 BIRJIA

Birjias of Jharkhand live in triangular or rectangular huts made up of bamboo, wood or mud. The huts of the Birjia tribe, usually, are devoid of windows: the huts have a small gate which is closed with a tati or a mat. The Birjia tribe possesses a patriarchal society: a Birjia family is usually a nuclear family with father as the head of the family. Moreover, the Birjia society is known as a monogamous society, yet the prevalence of bigamy cannot be overruled.

It goes without mention that they have a rural society where agriculture and forests play a vital role in the socio-economic life. Hence, Birjia economy is based on agricultural yields, as well as on hunting, fishing and labor.

2.4.6 CHERO

The Chero are one of the scheduled tribes of Jharkhand. In Jharkhand, Chero dwell in the districts of Ranchi, Santhal Pargana, Latehar and Palamu. Palamu seems to have a larger concentration of the Chero tribe in Jharkhand. Besides, the Chero at Jharkhand are also found in Bhojpur, Gaya, Champaran, Munger, Daltonganj, Patan, Lesliganj, Bhawanathpur, Rohtas, etc. It is noteworthy in this context that the Chero, also known as Cherwas or Cherus, was a martial group that annexed many new territories through war. They are said to be descendants of the Kshatriya lineage known as Chandravanshi.

2.4.7 GOND

On an all India basis the Gonds are the most popular tribal community (with a major part concentrated in Madhya Pradesh). Gond belongs to Palamu, Singhbhum and Ranchi districts. The Gond of Jharkhand, linguistically, belongs to the Dravidian race; but, racially, the Gond at Jharkhand hail from Proto- Australoid stock. They are usually forest fringe dwellers with strong family kinship based on love and affection. Marriage, birth are happy occasions. Women are the custodians of culture, norms and values in Gond society.

2.4.8 KHARIA

Kharia at Jharkhand belong to the Proto-Australoid group. The Kharia in Jharkhand are said to be the descendants of Nagvanshi Raja and are divided into three major sections namely Dudh Kharia, Dhelki Kharia and Hill Kharia. The Kharia of Jharkhand are one of the most primitive tribes that chiefly depend on the resources obtained from the forests of the territory. The Hill Kharia largely depend on roots, edible herbs, leaves, fruits, seeds, flowers, honey, wax, etc.; the Dhelki Kharia and Dudh Kharia depend on agriculture. The Kharias are basically centered round the hills and plains adjoining the hills. The settlement of the Kharia is dispersed throughout the different districts of Jharkhand. The houses of the Kharias are made up of straw, bamboo, mud, rope, etc. and are generally rectangular in shape. The houses, generally, possess a single room that contains a bedroom, kitchen and bathroom: we can find a cow-shed or pig stay attached to the house.

2.4.9 KHARWAR

Kharwar are found in the Latehar, Lohardaga, Ranchi, Hazaribagh, Chatra, Daltonganj, and Garhwa districts. The Kharwar of Jharkhand is also found in the Rohtasgarh district in Bihar. It is interesting to note that the Kharwar in Jharkhand are a group of traditional people who use the Khair grasses for various purposes. Due to the excessive use of the Khair leaves by them, the Kharwars are named so. A family is the unit of the Kharwar society which is generally nuclear in structure and comprises a husband, wife and their unmarried children because after marriage the children set up their own families. In fact, the concept of the joint family is completely absent from the Kharwar society.

2.4.10 MUNDA

The abode of the members of the extremely cherished Munda tribe is not confined to the borders of the state of Jharkhand. The Munda people have also penetrated into other beleaguering states of Orissa, Chhattisgarh, Bihar and West Bengal. The Mundas are one of the major tribal group of the state. Another Austro-Asiatic race ranked in the third position by way a population in Jharkhand. History suggests that they migrated here from north-western parts. Munda woman are very fond of ornaments. They mostly concentrated in Khunti area of Ranchi district. Their Language is Mundari. But according to some scholars they are basic people of here. In present day maximum numbers of Mundas lived in Ranchi, Gumla, Simdega and

Singhbhum. Head of Munda society is called as "Pahan" which is main head of the "Padha Society". In each religious place worship has done by Pahan. Their main God is "Singbonga". There are many gotra in munda society – Aaied, Kongadi, Gadi, Kerketta, Terom, Toppo, Dhanbar, Nag, Kachchhap, etc. Marriage between same gotras is not accepted. Their mother tongue is Mundari one of the major Ausro-Asiatic languages of India. They also speak Hindi. The Mundas are divided into totemic clans. They are patrilocal and agriculture is their traditional and primary occupation which they supplement by forest produce. Education has spread among them remarkably and many are employed in private and government organizations. Mundas are mainly nature worshippers. They also worship their ancestors, clans and village deities. Mundari folk songs and music are rich.

The abode of the members of the extremely cherished Munda tribe is not confined to the borders of the state of Jharkhand. The Munda people have also penetrated into other beleaguering states of Orissa, Chhattisgarh, Bihar and West Bengal. As a matter of fact, a handful of the Munda tribals have also been noted to have established their permanent domicile in Bangladesh as well.

The sphere where the Munda in Jharkhand bears a remarkable similitude with its contemporary tribes is mainly concerned with a conspicuous dialect and a unique life-style. This could be vividly illustrated from the fact that the lingo restricted to them is known as 'Mundari'. The legend that exists behind this extremely coveted and revered language elucidates that Mundari actually belonged to the Austro-Asiatic family of languages (Hoffman, 1950).

2.4.11 ORAON

The Oraons are the second largest tribal community of Jharkhand with over 10 lakh population. Majority of them reside in Ranchi and Hazaribagh areas. They speak 'Kurukh' belonging to a sub-group of Dravidian language family. They also speak Hindi and Sadri. The Oraons have several exoramous totemic clans and they use their clan names as surnames. They prefer to live in forest areas, land and forests being their main economic resources. They are mainly settled cultivators but depend on forest produces during the lean months. Educated Oraons are engaged in government and private jobs in large number. They have their own folk songs and folk tales. Both men and women participate in dance during festivals.

2.4.12 SANTHALS

With over 18 lakh population, the Santhals are the largest tribal group in the state, they dominate Jharkhand's tribal population. Their concentration is mainly in Dumka, Godda, Deoghar, Jamtara and Pakur districts of Santhal Parganas and East and West Singhbhum districts. They have a unique heritage of tradition, surprisingly sophisticated customs and tastes and lifestyles, and the most evocative of folk music, song and dance. Their mother tongue is Santhali, a language of Austro-Asiatic family. The Santhali language is elaborate, structured, richly endowed, with its own, recognized 'script', 'Alchiki', perhaps unmatched by any other tribal community, anywhere. Most of them also know Hindi or Bengali. The cultural refinement of the Santhals is reflected in their daily affairs – in the design, construction, color combinations, and the cleanliness and order of their homes. The drawings and motifs on the walls and the neatness of their courtyards, will shame many a swank, modern urban home (Boding, 1925).

2.5 TRIBES IN WEST BENGAL

As per the Constitution (Scheduled Tribes) Order, 1950, the following were listed as scheduled tribes in West Bengal: 1. Asur, 2. Baiga, 3. Badia, Bediya, 4. Bhumji, 5. Bhutia, Sherpa, Toto, Dukpa, Kagatay, Tibetan, Yolmo, 6. Birhor, 7. Birjia, 8. Chakma, 9. Chero, 10. Chik Baraik, 11. Garo, 12. Gond, 13. Gorait, 14. Hajang, 15. Ho, 16. Karmali, 17. Kharwar, 18. Khond, 19. Kisan, 20. Kora, 21. Korwa, 22. Lepcha, 23. Lodha, Kheria, Kharia, 24. Lohara, Lohra, 25. Magh, 26. Mahali, 27. Mahli, 28. Mal Pahariya, 29. Mech, 30. Mru, 31. Munda, 32. Nagesia, 33. Oraon, 34. Parhaiya, 35. Rabha, 36. Santal, 37. Sauria Paharia, 38. Savar, 39. Limbu (Subba), and 40. Tamang.

As per 2011 census scheduled tribes numbering 5,296,953 persons constituted 5.8% of the total population of the state. Santals constitute more than half (51.8%) of the total ST population of the state. Oraons (14%), Mundas (7.8%), Bhumij (7.6%) and Kora (3.2%) are the other major STs having sizeable population. Along with Santal, they constitute nearly 85% of the state's total ST population. The Lodhas, Mahalis, Bhutias, Bedias, and Savars are the remaining STs, and having population of one% or more. The rest of the STs are very small in population size. More than half of the total ST population of the state is concentrated in Medinipur, Jalpaiguri, Purulia,

TABLE 2.4   ST Population of West Bengal

S. No	Scheduled Tribe	Population	Percentage of the total ST population
1	Santal	2,280,540	51.8
2	Oraon	617,138	14
3	Munda	341,542	7.8
4	Bhumij	336,436	7.6
5	Kora	142,789	3.2
6	Lodha	84,966	1.9
7	Mahali	76,102	1.7
8	Bhutia	60,091	1.4
9	Bedia	55,979	1.3
10	Sabar	43,599	1

and Bardhaman districts. Of the remaining districts, Bankura, Malda, Uttar Dinajpur, and Dakshin Dinajpur have sizable ST population.

Tribes of West Bengal have occupied a large section of the total population of West Bengal. West Bengal state is the abode of numerous tribes who reside in the rural parts of the state. Their culture, religion, costumes, tradition have enriched the culture and tradition of West Bengal. Many of these tribes of West Bengal have adapted to diverse religious practices. Among them, Hinduism, with its subdivision of various castes and native tribes, has filled three fourths of the total tribal populace of the West Bengal state. Most of the people of the tribal groups of West Bengal speak in Bengali with their own localized accent. In fact these tribes are quite proud of their enriched culture and language. Variant dialects are also equally popular amongst these tribes of West Bengal. They are, in general, confined to the rural belt of the state. However, a small portion of this population has now moved to the urban belt, in search for employment and a better lifestyle.

2.5.1   ORAON

Oraon tribes of West Bengal are one of the biggest tribes in the whole of south Asia. These Oraon tribes converse with each other in Kurukh, a popular language belonging to the Dravidian family. Most of the Oraon tribal communities have taken up the profession of cultivation. Traditionally,

Oraons depended on the forest and farms for their ritual and economic livelihood, but in recent times, a few of them have become mainly settled agriculturalists. Small numbers of Oraons have migrated to the northeastern part of India.

2.5.2 MUNDA

Munda tribes are one of the largest tribes of India. Main languages spoken by these tribes include 'Munda' or 'Killi', Santali and Mundari. These Munda tribes are pious and religious minded, and mainly practice Hinduism.

The tribes of West Bengal are mostly farmers but many of them are engaged in some other occupations like carpentry, weaving, hunting, fishing, etc. Rice is the staple food of the tribal people of Bengal and sometimes they include fish, meat, chicken and fowls in their diet. Some of the tribes are adept in art and craft and their created items give evidence to the exclusive tribal arts. Tribes of West Bengal are famous all over the world for its proficiency in art and crafts. They are truly skilled in creating splendors to its outstanding works on carpentry, terracotta, drawings, and textile. Earthenware, brass and copper ware, needle works, wall-hanging, hand looms, fine muslin and silk clothes, wood statues, cane works, etc. are a couple of examples of handcrafts which have developed from the villages households of these tribes of West Bengal. Maximum of these craft products embellish the cottage industry of the state and have been spine of the economy of the rural provinces of the West Bengal state.

2.6 TRIBES OF RAJASTHAN

Rajasthan is home to various tribes who have very interesting history of origin, customs and social practices. Rajasthan tribals constitute around twelve percent of the total population of the state. The tribes of Rajasthan, India constitutes of mainly Bhils and the Minas. In fact, they were the original inhabitants of the area where Rajasthan stands now. Apart from these main tribes, there are also a number of smaller tribes in Rajasthan. However, all Rajasthan tribes share certain common traits, the variations being in their costumes, jewelry, fair and festivals, etc. The various tribes of Rajasthan are given in the following sub-sections.

2.6.1 BHILS

Bhills are the bow men of Rajasthan. The Bhils comprise of 39% of Rajasthan's tribal population and form an important group in the southern part of the state around Dungarpur, Udaipur and Chittorgarh. Their population dominated in the Banswara. Bhils are believed to be fine archers. The generic term derives from Bil (Bow) which describes their original talent and strength. Infect, Bhil bowmen even found a reference in the great Hindu epics Mahabharata and Ramayana. The Bhils were originally food gatherers. However, with the passage of time, they have taken up small-scale agriculture, city residence and employment. The major festivals of Bhils are the Baneshwar fair (held near Dungarpur) and Holi.

2.6.2 BISHNOIS

The Bishnoi community of Rajasthan have been identified as conservators. The Bishnois can be mostly seen in the western Rajasthan, especially in the Jodhpur and Bikaner areas. The Bishnois were the followers of the 15th century saint Guru Jambeshwar who due to a prolonged period of drought bade all his followers to protect all animal and plant life, since it seemed the only way to nurture nature. Ever since, their sanctity has extended to all trees and animals, and they do not allow either felling or hunting on their lands. They also believe that in their afterlife they will be reincarnated as deer, due to this the herds of deer can be seen roaming in their fields without fear. The Bishnois men are distinguished by their large white, turbans, while the women wear earth colors and have particularly ostentatious nose ring that establishes their identity. Cattle rearing and agriculture are their main occupation.

2.6.3 MINAS

Minas is the second largest tribal community of Rajasthan. It seems that the Minas may have been the original inhabitants of the Indus Valley civilization. They are militant defenders of Rajasthan. Minas have a tall, athletic build with sharp features, large eyes, thick lips and a light brown complexion. They are found dominating the regions of Shekhawati and eastern Rajasthan. Minas solemnize marriage in the younger years of the children.

2.6.4 GADULIYA LOHARS

Lohars are the nomadic blacksmiths of Rajasthan. The Gaduliya Lohars receives their name from their beautiful bullock carts (Gadis). These nomadic blacksmiths are said to have wandered from their homeland Mewar because of a pledge made to their acknowledged Lord Maharana Pratap. He was ousted from Chittorgarh by the Mughal emperor Akbar and The Gaduliya Lohars, a clan of warring Rajputs, vowed to re-enter the city only after the victory of Maharana Pratap.

2.6.5 GARASIAS

Garasias are the fallen Rajputs of Rajasthan and comprises only 2.7% of the Rajasthan tribals. These tribals have an interesting customs of marriage through elopement which usually take place at the annual Gaur fair held during the full moon of March. They inhabit mainly in Abu Road area of southern Rajasthan.

2.6.6 SAHARIYAS

Sahariyas, the jungle dwellers, is considered as the most backward tribe in Rajasthan and their name possibly derives from the Persian 'Sehr' (Jungle). The Sahariyas believed to be of Bhil origin, they inhabit the areas of Kota, Dungarpur and Sawai Madhopur in the southeast of Rajasthan. They are simple, illiterate and open to exploitation. Their main occupations include working as shifting cultivators, hunters and fishermen.

2.6.7 DAMORS

Belonging to the Dungarpur and Udaipur districts, Damors are mainly cultivators and manual laborers.

Rajasthan tribes also include the following: Meo and Banjara (the traveling tribes), Kathodi (found in Mewar region), Rabaris (cattle breeders, found in Marwar region), Sansi, Kanjar, Bhagalia, Bhil Gametia, Bhil Garasia, Bhil Kataria, Bhil Mama, Bhil Meena, Bhil people, Bhilala, Damor, Dhanka, Dholi Bhil, Dungri Bhil, Dungri Garasia, Garasia, Mewasi Bhil, Rawal Bhil, Tadvi Bhil and Vasava.

2.7   TRIBES OF BIHAR

The Scheduled Tribe (ST) population in the State of Bihar is 103,804,637 as per 2011 census, constituting 1.9% of the total population (1,336,573) of the State. The State has a total of twenty nine (29) Scheduled Tribes and all of them have been enumerated at 2011 census. The Scheduled Tribes are over-whelmingly rural as 94.6% of them reside in villages. District wise distribution of ST population shows that Katihar district has the highest proportion of STs (5.9%) followed by Jamui (4.8%), Banka (4.7%) and Purnia (4.4%). Sheohar district has the lowest proportion of the STs (0.01%), preceded by Darbhanga and Khagaria (0.03% each).

Bihar is home to a multitude of tribes that constitute the major chunk of social and culture map of Bihar. Before the year 2000, the number was even higher. But after the segregation of Jharkhand state, most of the tribes have moved to the Jharkhand. Most important and biggest of them all was the Santhal tribe. Their way of living is totally different from each other which make it a great experience to know each one of them. Most of them possibly migrated from Sub-Himalayan region. Like everywhere, tribals in Bihar as well earn their living from agriculture, including shifting cultivation and small cottage industries. Visiting one of their villages is truly an eye opening. Their houses are mud thatched with baked tiles for roofs. Most of the homes have their own gardens and farms from where the people obtain vegetables and necessary cereals. One can get to see a part of their culture and traditions on the walls of the houses itself. They are painted with different kinds of artwork and images. Every tribe has their own rituals, dances, festivals and music witnessing which helps understand and enjoy their culture better.

2.7.1   BATHUDI

Bathudi Tribe is among the most important tribes left in Bihar. They are the most colorful and artistic of all tribes in the state. Their homes which are ordinary mud thatched houses are made extraordinary with some exquisite multi colored flowery designs. The dressing style is also different from other with men usually preferring coarse dhoti of cotton and women choosing saris of different colors. Silver ornaments are also a must for Bathudi women. You will find most of the women sporting floral tattoos on their forehead or arms, which is a ritual of the tribe. The Bathudi people call this tattooing as Khada. The people of Bathudi tribe are mostly all Hindus and worship Hindu Gods and Goddesses. The common language spoken by Bathudi people is Bihari.

2.7.2 BINJHIA

The Binjhia tribe is known for their rich culture and heritage. You will find this tribe mostly residing near forests and hilly areas. Anthropologists consider Binjhia tribe to be the most advanced of all. It is evident from their homes which are built using tiles, wood, bamboo and are nicely and spaciously designed. Even the social structure is nicely laid out. A normal family consists of father, mother and their unmarried children. Father is the head of the family. Monogamy is more prevalent with widow remarriage being allowed. Binjhia tribe men wear dhoti, kurta, and ganji where as women dress up in sari, saya and blouse. Women also prefer wearing ornaments while going out.

2.7.3 BIRJIA

Birjia is one of the largest tribes in Bihar today. Earlier they were mainly resident of hilly areas but because of difficulties in agriculture, they moved down to plains. Most of the Birjia people are farmers and only few of them indulged in occupations like gathering, hunting, fishing, basketry, and working as daily labors. Their social structure is also similar to others with father being the head of the family and the main decision maker. Most of the Birjia tribe people practice Hinduism and worship Hindu Gods and Goddesses. Birjia folks also worship ancestral spirits for peace and prosperity.

2.7.4 CHIK BARAIK

Chik Baraik tribe is mainly found in the rural parts of Bihar. The most peculiar feature of this tribe is that you will not find any village or premises dedicated to them alone. They are considered very friendly and share space with other tribes. Chik Baraik is famous as a tribe of artists. They are involved in making cotton threads and cotton clothes. They also work as weavers, bird trappers, farmers and daily labors. The languages spoken by Chik Baraik people include Mundari, Sadani and Hindi.

2.7.5 *KHOND*

Khond tribes of Bihar are basically of Proto-australoid race and are considered as one of the scheduled tribes. Jharkhand, Singhbhum, Hazaribagh are the abode of this tribal community. As per the history, the Khond tribe is migrated from Orissa and is one of the major tribes of India. This tribal community linguistically belongs to Austro-Asiatic family. Their language is known as 'Kuvi' and is a type of Dravidian language. Khonds, also known as Kandhs, are an aboriginal tribe. They are divided in groups like hill Khonds or Kutia and plain-dwelling Khonds. They know the use of medicinal plants and treat diseases with the plants. The Khonds are a group of people who are tall and have some features of Aryan. They are considered as mixed Dravidian race. Khonds are divided in clans like Hansa, Beck, Hembram, Bedia, etc. The costumes of the male Khonds include dhoti, ganji and Shirt. The costumes of the women include saree and salwar kamiz dupatta. Ornaments play a major part in the costumes of the Khond women. They prefer to wear ornaments in neck, ear, finger, hair, nose, wrist and feet. The ornaments are basically made of bronze, shell, gold, brass, steel, nickel, seeds, thread, and gold and brass imitation. The Khonds follow animistic religion and they worship almost eighty-three gods. Some are also followers of Hinduism and Christianity. The name of their main deity is Sing Bonga. Their local deities are Thakur Dei, Thakur Deo, Burha Dei, Burha Deo, Borang Buru, Bhagbonga, and Tila bonga. Apart from worshipping their local deities, the Khonds worship goddess Durga, Bhagwati, goddess Lakshmi, Lord Shiva.

2.7.6 *PAHARI KORWA*

Pahari Korwa tribe occupies a major part of Bihar. The people of Pahari Korwa tribes practice cultivation, thereby, producing various crops. According to researches, these tribes belong to the Austro-Asiatic family group. If one divides the whole of the Pahari Korwa tribal community, one can find that there are two sub tribes, popularly known as Pahari Korwa and Dihari Korwa. The anthropologists have given a vivid description of their physical stature and form. According to them, these Paharia Korwa tribes have got medium to short height and have a dark brown or black skin. The society of the Paharia Korwa tribes is further segregated into 5 'totamistic clans', namely, Hansadwar, Samar, Edigwar, Ginnur and Renla. It has a rich heritage of culture and tradition, as depicted in their fairs, festivals, music and dancing forms. The villages that these Paharia Korwa tribes built are usually selected on the top of the hill. Sometimes they are being constructed in the fringes of the forest areas. Most of the families of Paharia Korwa follow the nuclear family structure. Just like any other tribal communities of Indian subcontinent, the religious practices of these Paharia Korwa is restricted to ancestral worship and to the worship of a handful of gods and deities.

2.7.7   SANTAL

Santal tribes reside in the serene regions of Singhbhum, Hazaribagh, Santhal Pragana, Giridih, Dhanbad, Ranchi and Bhagalpur, Purnea, Sharsha and Munger in Bihar.

Though a large part of tribal population has come under the state of Jharkhand, Bihar still boosts of most socially advanced and culturally rich tribes. Visit Bihar to witness a unique world of these tribes, uninfluenced by the outside ways and teaching us a lesson or two in how to live peacefully with the nature.

The most important festival for the tribes of Bihar is Sarhul, which commemorates the blossoming of Sal trees. Each and every tribal communities of Bihar fete this festival in the early days of spring time. Sal trees are worshipped in the sacred orchard. Different tribes have different ways of feting this festival. However, each one of these tribes of Bihar worships the 'spirit' of the Sal tree to try to find its blessings for a good harvesting.

Just like any other tribes, most of these tribes of Bihar practice cultivation including shifting cultivation. The most significant crop of these tribes of Bihar is paddy. On a daily basis, the meal of these tribes comprises of boiled cereals, millets. Amongst the delicacies include a curry of boiled vegetables or meat or any of the edible roots, and tubers nicely seasoned with salt and chilies. What is quite interesting to note is that milk and all the milk products are entirely missing from the menu of these tribes of Bihar. Marriage is also an important institution amongst all the tribes of Bihar state. Special rituals are feted. As soon as a child is born, he is in the beginning given a sip of milk of the goat before he is allowed to suck the mother's milk. Taboo is that goat's milk is never savored afterwards, as they thought that it is going to make the child quite argumentative. Religious nature of these tribes of Bihar is best reflected in the fact that there are diverse gods and goddesses, varying both in number and strength. For example, the Ho tribes have the tradition of appeasing only two village deities. These are namely, Desauli and Jahira Buru. According to the eminent anthropologists, some of these tribes of Bihar worship as many as 10 to 12 deities throughout the year. Invocation of spirits too is a popular phenomenon. However, most of the tribes revere the Singh Bonga as the Supreme Being.

2.8 TRIBES OF UTTAR PRADESH

2.8.1 AGARIYA

One of the Scheduled Tribes of India is the Agariya people who live primarily in the Uttar Pradesh and Madhya Pradesh states of India. During the years of the British rule, the ones who lived in and around Mirzapur were involved in the mining of iron. The languages that are spoken by the people of this tribe are Hindi, Agariya language and Chhattisgarhi.

2.8.2 KOL

Kols are mainly found in the Allahabad, Varanasi, Banda and Mirzapur districts, the Kol is the largest tribe in Uttar Pradesh. As stated in the history, this community migrated from the central parts of India almost five centuries ago. They are one of the Scheduled Castes available in UP. Divided into exogamous clans like Monasi, Rautia, Thaluria, Rojaboria, Bhil, Barawire, and Chero, they are followers of Hinduism and speak in Baghelkhandi dialect. Most of them do not have any land and depend upon the forest for the income. The leaves and firewood is collected by them and sold at the local markets.

2.8.3 KORWA

The Korwas, a Scheduled Tribe found in the Jharkhand, Chhattisgarh and Uttar Pradesh is economically and socially poor community. They receive many facilities from the Indian Government for their upliftment. In UP, the community is divided into four sub-groups, Dam Korwa, Agaria Korwa, Pahar Korwa and Dih Korwa who are further divided into Guleria, Huhar,

Haril, Leth, Mura, Munda and Pahari. They are isolated tribes and most of them are hunter gatherers. A few of them practice settled agriculture and are a part of the Hindu community. However, they worship their own tribal deity, Dih. Each settlement has a shrine of Goddess known as Diwai. Korwa people communicate in their mother tongue Korwa which is also known as Singli and Ernga alternatively. As a second language, Korwa people speak Chhattisgarhi and Sadri too.

2.8.4 SAHARIYA

These are the Scheduled Caste that is found in the Bundelkhand region. They are also referred to as the Banrawat, Rawat, Soarain and Banrakha. The name Sahariya originated from the Hindi word sahra which means the jungle. And, thus, Sahariya means the dwellers of the jungle. Many claim to have descended from Baiju Bhil, the worshipped or Shiva, the Hindu God and others trace the origin from Ramayana's Shabri. The division of the community is done in various gotras naming Lodhi, Sanauna, Solanki, Bagolia and more. The traditional occupation of the Sahariya community includes collecting honey, wood cutting, mining, making basket, breaking stones, etc. as they are majorly dependent upon the forests for their livelihood. Essentially followers of Hinduism, the Sahariyas also have numerous deities such as Gond Devi, Bhavani, Bijasur and Soorin.

2.8.5 THARU

Tharu tribe claim themselves to be the descendants of Ranas of Chittor. Dangurias are the dominant group of Gonda and Bahraich regions. The origin of this Tharu tribal community has got an important history to bank upon. After being thrown from the original homeland, they migrated to several other places including that of the state of Kerala. Thus as a natural consequence these Tharu tribes were displaced off their fertile lands and started working as bonded laborers in the fields of some of the affluent land employers. Tharu is considered as the biggest, traditional and primitive tribes in Uttar Pradesh. This tribe has mongoloid affinity. They are meat eaters and are also fond of liquor. Tharu society is patriarchal in nature. However, the women play a dominant role. They prefer joint family system. This tribe has strong traditional Panchayat organization in order to settle their disputes.

Divorce requires a social approval. They follow a monogamous marriage. As far as occupation of Tharu tribal community is concerned, they have tried their hands in practicing cultivation. They produce crops like barley, wheat, maize, and rice. They are also fond of rearing animals. Chickens, ducks, pigs, and goats are some of the common animals usually found in almost each and every household of Tharu village. Since several rivers flow through the region many of them have turned into fishermen. They have also adapted to the profession of hawkers. Agriculture is the main occupation of the Tharus. They are also experts in cattle herding, piggery and poultry. They are also engaged in carpentry, masonry, weaving and basketry.

ACKNOWLEDGEMENTS

The author is grateful to the Organizing Secretary, Deendayal Research Institute, Chitrakoot for the support in various ways.

KEYWORDS

- **Bhumij**
- **Ethnic communities**
- **Indo-Gangetic region and Central India**
- **Kora**
- **Munda**
- **Oraon**
- **Santal**

REFERENCES

Boding, P. O. (1925). The Santhal Medicine. Asiatic Society of Bengal.
Dube, B. K., & Bahadur, F. (1966). A study of the Tribal People and Tribal Areas of Madhya Pradesh. Tribal Research and Development Institute, Bhopal.
Dubey, K. C. (1970). The Dorlas of Bhopalpattam, Bastar District. *Bull. Tribal Res. Dev. Inst. Bhopal 8*, 1–12.
Elvin, V. (1939). The Baigas. Gyan Books Pvt. Ltd, New Delhi.
Elvin, V. (1947). Maria and their Ghotul. Oxford University Press, Bombay.

Grigson, W. (1949). The Maria Gond of Bastar. Oxford University Press, London.

Hoffman, J. H. (1950). Encyclopedia Mundarica. Government of Bihar.

Jain, S. K. (1981). Observations on Ethnobotany of the Tribals of Central India. In: Jain, S. K. (Ed.) Glimpses of Indian Ethnobotany. Oxford & IBH Publishing Co., New Delhi.

Pushpangadan, P., & Pradeep, P. R. J. (2008). A Glimpse at Tribal India: An Ethnobotanical Enquiry. Amity Institute for Herbal and Biotech Products Development, Peroorkada, Thiruvananthapuram.

Raizada, A. (1984). Tribal Development of in Madhya Pradesh-A Planning Perspective. Inter India Publication, New Delhi.

Russel, R. V., & Hira Lal (1919). The Tribes and Castes of the Central Provinces of India (Vols. 1–4). Cosmo Publication, New Delhi.

Sahoo, A. K., Pal, D. C., & Goel, A. K. (2013). A glimpse of the tribal and ethnobotanical diversity in India. *Ethnobotany 25,* 47–55.

Sikarwar, R. L. S. (1997). Ethnobotany of Sahariya Tribe, Madhya Pradesh. *Applied Botany Abstracts 17*(2), 129–140.

CHAPTER 3

ETHNOBOTANY OF INDUS VALLEY CIVILIZATION

K. V. KRISHNAMURTHY[1] and BIR BAHADUR[2]

[1]R&D, Sami Labs Ltd, Peenya Industrial Area, Bangalore–560058, India, E-mail: kvkbdu@yahoo.co.in

[2]Department of Botany, Kakatiya University, Warangal–506009, India, E-mail: birbahadur5april@gmail.com

CONTENTS

ABSTRACT

This chapter deals with the ethnobotanical aspects of Indus Valley Civilization, which is one of the oldest civilization of the world. The chapter introduces the basic features of this civilization and goes on to explain the involvement of Indus valley ethnic people in agriculture, domestication, crop plants cultivation, involvement of plants in medicinal and religious activities and trade. This chapter stresses that Indus Valley people were the first in Indian sub-continent to have introduced and perfected agriculture and technology. This civilization was also the one that helped in domestication and diversification of some crop species and acted as centers of these two phenomena.

3.1   INTRODUCTION

The Indus valley or Harappan Civilization is one of the very well-known ancient civilizations of the world and belongs to the pre-historic Indian sub-continent. The geographical stretch of this civilization is neither restricted to the Mohenjo-Daro- Harappan region from where it was first known (hence called Harappan Civilization) nor to Indus Valley, which is not the only or even the main river valley where this civilization grew and flourished. In fact, many people consider that it was centered around Sarasvati river, which got dried up long back due to climate change and which is now represented by a small river called Ghaggar (*see* Gupta, 1996). The geographical region of this civilization stretched from the Dasht Valley of the Makran coast in the West to Meerut and Saharanpur in the Upper Ganga-Yamuna Doab in the east (some even consider it even up to middle Ganga region), and from Jammu in the north to the Tapti river valley in the south. It also had a distinct presence as far south as upper Godavari River in Maharashtra and at Shortugai in North Afghanistan (Chakrabarti, 2004a). More than 300 sites are known so far covering an area of 1.5 million km^2 (Rao, 2008).

The most significant early publication on Indus Valley civilization was that of Marshall (1931), which emphasized, among others, the following three points: (i) the Civilization was deeply rooted in the Indian (pre-independence Indian) soil; (ii) it foreshadowed a number of features of India, particularly in religious beliefs, sculpture, crafts, etc., and (iii) in terms of India's "Vedic period" it was pre-vedic. The last aspect is often disputed by a number of scholars. The second most important book on this Civilization was by Wheeler (1953), which emphasized that this Civilization was started around 2500 B.C.E. and was destroyed around 1500 B.C.E. and that it had a comparatively later beginning than the Sumerian Civilization. However, both these statements are also disputed (see Frawley, 2010). Renfrew (1987) and Frawley (2010) even suggested that the Indus valley civilization is in fact Indo-Aryan civilization and was there even prior to the Indus Valley era.

Studies made subsequent to Wheeler in 1960s and 1970s revealed that this civilization can be recognized into four temporal phrases: Pre-Harappan (3500–2800 B.C.E.), Early Harappan (2800–2600 B.C.E.), Mature Harappan (2600–1900 B.C.E.) and Late Harappan (1900–1400 B.C.E.). The indigenous evolution of this civilization and its later transformation into subsequent cultures were also suggested (see more details in Possehl, 1993, 1999). Although these dates for this civilization are purely suggestive, some consider the earliest date for this civilization to about 8000 to 7000 years before the present (particularly in view of the excavations made in Baluchistan and Birrana and Rakhigarrhi in Harayana recently). We do not know with certainly who were the authors of this civilization but all available evidence, including skulls discovered, suggests that there was first a large proportion of aboriginal proto-Australoids (with dark skin), second the predominant Mediterranean type and third an occasional foreigner from the North-East India (Piggott, 1950). All Mediterraneans were speakers of Dravidian languages. One another unresolved issue relating to this civilization is the identity of the rulers and the political elites of the cities. There were neither palaces nor temples. Probably, instead of one social group with total control, the elites might have included merchants, shamans and individuals who controlled various resources (Kenoyer, 1998). Studies made after 1985 have brought to light several other aspects related to this civilization such as trade, farming, irrigation, crops, foods, art, etc. It was also brought to light that climate changes were the probable causes for the disappearance of this civilization. This chapter deals with the ethnobotany of the Indus Valley civilization (the usage "Harappan Civilization" is also employed interchangeably in the text), drawing information from very widely scattered archaeological and art sources, since there is an absence of deciphered writings of this civilization.

3.2 ETHNOAGRICULTURE

In India, the history of agriculture largely seems to begin with the practices of the inhabitants of Indus Valley civilization. The details of agriculture and its history can be obtained from various archaeological artifacts, as also from the art evidences drawn from these artifacts such as pottery, sherds, seals, etc. (Tiwari, 2008). Although this civilization is supposed to be an urban civilization, agriculture was the mainstay of life. The archaeological remains of large granaries of Harappa and Mohenjo-Daro prove that agricultural activities systematically developed fully (Ghosh, 1989). The Baluchistan/Mehrgarh sequence took India's wheat-barley-cattle-sheep-goat agricultural profile back in history to around 7000 B.C.E. The Pre-Harappan Cultures of Amri-Nal, Quetta, Kulli, Zhob, Kotdiji/ Sothi and Banawali evolved out of agricultural practices and developed into a complex pattern of settled life. These facts established Indus Valley Civilization area as an independent region of early agriculture (Tiwari, 2008).

The Harappan Civilization people appeared to have interacted with the hunter-gatherers, pastoral nomadics and incipient farmers who were already present there before them and this interaction led to the growth of farming settlements, as well as to the introduction of irrigation as an important component of agriculture. Also, these people used the so-called Persian wheel for drawing water for agricultural (as well as for domestic) purposes. The fertile alluvial soil, heavy monsoon rainfall at that time and the free and plentiful availability of water from various rivers were highly conductive for the cultivation of different agricultural and horticultural crops. There are strong evidences that seem to indicate that the economy of the people of this civilization was built on agricultural practice. They also knew of flood irrigation and well-irrigation. Excavated evidence across the Harappan civilization area indicate that cultivators developed complex and extensive system of dams and canals. These farmers frequently used contouring, bunding, terracing and benching for water management. They had used in early phases of civilization only toothed harrow for plowing and did not know about the plow. However, in later phases of civilization they had used plows. The archaeological site of Kalibangan has yielded the first evidence of a plowed field in the Indian sub-continent, whose plow marks are identical with those in the mustard-and horse gram cultivating fields of the same area today (Chakrabarti, 2004b). The archaeological finding also includes a sickle which was probably used for harvesting wheat, barley and rice.

The agricultural geography of the area across which Indus Valley archaeological sites are dispersed, however, varies greatly and perhaps the types of agricultural practices were also quite varied. It also varied according to the crops cultivated.

3.3   ETHNODOMESTICATION OF CROP SPECIES

It is generally known that there are distinct centers of origin of cultivated plants in the world (originally 8 and subsequently 12 centers were recognized). It is also known that there are centers of diversification of these domesticated taxa. India is recognized as one of the major centers of origin and diversification of crops (*see* Krishnamurthy, 2003). In India, the Indus Valley Civilization people must have been involved in these two processes as they were historically one of the original peoples who initiated and perfected the art and science of agriculture. This is also supported by the fact that wild as well as cultivated species/varieties have been found in the excavated sites for some of the crop plants. The use of wild plants was followed chronologically by their domesticated derivatives.

The origin of wheat-barley cultivation for the Indus Valley Civilization area can be traced to Baluchistan. In case of barley (*Hordeum vulgare*) there is a clear evidence of transition from the wild state to domestication. The earlier available data of Mehrgarh site (5000–4000 B.C.E.) provide evidence of barley cultivation. The remains of barley belonging to two-rowed and six-rowed types were found (Asthana, 1985). There are evidences that the wild ancestor of barley, *H. spontaneum*, got spread to S. Asia, mainly the Indus valley civilization area, from about 8000 ^{14}C years ago. Remains of *H. spontaneum* have been found in the Harappan area (Vishnu-Mittre and Savithri, 1982), along with *H. vulgare* var. *nudum* and *H. vulgare* var. *hexastichum*. Therefore, the Indus Valley Civilization area might have served as a center of diversification of barley and the ethnic people of this civilization must have played an important role in this diversification. In the case of wheat, all the known wheat species belonging to the Eincorn (*Tritium monococcum*), Emmer (*T. dicoccum*), Macaroni (*T. durum*), Bread (*T. aestivum*) and Compact (*T. compactum* and *T. sphaerocarpum*) categories were obtained from archaeological sites of Harappan civilization. Most of these got spread from the Fertile Crescent region of Far East to the Harappan region. Both, the compact species were very characteristic of this civilization. Therefore, the Indus Valley region must have been an essential center of diversification of wheat species.

All the present day rice varieties essentially belong either to the *japonica* group with smaller grains or to the *indicia* group with longer grains. Both these groups were known to have been domesticated. The *indica* group is believed to have been domesticated in North India and this must have happened during the Indus Valley cultivation period (or even a little earlier). The wild ancestor is believed to be *Oryza rufipogon*, while the cultivated species is *Oryza sativa*. Charred remains of rice from Chopani Mando dated to 9000–8000 B.C.E. represents a wild variety of rice (Agrawal and Kharakwal, 2002). Subsequently, both wild and cultivated rice, for example, have been identified in the rice impressions in burnt and plaster pieces at Hulas, a Harappan site. Thus, wild and cultivated rice species simultaneously had existed and were used by Harappan people. Domesticated rice was cultivated in the central Ganga region in the Belar Valley bordering Allahabad as early as 5000 B.C.E., if not earlier. Evidence of cultivated rice, along with its wild progenitor has been found in the year 2002 and these were embedded in the core of a number of pot sherds from Lahuradeva in Sarayupar area (5298 ± 60 and 4196 ± 90 B.C.E.) in the Ganges valley. Rice also occurred in the early Harappan context in Balu and Kunal in Haryana and at Damdama in Uttar Pradesh around 5000 B.C.E. In the mature Harappan Context it occurred at Harappa and in Gujarat. Thus, we cannot assume that the Indus Civilization people learnt rice domestication/cultivation via diffusionary medavism from China (Srivastava, 2008) as it was believed earlier. The diffusion of cultivation of rice, North West in competition to wheat-barley in around 2800–2700 B.C.E., as we have evidence of cultivated rice in Haryana around this time.

Millets such as sorghum, and ragi are found in the Early Harappan Stage of Harappa in Punjab and Banawali in Haryana. In Mature Harappan levels they occur widely (Singh, 2008b). Thus, this region may be one of the three centers of origin of for at least some of the millets (Weber, 1991a, b).

Cotton was another plant that was domesticated by the Indus Valley people. The species of cotton domesticated was *Gossypium arboreum*, the tree-cotton. This is found wild in many parts of Gujarat, which was one of the main sites of Indus Valley civilization. Remnants of both wild and cultivated tree-cotton remains were found in the archaeological remains.

3.4 ETHNIC CROPS

A rich alluvial soil provided the base for a broad-based subsistence pattern in most of the Indus Valley Civilization area. Several crops were known to have been cultivated during this civilization (Weber, 1991a, b; Chakrabarti, 1999). This fact is evident from archaeological excavations made at several sites. According to Weber (1991b) the Harappan people were growing a number of crops in both summer and winter cultivations. It is unlikely that the same package of crops was grown throughout various ecologies and periods. The major staple crops were barley, wheat, rice and millets. However, in the Sind region rice, various millets, wheat and barley were the staple crops. These four crops were cultivated in most regions of Harappan civilization, particularly in the later phases.

The cultivated wheats recovered were *Triticum dicoccum*, *T. durum*, *T. aestivum*, *T. compactum* and *T. sphaerocarpum; T. monococcum* was perhaps restricted to Baluchistan-Mehrgarh region. Wheat belonging to all these species was the staple food of the Indus Valley people (Rao, 2008; Sharma, 2008).

Barley from Mohenjo-Daro and Harappa were varieties of *Hordeum vulgare*: *nudum* and *hexastichum*. However, the progenitor wild species of barley, *H. spontaneum* was also perhaps used in this region (Vishnu-Mittre and Savithri, 1982).

Rice was another equally important staple crop of Harappan people. As already mentioned, evidence of cultivated rice has been formed from several archaeological sites, the most important being from Sarayupar area in the Ganges valley and from Koldihwa, around 5th to 4th millennium B.C.E. An analysis of rice husks by Vishnu-Mittre and Te-Tzu Chang revealed that they belonged to the domesticated *Oryza sativa* group (dating back to 9000–8000 years from the present) (Singh, 2008a). Hence, Wenming (2002) was correct/right to conclude that rice agriculture originated in India no later than 9000 years back from the present.

As already mentioned, millets like sorghum and ragi are found in the Early Harappan stage in Punjab and Banawali in Haryana, while in Mature Harappan level they occur widely (Singh, 2008b). Other millets known from Indus Culture were pearl millet, foxtail millet (Italian millet) and Kodo millet. Job's Tears (*Coix lacrryma-jobi*) was known from Kuntas site.

Among other crops were a variety of pulses. These include *Cicer arietinum, Lablab purpureus, Vicia faba, Lathyrus sativus, Macrotyloma uniflorum, Lens culinaris, Vigna radiata, Pisum sativum, Cajanus cajan, Vigna mungo* and *Pisum arvensis.*

There were oil seeds like *Sesamum indicum, Linum usitatissimum, Brassica juncea* and cotton. There were reports of castor but there are others who question its presence at that period of time (Swamy, 1973).

Among the fruits cultivated by Indus Valley Civilization people were grapes, pomegranate, dates, banana, melon and jujube. The remnants of almonds and walnuts were also found. *Emblica* fruits in carbonized form were available in the Pre-Harappan Banawali. There were also a few vegetables.

The fiber crops used by Harappan people were cotton, silk cotton and linseed. Cotton was an important crop that was used from both wild and cultivated *Gossypium arboreum*. Evidence of cotton cultivation was found in Hulas site. Cotton from Mohenjo-Daro was available in the form of fragments of cloth and string. Cotton was perhaps used even earlier to 5000 B.C.E.

A number of tree crops were utilized, both as fuel and as construction/carpentry items. The identified trees from excavated pieces of wood charcoal include deodar (*Cedrus deodara*), teak (*Tectona grandis*), sisam (*Dalbergia sissoo*), sal (*Shorea robusta*) and babul (*Acacia* sp.). The oldest record of wood use in Indian region is from Harappa and Harappan Gujarat. In the former region two woods were found to have been used in coffin: deodar and rose wood (*Dalbergia latifolia*). Other woods found here were *Ziziphus mauritiana* (jujube) in a wooden mortar (for pounding grains). The charred timbers recovered from Lothal in Gujarat were species of *Acacia* and *Albizia, Tectona grandis, Adina cordifolia* and *Soymida febrifuga* (Chowdhuri, 1971). Wooden tools and implements were found in Burzhom in Kashmir. Remains of *Holarrhena antidysenterica* were found in Hastinapur. Remnants of *Tamarix dioica* was found in Kalibangan and Rohira, while remains of *Capparis aphylla, Manilkara hexandra, Lawsonia inermis* and *Prosopis spicigera* were found in Rohira. Thus, the credit for wood work and carpentry in Ancient India certainly belongs to Indus Valley people.

The chief archaeobotanical data of Indus Valley civilization come from pollen, seeds, charcoal, plant impression and phytoliths. Weber (1991b) has derived the relative abundance value for the periods belonging to 2400–2034 B.C.E. (of the Harappan Valley area) from the three principal means of measuring the archaeobotanical record: percentage of a plant's remains, its ubiquity and its density. He then constructed a three-tiered hierarchical model of plant use in an inverted pyramidal form. The high density, highest percentage and great abundance tier included wheat and barley, the median density, percentage and abundance tier included *Lens, Cicer, Pisum* and *Lathyrus*. The low density and percentage level included *Ziziphus, Trianthema, Chenopodium* and many weeds and wild plants.

3.5 ETHNIC FOOD

Food is one of the most essential needs of mankind of all cultures including the Indus valley culture and agriculture is the prime source of food for people. Barley, wheat, rice and millets formed the staple food but these people also used pulses like *Phaseolus mungo*, *Cicer arietinum*, *Lablab purpureus*, *Vigna radiata* and other legumes as mentioned in the previous section. They also used a number of fruits and oil crops. Since the civilization was essentially urban, agriculture was the main source of these foods, and foraging was extremely limited.

Architectural structures described as granaries have come to light from Mohenjo-Daro, Harappa and Lothal. These are huge store houses. In Lothal 65 terracotta sealing with impressions of reed, woven fiber, matting and even twisted cords have been found in the granary, which suggest that "packages of goods were stored in bulk in these granaries. However, the storage containers range from pits lined with lime plaster for storing wheat/barley) discovered from the Kalibanga sites (Madhubala, 2004) to pots of various sizes and shapes. The pottery included storage jars, perforated cylindrical jars, beakers, bowls, shallow dishes, dish on stand, cups, pedestalled cups, plates, jugs, etc. (Tiwari, 2008). A few grinding stones were also discovered.

Interesting evidence regarding cooking practice is also revealed by the presence of both underground and overground varieties of mud ovens inside the houses in Kalibangan site. These resembled the present-day tandoories of Punjab and Rajasthan. Kalibangan residents were mainly wheat-eaters. A terracotta figurine from Mohenjo-Daro kneading flour (probably of wheat) has been described. A few saddle querns, cylindrical rollers. Platform for pounding grains and pestle-mortars have also been found. The querns found in the excavations show how the grinding of grains was done on an extensive scale. All these archaeological finds form important evidences of various activities such as storing, processing, cooking and preserving the different agricultural products. They also give indirect evidences of the culinary art and esthetics of their times to handle a variety of foods and beverages.

3.6 PLANTS IN RELIGION AND RITUALS

An issue that had received very little/scant attention is the nature of the Harappan religion. There is no evidence of a centralized or state religion in the Indus Valley civilization. However, it has been known that the religion of the ancient Indians was Animism and its improved version Totemism. Hence, it is highly probable that it is so for the Indus Valley people also. Trees, other plants, snakes and animals like bulls were worshipped as totem objects. There is also evidence for some form of phallic worship.

Hinduism in its medieval or modern forms was not present in the Indus Valley Civilization. But yet, some major elements of Hinduism seem to be present in the archaeological finds. Thus, we can conclude that these people have contributed a great many elements of paramount importance to Hinduism. According to Piggott (1950) prehistoric Hindu society owed more to Harappans than it did to Aryans. The religion of the Atharva Veda resembles that one so closely that one is forced to the conclusion that it was taken over from the people of Indus Valley civilization. The Atharva Veda was shown to be entirely different from Rig Veda, but represents a much more primitive stage of thought and is a book of spells and incantations appealing to the demon-world and teems with notions about witchcraft current among lower grades of the population (Macdonnel, 1925). It is possible from the archaeological finds to trace some of the major elements of later Indian religious aspects, including tree worship. A number of plants used in the worship of gods and goddesses, especially earth goddess and Siva-Parvati worship seem to have been first tried in this civilization. It is most likely that phallus worship later gave rise to the Sivalinga concept.

The major evidences for tree/plant worship of the Indus valley civilization people are the following: Fertility symbols such as vegetation growing from the womb of a female figure, the representation on a seal of a supplicant kneeling before a figure perched in a tree and a deity being shown in the branches of the sacred pipal tree. Religious importance and prominence was given to pipal leaves.

Religious practices of Indus Valley civilization people included purification rites, magic and fire rituals. The ceremonial platform of Kalibangan might have been used for religious practices and rituals. Asceticism also seems to have been practiced by the Harappan individuals. Several seals show a traditional yogic *asana* (=posture). These representations strongly suggest shamans, who were also medicine men. The robed "priest" stone-figure discovered at Mohenjo-Daro reminds us of these shamans (Chakrabarti, 2004a). If the Indus ascetics were really shamans, then they performed ritualistic magical healing which are the principal functions of shamans throughout the world's primitive cultures (see some more details in the next section)

(During Caspers, 1993). The cultural artifacts symbolize the multiple uses of crops and parts of plants such as husks, corns, dried organs, seeds, fruits, etc. which are used for various religious (and decoration) purposes.

In Mehrgarh site the dead were buried with considerable quantities of funerary offerings comprising animal sacrifices (a strong aspect of animism) and utilitarian objects such as tools, baskets, grinding-stones, grains and many types of ornaments.

3.7 ETHNOMEDICINE

According to experts on Indus valley civilization Harappan people believed that disease is the result of malevolent influences exercised by a spirit, god/goddess or supernatural being. Thus, disease is a magical or magico-religious rather than a natural phenomenon. Like the medicine of the great civilizations of Egypt and Mesopotamia, the healing system of the Harappan culture was inextricably connected with the culture's religious beliefs and practices and probably based predominantly on magic (Gajjar, 1971). Since diseases are attributed to supernatural causes, they are treated by magic incantations and other rituals. To ward off diseases the Harappan ascetics or shamans use charms, amulets and talismans. They performed ritualistic magical healings. The shamans heal the patients' diseases by means of rituals including such elements as ecstatic dances, magical flights, the use of potent herbs and amulets, the recitation of incantations and exorcisms. Thus, the medical system of this civilization can be correctly described as primitive. These medical beliefs must have been emanated and popularized in the Harappan civilization (Das Gupta, 1932), but we do not have direct evidence.

It is strongly believed that the Harappan culture people and their shamanism, used, in the healing rituals, among other things, plants. In this healing process plants were not only worshipped but were also used as medicines. This is indicated by the huge number of plants and their remains found in many archaeological sites of Indus Valley Civilization. For example, at the Late Harappan site of Surkotada in Kutch charred lumps of carbonized seeds were found in earthen pots (dated approximately 1970–1600 B.C.E.). There were 574 seeds, majority of which belong to wild plants and 7% to cereals that included millets. The wild species include grasses (257 seeds), sedges, Chenopodiaceae, Amaranthaceae, Polygonaceae and *Euphorbia* species. At least some of these wild plants might be of medicinal value, while others may be of nutraceutical value. In Rojoli site more than 10,000 seeds belonging to 70 different plants were recovered and out of which many have medicinal properties.

In the Pre-Harappan Banawali, carbonized remains show the fruits of south Indian soap nut tree (*Sapindus emarginatus*), amla (*Phyllanthus emblica*) and pods of Shikakai (*Acacia concinna*). These indicate ample testimony to the richness of experience and understanding of the properties of plant products which enabled pre-Harappans to prepare the herbal shampoo (Saraswat et al., 2000) and hair-care products such as herbal detergents.

3.8   TRADE IN ETHNOBOTANICALS

The Indus Valley civilization makes the first clearly discernible phase of India's trade. There were both external and internal trades, the former involving regions not belonging to this civilization. Internal trade involved different regions of the civilization. There were both land and sea trades. There was an increasing integration of coastal communities with inland trading circuits as well as expansion of the maritime networks themselves (*see* Kenoyer, 1997). The regions important for the seafaring activities of the Harappans were the Makran and Indus coasts, Kutch and Savrashtra.

Scholars working on the archaeological sites of Tepe Yahaye and She-i-Sokta in Iran emphasized that the Mesopotamian culture moved eastward and proposed that long-distance trade between the Harappan civilization and Mesopotamia was the most important factor behind the Harappan civilization genesis. Baluchistan, a very important Harappan civilization site, had clearly defined trade routes to Iran, Afghanistan and Punjab along with an equally clear network of internal trade routes (Chakravarti, 2004b; Lahiri, 1992). Ratnagar (2000) had also pointed out that Harappan people had trade with Mesopotamia, at least around 2000 B.C.E., if not earlier.

The trade was largely in cotton textiles, dyes such as indigo and miscellaneous plant products. At Mohenjo-Daro two silver vases wrapped in red-eyed cotton cloth were found; Fabric impressions on pottery or the reverse of terracotta seals are relatively larger in number. Lothal yielded impressions of woven cloth on the reverse of terracotta sealings baked in a fire at the site of a warehouse. There is no information on trade in grains, but there were movements of them from regions of plenty to regions of scarcity. Goods meant for trade are transported by pack bullocks and carts. There are many terracotta representations of carts and boats. Traders traveled along caravans and sea routes.

Mohenjo-Daro is considered as a major pilgrimage/sacred/trading city. It had a great bath and granary. The agricultural products were collected here for religious offerings as well as for trading (Kondo et al., 1995).

KEYWORDS

- ancient trade routes
- ethnic crops
- ethnoagriculture
- ethnodomestication
- Harappan civilization
- Mohenjo-Daro

REFERENCES

Agarwal, D. P., & Kharakwal, J. S. (2002). South Asian Prehistory Aryan Books International, New Delhi, India

Asthana, S. (1985). Pre-Harappan Cultures of India and Borderlands. Books & Books, New Delhi, India

Chakrabarti, D. K. (1999). India: An Archaeobotanical history. Oxford Univ. Press, New Delhi, India

Chakrabarti, D. K. (2004a). Introduction. In: Chakrabarti, D. K. (ed.). Indus Civilization sites in India. New Discoveries. Marg Publications, Mumbai, India. pp. 6–22.

Chakrabarti, D. K. (2004b). Prelude to the Indus Civilization. Chakrabarti, D. K. (ed.). Indus Civilization sites in India. New Discoveries. Marg Publications, Mumbai, India. pp. 23–28.

Chowdhuri, K. A. (1971). Botany in the Medieval Period from Arabic and Persian sources. In: Bose, D. M., Sen, S. N., & Subbarayappa, B. V. (eds.). A concise History of Science in India. Indian National Science Academy, New Delhi, India. pp. 392–400.

Das Gupta, S. (1932). A history of Indian Philosophy Vols. I & II. Cambridge Univ. Press, New Delhi, India.

During Caspers, C. L. (1993). Another face of the Indus Valley magico-religious system. In: Gail, A. J., & Mevissen, G. J. R. (eds.). South Asian Archaeology. Franz Steiner Verlag, Stuttgart. pp. 65–86.

Frawley, D. (2010). Gods, Sages and Kings. (reprinted) Motilal Banarsidass Publishers Pvt. Ltd. Delhi, India

Gajjar, I. N. (1971). Ancient Indian Art and the West. D. B. Taraporevala Sons & Co. Pvt. Ltd., Bombay.

Ghosh, A. (1989). An Encyclopedia of Indian Archaeology. Munshiram Manoharlal Publishers Pvt. Ltd. New Delhi.

Gupta, S. P. (1996). The Indus-Sarasvati Civilization: Origins, Problems and Issues. Pratibha Prakashan, Delhi, India.

Kenoyer, J. M. (1997). Trade and technology of the Indus Valley: New insights from Harappa, Pakistan. *World Archaeology 29*, 262–280.

Kenoyer, J. M. (1998). Ancient cities of the Indus Civilization. American Institute of Pakistan Studies, Karachi, Pakistan.

Kondo, R., Ichikawa, A., & Morioka, T. (1995). Taking a bath in Mohenjo-Daro. In: Allchin, R., & Allchin, B. (eds.). South Asian Archaeology. Vol. I. Oxford & IBH, New Delhi, India. pp. 127–137.

Krishnamurthy, K. V. (2003). Textbook of Biodiversity, Science Publishers, Enfield, USA.

Lahiri, N. (1992). The Archaeology of Indian trade Routes (up to *c* 200 B.C.). Oxford University Press, Delhi.

Madhu Bala. (2008). Kalibangan. In: Chakrabarti, D. K. (Ed.). Indus Civilization sites in India. New Discoveries. Marg Publications, Mumbai, India. pp. 34–43.

Marshall, J. (ed.) 1931. Mohenjo-Daro and the Indus Civilization. 3 Vols., Arthur Probsthan, London.

Mcdonnel, W. (1925). A History of Sansrit Literature Heinemann Ltd., London.

Piggott, S. (1950). Prehistoric India. Penguin Books. London.

Possehl, G. L. (ed.) (1993). Harappan Civilization: A Recent Perspective. 2nd Edition. Oxford & IBH, New Delhi, India

Possehl, G. L. (1999). The Indus Age: The Beginnings. University of Pennsylvania Press, Philadelphia, USA.

Rao, S. R. (2008). Agriculture in the Indus Civilization. In: Gopal, L., & Srivastava, V. C. (Eds.). History of Science, Philosophy History Agriculture in India (Up to *C* 1200 A.D.). Centre for Studies in Civilization. New Delhi, India. pp. 171–202.

Ratnagr, S. (2000). The End of the Great Harappan Tradition. Manohar Publishers and Distributors, New Delhi, India.

Renfrew, C. (1987). Archaeology and Languages. Cambridge Univ. Press, New York.

Saraswat, K. S., Srivastava, C., & Pokharia, A. K. S. (2000). VI. Palaeobotanical and pollen analytical investigations. In: Bisht, R. S., Dorje, C., & Baneji, A. (eds.) Indian Archaeology 1993–94: A review. Archaeological Survey of India, New Delhi, India. pp. 143–145.

Sharma, B. D. (2008). The origin and history of wheat in Indian Agriculture. In: Gopal, L., & Srivastava, V. C. (eds.). History of science, philosophy history, agriculture in India (Up to *C* 1200 A.D.). Centre for Studies in Civilization. New Delhi, India. pp. 126–142.

Singh, P. (2008a). Origin of agriculture in the Middle Ganga plain. In: Gopal, L., & Srivastava, V. C. (eds.). History of science, philosophy, history, agriculture in India (Up to *C* 1200 A.D.). Centre for Studies in Civilization. New Delhi, India. pp. 3–18.

Singh, P. (2008b). History of millet cultivation in India. In: Gopal, L., & Srivastava, V. C. (eds.). History of science, philosophy, history, agriculture in India (Up to *C* 1200 A.D.). Centre for Studies in Civilization. New Delhi, India. pp. 107–119.

Srivastava, V. C. (2008). Introduction. pp. XXIX–XXXIV. In: Gopal, L., & Srivastava, V. C. (eds.). History of science, philosophy, history, agriculture in India (Up to *C* 1200 A.D.). Centre for Studies in Civilization. New Delhi, India.

Swamy, B. G. L. (1973). Sources for a history of plant science in India. I. Epigraphy. *Indian J. Hist. Sci. 8,* 61–98.

Tiwari, P. (2008). History of agriculture as reflected in the art of India. In: Gopal, L., & Srivastava, V. C. (eds.). History of science, philosophy, history, agriculture in India (Up to *C* 1200 A.D.). Centre for Studies in Civilization. New Delhi, India. pp. 784.

Vishnu-Mittre & Savithri, R. (1982). Food Economy of the Harappans. In: Possehl, G. L. (ed.). Harappan Civilization. Oxfords & IBH, New Delhi, India. pp. 205–221.

Weber, S. A. (1991a). Plants and Harappan subsistence: An example of stability and change from rojdi. Oxford & IBH Publishing Co., New Delhi, India.

Weber, S. A. (1991b). Plants and Harappan subsistence. Westview Press, Boulder, USA.

Wenming, Y. (2002). The origins of rice agriculture pottery and cities. In: Yasuda, Y. (Ed.). The origins of pottery and Agriculture. Roli Books, New Delhi, India, pp. 151–156.

Wheeler, M. (1953). The Indus Civilization. Cambridge University Press, Cambridge, UK.

CHAPTER 4

ETHNIC FOOD PLANTS OF INDO-GANGETIC PLAINS AND CENTRAL INDIA

SUDIP RAY

Department of Botany, PMB Gujarati Science College, Indore – 452001, Madhya Pradesh, India, E-mail: sudbot@yahoo.com, raysudip8@gmail.com

CONTENTS

ABSTRACT

Present study reports 483 species of edible plants which are distributed in 281 genera and 118 families. Among 118 families, Agaricaceae, Diplocystaceae (Fungi), Adiantaceae, Marsileaceae, Parkeriaceae, Polypodiaceae, Thelypteridaceae (Pteridophyta), Cycadaceae, Gnetaceae, Pinaceae (Gymnospermae) are from other groups while 109 families belong to Angiospermae. Ten dominant families which are consumed for edible foods in Indo-Gangetic plains and Central India are Leguminosae (25), Poaceae (15), Moraceae (12), Amaranthaceae (12), Rubiaceae (12), Rhamnaceae (9), Cucurbitaceae (9), Amaryllidaceae (9), Euphorbiaceae (9), and Tiliaceae (8). A total of 30 families represented with single species and monogeneric. Of the total taxa recorded for wild edible foods, 31% are trees, 8% shrubs, 10% climbers and 51% are herbs. In all the edible plants, leaves and petiole of 131 plant species, fruits of 191 plant species (including unripe and ripe), underground parts of 61 plant species (including rhizome, tubers, corms, stolon, bulbs, root and root stock), seeds of 62 plant species, flowers of 23 plant species, stems and shoots of 23 plant species and whole plants of 11 plant species are consumed by ethnic people of Indo-Gangetic plains and Central India.

4.1   INTRODUCTION

Wild resources have been utilized as food by human beings since very ancient time. People were dependent on plants and plant products to meet their basic need for food, shelter and medicine. India is a rich repository of cultural heritage for varied ethnic groups and a healthy tradition of folks practices of utilization of wild plants (Kumar and Mishra, 2011). Of the total floristic wealth of about 20,000 species of angiosperms available in India, about 600 fall in the edible category for use directly or indirectly as food stuffs (Watt, 1971). Arora (1991) reported 3900 plant species used by tribals as food out of 45,000 wild plants in India. Most of the people inhibiting in the rural and remote areas of India are economically very poor and depend on noncultivated wild plants for food. These poor inhabitants spend most of their valuable time in collecting wild edible plants (Pieroni, 2001). Millions of the people in many developing countries do not have enough food to meet their daily requirement and a further more people are deficient in one or more micronutrients (FAO, 2004).

Rakesh et al. (2004) viewed that the nutritional value of traditional wild plants is higher than several known common vegetables and fruits. The folk selection was based on local needs, customs, preferences and habits (Arora, 1994). The tribes who still live in their undisturbed forest areas and having the traditional food habits like consumption of large variety of seasonal foods have been observed to be healthy and free from most of the diseases (Anonymous, 1995).

Indo-Gangetic plain is known as Indus Ganga. The area dealt with 1,96,000 sq. km. fertile plain encompassing most of northern and eastern India. It is bounded on the north and north east by a portion of the main chain of the western Himalaya, feed its numerous river and are the source of the fertile alluvium deposited across the region by two river system. The southern edge of the plain is marked by the Chhotanagpur plateau. Eastern part extends up to Bengal. On the south and south west of the boundary follow watershed from which all the river of the west drains into Ganges and Yamuna. The watershed extends along the northern slopes of the numerous groups of hills, Vindhyan mountains which separate the Gangetic plain from Narmada valley. The large piece of the country lying to the south west of Gangetic plain proper includes a portion of Baghel khand to Central India, also Bundelkhand, the Malwa plateau, Eastern Rajputana a small piece of the western Punjab and in the neighborhood of Delhi. Politically the Indo-Gangetic region includes plains of Punjab, Haryana, Rajasthan, Uttar Pradesh, Bihar, plains of West Bengal while Central India includes the state of Madhya Pradesh and Chhattisgarh. Indian subcontinent is inhabited by over 53 million tribes belonging to 550 different communities under 227 ethnic groups (Anonymous, 1995). Two third of tribal population is concentrated in Madhya Pradesh, Orissa, Bihar, Gujarat and Rajasthan states of India. The tribal communities inhabiting in the hills and remote forest areas of Indo-Gangetic and Central region are as follows:

- **Central India:** Baigas, Bharia, Bhils, Bhilalas, Pateliya, Gond, Korku, Kols, Marias, Maria gonds, Majhi, Oraon, Pradhans, Sahariya, Kharia, Kawarl, Pardhi, Bahelia.
- **Bihar:** Lohar, Oraon, Parhaiya, Baiga, Khond, Munda, Santal, Tharu, Kol.
- **Punjab:** Gujjar, Jat, Khatri, Bhanjra.
- **Rajasthan:** Banjara, Bhil, Tadvi, Bhilala, Damor, Garasia, Meena, Sahariya, Kathodi.
- **Uttar Pradesh:** Baksa, Bhotia, Bhuinya, Kurmi, Kharwar, Tharu.

- **West Bengal:** Chik baraik, Kheria, Korwa, Lodha, Moonda, Oraon, Rabha, Rajbanshi, Gond, Santal, Birhor.

4.2   REVIEW OF LITERATURE

Traditional knowledge of local people on wild edible plants has contributed a lot to the society and proved to be effective up to this day (Lee et al., 2008). Studies on wild food plants have been carried out by several workers from various parts of the country (Jain, 1963; Radhakrishna et al., 1996; Arora and Pandey, 1996). Several workers conducted studies on wild edible plants in Indo-Gangetic region and Central India (Duthie, 1903–1929; Haines, 1963; Prain, 1961; Verma, 1993; Jain and Tiwari, 2012). Significant works have been made on wild food plants in Central India (Dwivedi and Singh, 1984; Maheshwari, 1990; Pandey and Oomachan, 1992; Verma, 1993; Shukla, 1996).

Roy and Rao (1957) have studied the nutritional aspects of the diet of Marias and some of the wild plants eaten by Gonds and Santhals of Central India. Katewa et al. (2000) enlisted 48 wild plants which are used for food from the Aravalli hills of south east Rajasthan. Workers like King (1969), Kanodia and Gupta (1968) and Bhandari (1974) have documented famine food plants through personal collection of information from the desert tribes men and women in western Rajasthan. Considerable works on wild edible plants for food from Rajasthan have been made (Singh and Singh, 1981; Sebastein and Bhandari, 1990; Joshi and Awasthi, 1991; Singh and Pandey, 1998).

Rai (1992) studied ethnobotany of Gond tribes of Seoni district and mentioned plants used as food and general uses. Authors have enlisted 19 edible and other useful plants which are used by Gond tribes in Seoni district, Madhya Pradesh. Mukherjee and Ghosh (1992) recorded 132 useful plant species of Birbhum district, West Bengal which are distributed under 113 genera. Oomachan and Masih (1988) reported 28 wild plant species which are consumed as food or vegetables by aboriginals of Vindhyan plateau, an important region of Central India. Dwivedi and Pandey (1992) documented 28 wild plants which are consumed as food or vegetables by aboriginals of Vindhyan plateau, Central India.

Das (1992) documented 31 wild plants including two Pteridophytes from Midnapur district, West Bengal which are used as food resources during drought and floods. Sahu (1996) studied on life support promising food plants among aboriginals of Bastar in Madhya Pradesh. These are useful in famine and scarcity conditions or floods. Patole and Jain (2002) reported 45 edible plants from Pachmarhi hill biosphere reserve which are used by ethnic people. Kumar (2003) carried ethnobotanical study on wild edible plants of Surguja District of Chhattisgarh, MP and recorded 116 plant species consumed as food by tribals. Out of them 58 species are used as vegetable, 47 as fruits and 32 species for other purposes such as spices and condiments, sauces, etc. Khan et al. (2008) worked on certain ethnobotanical information on food and medicinal plants of Rewa division of Madhya Pradesh and reported 50 plant species used by tribals and rural people, of which 24 plants are consumed as food and vegetables during scarcity of food and famine. Some wild edible plants used by tribals of Nimar region in Madhya Pradesh have been reported (Satya, 2006; Mishra, 2010; Sisodiya, 2012). Bandyopadhyay and Mukherjee (2009) recorded 125 wild edible plants of Koch Bihar district, West Bengal which are distributed in 102 genera and 54 families.

A good number of research works have been published on wild edible plants in West Bengal (Jain and De, 1964; Maji and Sikadar, 1982; Das, 1999; Mukherjee and Ghosh, 1992). Kumar et al. (2010) reported 37 wild edible plants which are consumed as food.

Jadhav (2011) documented 58 wild plants used as source of food by Bhil tribe of Ratlam district, Madhya Pradesh. The inhabitants consume 27 plant species as edible vegetables and 31 species for edible fruits. Jain and Tiwari (2012) published research paper on nutritional value of some traditional edible plants used by tribal communities during emergency with reference to Central India. Fatma and Pan (2012) conducted exploration work on wild edible plants of Bihar, India and reported 253 wild edible plant species which are distributed in 86 families. Bandyopadhyay et al. (2012) reported 99 wild edible plants of Howrah district, West Bengal which are distributed in 77 genera belonging to 53 families. Banerjee et al. (2013) studied wild edible plant species consumed by different tribal communities inhabiting in the Bankura district, West Bengal. A total of 50 plant species belonging to 38 families and 48 genera were reported from study area. Singh and Ahirwar (2015) recorded 34 wild edible species which provide food and vegetables to inhabiting tribals of Chanda forest, Dindori district, Madhya Pradesh.

Kapale et al. (2013) studied traditional food plants used by Baiga tribes residing in Amarkantak – Achanakmar biosphere, Central India and enlisted 26 vegetable plants. Dwivedi et al. (2014) documented 58 wild plant species which are used as supplementary source of food by tribals-socially poor communities of district Sonbhodra in Uttar Pradesh. Kaur and Vaishistha (2014) made ethnobotanical survey of Karnal district, Haryana and enumerated some wild food plants.

Sandya and Ahirwar (2015) carried out ethnobotanical studies of some wild edible plants of Jaitpur forest, Shahdol district, Madhya Pradesh, Central India and reported 34 wild edible species which are utilized as food supplement in large scale by different tribes. Singh and Ahirwar (2015) documented 38 wild edible plants of Bandhavgarh National Park, Umaria district, Madhya Pradesh. This study reports that fruits of 20 plant species are edible and tender shoot of 19 plants are used as vegetables. Alawa and Ray (2016) made ethnobotanical survey in remote forest areas of Dhar district, Madhya Pradesh and focused on 32 wild plants which are used as vegetables. These plants are distributed in 31 genera and 23 families. Ray and Sainkhediya (2016) studied wild edible plant resources of Satpura hill ranges, Harda district, Madhya Pradesh and documented 59 wild plants which are used as edible food by tribals inhabiting in the area. Of the 59 plants, 23 plants are utilized as edible fruits, 36 plants are used as vegetables and seeds of 7 plants are used as food during famine or drought.

A list of food plants of Indo-Gangetic plains and Central India is given in Table 4.1 along with edible part, mode of preparation and references.

4.3   GROWTH FORM ANALYSIS

There are mainly four types of growth forms of edible plants used by tribals in the area which includes herbs (248), shrubs (39), trees (150) and climbers (49). Herbs make up the highest proportion (51%) of the total edible species followed by tree (31%), climber (10%) and shrubs (8%) respectably.

4.4   EDIBLE PARTS ANALYSIS

The tribals of India raised a number of agricultural crops. Most of them practice, settled agriculture now. But they do supplement their food with a number of wild edible plants particularly in times of scarcity. Depending upon the nature of different species, the tribals consume fruits, seeds or grains, leaves, roots, tubers, barks, flowers or sometime as whole plants. The whole

TABLE 4.1 Ethnic Food Plants of Indo-Gangetic Plains and Central India

S. No.	Name of plant	Edible part	Mode of preparation	Reference
1.	*Abelmoshus ficulneus* (L.) Wt.	Fr	Young fruits used as vegetables	Alawa and Ray, 2016; Kumar, 2003
2.	*Abelmoschus moschatus* Medic.	Fr	Fruits edible	Bandyopadhyay and Mukherjee, 2009; Kumar, 2003
3.	*Abutilon indicum* (L.) Sweet	L, Fr	Leaves are cooked as vegetable. Ripe fruits eaten	Bandyopadhyay and Mukherjee, 2009
4.	*Acacia catechu* (L.) Willd.	Gum	Taken with betel leaf	Bandyopadhyay and Mukherjee, 2009
5.	*Acacia nilotica* (L.) Delile	Gum	Gum edible	Kumar, 2003; Upadhyay and Chauhan, 2003
6.	*Achyranthes aspera* L.	L	Leaves used as vegetable	Kumari and Kumar, 2000; Singh and Ahirwar, 2013; Sandya and Ahirwar, 2015
7.	*Acorus calamus* L.	Rh	Rhizome eaten	Kapale et al., 2013
8.	*Adansonia digitata* L. (Plate 2B)	Fr	Fruits edible	Jadhav, 2011; Kumar, 2003
9.	*Adenanthera pavonina* L.	L	Leaves edible	Fatma and Pan, 2012
10.	*Aegle marmelos* (L.) Corr.	Fr	Fruit pulp and juice are edible	Katewa et al., 2000; Jain et al., 2012; Jadhav, 2011; Singh and Ahirwar, 2013; Katewa et al., 2012
11.	*Aerva lanata* Juss.	L	Leaves edible	Fatma and Pan, 2012
12.	*Aeschynomone aspera* L.	L	Leaves eaten	Fatma and Pan, 2012
13.	*Agaricus campestris* L.	WP	Fruiting body edible	Kumar, 2003; Banerjee et al., 2013
14.	*Agave americana* L.	L	Leaf sliced and eaten	Alawa and Ray, 2016
15.	*Agave cantula* Roxb.	Fl, Ts	Flowers and fruits cooked and eaten	Kumar, 2003
16.	*Alangium salvifolium* (L.f.) Wang	Fr	Fruit pulp edible	Fatma and Pan, 2012

TABLE 4.1 (Continued)

S. No.	Name of plant	Edible part	Mode of preparation	Reference
17.	*Albizia procera* (Roxb.) Benth.	Fr	Fruit edible	Fatma and Pan, 2012; Kumar et al., 2010
18.	*Allium ampeloporasum* L.	Bulbil	Bulbils eaten	Fatma and Pan, 2012
19.	*Allium ascalonicum* L.	Bulbil	Leaves edible	Dwivedi et al., 2014; Fatma and Pan, 2012
20.	*Allium corolianum* DC.	L	Leaves edible	Fatma and Pan, 2012
21.	*Allium schoenoprasum* L	Bulbil	Leaves edible	Fatma and Pan, 2012
22.	*Allium tuberosum* Rottl. ex Spreng.	Bulbil	Leaves edible	Fatma and Pan, 2012
23.	*Allophyllus serratus* (Roxb.) Radlk.	Fr	Ripe fruits eaten raw	Bandyopadhyay and Mukherjee, 2009
24.	*Alocasia indica* Schott.	Rh	Rhizomes are used as vegetable	Das, 1999; Fatma and Pan, 2012
25.	*Alocasia macrorrizos* (L.) G. Don	L, R	Rootstock and leaves are used as vegetable	Bandyopadhyay and Mukherjee, 2009; Fatma and Pan, 2012
26.	*Aloe vera* (L.) Burm. f.	L	Leaves used as food	Jadhav, 2011
27.	*Alpina galanga* (L) willd	Rh, S, Fl	Leaves, seeds and flowers edible	Fatma and Pan, 2012
28.	*Alternanthera ficoides* R.Br. ex Roem.	L	Leaves are used as vegetable	Das, 1999
29.	*Alternanthera paronychioides* St.Hill	L	Leaves are cooked	Singh and Ahirwar, 2013; Sandya and Ahirwar, 2015
30.	*Alternanthera philoxeroides* (Mart.) Griseb.	L	Tender leafy twigs are used as vegetable	Kumari and Kumar, 2000; Fatma and Pan, 2012; Bandyopadhyay and Mukherjee, 2009
31.	*Alternanthera pungens* Kunth	L	Leaves are chewed	Jain et al., 2012; Kapale et al., 2013
32.	*Alternanthera sessilis* (L.) R. Br. ex DC.	L	Young leaves are used as vegetable.	Das, 1999; Kumari and Kumar, 2000; Jadhav, 2011

S. No.	Name of plant	Edible part	Mode of preparation	Reference
33.	*Alternanthera triandra* Lamk.	L	Leaves are eaten as vegetable	Fatma and Pan, 2012
34.	*Amaranthus caudatus* L.	L	Leaves eaten as vegetable	Jadhav, 2011; Fatma and Pan, 2012
35.	*Amaranthus cruentus* L.	S	Seeds are used as grains	Alawa et al., 2016
36.	*Amaranthus hybridus* L.	L	Leaves are eaten	Kapale et al., 2013
37.	*Amaranthus paniculatus* L.	L	Leaves are eaten	Kapale et al., 2013
38.	*Amaranthus polygamus* L.	L	Leaves used as vegetable	Fatma and Pan, 2012
39.	*Amaranthus spinosus* L.	L, Tw	Leaves and young twigs eaten as vegetable.	Jain, 1963; Kumari and Kumar, 2000; Kumar, 2003; Banerjee et al., 2013; Sandya and Ahirwar, 2015; Alawa and Ray, 2016
40.	*Amaranthus tricolor* L.	St, L	Tender shoot and leaves cooked and eaten	Bandyopadhyay and Mukherjee, 2009; Fatma and Pan, 2012
41.	*Amaranthus viridis* L.	S	Seeds used as grains	Dwivedi et al., 2014; Ray and Saikhediya, 2016; Kumar, 2003
42.	*Amorphophallus bulbifer* (Roxb.) Blume	C	Corm boiled and eaten	Ray and Saikhediya, 2016
43.	*Amorphophallus konkanensis* Hett et al.	C	Corm boiled and eaten	Ray and Saikhediya, 2016
44.	*Amorphophallus margaritifer* (Roxb.) Kunth	S, St	Stem and seeds used as vegetables	Jain, 1963; Fatma and Pan, 2012
45.	*Amorphophallus peoniifolius* (Dennst.) Nicolson.	C	Corm boiled sliced and used for food preparation	Das, 1999; Bandyopadhyay and Mukherjee, 2009; Kapale et al., 2013; Dwivedi et al., 2014; Jadhav, 2011; Fatma and Pan, 2012
46.	*Ampellocissus barbata* (Wall.) Planch.	Fr, C	Ripe fruits eaten raw. Corms boiled, sliced and eaten with salt	Fatma and Pan, 2012, Bandyopadhyay and Mukherjee, 2009

TABLE 4.1 (Continued)

S. No.	Name of plant	Edible part	Mode of preparation	Reference
47.	*Ampelocissus latifolia* (Vahl) Planch.	Fr	Fruits are eaten raw	Katewa et al., 2000; Kumar, 2003; Sandya and Ahirwar, 2015
48.	*Ampelocissus tomentosa* (Heyne ex Roth) Roth	Fr	Fruit is edible	Kumar, 2003
49.	*Ampelopteris prolifera*	Fr	Young fruits eaten as vegetables	Bandyopadhyay and Mukherjee, 2009
50.	*Ananas comosus* (L.) Merr.	Fr	Ripe fruits edible	Bandyopadhyay and Mukherjee, 2009
51.	*Andrographis paniculata* Nees	L	Young leaves with other vegetables used in food preparation	Katewa et al., 2000; Fatma and Pan, 2012
52.	*Andropogon halepensis* (L.) Brot	S	Grains eaten	Fatma and Pan, 2012
53.	*Andropogon squarrosus* L.f	S, R	Grains and dried roots eaten as food	Fatma and Pan, 2012
54.	*Anethum graveolens* L.	WP	Leaves eaten	Jadhav, 2011; Fatma and Pan, 2012
55.	*Annona reticulata* L.	Fr	Fruits are eaten raw	Jadhav, 2011; Banerjee et al., 2013
56.	*Annona squamosa* L.	Fr	Fruits are eaten raw	Sandya and Ahirwar, 2015; Katewa et al., 2000; Dwivedi et al., 2014
57.	*Anogeissus latifolia* (Roxb. ex DC.) Wall. ex Guillem. & Perr.	Fr	Ripe fruits are edible	Fatma and Pan, 2012
58.	*Anthocephalus chinensis* (Lamk.) A.Rich. ex Walp.	Fr	Fruits eaten	Kumar, 2003
59.	*Antidesma diandrum* Retz.	L	Leaves pounded with rice, some gram flour is mixed with it a curry is made	Jain, 1963; Kumari and Kumar, 2000; Kumar, 2003
60.	*Antidesma ghaesembilla* Gaertn.	Fr	Ripe fruits eaten raw	Bandyopadhyay and Mukherjee, 2009

S. No.	Name of plant	Edible part	Mode of preparation	Reference
61.	*Aponogeton monostachyon* L.	Fr	Fruits eaten	Fatma and Pan, 2012
62.	*Ardisia humilis* Vahl	L	Leaves eaten	Fatma and Pan, 2012
63.	*Argyreia nervosa* (Burm.f.) Bojer	L	Leaves used as vegetables	Fatma and Pan, 2012; Alawa and Ray, 2016
64.	*Arisaema tortuosum* Schott.	L, Ts	Leaves and tubers eaten	Jain, 1963; Kumar, 2003
65.	*Artocarpus heterophyllus* Lam.	Fr	Young unripe fruits eaten as vegetables	Jadhav, 2011
66.	*Asphodelus tenuifolius* Cav.	Rh	Rhizomes are eaten	Fatma and Pan, 2012
67.	*Astraeaeus hygrometricus* (Pers) Morg	WP	Cooked as vegetable	Kumar, 2003
68.	*Atylosia scarabaeoide* (L.) Benth.	S	Seeds eaten	Alawa and Ray, 2016
69.	*Avena sterlis* subsp *ludoviciana* (Durieu) Gillete & Magne	S	Seeds are used as grains	Mishra, 2010
70.	*Averrhoa bilimbi* L.	Fr	Ripe fruits eaten	Fatma and Pan, 2012
71.	*Averrhoa carambola* L.	Fr	Ripe fruits eaten	Fatma and Pan, 2012; Banerjee et al., 2013
72.	*Azadirachta indica* Juss.	L	Young leaves fried and eaten with rice	Katewa et al., 2000; Kumar et al., 2010; Bandyopadhyay and Mukherjee, 2009; Banerjee et al., 2013
73.	*Bacopa monnieri* (L.) Pennel	L	Leaves fried and eaten as useful food	Banerjee et al., 2013; Das, 1999
74.	*Bambusa arundinacea* (Retz.) Willd.	Sh	Tender shoots used as vegetable	Bandyopadhyay and Mukherjee, 2009; Kapale et al., 2013
75.	*Bambusa balcooa* Roxb.	Sh	Tender shoots used as vegetable	Bandyopadhyay and Mukherjee, 2009
76.	*Bambusa bambosa* Roxb.	L	Leaves eaten	Fatma and Pan, 2012
77.	*Bambusa tulda* Roxb.	L	Leaves eaten	Fatma and Pan, 2012

TABLE 4.1 (Continued)

S. No.	Name of plant	Edible part	Mode of preparation	Reference
78.	*Bambusa vulgaris* Sehrad.	Sh	Tender shoots cooked as vegetables	Bandyopadhyay and Mukherjee, 2009
79.	*Basella rubra* L.	L	Leaves eaten	Dwivedi et al., 2014; Fatma and Pan, 2012
80.	*Bauhinia malabarica* Roxb.	St	Bark chips added in local wine	Jain, 1963
81.	*Bauhinia purpurea* L.	Fl, R, Flb	Cooked as vegetable	Kumari and Kumar, 2000; Fatma et al., 2012; Dwivedi et al., 2014; Bandyopadhyay and Mukherjee, 2009
82.	*Bauhinia racemosa* Lanill.	Fl, S	Flower and seeds eaten	Fatma and Pan, 2012
83.	*Bauhinia retusa* Roxb.	L	Leaves as vegetable	Kumari and Kumar, 2000
84.	*Bauhinia semla* Wunderlin	Fr	Boiled fruit are eaten	Kumar, 2003
85.	*Bauhinia vahlii* Wight & Arn.	S, Fl	Seeds and flowers eaten.	Singh and Ahirwar, 2013; Fatma and Pan, 2012; Kumar et al., 2010; Jain et al., 2012
86.	*Bauhinia variegata* L.	Fl	Flowers used as vegetable	Fatma and Pan, 2012; Kapale et al., 2013; Dwivedi et al., 2014; Sandya and Ahirwar, 2015
87.	*Begonia picta* Sm.	L	Leaves cooked as vegetable	Kumar, 2003; Kumari and Kumar, 2000
88.	*Benincasa hispida* (Thunb.) Cogn.	Fr	Fruits used as vegetable	Fatma and Pan, 2012; Bandyopadhyay and Mukherjee, 2009; Dwivedi et al., 2014
89.	*Benthamidia capitata* (Wall.) H.Hara	Fr	Fruits edible	Fatma and Pan, 2012; Jain et al., 2012; Katewa et al., 2000
90.	*Berberis asiatica* Roxb. ex DC.	L	Leaves a cut into small pieces and eaten	Fatma and Pan, 2012
91.	*Beta vulgaris* L.	L	Leaves eaten as vegetables	Fatma and Pan, 2012

S. No.	Name of plant	Edible part	Mode of preparation	Reference
92.	*Boerhavia diffusa* L.	L	Leaves as vegetable	Das, 1999; Kumari and Kumar, 2000; Bandyopadhyay and Mukherjee, 2009; Banerjee et al., 2013
93.	*Boerhavia repanda* Willd.	L	Leaves used as vegetable	Jain, 1963; Kumar, 2003
94.	*Bombax ceiba* L.	S	Seeds edible	Kumar, 2003; Kumar et al., 2010; Jadhav, 2011; Fatma and Pan, 2012
95.	*Borassus flabelifer* L.	Fr	The pulp of the fruit is eaten after roasting	Das, 1999; Jain, 1963
96.	*Boswellia serrata* Roxb.	S	Seeds edible	Kumar et al., 2010
97.	*Brassica juncea* (L.) Coss	WP	Whole plant cooked as vegetable. Edible oil extracted from seeds	Banerjee et al., 2013
98.	*Breynia rhamnoides* Muell. Arg.	Fr	Ripe fruits eaten raw	Bandhopadhy and Mukherjee, 2012
99.	*Bridelia retusa* Spreng.	Fr	Eaten raw	Bandyopadhyay and Mukherjee, 2009
100.	*Bridela stipularis* Blume	Fr	Eaten raw by children	Jain, 1963; Bandyopadhyay and Mukherjee, 2009
101.	*Bryonopsis lacinosa* (L.) Naud.	L, Fr	Fruits eaten after cooking, leaves cooked as vegetables	Bandyopadhyay and Mukherjee, 2009
102.	*Buchnania cochinchinensis* (Lour.) Almeida	S, Gum	Seeds eaten. Gum is edible and relished by children.	Upadhyay and Chauhan, 2003; Fatma and Pan, 2012; Kumar et al., 2010
103.	*Butea monosperma* (Lam.) Taub.	Fl	Flowers eaten as vegetable	Fatma and Pan, 2012; Kumar et al., 2010; Sandya and Ahirwar, 2015; Bandyopadhyay and Mukherjee, 2009
104.	*Butomopsis latifolia* Kunth	L	Leaves eaten as vegetable	Kumar, 2003

TABLE 4.1 (Continued)

S. No.	Name of plant	Edible part	Mode of preparation	Reference
105.	*Cajanus sarabaeoides* (L.) Du Petit	S	Seeds cooked and eaten	Kumar, 2003
106.	*Callicarpa macrophylla* Vahl	Fr	Fruit edible	Bandyopadhyay and Mukherjee, 2009; Fatma and Pan, 2012
107.	*Canavalia virosa* Wight & Arn.	Fl	Flowers edible	Fatma and Pan, 2012
108.	*Canthium parviflorum* Roxb.	Fr	Ripe fruits eaten raw	Kaur and Vashistha, 2014
109.	*Capparis deciduas* (Forssk.) Edgew.	Fr	Ripe fruit eaten	Sisodiya, 2012
110.	*Capparis zeylanica* L.	Fr	Fruits are eaten	Alawa and Ray, 2016
111.	*Carica papaya* L.	Fr	Unripe fruit as vegetable and ripe edible	Jadhav, 2011
112.	*Carissa spinarum* L.	Fr	Ripe fruits eaten	Katewa et al., 2000; Fatma and Pan, 2012; Singh et al., 2015
113.	*Carum roxburghianum* (DC.) Kurz	S	Seeds edible	Fatma and Pan, 2012
114.	*Caryota urens* L.	Fr, Sap	The sap is fermented to make wine. Dried fruits chewed as betel nut	Jain, 1963; Bandyopadhyay and Mukherjee, 2009
115.	*Casearia graveolens* Dalz.	S	Seeds yield edible oil	Jain, 1963
116.	*Casearia tomentosa* Roxb.	S	Seeds edible	Fatma and Pan, 2012
117.	*Cassia fistula* L.	Fl	Flowers eaten	Jadhav, 2011
118.	*Cassia occidentalis* L.	L	Vegetable	Kumari and Kumar, 2000

S. No.	Name of plant	Edible part	Mode of preparation	Reference
119.	*Catunaregum spinosa* (Thunb.) Tirv.	Fr	Fruits edible	Kumar, 2003
120.	*Caturegum uliginosa* (Retz.) Siva	Fr	Unripe fruits eaten	Alawa and Ray, 2016, Fatma and Pan, 2012
121.	*Cayratia auriculata* (Wall.) Gamble	L	Young leaves cooked and eaten as vegetable	Kumar, 2003
122.	*Celastrus paniculata* Willd.	Fr	Ripe fruits eaten raw and young fruits cooked as vegetables	Jain, 1963; Alawa and Ray, 2016
123.	*Celosia argentea* L.	L	Leaves used as vegetable	Jain, 1963; Katewa et al., 2000; Kumari and Kumar, 2000; Kumar, 2003
124.	*Celtis australis* L.	Fr	Fruits edible	Kaur and Vashistha, 2014
125.	*Centella asiatica* (L.) Urb.	L	Leaves cooked as food	Katewa et al., 2000; Kumari and Kumar, 2000; Kumar, 2003; Bandyopadhyay and Mukherjee, 2009; Dwivedi et al., 2014
126.	*Ceratopteris thalictriodes* (L.) Brongn.	L	Leaves cooked as vegetable	Fatma and Pan, 2012; Bandyopadhyay et al., 2012
127.	*Ceriscoides turgida* (Roxb.) Tirveng.	Fr	Unripe fruits eaten as vegetables	Kumar et al., 2010
128.	*Ceropogia bulbosa var. bulbosa*	St, B, L	Bulbs eaten raw, unripe fruits used for making pickles. Leaves as vegetable	Bandyopadhyay and Mukherjee, 2009; Jadhav, 2011; Alawa and Ray, 2016
129.	*Ceropogia bulbosa var. lushii* Roxb	L	Bulbs consumed raw	Katewa et al., 2000
130.	*Ceropegia hirsuta* Wt. & Arn.	Ts	Tubers cooked and eaten as vegetable	Kumar, 2003

TABLE 4.1 (Continued)

S. No.	Name of plant	Edible part	Mode of preparation	Reference
131.	*Cheilocostus speciosus* (J Koenig) C. D. Specht	Rh	Rhizome is edible	Kapale et al., 2013
132.	*Chenopodium album* L.	L	Leaves consumed as vegetable	Jadhav, 2011; Sandya and Ahirwar, 2015; Singh et al., 2015
133.	*Chlorophytum arundinaceum* Baker	Fr	Flowers eaten	Ray and Saikhediya, 2015; Fatma and Pan, 2012
134.	*Chlorophytum borivilianum* Sant. & Fern.	R, Ts	Root and tubers consumed raw with ghee and cane sugar in the form of ladoos for vitality	Kumar, 2003; Katewa et al., 2000
135.	*Chlorophytum tuberosum* Baker	R, Ts	Root tubers boiled and eaten	Fatma and Pan, 2012; Katewa et al., 2000; Jadhav, 2011
136.	*Cinnamomum tamala* Nees & Eberm.	L	Leaves used as vegetable	Bandyopadhyay and Mukherjee, 2009
137.	*Cissus adnata* Roxb.	L	Cooked as vegetable	Kumari and Kumar, 2000; Bandhopadhyay and Mukherjee, 2009; Fatma and Pan, 2012
138.	*Cissus quadrangularuis* L.	Fr	Fruits edible	Jadhav, 2011; Fatma and Pan, 2012
139.	*Cissus repanda* (Wight & Arn.) Vahl	St	Watery sap of the plant is drunk.	Jain, 1963
140.	*Cissus repens* Lamk.	Fr	Fruits edible	Fatma and Pan, 2012
141.	*Citrullus colocynthis* (L.) Schrad.	Fr	Fruits eaten raw	Ray and Saikhediya, 2016
142.	*Cleome chelidoni* L.f.	S	Seeds edible	Kumar, 2010
143.	*Cleome gynandra* (L.) DC.	L	Leaves eaten	Fatma and Pan, 2012
144.	*Cleome monophylla* L.	L, Fr	Leaves as vegetable, fruits edible	Kumari and Kumar, 2000; Kumar, 2003; Fatma and Pan, 2012

S. No.	Name of plant	Edible part	Mode of preparation	Reference
145.	*Cleome pentaphylla* (L.) DC.	L	Leaves cooked and eaten	Mishra, 2008; Fatma and Pan, 2012
146.	*Cleome viscosa* L.	L	Leaves as vegetable	Kumari and Kumar, 2000
147.	*Clerodendrum indicum* (L.) Kuntze	L	Tender leaves eaten after cooking	Dwivedi et al., 2014
148.	*Coccinia grandis* (L.) Voight	Fr	Used as vegetable	Katewa, 2000; Singh and Ahirwar, 2013; Banerjee et al., 2013; Dwivedi et al., 2014
149.	*Coix lacryma-jobi* L.	S	Grains eaten as cereals	Jain, 1963
150.	*Colocasia esculenta* (L.) Schott	L, Pt, Stolon	Leaves, petiole and stolons serve as very good vegetable	Das, 1999; Bandyopadhyay and Mukherjee, 2009; Jadhav, 2011
151.	*Combretum nanum* Buch-Ham.	S	Seeds eaten	Jain, 1963
152.	*Commelina benghalensis* L.	Fl, L	Flowers and leaves used as vegetable	Kumari and Kumar, 2000; Banerjee et al., 2013; Fatma and Pan, 2012; Jadhav, 2011, Bandhopadhay and Mukherjee, 2009
153.	*Commelina obliqua* Buch-Ham. ex D. Don	L, WP	Leaves and sometime whole plant eaten as vegetable.	Kumar, 2003
154.	*Corchorus capsularis* L.	St, L	Tender shoot and leaves eaten as vegetable	Das, 1999; Bandyopadhyay and Mukherjee, 2009
155.	*Corchorus olitorius* L.	L	Leaves used as vegetable	Jadhav, 2011; Fatma and Pan, 2012
156.	*Corchorus trilocularis* Lamk.	L	Leaves used as vegetable	Singh and Ahirwar, 2013; Sandya and Ahirwar, 2015
157.	*Cordia dichotoma* G.Forst.	Fr	Ripe fruits eaten raw and for making pickles	Kumari and Kumar, 2000; Kumar, 2010; Jadhav, 2011

TABLE 4.1 (Continued)

S. No.	Name of plant	Edible part	Mode of preparation	Reference
158.	Corylus jacquemontii Decne	Fr	Fruits edible	Kaur and Vashistha, 2014
159.	Cosmostigma racemosa (Roxb.) Wight	Infl	Used as vegetable	Fatma and Pan, 2012
160.	Crateva religiosa Forst. f.	Fr, L	Fruits and leaves eaten	Fatma and Pan, 2012
161.	Crinum defixum Ker Gawl.	Rh	Rhizomes eaten	Fatma and Pan, 2012
162.	Crinum latifolium L.	L	Used as vegetable	Alawa and Ray, 2016
163.	Croton tiglium L.	S	Seeds yield edible oils	Mishra, 2008; Fatma and Pan, 2012
164.	Cryptocoryne retrospiralis Kunth	WP	Whole plant is commonly eaten as vegetable	Jain, 1963
165.	Cucumis callosus (Rottl.) Cogn.	Fr	Fried and dried slices of fruits used as vegetable	Dwivedi et al., 2014
166.	Cucumis melo L. var. agristis Naud.	L	Leaves are cooked and eaten as vegetable	Kumar, 2003
167.	Cucumis prorphetarum L. (Plate 1B)	Fr	Unripe fruits used as vegetable	Alawa and Ray, 2016
168.	Curculigo orchioides Gaertn.	R	Roots edible	Fatma and Pan, 2012 Kumar, 2003
169.	Curcuma amada Roxb.	Rh	Rhizomes used as flavoring agents	Banerjee et al., 2013
170.	Curcuma angustifolia Roxb.	T	Tubers used as vegetable	Jain, 1963; Fatma et al., 2012; Ray and Saikhediya, 2016
171.	Curcuma aromatica Salisb.	Sh, L	Shoots and leaves used as vegetable	Singh and Ahirwar, 2013;
172.	Curcuma decipiens Dalzell	Rh	Rhizome eaten	Ray and Saikhediya, 2016

S. No.	Name of plant	Edible part	Mode of preparation	Reference
173.	*Curcuma leucorhiza* Roxb.	T	Tubers are cooked and eaten	Banerjee et al., 2013
174.	*Curcuma rectinata* Roxb.	T	Tubers boiled and eaten	Banerjee et al., 2013
175.	*Cyamopsis tetragonoloba* (L.) Taub.	Fr	Tender pods used as vegetable	Fatma and Pan, 2012; Singh et al., 2013
176.	*Cyanotis cristiata* (L.) D.Don.	WP	Whole plant cooked and eaten as vegetable	Kumar, 2003
177.	*Cycas pectinata* Griff.	St	Pith and outer soft tissue are eaten	Fatma and Pan, 2012; Banerjee et al., 2013
178.	*Cyperus esculenta* L.	Rh	Rhizomes eaten	Singh et al., 2015
179.	*Cyperus rotundus* L.	Rh	Rhizomes cooked	Fatma and Pan, 2012
180.	*Cyphostemma auriculatum* (Roxb.) P. Singh & V Shetty	Fr	Fruits edible	Fatma and Pan, 2012
181.	*Dalbergia lanceolaria* L.f.	L	Leaves used as vegetable	Kumar et al., 2010
182.	*Dendrocalamus hamiltonii* Nees & Arn. ex Murno	Sh	Young shoots boiled and eaten	Fatma and Pan, 2012
183.	*Dendrocalamus strictus* (Roxb.) Nees	S	Seeds edible	Kumar et al., 2010; Singh, 2015; Alawa and Ray, 2016
184.	*Dichanthium annulatum* (Forsk.) Stapf	S	Grains eaten	Kaur and Vashistha, 2014
185.	*Digera muricata* (L.) Mart.	L	Leaves cooked as vegetables	Jain, 1963; Katewa, 2000; Jadhav, 2011
186.	*Dillenia indica* L.	Calyx	Used in making pickles	Jain, 1963; Bandyopadhyay and Mukherjee, 2009
187.	*Dillenia pentagyna* Roxb.	Fr, Fl	Flowers and fruits edible	Jain, 1963; Kumar et al., 2012; Fatma and Pan, 2012

TABLE 4.1 (Continued)

S. No.	Name of plant	Edible part	Mode of preparation	Reference
188.	*Dioscorea alata* L.	Ts	Tubers kept in running water overnight and then used as vegetable	Katewa et al., 2000; Jadhav, 2011; Bandyopadhyay and Mukherjee, 2009
189.	*Dioscorea bulbifera* L.	Ts	Used as vegetables	Katawa, 2000; Jadhav, 2011
190.	*Dioscorea daemona* Roxb.	Ts	Tubers are kept in running water overnight and then used as vegetables.	Katewa et al., 2000
191.	*Dioscorea esculenta* Burkill	Ts	Tuber cooked as vegetable	Bandyopadhyay et al., 2012
192.	*Dioscorea globosa* Roxb.	Ts	Tuber eaten	Kapale et al., 2012
193.	*Dioscorea hamiltonii* Hook. f.	Ts	Tubers eaten after processing	Jain, 1963
194.	*Dioscorea hispida* Dennst	Ts	Tubers eaten after processing	Bandyopadhyay and Mukherjee, 2009; Jain, 1963
195.	*Dioscorea oppositifolia* L.	Ts	Tubers cut into pieces, small cubes, slices boiled. Boiled slices eaten	Fatma and Pan, 2012; Bandyopadhyay and Mukherjee, 2009
196.	*Dioscorea pentaphylla* L.	Ts	Tuber eaten	Jain, 1963; Fatma and Pan, 2012
197.	*Dioscorea puber* Bl.	Ts	Tuber eaten	Jain, 1963; Kapale et al., 2012
198.	*Dioscorea wallichi* Hook. f.	Ts	Boiled tubers cooked as vegetable	Jain, 1963; Bandyopadhyay, 2012
199.	*Diospyros exsculpta* Buch.- Ham.	Fr	Ripe fruits eaten raw	Katewa et al., 2000; Jadhav, 2011; Banerjee et al., 2013; Singh and Ahirwar, 2013; Kumar, 2013
200.	*Diospyros kaki* L.	Fr	Fruits edible	Fatma and Pan, 2012
201.	*Diospyros malabarica* (Desr.) Kostel	L	Used as vegetable	Banerjee et al., 2013

S. No.	Name of plant	Edible part	Mode of preparation	Reference
202.	*Diospyros melanoxylon* Roxb. (Plate 1G)	Fr	Ripe fruits eaten raw	Katewa et al., 2000; Kumar et al., 2010; Singh and Ahirwar, 2013; Sandya and Ahirwar, 2015; Ray and Saikhediya, 2016
203.	*Diospyros montana* Roxb.	Fr	Ripe fruits eaten raw	Kumar et al., 2010
204.	*Diospyros peregrina* (Gaertn.) Gurke	S	Seeds edible; Fruits eaten raw	Jain, 1963; Fatma and Pan, 2012
205.	*Diplazium esculatum* (Retz.) C. Presl.	L	Young leaves cooked	Banerjee et al., 2013
206.	*Diplazium frondosum* (Clarke) Christ.	Sh	Young shoots edible	Bandyopadhyay and Mukherjee, 2009; Kumar, 2010
207.	*Diploknema butyracea* (Roxb.) Lam.	Fr, S, Fl	Fruits, seeds and flowers edible	Kumar et al., 2010
208.	*Dolichos briflorus* L.	S	Seed grains eaten	Fatma et al., 2012
209.	*Dregea volubilis* (L.f.) Benth ex Hook.f.	Fr	Unripe fruits eaten	Jadhav, 2011; Fatma and Pan, 2012
210.	*Drimia indica* (Roxb.) Jessop	L	Leaves used as vegetable	Alawa and Ray, 2016
211.	*Duchesnea crysantha* (Zoll. & Moritzi) Miq.	Fr	Ripe fruits eaten	Bandyopadhyay and Mukherjee, 2009
212.	*Echinochloa crus-galli* (L.) P. Beuv.	Fr	Fruits boiled and taken with salt	Bandyopadhyay and Mukherjee, 2009
213.	*Echinochloa frumentacea* L.	S	Seeds used as grains	Fatma and Pan, 2012
214.	*Eclipta prostrata* L.	L	Cooked as vegetable	Banerjee et al., 2013
215.	*Ehretia acuminata* R.Br.	Fr	Ripe fruits eaten raw	Jadhav, 2011
216.	*Ehrertia leavis* Roxb.	Fr	Fruits eaten raw	Kumar et al., 2010; Fatma and Pan, 2012

TABLE 4.1 (Continued)

S. No.	Name of plant	Edible part	Mode of preparation	Reference
217.	*Elaeocarpus serratus* L.	Fr	Unripe fruits used for making pickles	Bandyopadhyay and Mukherjee, 2009
218.	*Eleocharis dulcis* (Burm. f.) Trin. ex Hench.	Rh	Rhizomes used as vegetables	Ray and Saikhediya, 2016
219.	*Eleusine indica* L.	S	Seeds eaten as grains	Bandyopadhyay and Mukherjee, 2009; Mishra, 2008
220.	*Eleusine aegiptia* (L.) Roxb.	S	Seed flours eaten during drought.	Fatma and Pan, 2012
221.	*Eleusine coracana* (L.) Gaertn.	S	Seeds consumed during famine	Ray and Saikhediya, 2016
222.	*Emilia sonchifolia* (L) Dc	L	Leaves cooked as vegetable	Kumar, 2003
223.	*Enhydra fluctuans* Lour.	WP	Cooked as vegetable	Bandyopadhyay and Mukherjee, 2009; Banerjee et al., 2013
224.	*Ensete superbum* (Roxb.) Cheesman (Plate 1D)	Rh, Fr	Rhizomes boiled and eaten. Unripe fruits used as vegetable	Mishra, 2008; Alawa et al., 2016
225.	*Erioglossum rubignosum* Roxb.	Fr	Fruits eaten	Fatma and Pan, 2012
226.	*Eruca sativa* Mill.	S, WP	Young plants used in salad and as vegetables. Seed oil used in pickles	Kaur and Vashistha, 2014
227.	*Erycibe paniculata* Roxb.	Fr	Fruits edible	Fatma and Pan, 2012
228.	*Eryngium foetidum* L.	L	Used as flavoring agents	Bandyopadhyay and Mukherjee, 2009
229.	*Erythrina indica* Lamk.	S	Seeds consumed	Fatma and Pan, 2012
230.	*Eulophia herbacea* Lindl.	Ts	Tubers boiled and eaten with flowers of *Madhuca longifolia*	Kumar, 2003

S. No.	Name of plant	Edible part	Mode of preparation	Reference
231.	*Eulophia nuda* Lindl.	Ts	Tubers boiled and eaten	Kumar, 2003
232.	*Euphorbia elegans* Spreng.	WP	Whole plant is used as vegetable	Jain, 1963
233.	*Euphorbia hirta* L.	L	Leaves as vegetable	Kumari and Kumar, 2000
234.	*Euphorbia longam* Steud.	S	Seeds edible	Banerjee et al., 2013
235.	*Euphorbia prostrata* Ait.	WP	Whole plant is used as vegetable	Jain, 1963
236.	*Fagaria oxyphylla* (Edgew.) Engler	Fr	Fruits edible	Kaur and Vashistha, 2014
237.	*Fagopurum esculentum* Moench.	L, Fr	Leaves and fruits cooked as vegetable	Bandyopadhyay and Mukherjee, 2009
238.	*Ficus auriculata* Lour.	Fr	Fruits eaten raw and cooked as vegetables	Singh and Ahirwar, 2013; Kumar et al., 2010
239.	*Ficus benghalensis* L.	Fr	Ripe fruits eaten raw	Kumar, 2003; Jadhav, 2011; Fatma and Pan, 2012; Singh and Ahirwar, 2013
240.	*Ficus cunia* Buch.	Fr	Ripe figs eaten	Jain, 1963
241.	*Ficus geniculata* Kurz	L	Leafy vegetable	Kumari and Kumar, 2000
242.	*Ficus hispida* L.	Fr	Receptacles are cooked as vegetables	Bandyopadhyay and Mukherjee, 2009; Kumar, 2010; Banerjee et al., 2013
243.	*Ficus infectoria* Willd.	L, Fr	Leafy vegetable, fruits edible	Kumari and Kumar, 2000; Fatma and Pan, 2012
244.	*Ficus palmata* Forssk.	Fr	Fruits edible	Tiwari et al., 2010
245.	*Ficus racemosa* L.	Fr	Eaten raw as famine food	Katewa, 2000; Kumar, 2003; Jadhav, 2011; Bandyopadhyay and Mukherjee, 2009; Kumar et al., 2010; Singh and Ahirwar, 2013
246.	*Ficus religiosa* L.	Fr	Ripe fruits eaten raw	Jain, 1963; Kumar, 2003; Fatma and Pan, 2012; Singh and Ahirwar, 2013

TABLE 4.1 (Continued)

S. No.	Name of plant	Edible part	Mode of preparation	Reference
247.	*Ficus rumphii* Blume	Fr	Eaten as vegetables	Fatma and Pan, 2012; Tiwari et al., 2010
248.	*Ficus semicordata* Buch-Ham. ex Sm.	Fr	Fruits eaten raw	Fatma and Pan, 2012
249.	*Ficus virens* Aiton	Fr	Fruits eaten	Ray et al., 2016
250.	*Flacourtica indica* (Burm.f.) Merr.	Fr	Eaten raw	Katewa et al., 2000
251.	*Flacourtia jangomas* (Lour.) Raeusch.	Fr	Ripe fruits eaten	Bandyopadhyay and Mukherjee, 2009
252.	*Flueggea virosa* (Roxb ex Willd.) Baill.	Fr	Ripe fruits eaten	Bandyopadhyay and Mukherjee, 2012
253.	*Garcina cowa* Roxb. ex DC.	Fr	Fruits eaten raw	Fatma and Pan, 2012
254.	*Garcina xanthochymus* Hook. f.	Fr	Fruits eaten raw	Fatma and Pan, 2012
255.	*Gardenia companulata* Ross	Fr	Fruits eaten raw	Fatma and Pan, 2012
256.	*Gardenia gummifera* L.f.	Fr	Fruits edible	Mishra, 2008; Fatma and Pan, 2012
257.	*Gardenia latifolia* Soland.	Fr	Fruits eaten raw	Fatma and Pan, 2012; Kumar et al., 2010, Kumar, 2003
258.	*Gardenia resinifera* Roth	Fl	Flowers eaten	Kumar, 2003
259.	*Garuga pinnata* Roxb.	Fr	Fruits edible	Jadhav, 2011; Fatma and Pan, 2012
260.	*Geodorum densiflorum* (Lam.) Schl. (Figure 1E)	Pseudobulb	Pseudobulb consumed with honey	Alawa and Ray, 2016
261.	*Glinus lotoides* L.	L	Leafy twigs cooked as vegetables	Bandyopadhyay and Mukherjee, 2009
262.	*Glinus oppositifolius* (L.) A.DC.	L	Fruits edible	Banerjee et al., 2013
263.	*Gloriosa superba* L.	R	Roots edible	Fatma and Pan, 2012

S. No.	Name of plant	Edible part	Mode of preparation	Reference
264.	*Gmelina arborea* Roxb.	Fr	Fruits edible	Singh and Ahirwar, 2013; Sandya and Ahirwar, 2015; Kumar et al., 2010
265.	*Graptophyllum pictum* Griff.	L	Leaves eaten	Fatma and Pan, 2012
266.	*Grewia abutilifolia* Roxb. ex DC.	Fr	Fruits edible	Jain, 1963; Ray et al., 2016
267.	*Grewia asiatica* L.	F	Ripe fruits eaten raw	Bandyopadhyay and Mukherjee, 2009; Tiwari, 2010; Banerjee et al., 2013
268.	*Grewia damine* Gaertn.	Fr	Fruits edible	Fatma and Pan, 2012
269.	*Grewia flavescens* Juss.	Fr	Fruits are eaten raw	Kumar, 2003; Fatma and Pan, 2012
270.	*Grewia hainesiana* Hole	Fr	Fruits edible	Fatma and Pan, 2012
271.	*Grewia hirsuta* Vahl	Fr	Eaten raw	Jain, 1963; Kumar, 2003; Singh and Ahirwar, 2013; Katewa et al., 2000; Dwivedi et al., 2014
272.	*Grewia sapida* Roxb. ex DC.	Fr	Fruits edible	Ray et al., 2016; Fatma and Pan, 2012
273.	*Grewia selerophylla* Roxb.	Fr	Fruits eaten raw	Fatma and Pan, 2012
274.	*Grewia tenax* (Forssk.) Fiori (Plate 1F)	Fr	Fruits are edible	Ray and Saikhediya, 2016
275.	*Grewia tillifolia* Vahl	Fr	Fruits edible	Jain, 1963; Kumar et al., 2010; Jadhav, 2011; Bandyopadhyay and Mukherjee, 2009
276.	*Guazuma ulmifolia* Lam.	Fr	Ripe fruits eaten raw	Bandyopadhyay and Mukherjee, 2009
277.	*Guizotia abyssynica* (L.) Cass.	Fr	Roasted	Bandyopadhyay and Mukherjee, 2009
278.	*Gymnema sylvestre* R.Br.	L	Leaves eaten	Fatma and Pan, 2012
279.	*Habenaria marginata* Colebr.	Ts	Tubers cooked and eaten as vegetable	Kumar, 2003
280.	*Hedyotis scandens* Roxb.	L	Cooked as vegetable	Bandyopadhyay and Mukherjee, 2009
281.	*Helitropium ovalifolium* Forssk.	L	Leaves eaten as vegetable	Kapale et al., 2013

TABLE 4.1 (Continued)

S. No.	Name of plant	Edible part	Mode of preparation	Reference
282.	*Hibiscus cancellatus* Roxb.	R	Roots eaten	Fatma and Pan, 2012
283.	*Hibiscus crinitus* G.Don	R	Roots eaten after frying or boiling	Jain, 1963
284.	*Hibiscus rugosus* Mast.	R	Tuberous roots boiled and their rind is removed and inner part is eaten	Jain, 1963
285.	*Hibiscus sabdariffa* L.	L, Fr	Fruits eaten. Leaves used as vegetable	Kumar, 2003; Jadhav, 2011; Fatma and Pan, 2012
286.	*Holboelia latifolia* Wall.	Fr	Fruits edible	Kaur and Vashistha, 2014
287.	*Holoptelea integrifolia* (Roxb.) Planch	S	Eaten by children	Katewa et al., 2000; Kumar et al., 2010; Jadhav, 2011
288.	*Hygrophila auriculata* (Schum.) Haines	L	Leaves eaten	Katewa et al., 2000; Bandyopadhyay and Mukherjee, 2009
289.	*Hymenodictyon excelsum* Wall.	L	Leaves cooked as vegetable	Jain, 1963
290.	*Impatiens balsamina* L.	L	Leaves eaten	Fatma and Pan, 2012
291.	*Indigofera cassioides* Forssk.	Fl	Flowers used as vegetable	Kumar, 2003; Singh and Ahirwar, 2013
292.	*Indigofera gerardiana* Wall. ex Baker	Fr	Fruits eaten	Kumar et al., 2010
293.	*Indigofera pulchella* Roxb.	Fl	Flowers eaten raw	Jain, 1963
294.	*Iphigenia indica* (L.) Kunth	B	Used as vegetable	Katewa et al., 2000
295.	*Ipomoea aquatica* Forssk. (Plate 1A)	L, Sh	Leaves and young tender shoots cooked as vegetable	Jain, 1963; Kumari and Kumar, 2000; Bandyopadhyay, and Mukherjee, 2009; Kumar et al., 2010; Fatma and Pan, 2012

S. No.	Name of plant	Edible part	Mode of preparation	Reference
296.	*Ipomoea batatas* (L.) Lam.	R	Roots used as vegetable	Banerjee, 2013
297.	*Ipomoea nil* (L.) Roth	L	Leaves eaten	Fatma and Pan, 2012
298.	*Ipomoea reptans* Forssk.	Sh	Tender shoots eaten	Fatma and Pan, 2012
299.	*Ixora parvifolia* Vahl	Fr	Fruits edible	Fatma and Pan, 2012
300.	*Lablab purpureus* (L.) Sweet	Fr, S	Pods used as vegetable. Seeds used as pulses	Dwivedi et al., 2014
301.	*Lagerstroemia parviflora* Roxb.	L	Leaves cooked as vegetables	Katewa et al., 2000; Fatma and Pan, 2012
302.	*Lantana camara* L.	Fr	Fruits edible	Jadhav, 2011
303.	*Lasia spinosa* L.	L	Used as vegetable	Bandyopadhyay and Mukherjee, 2009
304.	*Leea asiatica* (L.) Rids.	Fr	Fruits eaten	Kumar, 2003
305.	*Leea macrophylla* Roxb. ex Hornem	L	Leaves eaten	Jain, 1963; Fatma and Pan, 2012
306.	*Lepisanthes rubiginosa* (Roxb.) Leenh.	L	Young leaves vegetable	Bandyopadhyay and Mukherjee, 2009
307.	*Leptadenia pyrotechnica* (Forssk.) Decne	Fr	Fruits used as vegetable	Katewa et al., 2000; Fatma and Pan, 2012
308.	*Leucas aspera* (Willd.)Link	L	Leaves cooked and eaten	Kumar, 2003
309.	*Leucas cephalotes* (Roth) Spreng.	L	Leaves cooked	Kumar, 2003; Fatma and Pan, 2012; Jadhav, 2011
310.	*Leucas lavandulaefolia* Sm.	L, T	Leafy twig cooked as vegetable	Bandyopadhyay and Mukherjee, 2009
311.	*Leucas montana* (Roth) Spreng.	L	Leaves eaten	Fatma and Pan, 2012
312.	*Limonia acidissima* L.	F	Ripe fruits eaten raw and also used in making pickles	Bandyopadhyay and Mukherjee, 2009; Jadhav, 2011; Singh and Ahirwar, 2013; Fatma and Pan, 2012

TABLE 4.1 (Continued)

S. No.	Name of plant	Edible part	Mode of preparation	Reference
313.	*Lonicera angustifolia* Wall. ex DC.	Fr	Fruit edible	Fatma and Pan, 2012
314.	*Luffa cylindrica* (L.) M. J.Roem	Fr	Young fruits used as vegetables	Banerjee et al., 2013
315.	*Madhuca longifolia* (L.) Macbride (Plate 1H)	Fr	Fruits edible	Jain, 1963; Kumar, 2003; Singh and Ahirwar, 2013; Kumar et al., 2010; Banerjee et al., 2013
316.	*Maesua indica* (Roxb) A.DC.	Fr	Ripe fruits eaten raw	Bandyopadhyay and Mukherjee, 2009
317.	*Mahonia borealis* Takeda	Fr	Fruits edible	Kaur and Vashistha, 2014
318.	*Malva verticillata* L.	L	Tender leafy twigs cooked as vegetables	Bandyopadhyay and Mukherjee, 2009
319.	*Mangifera indica* L.	Fr, S	Ripe fruits edible. Unripe fruits used for vegetables and pickles.	Kumar, 2003; Banerjee et al., 2013; Kumar et al., 2010; Jadhav, 2011; Jain et al., 2012
320.	*Manihot utilissima* Pohl	Ts	Tuber boiled and eaten	Fatma and Pan, 2012
321.	*Manilkara hexandra* (Roxb.) Dub. (Plate 1 I)	Fr	Fruits edible	Fatma and Pan, 2012; Jadhav, 2011
322.	*Maranta arundinacea* L.	Ts	Tubes are boiled and eaten	Fatma and Pan, 2012
323.	*Marsedenia hamiltoniana* Wight & Arn.	Fr	Fruits eaten	Fatma and Pan, 2012
324.	*Marsilea minuta* L.	L	Cooked as vegetable	Kumari and Kumar, 2000; Banerjee et al., 2013; Bandyopadhyay and Mukherjee, 2009
325.	*Martynia annua* L.	Fr	Cotyledon edible	Jadhav, 2011
326.	*Medicago sativa* L.	L	Leaves cooked as vegetable	Jadhav, 2011

S. No.	Name of plant	Edible part	Mode of preparation	Reference
327.	*Melastoma malabathricum* L.	S	Seeds and placenta edible	Fatma and Pan, 2012; Bandyopadhyay and Mukherjee, 2009
328.	*Melilotus alba* Desr.	L	Leaves cooked as vegetables	Bandyopadhyay et al., 2012
329.	*Melochia corchorifolia* L.	L	Tender leaves cooked as vegetables leaves eaten	Bandhyopadhyay, 2012; Fatma and Pan, 2012
330.	*Memecyclon edule* Roxb.	Fr	Berries eaten	Fatma and Pan, 2012
331.	*Mentha viridis* L.	L	Young leaves used as vegetables	Banerjee et al., 2013
332.	*Merremia umbellata* Hall	L	Young leaves cooked with rice and eaten.	Jain, 1963
333.	*Meyna spinosa* Roxb. ex Link	Fr	Ripe fruit eaten raw	Bandyopadhyay and Mukherjee, 2009
334.	*Michelia champaca* L.	Fl	Flowers edible	Fatma and Pan, 2012
335.	*Miliusa velutina* (Dunnel) Thom	Fr	Ripe fruits eaten raw	Fatma and Pan, 2012
336.	*Mimosops elengi* L.	Fr, L	Ripe fruits and tender leaves eaten raw	Katewa et al., 2000; Alawa and Ray, 2016
337.	*Molluga spergula* L.	L	Leaves used as vegetable	Banerjee et al., 2013
338.	*Momordica charantia* L. (Plate 1C)	Fr	Unripe fruits eaten as vegetable	Kumar, 2003
339.	*Momordica cochinchinensis* (Lour.) Spreng.	Fr	Unripe fruits cooked as vegetables	Bandyopadhyay and Mukherjee, 2009
340.	*Momordica dioca* Roxb.	Fr	Fruits eaten as vegetables	Singh and Ahirwar, 2013; Bandyopadhyay et al., 2012
341.	*Moringa oleifera* Lam.	Fl	Flower and fruits eaten	Jadhav, 2011; Singh et al., 2015
342.	*Moringa pterigosperma* Gaertn.	Fr	Unripe fruits used as vegetable	Fatma and Pan, 2012
343.	*Morus alba* L.	Fr	Fruits and cotyledons edible	Alawa and Ray, 2016

TABLE 4.1 (Continued)

S. No.	Name of plant	Edible part	Mode of preparation	Reference
344.	*Morus indica* L.	Fr	Fruits edible	Fatma and Pan, 2012
345.	*Morus serrata* Roxb.	Fr	Fruits edible	Jadhav, 2011
346.	*Mukia madaraspatana* (L.) M.Roem.	Fr	Fruits edible	Jadhav, 2011
347.	*Murraya koengii* (L.) Spreng.	L	Leaves are used as flavoring agent	Bandyopadhyay and Mukherjee, 2009; Banerjee et al., 2013
348.	*Musa balbisiana* Colla	Fr	Unripe fruits cooked as vegetables	Badyopadhyay, 2009; Dwivedi et al., 2014
349.	*Nasturtium officinale* R.Br.	L	Leaves edible	Banerjee et al., 2013
350.	*Nelumbo nucifera* Gaertn.	Fr, Pt	Petioles used as vegetable and ripe fruits eaten raw	Katewa et al., 2000; Fatma and Pan, 2012; Dwivedi, 2016
351.	*Nigella sativa* L.	S	Seeds eaten	Fatma and Pan, 2012
352.	*Nymphaea alba* L.	Rh	Rhizome eaten	Fatma and Pan, 2012
353.	*Nymphaea nouchali* Burm. F.	Pd	Peduncles cooked as vegetable	Bandyopadhyay and Mukherjee, 2009; Fatma and Pan, 2012
354.	*Nymphaea pubescens* Willd.	Pd	Peduncles cooked as vegetables	Bandyopadyay and Mukherjee, 2009; Fatma and Pan, 2012
355.	*Nymphaea rubra* Roxb. ex Salisb.	Pt	Petioles cut into pieces and cooked	Banerjee et al., 2013
356.	*Nymphaea stellata* Willd.	Pd	Peduncles cooked as vegetable	Bandhopadhyay and Mukherjee, 2009; Fatma and Pan, 2012
357.	*Nympoides hydrophylla* (Lour.) Kuntze	S, Fl	Leaves and flowers eaten as vegetables	Fatma and Pan, 2012

S. No.	Name of plant	Edible part	Mode of preparation	Reference
358.	*Nymphoides indica* (Roxb.) Kuntze	S, Fl	Seeds and flowers eaten	Fatma and Pan, 2012
359.	*Ocimum americanum* L.	S	Seeds cooked in water and then consumed with curd	Katewa et al., 2000; Bandyopadhyay et al., 2012
360.	*Ophioglossum reticulatum* L.	L	Leaves as vegetable	Kumari and Kumar, 2000
361.	*Opuntia dillenii* (Ker Gawl.) Haw.	Fr	Fruits edible	Fatma and Pan, 2012; Banerjee et al., 2013
362.	*Oroxylum indicum* (L.) Vent.	Fr, S	Fruits and seeds eaten	Jain, 1963; Fatma and Pan, 2012; Bandyopadhyay and Mukherjee, 2009
363.	*Ottelia alismoides* Pers.	Rh	Rhizomes boiled and eaten	Fatma and Pan, 2012
364.	*Ougeinia oogeinensis* (Roxb.) Hochr.	Fl	Flowers cooked and eaten as vegetable	Kumar, 2003
365.	*Oxalis corymbosa* DC.	B	Eaten raw and it also used for pickles	Bandyopadhyay and Mukherjee, 2009; Fatma and Pan, 2012
366.	*Oxalis corniculata* L.	L	Leaves cooked as vegetable	Kumari and Kumar, 2000; Jain et al., 2012; Fatma and Pan, 2012
367.	*Pachyrhizus angulatus* L.C Rich ex DC.	L	Leaves eaten	Fatma and Pan, 2012
368.	*Paederia foetida* L.	L	Leaves as vegetables	Banerjee et al., 2013; Fatma and Pan, 2012; Bandyopadhyay and Mukherjee, 2009
369.	*Panicum sumatrense* Roth	S	Seeds taken as grains	Mishra, 2010; Sisodiya, 2012
370.	*Parthenocissus semicordata* (Wall.) Planch.	Fr	Fruits edible	Kaur and Vashistha, 2014
371.	*Paspalum scrobiculatum* L.	S	Leaves eaten	Fatma and Pan, 2012
372.	*Pennisetum americanum* (L.) K. Schum.	S	Seeds as food grains	Sisodiya, 2012

TABLE 4.1 (Continued)

S. No.	Name of plant	Edible part	Mode of preparation	Reference
373.	*Pereskia belo* L.	Fr	Seed grain used	Fatma and Pan, 2012
374.	*Pergularia extensa* (Jacq.) N. E.Br.	Fl, Fr	Flower and fruits as spice	Kumar, 2003
375.	*Persicara barbata* (L.) H.Hara (Syn.: *Polygonum barbatum* L.)	L	Leaves as vegetable	Katewa et al., 2000; Kumari and Kumar, 2000
376.	*Persicaria chinensis* (L.) H Gross	L	Used as vegetables	Bandyopadhyay and Mukherjee, 2009
377.	*Persicaria glabra* (Willd.) M.Gomez (Syn.: *Polygonum glabrum* Willd.)	L	Leaves used as vegetable	Kumari and Kumar, 2000; Mishra, 2010
378.	*Peucedanum nagpurense* (Clarke) Prain	Fr	Fruits used as vegetables	Fatma and Pan, 2012
379.	*Phaseolus aconitifolius* Jacq.	S	Seeds eaten grain	Fatma and Pan, 2012
380.	*Phaseolus calcaratus* Roxb.	S	Seeds edible	Fatma and Pan, 2012
381.	*Phlogacanthus thyrisiflorus* (Roxb.) Nees	Fl	Used as vegetables	Bandyopadhyay and Mukherjee, 2009
382.	*Phoenix acaulis* Roxb.	Fr	Ripe fruits are eaten	Dwivedi et al., 2014
383.	*Phoenix humulus* Royle	Fr	Ripe fruits eaten	Jain, 1963
384.	*Phoenix sylvestris* Roxb.	Fr	Ripe fruits eaten raw	Jadhav, 2011; Katewa et al., 2000; Bandyopadhyay and Mukherjee, 2009
385.	*Phyllanthus emblica* L.	Fr	Unripe fruits are used for pickles and fruits are edible	Kumar, 2003; Bandyopadhyay and Mukherjee, 2009; Jadhav, 2011; Sandya and Ahirwar, 2015
386.	*Physalis minima* L.	Fr	Fruits edible	Banerjee et al., 2013; Jadhav, 2011; Dwivedi et al., 2014

S. No.	Name of plant	Edible part	Mode of preparation	Reference
387.	*Phytolaca acinosa* L.	L	Leaves edible	Bandyopadhyay and Mukherjee, 2009; Jadhav, 2011
388.	*Pimpinella diversifolia* DC.	Fl	Flowers as spice	Kumar, 2003
389.	*Piper longum* L.	Fr	Fruits edible	Alawa and Ray, 2016
390.	*Pithecellobium dulce* (Roxb.) Benth.	Fr	Eaten raw as vegetables and ripe fruits edible	Katewa et al., 2000; Bandyopadhyay and Mukherjee, 2009; Jadhav, 2011
391.	*Plesmonium margaritiferum* (Roxb.) Schott	S, St	Stem and seeds used as vegetable	Jain, 1963; Fatma and Pan, 2012
392.	*Plumbago zeylanica* L.	Fr	Young leaves used as vegetables	Alawa and Ray, 2016
393.	*Polyalthia cerasioides* (Roxb.) Bedd.	Fr	Fruits eaten	Fatma and Pan, 2012
394.	*Polyalthia suberosa* (Roxb.) Thwaites	Fr	Ripe fruits eaten raw	Bandyopadhyay and Mukherjee, 2009
395.	*Polycarpon prostratum* (Forssk.) Aschers	Fr	Ripe fruits eaten raw	Bandyopadhyay and Mukherjee, 2009
396.	*Polygonatum multiflorum* (L.) All.	Fr	Fruits eaten	Banerjee et al., 2013
397.	*Polygonum alatum* L.	L	Leaves used as vegetables	Fatma and Pan, 2012
398.	*Polygonum plebeium* R.Br.	L	Leaves cooked as vegetable	Das, 1999; Kumari and Kumar, 2000;
399.	*Polygonum serrulatum* Lag.	L	Leaves eaten as vegetable	Kumar, 2003
400.	*Portulaca oleracea* L.	L, St	As vegetable	Kumari and Kumar, 2000; Fatma and Pan, 2012; Bandyopadhyay and Mukherjee, 2009; Katewa et al., 2000
401.	*Portulaca quadrifida* L.	L	Consumed as vegetable	Kumari and Kumar, 2000; Jadhav, 2011; Banerjee et al., 2013; Fatma and Pan, 2012

TABLE 4.1 (Continued)

S. No.	Name of plant	Edible part	Mode of preparation	Reference
402.	*Prosopsis cinerea* (L.) Druce	Fr	Pods used as an ingredient	Katewa et al., 2000
403.	*Prunus persica* (L.)Stokes	Fr	Fruits eaten	Kumar, 2003
404.	*Psidium guajava* L.	Fr	Ripe fruits edible	Jadhav, 2011
405.	*Psydrax dicoccos* L.	Fr	Fruits eaten raw	Fatma and Pan, 2012
406.	*Pterocarpus marsupium* Roxb.	L	Seeds edible	Fatma and Pan, 2012; Kumar et al., 2010
407.	*Pueraria tuberosa* DC.	S	Seed, flower and roots edible	Kumar, 2003; Fatma and Pan, 2012; Singh et al., 2015
408.	*Punica granatum* L	Fr	Fruits edible	Fatma and Pan, 2012
409.	*Randia dumentorum* Lamk.	Fr	Ripe fruits eaten	Fatma and Pan, 2012
410.	*Randia uliginosa* (Retz) Poir.	Ts	Tuber eaten	Fatma and Pan, 2012
411.	*Rhus parviflora* Roxb.	Fr	Fruits edible	Kumar, 2003; Alawa, 2016
412.	*Rhus semialala* Murray	Fr	Eaten raw	Fatma and Pan, 2012
413.	*Rivea hypocrateriformis* (Desr.) Choisy	Fr	Drupes eaten	Kumar, 2003; Fatma and Pan, 2012
414.	*Rubus ellipticus* Sm.	L	Leaves eaten	Fatma and Pan, 2012
415.	*Rumex dentalus* L.	L	Used as vegetables	Bandyopadhyay and Mukherjee, 2009
416.	*Rumex hastatus* Don	L	Leaves edible	Dwivedi et al., 2014
417.	*Rumex vesicarius* L.	L	Leaves cooked as vegetable	Kumari and Kumar, 2000
418.	*Sacciolepis interupta* (Willd.) Stapf	S	Seeds eaten as grains	Jain, 1963
419.	*Sageratia filiformis* (Roem. & Schult.) G.Don	Fl, Fr	Flowers and fruits eaten	Katewa et al., 2000

S. No.	Name of plant	Edible part	Mode of preparation	Reference
420.	*Sagittaria sagitiifolia* L.	Ts	Starchy tubers edible	Das, 1999; Fatma and Pan, 2012
421.	*Salvadora persica* L.	Fr	Ripe fruits eaten	Bandyopadhyay and Mukherjee, 2009; Kumar et al., 2010
422.	*Santalum album* L.	Fr	Ripe fruits edible	Kumar et al., 2010
423.	*Saponaria vaccaria* L.	S	Seed oil edible	Fatma and Pan, 2012
424.	*Schefflera venulosa* (Wight & Arn.) Harms	Fr	Fruits eaten	Kumar, 2003
425.	*Schleichera oleosa* (Lour.) Oken	S	Seeds edible	Jain, 1963; Kumar et al., 2010; Sandya and Ahirwar, 2015
426.	*Semecarpus anacardium* L.f.	L, Fl	Young leaves cooked, Flowers eaten	Jain, 1963; Kumar, 2003; Fatma and Pan, 2012; Dwivedi et al., 2014; Alawa and Ray, 2016
427.	*Senna obtusifolia* (L) H. S. Irwin & Barneby	L	Used as vegetable. Ripe fruits edible used pickles	Jadhav, 2011
428.	*Senna siamea* (Lam.) H. S. Irwin & Barneby	Fl	Flowers cooked as vegetable	Fatma and Pan, 2012
429.	*Senna sophera* (L.) Roxb.	L	Cooked as vegetable	Bandyopadhyay and Mukherjee, 2009
430.	*Senna tora* (L.) Roxb. (Syn.: *Cassia tora* L.)	L	Cooked as vegetable	Bandyopadhyay and Mukherjee, 2009; Jadhav, 2011; Kumar, 2010
431.	*Sesbania grandiflora* (L.) Poir.	Fl	Flowers fried and eaten	Jadhav, 2011; Banerjee, 2013
432.	*Setaria glauca* (L.) P. Beuv.	S	Grains boiled and consumed as food.	Kaur and Vashistha, 2014
433.	*Setaria italica* (L.) P. Beauv.	S	Seeds used as grains	Fatma and Pan, 2012
434.	*Shorea robusta* Gaertn. F.	S	Seed oil edible	Kumar, 2003; Fatma and Pan, 2012; Singh et al., 2015

TABLE 4.1 (Continued)

S. No.	Name of plant	Edible part	Mode of preparation	Reference
435.	*Sida veronicifolia* Roxb.	L	Leaves cooked	Fatma and Pan, 2012
436.	*Similax ovalifolia* Roxb.	L	Cooked as vegetable	Bandyopadhyay and Mukherjee, 2009
437.	*Smilax prolifera* Roxb.	Fr	Young fruits cooked as vegetable	Jain, 1963
438.	*Smilax zeylanica* L.	L, Rh	Leaves and rhizomes eaten	Fatma and Pan, 2012
439.	*Smithia conferta* Sm.	L	Leaves cooked as vegetable	Jain, 1963
440.	*Solanum myriacanthum* Dunni	Fr	Unripe fruits cooked as vegetables	Fatma and Pan, 2012
441.	*Solanum nigrum* L.	Fr	Fruits edible	Jain, 1963; Kumar et al., 2010; Jadhav, 2011; Banerjee et al., 2013
442.	*Solanum torvum* Sw.	Fr	Fruits eaten	Bandyopadhyay and Mukherjee, 2009
443.	*Solanum xanthocarpum* Schrad.	F	Young fruits cooked as vegetable	Jadhav, 2011; Fatma and Pan, 2012; Ray and Saikhediya, 2016
444.	*Solena amplexicaulis* (Lam) Gandhi	L, Fr	Leaves and unripe fruits used as vegetables	Kumar, 2003; Bandyopadhyay and Mukherjee, 2009
445.	*Sphaeranthus indicus* L	L	Young leaves eaten	Jain, 1963
446.	*Spilanthes oleracea* L.	L	Leaves cooked as vegetable	Bandyopadhyay and Mukherjee, 2009
447.	*Spondias pinnata* (L.) Kurz	F	Unripe and ripe fruits eaten raw and for making pickles	Jain, 1963; Bandyopadhyay and Mukherjee, 2009; Dwivedi et al., 2014
448.	*Sterculia foetida* L.	S	Seeds roasted	Fatma and Pan, 2012
449.	*Sterculia urens* Roxb.	Fr, S, Gum	Eaten as vegetable and gum also obtained	Katewa et al., 2000
450.	*Sterculia villosa* Roxb. ex Smith	S	Roasted seeds eaten	Bandyopadhyay and Mukherjee, 2009

S. No.	Name of plant	Edible part	Mode of preparation	Reference
451.	*Symplocos racemosa* Roxb.	L	Young leaves cooked and eaten as vegetable	Kumar, 2003
452.	*Syzigium cumini* (L.) Skeels	Fr	Fruits edible	Kumar et al., 2010; Jadhav, 2011
453.	*Syzigium nervosum* DC.	Fr	Ripe fruits eaten	Kumar, 2003
454.	*Syzigium samarangense* (Blume) Merr. & L. M.Perry	Fr	Fruits edible	Kumar et al., 2010
455.	*Tacca leontopetaloides* (L.) Kuntz	Ts	Tubers macerated and eaten	Jain, 1963; Fatma and Pan, 2012
456.	*Tamarindus indica* L.	Fr, Fl	Fruits edible	Bandyopadhyay and Mukherjee, 2009; Kumar et al., 2010; Jadhav, 2011; Sandya and Ahirwar, 2015
457.	*Terminalia alata* Heyne ex Roth	Gum	Gum edible	Kumar, 2003
458.	*Terminalia bellerica* (Gaertn.) Roxb.	Fr	Fruits edible	Jadhav, 2011; Kumar, 2010; Singh et al., 2015
459.	*Tetrastigma bracteolatum* (Wall.) Planch.	Fr	Ripe fruits eaten	Bandyopadhyay and Mukherjee, 2009
460.	*Tetrastigma leucostapylum* (Dennst.) Balakr.	Fr	Ripe fruits eaten	Bandyopadhyay and Mukherjee, 2009
461.	*Trapa natans* L.	Fr	Eaten raw	Fatma and Pan, 2012; Alawa and Ray, 2016
462.	*Trema politoria* Planch.	Fr, Gum	Fruit kernel and gums eaten	Fatma and Pan, 2012
463.	*Tribulus terrestris* L.	Fr, L	Used as vegetables	Katewa et al., 2000
464.	*Trichosanthes anguina* L.	Fr	Eaten as vegetables during famine	Bandyopadhyay and Mukherjee, 2009
465.	*Trigonella corniculata* L.	L	Young leaves as vegetables	Banerjee et al., 2013

TABLE 4.1 (Continued)

S. No.	Name of plant	Edible part	Mode of preparation	Reference
466.	*Typha angustata* Chaub.	Fr, Fl	Flowers and fruits eaten	Fatma and Pan, 2012
467.	*Typhonium trilobatum* Schott	L, Pt	Leaves and petioles cut into pieces, steamed and fried with kalijira (kalaunji) during famine	Das, 1999; Bandyopadhyay and Mukherjee, 2009
468.	*Vangueria pubescens* Kurz	Fr, L	Fruits and leaves cooked as vegetable	Fatma and Pan, 2012
469.	*Vangueria spinosa* Roxb.	L	Leaves as vegetable	Kumari and Kumar, 2000
470.	*Ventilago denticulata* Willd.	S	Seeds yield an edible oil	Jain, 1963
471.	*Ventilago madaraspatana* Gaertn.	S	Seeds eaten	Fatma and Pan, 2012; Ray and Saikhediya, 2016
472.	*Vernonia cinerea* Less.	L	Leaves consumed as vegetable	Fatma and Pan, 2012
473.	*Vicia faba* L.	Fr, S, L	Leaves and pods used as vegetable. Seeds used as pulses.	Dwivedi et al., 2014
474.	*Vigna catjang* (Burm. f.) Wall.	Fr	Fruits edible	Fatma and Pan, 2012
475.	*Vitex negundo* L.	S	Seeds edible	Katewa et al., 2000
476.	*Wattakaka volubilis* (L.) Stapf	L	Leaves cooked to prepare vegetable	Alawa and Ray, 2016

S. No.	Name of plant	Edible part	Mode of preparation	Reference
477.	*Woodfordia fruticosa* (L.) Kurz	Fl	Leaves used as vegetable	Jadhav, 2011; Alawa and Ray, 2016
478.	*Xanthium strumarium* L.	L, St	Young leaves and stem used as vegetable	Banerje et al., 2013
479.	*Xylia xylocarpa* (Roxb.) Taub.	S	Seeds roasted in an earthen pot and eaten	Jain, 1963
480.	*Zehneria umbellata* Thwaites	Rh	Rhizomes used for vegetable	Banerjee et al., 2013
481.	*Zingiber zerumbet* (L.) Roscoe	Rh	Rhizomes edible	Fatma and Pan, 2012
482.	*Ziziphus jujuba* Mill.	Fr	Unripe fruits used as vegetable	Singh et al., 2015; Katewa et al., 2000
483.	*Ziziphus mauritiana* Lamk.	Fr	Fruits edible	Jadhav, 2011; Singh and Ahirwar, 2013
484.	*Ziziphus nummularia* (Burm.f.) Wight & Arn.	Fr	Fruits edible	Jadhav, 2011; Dwivedi et al., 2014
485.	*Ziziphus oenoplia* (L.) Mill.	Fr	Fruits edible	Bandyopadhyay and Mukherjee, 2009
486.	*Ziziphus rugosus* (L.) Mill.	Fr	Fruits edible	Fatma and Pan, 2012, Dwivedi et al., 2014
487.	*Ziziphus xylopyrus* (Retz.) Willd.	Fr	Fruits edible	Jadhav, 2011; Alawa and Ray, 2016

plants or their products are used variously such as vegetables, raw fruits, nuts, beverages or drinks, pickled or, oilseed, grains or condiments.

The nutritional aspects of the diets of Marias have been documented by Roy and Rao (1957). It is revealed from the study that out of 483 edible plants recorded, most of the plants are significant for fruits (191), followed by leaves (131), underground parts (61), seeds (67), flower (23), stem and shoot (23), whole plant as vegetable (10), bulbils (4) and others (17).

4.4.1   STEM

Stems and young tender shoots of 23 plant species are used for food. Young shoots of *Bambusa arundinacea, Dendrocalamus hamiltonii, Enhydra fluctuans* and tender twig of *Ipomoea aquatica* (Plate 4.1A) are cooked as vegetable. Pith and outer soft tissue of *Cycas pectinata* are eaten (Fatma and Pan, 2012).

4.4.2.   UNDERGROUND PARTS

Underground parts of plant are consumed as vegetables. In these vegetables the food is stored in underground parts and important source of starch. The storage organs may be true roots, rootstock or modified stem and root like rhizome, corms, tubers and bulbs (Pandey, 2006).

There are 61 wild plants species are known to be important for underground parts in the area under review of which rhizome (19), tubers (21), root and rootstock (13), bulbils (4) stolons (2), bulbs (2), corms (4) of these plants are gathered and consumed by tribals.

Rhizomes, corms or tubers are eaten raw or cooked after repeated washing to remove the bitterness and pungency. Tribal men generally dig out underground parts (Arora and Pandey, 1996). Tubers of *Alocasia indica, Colocasia esculenta, Typhonium trilobatum, Momordica dioca, M. cochinchinensis, Randia uliginosa, Tacca leuntopetaloides, Dioscorea hispida, D. esculenta, D. alata, D. globosa, D. bulbifera, D. pentaphylla, Amorphophallus konkanensis, A. campanulatus* are boiled and taken as staple food during scarcity of food.

Rhizome and corms of *Costus speciosus, Curcuma amada, Cyperus esculenta, Nelumbo nucifera, Musa balbsiana, Alpina glanga, Chlorophytum tuberosum, Asphodelus tenuifolius, Amorphophallus campanulatus* are

(A) *Ipomoea aquatica* Forssk.   (B) *Cucumis prophetarum* L.   (C) *Momordica dioica* Roxb. ex Willd.

(D) *Ensete superbum* (Roxb.) Cheesm.   (E) *Geodorum densiflorum* (Lam.) Schl.

(F) *Grewia tenax* (Forssk.) Fiori   (G) *Diospyros melanoxylon* Roxb.

(H) Flowers of *Madhuca longifolia* (L.) Macbr.   (I) *Manilkara hexandra* (Roxb.) Dub.

PLATE 4.1

cooked as vegetable in Bihar, West Bengal and Madhya Pradesh. Rhizome of *Curcuma lucca* is used for starch source. Bulbs of *Geodorum densiflorum* (Plate 4.1E) and *Chlorophytum borivilianum* are dug out and eaten by tribal women to increase strength in Dhar district, Western Madhya Pradesh (Alawa, 2013). Stolons of *Elaeocarpus dulcis, Colocasia esculenta, Typhonium trilobatum* are taken as vegetable.

4.4.3 LEAVES

Vegetable is usually applied to edible plants which store up reserve food in roots, stems, leaves which are eaten, cooked or raw as salad. The nutritive value of vegetables is high due to presence of indispensable minerals, salts and vitamins. Wild vegetables provide adequate number of crude fibers, fats, carbohydrate, proteins, mineral elements like Ca, Na, K, Fe, Mg, Mn, Cu, Zn in addition to vitamins (Gogoi and Kalita, 2014).

There are 175 wild plant species of which leaves of 131 plants and unripe green fruits of 44 plant species are collected and eaten raw or cooked as vegetables. Young tender fronds of 3 pteridophyta plant species are eaten as vegetables in Madhya Pradesh, Uttar Pradesh and Bihar.

Diplazium esculentum is common edible species in West Bengal. *Ceratopteris thalictroides, Adiantum caudatum* and *Marsilea minuta* are consumed as green vegetable in Madhya Pradesh and Chhattisgarh.

Leaves of *Colocasia esculenta* are widely used as vegetable in the all the regions of Indo-Gangetic and Central India (Bandyopadhyay et al., 2012; Kapale, 2013). Young leaves of *Cycas pectinata* are eaten as vegetable in West Bengal (Bandyopadhyay and Mukherjee, 2009).

Leaves of *Amaranthus viridis, A. caudatus, A. spinosus, A tricolor, Basella rubra, Chenopodium album, Ipomoea aquatica, Nasturtium officinale, Senna tora, Portulaca quadrifida* are very commonly consumed in Madhya Pradesh, West Bengal, Bihar and Chhatishgarh (Jadhav, 2011; Das, 1999).

Leaves of *Ceropegia bulbosa, Cassia obtusifolia, Celosia argentea, Digera muricata, Hygrophila auriculata, Lagerstroemia parvifolia, Polygonum barbatum* are used as leaf vegetable in Rajasthan (Katewa et al., 2000). Leaves of *Trigonella polycrata, Phytolacca acinosa, Plantago major, Rumex hastatus, Trianthema portulacastrum* are used as vegetable in Uttar Pradesh (Joshi and Tiwari, 2000). Leaves of *Cissus adnata,*

Hedyotis scandens, *Paedaria scandens*, *Persicaria chinensis*, *Rumex dentatus* and *Typhonium trilobatum* are cooked as vegetable in Bihar and adjacent remote areas. Leaves *of Leea asiatica*, *Argyreia strigosa*, *Rivea hypocrateriformis* are used as vegetable during drought and scarcity of food in Nimar region of Madhya Pradesh (Mishra, 2010; Satya, 2006). Leaves of *Basella rubra* and *Cassia obtusifolia* are roasted and eaten in Madhya Pradesh, Bihar and West Bengal (Bandyopadhyay et al., 2012; Fatma and Pan, 2012).

Leaves of *Cinnamomomum tamala*, *Murraya koenigii*, *Curcuma amada* and *Zingiber zerumbet* are used as flavoring agents (Banerjee et al., 2013). Petiole and peduncle of *Nymphaea pubescens*, *N. rubra*, *N. nouchali* and *N. stellata* are peeled, cut into pieces, steamed and cooked to prepare food (Bandyopadhyay, 2012; Basu and Mukherjee, 1996). Kumar et al. (2013) gave an account of 21 leafy vegetables supplemented to malnutrition among the tribals of Jharkhand.

4.4.4   FLOWERS

A few wild species are important for their edible flowers, buds, and inflorescence. Flowers of 23 plant species are reported to be collected and eaten by tribals in the Indo-Gangetic and Central India. In Madhya Pradesh, Bihar, Orissa and adjoining tracts of peninsular India, the tribal collect flowers of *Madhuca indica* which constitute important article of food being eaten raw or cooked (Singh and Arora, 1978). Roy and Rao (1959) studied the chemical composition of flowers and reported percentage of alcohol content in liquor and distilled spirit of *Madhuca longifolia* as 4.4% and 19.58%, respectively.

Flowers of *Sesbania grandiflora*, leaves of *Coleus anthelmintica* and *Eichhornia crassipes* are fried and eaten. Flowers and leaves of many wild plants are used as supplementary vegetables in Dharbhanga district, Bihar (Jha et al., 1996). Flowers of *Chlorophytum arundinaceum*, *Nymphoides hydrophylla*, *Woodfordia fruticosa*, *Dregea volubilis*, *Indigofera cassioides*, *Rhododendron arboreum*, *Viola canescens*, *Butea monosperma*, *Oroxylum indicum*, *Cassia fistula*, *Madhuca indica*, *Moringa oleifera* are plucked and eaten. Flowers and floral buds of *Bauhinia variegata* are sold in the market for its edible value.

4.4.5   UNRIPE FRUITS

The unripe green fruits of 44 plants species are used as vegetables and in some cases used as pickles. Commonly consumed plants in study area are *Ensete superbum* (Plate 4.1D), *Artocarpus heterophyllus, Cerescoides turgida, Cordia dichotoma, Carica papaya, Elaeocarpus serratus, Mangifera indica, Solanum xanthocarpum, Cassia mimosoides, Cucumis prophetarum* (Plate 4.1B), *Momordica balsimina, M. cochinchinensis, M. dioica* (Plate 4.1C), *Leptadenia reticulata, Coccinia grandis, Spondias pinnata, Zehneria umbellata* and *Hibiscus sabdariffa.*

Receptacles of *Ficus rumphii, F. racemosa, F. hispida, F. locar* are cooked as vegetable. Leaves, flower and fruits of *Moringa oleifera* and *M. pterigosperma* are cooked as vegetable.

4.4.6   RIPE FRUITS

Morphologically a fruit is the seed-bearing portion of the plant, and consists of the ripened ovary and its contents. Ripe fruits of 147 plant species are edible in the study area and usually fruits of 118 plant species are eaten without cooking by local communities because of containing appreciable amount of nutrients and energy. Although nutritive value of wild fruit is little known but it is apparent from the study that some of the fruits are rich in protein, minerals and carbohydrate. Fruits of *Aegle marmelos, Feronia limonia* and *Ziziphus rugosus* are rich in protein. Fat content is more in *Gardenia latifolia, Spondias pinnata, Ficus* spp. where as mineral content is found to be more in *Carissa congesta, Ziziphus rugosus* and *Ficus* spp. Apart from this it is reported that fruits of *Spondias pinnata* and *Artocarpus lacucha* are rich in Vitamin A (Aykroyd, 1956). Thus fruits are useful food supplements for ethnic people.

Edible kind of fruits are mainly obtained from families Arecaceae, Annonaceae, Anacardiaceae, Capparaceae, Euphorbiaceae, Rosaceae, Rubiaceae, Rutaceae, Myrtaceae, Tiliaceae and Rhamnaceae. Ripe fruits of *Aegle marmelos, Grewia asiatica, G. subinequalis, G. tiliaefolia, G. flavescens, G. tenax, G. sclerophylla, G. hirsuta, Annona squamosa, A. reticulata, Morus alba, Averrhoa carambola, Rubus ellipticus, Artocarpus lacucha, Borassus flabellifer, Phoenix sylvestris, Carissa carandas, Citrullus colocynthis, Diospyros melanoxylon* (Plate 4.1G), *Manilkara hexandra, Mimusops elengi, Miliusa velutina, Meynna spinosa, Pereskia velo, Randia dumentorum, Tamarindus indica* and *Phyllanthus emblica* are consumed as fruits.

Fruits of *Artocarpus heterophyllus, Feronia limonia, Cordia dichotoma, Carissa opaca, Spondius pinnata* and *Mangifera indica* are used for making pickles. Calyces of *Dillenia indica* are used for making pickles in West Bengal (Bandyopadhyay, 2012).

4.4.7 SEEDS

Seed grains of *Echinochola colonum, Eleusine indica, Paspalum scrobiculatum, Fagopyrum cymosum, Panicum millaceam, P. sumatrense, Dactyloctenium sindicum, Avena ludoviciana, Oryza rufipogon, Eleusine coracana* are boiled and taken with salt during famine and drought. A total of 67 wild plant species possess edible seeds which are consumed during famine as scarcity of food. Seeds of *Indigofera glandulosa, Indigofera cordifolia, Grewia tenax* (Plate 4.1F), *Artocarpus heterophyllus* are eaten. In the hilly tracts of Central India, Bihar the tribal inhabitant of remote forest area collect ripe and unripe seeds of *Bauhinia vahlii, Mucuna prurita* which are boiled, roasted and eaten. Seeds of *Nymphaea* and *Nelumbo* species are eaten raw from the ripe carpel. Seeds of *Buchnania cochinchinensis* are edible and sold in the market. Seeds of *Oroxylum indicum* are pounded and made into flour in time of scarcity (Jain, 1963.)

4.5 OTHERS

4.5.1 GUMS

Gum exudes from the stumps of pruned branches and other scars of trees. The gum forms with water dark colored tasteless mucilage. Gums of *Acacia catechu, Sterculia villosa, Holoptelea integrifolia, Boswellia serrata, Trema politoria* and *Trema orientalis* are consumed and especially given to tribal women after delivery to increase strength. Gum of *Acacia nilotica, Cochlospermum religiosum* and *Buchnania cochinchinensis* is edible (Upadhyay and Chouhan, 2003).

4.5.2 INDIGENOUS DRINK

The tribals prepare country liquor from flowers of *Madhuca indica*, grains of *Oryza sativa*, fruit of *Ziziphus jujuba*, stem juice of *Saccharum officinarum*, fruits of *Ficus glomerata*, fruits of *Pheonix sylvestris*, fruits of *Borassus fla-bellifer*, *Manilkara hexandra* and *Citrus limon*. This local wine is distributed and drunk in every tribal festival and marriage ceromony. Tribals collect the sugary juice from *Phoenix sylvestris* and *Borassus flabellifer* early in the morning and taken as soft drink. The liquor prepared from *Madhuca longifo-lia* (Plate 4.1H) is the commonest beverage consumed in varying quantities by all the tribals of Bastar in Madhya Pradesh. Salphi obtained from *Caryota urens* is common in northern parts of Bastar district (Jain, 1963).

4.6 PLANTS THAT IMPROVE THE ECONOMY

Most of the tribals and rural people are very poor economically and depend on non cultivated wild food plants for food. Tribals and villagers collect twigs, flowers, fruits, underground parts and other plant parts from forest for their own consumption and sometimes sell these in the local village market for earning money (Plate 4.2A). The plants like *Annona squamosa, Artocarpus lacucha, Artocarpus heterophyllus, Colocasia esculenta, Phyllanthus emblica, Trianthema portulacastrum, Typhonium trilobatum, Madhuca lon-gifolia, Mimusops elengi, M. hexandra* (Plate 4.1I), *Diospyros melanoxylon, Grewia tiliaefolia, G. subinequalis, Helicteres isora, Terminalia bellerica, Tamarindus indica, Buchnania cochinchinensis,* and *Ziziphus jujuba* are directly useful for economic upliftment of tribals of the area.

4.7 DISCUSSION

It is revealed from the present review that utilization of plants generally depends upon the availability of these plants in forests. Vegetables are regu-larly eaten by tribals, either cooked or separate preparation. They may be leafy vegetable or non leafy and tuberous. Mostly leaves, fruits, tuber, flow-ers, rhizome, inflorescence, stem, seeds or sometime whole plants are used as supplementary foods. Analytical study proves that the plants used by trib-als as food rich in nutritional property (Jain, 1963). Several time, plant parts are used as staple food while some are used at the time of scarcity like fam-ine, drought, etc. Most of the edible fruits are eaten as raw, which can pro-vide essential supplements of vitamins and minerals. It is the sweetest pulp or the fleshy palatable pericarp of ripe berries, drupe or nuts that is generally

(A) Tribal woman selling fruits of *Adansonia, Tamarindus* and *Ziziphus*

(B) Fruits of *Adansonia digitata* L.

PLATE 4.2

consumed. Tribals consume sufficient amount of fiber food in their diet, hence constipation problem is rarely found.

The method of wild plant collection is very easy. Ethnic people are very much familiar and careful regarding selection of fruit. Usually they avoid eating unripe fruit in the forest because they know well that eating of immature unripe fruits may cause harm or even death. Some unripe fruits are poisonous but same fruits are edible on ripening. They learnt it by experience or from their ancestors passing it from generation to generation. Most of tribal women collect leaves and flower from natural forests (Prasad and Bhatnagar, 1991) and men would dig out tuberous food (Arora and Pandey, 1996). Leaves, flowers, fruits are plucked whereas underground parts like rhizome tubers, corms are dugout. They do not dug out complete underground parts of plant and leave some part for propagation in future. Mostly underground parts are gently washed and boiled, then eaten. Tubers are largely eaten after processing. Tubers are boiled in water and their skin is removed. They are sliced into thin pieces. The acrid content of the tuber chips is washed away. This fresh chips can be cooked like rice or with rice (Jain, 1963).

4.8   CONCLUSION

Wild edible plants are closely linked with socioeconomic condition of tribal people. Most of these wild forms if cultivated in large scale may provide good economy. Increased over exploitation of wild edible plants may cause threat to certain species. Sustainable use of wild plant resources may solve the food crisis of the region. Conservation efforts are required by plantation and protection of these plants with active participation of local people. Detailed investigations on nutritional profile of all the reported wild edible food plant species are required before introducing these wild plant resources. It is hoped that wild edible plants would definitely fulfill the food crisis of 21st century particularly in the developing nations.

ACKNOWLEDGEMENT

Author is very much grateful to Dr. C. M. Solanki, Retd. Professor and Ex Head, Department of Botany, PMB Gujarati Science College, Indore and Dr. V. B. Diwanji, Retd. Professor and Ex Head, Dept of Botany, Holkar Science College, Indore for their valuable suggestions, encouragement and valuable insights. Thanks are due to Jeetendra Sainkhediya, Research Scholar for help in data collection. The Author also thanks the Principal, Head and all teaching staff members of Botany Department, PMB Gujarati Science College, Indore for their cooperation.

KEYWORDS

- **Bhil**
- **ethnic food plants**
- **ethnic people**
- **fruits**
- **Marias**
- **Santhal**
- **vegetables**

REFERENCES

Alawa, K. S. (2013). Ethnobotany of Dhar district, India. PhD Thesis (Unpublished) DAVV, Indore, India

Alawa, K. S., & Ray, S. (2016). Some wild vegetable plants used by tribals of Dhar district, India. *Indian J. Applied and Pure Biol., 31*(1), 65–69.

Anonymous (1995). Ethnobiology in India. A status report, All India coordinated research project on Ethnobiology. Ministry of Environment and Forests, Government of India, New Delhi, India.

Arora, R. K. (1991). Conservation and Management concept and Approach in Plant Genetic resources. In: Paroda R. S., & R. K. Arora (eds.), IBPGR, Regional office South and South east Asia New Delhi, p. 25.

Arora, R. K. (1994). Ethnobotanical studies on plant genetic resources – national efforts and concern. *Ethnobotany 7,* 125–136.

Arora, R. K., & Pandey (1996). Wild edible plants of India. National Bureau of Plant Genetic Resources, New Delhi, India.

Aykroyd, W. R. (1956). Nutritive value of Indian foods and the planning of satisfactory diets. Health. ICMR Bull. Govt. of India, Delhi, India.

Bandyopadhyay, S., Haldar, D., & Mukherjee, S. (2012). A census of wild edible plants from Howrah district, West Bengal, India. *Proceeding of UGC sponsored National Seminar Plant Research Science in Human Welfare,* pp. 14–25.

Bandyopadhyay, S., & Mukherjee, S. (2009). Wild edible plants of KuchBihar district, WB, *Natural product Radiance 8*(1), 64–72

Banerjee, A., Mukherjee, A., & Sinhababu, A. (2013). Ethnobotanical documentation of some wild edible plants in Bankura district, WB, India. *J. Ethnobiol. Traditional Medicine 120*, 585–590.

Basu, R., & Mukherjee (1996). Food plants of the tribes in Paharias of Purulia. *Advances in Plant Science 9* (2), 209–210.

Bhandari, M. M. (1974). Native resources used as famine food in Rajasthan. *Econ. Bot. 28*, 73–81.

Das, D. (1999). Wild food plants of Midnapur, WB during drought and famine. *J. Econ. Taxon. Bot. 23*, 539–547.

Duthie, J. F. (1903–1929). Flora of upper Gangetic Plain and adjacent Siwalik and Sub-Himalayan tracts, 3 Vol., BSI publication, Calcutta, India.

Dwivedi, S. N., & Pandey, A. (1992). Ethnobotanical studies on wild and indigenous species of Vindhyan plateau. *Maxpaceous flor*a, pp. 144–148.

Dwivedi, S. N., & Singh, H. (1984). Ethnobotany of Kols of Rewa division, Madhya Pradesh. *Proct. Natl. Sem. Envt. EPCO.* II 37–44.

Dwived, S. V., Anand, R. K., Singh, M. P., & Mishra, P. K. (2014). Studies on lesser known food plants used by tribals – socially poor communities of district Sonbhadra in Uttar Pradesh. *Int. J. Plant Sci. 9(1)*, 248–251.

FAO (2004). The state of food insecurity in the world food summit 2nd millennium developmental goals. Annual reports, Rome, Italy.

Fatma, N., & Pan, T. K. (2012). Checklist of wild edible plants of Bihar, India. *Our Nature, 1*, 233–241.

Gogoi, P., & Kalita, J. C. (2014). Proximate analysis and minerals components of some edible medicinally important vegetables of Kamrup district of Assam, India. *Int. J. Pharma Biosci. 5*(4), 451–457.

Haines, H. H. (1961). The Botany of Bihar and Orissa. 3 Vol. (Rep.). BSI Publication, Calcutta.

Jadhav, D. (2011). Wild plants used as a source of food by the Bhil tribe of Ratlam district (M.P.). *J. Econ. Taxon. Bot. 35* (4), 707–710.

Jain, A. K., & Tiwari, P. (2012). Nutritional value of some traditional edible plants used by tribal communities during emergency with reference to Central India. *Indian J. Trad. Knowl. 11*(1), 51–57.

Jain, S. K. (1963). Wild plant foods of tribals of Bastar. *Proc. Nat. Inst. Ind. 30 B*, 56–80.

Jain, S. K., & De, S. N. (1964). Some less known plant foods among the tribals of Purulia (WB). *Sci. & Cult. 30*, 285–286.

Jha, V., Mishra, S., Gupta, A. N., & Jha, A. (1996). Leaves and flowers utilized as supplementary vegetable in Dharbhanga (north Bihar) and their ethnobotanical significance. *J. Econ. Taxon. Bot. Addl Ser 12*, 395–402.

Joshi, P., & Awasthi. (1991). A life support plant species used in famine by tribals of Aravallis. *J. Phytol. Res. 4(2)*, 193–196.

Kanodia, K. C., & Gupta, R. K. (1968). Some useful and interesting supplementary food plants of the arid region. *J. D. Agric. Trop. Bot. Appl. 15*, 71–74.

Kapale, R., Prajapati, A. K., Napit, R. S., & Ahirwar, R. K. (2013). Traditional food plants of Baiga tribals. A survey study in tribal villages of Amarkantak Biosphere, Central India. *Indian J. Sci. Res. Tech. 1*(2), 27–30.

Katewa, S. S., Nag, A., & Guria, B. D. (2000). Ethnobotanical studies on wild plants for food from the Aravalli hills of south east Rajasthan. *J. Econ. Taxon. Bot. 23*(2), 259–264.

Kaur, R., & Vashistha, B. D. (2014). Ethnobotanical studies on Karnal district, Haryana, India. *Int. Res. J. Biol. Sci., 3*(8), 46–55.

Khan, A. A., Agnihotri, S. K., Singh M. K., & Ahirwar, R. K. (2008). Enumeration of certain Angiospermic plants used by Baiga tribe for conservation of plants species. *Plant Arch., 1*(8), 289–291.

King, G. (1969). Famine foods of Marwar. *Proc. Asiat. Soc. Beng 38,* 116–122.

Kumar, A., & Mishra, R. N. (2011). Computerized based taxonomy in the identification of ethnomedicinal plants of Shakumbhari Devi of Shivalik hills. *J. Indian Bot. Soc. 90,* 244–250.

Kumar, S., Kumari, B., & Goel, A. K. (2013). Study of leafy vegetables supplemental to malnutrition among tribals in Jharkhand. *Ethnobotany 25,* 135–138.

Kumar, S., Yadav, S., & Yadav, D. K. (2010). Study on biodiversity and edible bioresources of Betla National Park, Palamu, Jharkhand, India. *J. Econ.Taxon. Bot. 34*(4), 725–729.

Kumar, V. (2003). Wild edible plants of Surguja district of Chhattisgarh state, India, *J. Econ. Taxon. Bot 27*(2), 272–282.

Kumari, B., & Kumar, S. (2000). A check list of some leafy vegetables used by tribal in and around Ranchi, Jharkhand. *Zoos' Print J. 16,* 442–444.

Maheshwari, J. K. (1990). Recent ethnobotanical researches in Madhya Pradesh. *S. E. B. S. News Let.* (1–3), 5.

Maji, S., & Sikdar, J. K. (1982). A taxonomic survey and systematic census on the edible wild plants of Midnapur district, WB. *J. Econ. Taxon. Bot. 3,* 717–737.

Mishra, S. (2010). Ethnobotany of Korku, Gond and Nihal tribes of East Nimar, MP. PhD Thesis unpublished, DAVV, Indore.

Mukherjee, C. R., & Ghosh, R. B. (1992). Useful plants of Birbhum district, W. B. *J. Econ. Taxon. Bot, Addl. Series, 10,* 83–95.

Oomachan, M., & Masih, S. K. (1988). Multifarious uses of plants by the forest tribals of M. P. wild edible plants. *J. Trop. Forestry, 11,* 163–169.

Pandey, A., & Oomachan, M. (1992). Studies on less known wild food plants in rural and tribal areas around Jabalpur. *Indian J. Pure Appl. Biol. 7*(2), 129–136.

Patole, S. N., & Jain, A. K. (2002). Some edible plants of Panchmarhi Biosphere Reserve (M. P.). *Ethnobotany 14,* 48–51.

Pieroni, A. (2001). Evaluation of the cultural significance of wild food and botanicals traditionally consumed in north western Tuskey, Italy. *J. Ethnobiol, 21,* 89–104.

Prain, D. (1961). Bengal plants, 2 vols, BSI publication, Calcutta, India.

Prasad, R., & Bhatanagar, P. (1991). Wild edible products in the forests of Madhya Pradesh. *J. Tropical forestry 7*(3), 210–218.

Radhakrishnan, K., Panduranngan, A. G., & Pushpangadan, P. (1996). Ethnobotany in wild edible plants of Kerala, India. In: Jain, S. K. (ed.). Ethnobiology in Human welfare. Deep publications, New Delhi, India, pp. 48–51.

Rai, M. K. (1992). Observation on the ethnobotany of Gond tribe of Seoni district, plants used as food. *J. Econ. Taxon. Bot. Addl Series 10,* 281–283.

Rakesh, K. M., Kottapalli, S. R., Krishna, G. S. (2004). Bioprospecting of wild edibles for rural development in the central Himalayan mountains of India. *Mountain Res. Development. 24(2),110–113*

Ray, S., & Saikhediya, S. (2016). Wild edible plant resources of Harda district, M. P. *J. Biol. Pharmaceut. Chem. Res. 3* (1), 1–3.

Roy, J. K., & Rao, R. K. (1957). Investigations on the diet of the Maria of Bastar district. *Bull Dep. Anthrop. Govt. India 6,* 33–45.

Roy, J. K., & Rao, R. K. (1959). Mahua spirit and the chemical composition of the raw material (Mahua flowers). *J. Inst. Chem. 31,* 64.

Sahu, T. R. (1996). Life support promising food plant among aboriginals of Bastar (MP), India. In: S. K. Jain (ed.) Ethnobiology in human welfare. Deep publications, New Delhi, India, pp. 26–30.

Sandya, G. S., & Ahirwar, R. (2015). Ethnobotanical studies of some wild edible plants of Jaitpur forest district Shadol, M. P. Central India. *Inter. J. Pharm. Pharmaceut. Res. 4,* 282–282.

Satya, V. (2006). Ethnomedicinal studies with a particular reference to their conservational based religious and cultural ceremonies in tribal belt of west Nimar region of MP. PhD thesis (unpublished), DAVV, Indore, India.

Sebastian, M. K., & Bhandari, M. M. (1990). Edible wild plants of the forest areas of Rajasthan *J. Econ. Taxon. Bot. 14*(3), 689–694.

Shukla, K. M. L. (1996). Ethnobotanical studies on the tribals of Bilaspur district with special references to Korwa tribe. PhD thesis, A.P.S. University, Rewa (M.P.), India.

Singh, G., & Ahirwar, R. (2015). An ethnobotanical survey for certain wild edible plants of Chanda forest district, Dindori, Central India. *Int. J. Sci. Res., 4*(2), 1755–1757.

Singh, G. K., & Ahirwar, R. (2015). Documentation of some ethnobotanical wild edible plants of Badhavgarh National Park, District Umeria, Madhya Pradesh, India. *Int. J. Curr. Microbiol. Appl. Sci. 4*(8), 459–463.

Singh, H. B., & Arora R. K. (1978). Wild plants of India, NBPGR, IARI, New Delhi, India.

Singh, V., & Singh, P. (1991). Wild edible plants of eastern Rajasthan. *J. Econ. Taxon. Bot. 2,* 197–207.

Singh, V., & Pandey, R. P. (1998). Ethnobotany of Rajasthan. Scientific publishers, Jodhpur, India.

Sisodiya, S. (2012). Study of floristic composition and phytoresources of Barwani district, MP, Thesis unpublished, DAVV, Indore, India.

Upadhyay, R., & Chauhan, S. V. S. (2003). Ethnobotanical uses of plant gums by the tribals. *J Econ. Taxon. Bot. 27*(3).

Verma, P. (1993). Ethnobotanical studies on the tribals of Shadol district with special reference to Amarkantak. PhD Thesis, A. P. S. University, Rewa, M.P., India.

Watt, G (1971). A dictionary of the economics products of India (reprinted). Cosmo Publication, Delhi, India.

CHAPTER 5

ETHNOMEDICINAL PLANTS OF THE INDO-GANGETIC REGION AND CENTRAL INDIA

CHOWDHURY H. RAHAMAN,[1] SUMAN K. MANDAL,[1] and T. PULLAIAH[2]

[1]*Department of Botany, Visva-Bharati University, Santiniketan–731235, West Bengal, E-mail: habibur_cr@yahoo.co.in*

[2]*Department of Botany, Sri Krishnadevaraya University, Anantapur–515003, Andhra Pradesh, India, E-mail: pullaiah.thammineni@gmail.com*

CONTENTS

ABSTRACT

In this chapter a detailed review has been made on the ethnomedicinal plant resources used by the ethnic people reside in Indo-Gangetic region and

Central India. Here, 244 published literatures have been consulted from different states of the said region of India. A total of 528 plant species under 112 families have been enlisted which are used by the tribal communities for a wide range of medicinal purposes. In most of the cases leaves are used followed by underground parts, fruits, seeds, etc. The recorded species have been enumerated in tabular form with its updated scientific name, popular synonym, plant parts used, diseases and ailments cured, and allied references. The vast phytoresources of this part of India need further scientific studies for bioprospecting of natural products and also for conservation of important ethnomedicinal species.

5.1   INTRODUCTION

The plant people relation dates back to the very early period of human civilization. With the gradual development of human civilization, dependence of primitive men increased more and more on their surrounding plant resources, not only for food, but also for fodder, fuel, drug and shelter. The knowledge of curing diseases and ailments was developed as a system of health care through trial and error over a long period of time. Like any other traditional knowledge, this system of knowledge on health care is also transmitted orally from one generation to another (Rahaman, 2015). Such age-old healthcare systems have been developed by the ancient people in different corners of the world where they are living in close association with the nature. Since time immemorial, for the treatment of ill health plant wealth have been in use among different ethnic and traditional communities throughout the world including India. This is the basis of ethnomedicinal study today. Ethnomedicine is an important branch of ethnobotany which deals with the traditional practices involved in using biological resources including plants by the ethnic people for curing the diseases and ailments of their own and their domesticated animals. Medicinal plants which are used in preparation of remedies by the ethnic communities are generally known as ethnomedicinal plants. India is a country of rich biodiversity and is endowed with her great cultural heritage. In India, there are about 550 tribal communities covered under 227 ethnic groups residing in about 5000 villages in and around different forest and vegetation types (Pushpangadan, 2002).

Studies in folk medicine have been started in India long before the Indian ethnobotany. The Bodding's work on Santhal medicine is considered as a pioneering work in this field of folk medicine (Bodding, 1925, 1927). Another earliest work in the field of ethnomedicine had been made by Bressers (1951) on tribal medicine of Ranchi district, then Bihar state (Jharkhand). Subsequently, a large number of research works exclusively on folk or ethnomedicine have been documented from almost each state of India. There are plenty of references on folk remedies or ethnomedicinal plants along with other uses of plant resources in different ethnobotanical literatures of India that published in the form of book, research article or technical report (Saxena and Dutta, 1975; Jain and Borthakur, 1980; Saxena, 1986; Singh and Pandey, 1998; Rai et al., 2004; Ghosh et al., 2011; Rahaman and Pradhan, 2011; Rahaman and Karmakar, 2015). Many of the published articles on ethnobotany as well as ethnomedicine have been reviewed, scientifically complied and finally, it has been published in the form of a book or review article, but no such compilation work exclusively on ethnomedicine or ethnomedicinal plant resources from Indo-Gangetic region and central part of India has so far been documented (Jain, 1991; Dutta and Dutta, 2005; Choudhary et al., 2008; Mao et al., 2009; Ekka, 2011a; Sharma and Thokchom, 2014).

In this context, present work has been undertaken to review a wide range of published literature on ethnobotany as well as ethnomedicine from the Indo-Gangetic region and Central India, to highlight the ethnomedicinal plant resources of the area and their therapeutic uses. It is a humble step towards preparing a complied huge database on ethnomedicinal knowledge and related medicinal plant resources of this geographic area of India.

5.2 MATERIALS AND METHODS

Indo-Gangetic region and Central India are the two vast regions of India which are inhabited by a large number of ethnic communities and traditional knowledge of those ethnic communities has so far been documented by many researchers. A perusal of about 244 published literatures on ethnomedicinal plants from the states of Rajasthan, Punjab, Haryana, Uttar Pradesh, Bihar, Jharkhand, West Bengal, Madhya Pradesh and Chhattisgarh has been performed for present review (Figure 5.1). During review, all the data regarding plant parts used, disease cured and scientific name of the plant species have critically been checked and presented scientifically. Special care has been taken to update the nomenclature of each taxon and in many cases old nomenclature of the species has properly been updated. It was also noticed that a particular taxon published in different papers as separate taxa with its different old synonyms and here in all the cases, the old synonyms

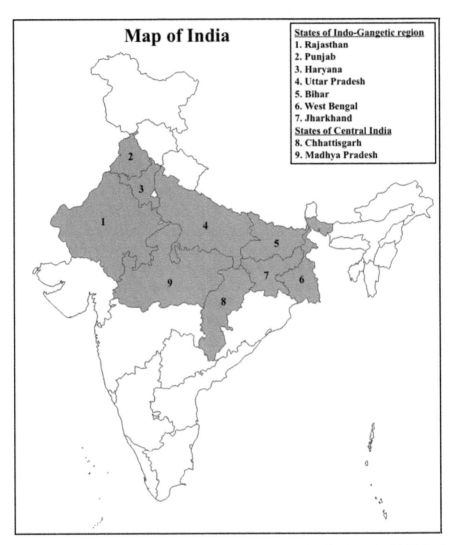

FIGURE 5.1   Map of the Indo-Gangetic region and Central India (from the highlighted States literature survey was made).

have been merged into a valid name i.e., separate taxa with old synonyms have been put together in one valid taxon. For checking the valid scientific names of all the recorded plant species, the nomenclature for the plant taxa of "The Plant List" have been followed (www.theplantlist.org). Finally, all the recorded ethnomedicinal plants have been enumerated in tabular form with its updated scientific name, popular synonym, its parts used, diseases and ailments cured and allied references (Table 5.1).

TABLE 5.1 Ethnomedicinal Plants of Indo-Gangetic Region and Central India

S. No.	Name of the plant species	Plant parts used	Ethnomedicinal uses	References
1.	*Abelmoschus esculentus* (L.) Moench [Syn.: *Hibiscus esculentus* L.]	Roots	Burning sensation during urination, increase sexual potency, urinogenital problems in males	Khanna, 2002; Jadhav, 2006a, 2011
2.	*Abelmoschus moschatus* Medik. Figure 5.2A.	Fruits, roots	Male sterility, hematuria, gout, rheumatism, aphrodisiac	Tirkey, 2004; Malviya et al., 2011
3.	*Abroma augusta* (L.) L.f.	Roots	Tuberculosis, pain in backbone	Tirkey, 2004
4.	*Abrus precatorius* L.	Leaves, seeds, roots	For abortion and antifertility in women, diarrhea, blisters in mouth, skin diseases, fever, sciatica, asthma, nervous disorders, baldness, sedative, abortifacient, leucorrhoea, coma, menstrual disorders, aphrodisiac	Sahu, 1982; Sahu et al., 1983; Khanna, 2002; Katewa et al., 2003; Tirkey, 2004, 2006; Katewa and Galav, 2005; Bandhyophyay and Mukherjee, 2006; Mishra, 2008; Mairh et al., 2010; Shukla et al., 2010; Malviya et al., 2011; Singh et al., 2014; Pandey et al., 2015
5.	*Abutilon indicum* (L.) Sweet	Roots, leaves, whole plant	Gonorrhea, piles, urethritis, dysentery, urinary troubles, gout, leucorrhoea, toothache, diuretic, kidney stones, dental problems, mouth diseases, aphrodisiac, analgesic, diuretic	Sahu, 1982; Katewa et al., 2003; Kumar et al., 2003; Tirkey, 2004; Jadhav, 2006b; Kumar and Chauhan, 2006; Mishra, 2008; Jain et al., 2009; Prachi et al., 2009; Malviya et al., 2011; Singh et al., 2012; Singh et al., 2014; Kumar et al. 2015
6.	*Acacia catechu* (L.f.) Willd.	Bark, gum, roots, flowers	Tonic for pregnant ladies, skin disease, jaundice, diarrhea, sore throat, gonorrhea, anti-inflammatory, rheumatism	Mishra, 2008; Shukla et al., 2010; Anand et al., 2013; Quamar and Bera, 2014; Sharma and Kumar, 2011; Kumar et al., 2015
7.	*Acacia concinna* (Willd.) DC.	Fruits	Hair tonic	Tirkey, 2004

A. *Abelmoschus moschatus* Medik.

B. *Anogeissus latifolia* (Roxb. ex DC.) Wall. ex Guill. & Perr.

C. *Argyreia nervosa* (Burm.f.) Boj.

D. *Barleria prionitis* L.

E. *Careya arborea* Roxb.

F. *Casearia tomentosa* Roxb.

FIGURE 5.2

TABLE 5.1 (Continued)

S. No.	Name of the plant species	Plant parts used	Ethnomedicinal uses	References
8.	*Acacia jacquemontii* Benth.	Bark, gum	Abortifacient, snake bite, kidney & renal disorders, asthma, sore in mouth	Choudhary et al., 2009
9.	*Acacia leucophloea* (Roxb.) Willd.	Shoots, bark	Rheumatic pain, diarrhea	Kumar and Chauhan, 2006; Quamar and Bera, 2014
10.	*Acacia nilotica* (L.) Del. ssp. *indica* (Benth.) Brenan [Syn.: *A. arabica* (Lam.) Willd.]	Bark, gum, roots, leaves, stem, fruit, flower, seeds	Premature ejaculation, diarrhea, dysentery, mouth ulcer, toothache, cold, whooping cough, tonic, piles, wounds, snakebite, swelling, tooth ache, bleeding in gum, ulcers, aphrodisiac, bone fracture	Sahu et al., 1983; Khanna, 2002; Pandey and Verma, 2002; Kumar et al., 2003; Kumar and Chauhan, 2006; Sharma and Kumar, 2007; Mishra, 2008; Jadhav, 2006b, 2008; Jain et al., 2009; Kala, 2009; Shukla et al., 2010; Malviya et al., 2011; Mondal and Rahman, 2012; Singh et al., 2012; Quamar and Bera, 2014
11.	*Acacia senegal* (L.) Willd.	Gum	Inflammation, sore nipples, burns, nodular leprosy	Jain et al., 2009
12.	*Acalypha ciliata* Forssk.	Leaves	Ulcers	Jain et al., 2009
13.	*Acalypha indica* L.	Leaves	Bed sores, ringworm infection, scabies, eczema, chest pain	Upadhyay and Singh, 2005; Jain et al., 2009; Shukla et al., 2010; Singh et al., 2012
14	*Acalypha lanceolata* Willd.	Leaves	Boils, sores, swellings	Jain et al., 2009
15.	*Acampe praemorsa* (Roxb.) Blatt. & McCann [Syn.: *Saccolaabium papillosum* Lindl.]	Whole plant	Cracked bone	Tirkey, 2004
16.	*Acanthospermum hispidum* DC.	Leaves	Skin diseases	Shukla et al., 2010

TABLE 5.1 (Continued)

S. No.	Name of the plant species	Plant parts used	Ethnomedicinal uses	References
17.	*Achyranthes aspera* L.	Roots, inflorescence, whole plant, leaves, stem, seeds	Stop bleeding after abortion, fever, for easy delivery, menstruation, dog bite, snake bite, cough, leucorrhoea, fever, pneumonia, kidney stone, stomach pain, jaundice, tooth ache, piles, anemia, itching, sores, pyorrhea, skin disease, asthma, bronchitis, piles, gynecological disorders, dysentery	Sahu, 1982; Maheshwari et al., 1986; Jain and Patole, 2001; Kumar et al., 2003; Maliya, 2004; Upadhyay and Singh, 2005; Jadhav, 2006b; Kumar and Chauhan, 2006; Rai, 2006; Sharma and Kumar, 2007; Mishra, 2008; Jadhav, 2008a; Jain et al., 2009; Kala, 2009; Shukla et al., 2010; Alawa and Ray, 2012; Jain et al., 2012; Singh et al., 2012; Tomar et al., 2012; Sahu et al., 2014; Singh et al., 2014; Maity et al., 2015
18.	*Acmella paniculata* (Wall. ex DC.) R. K.Jansen [Syn.: *Spilanthes paniculata* Wall. ex DC.]	Stem, flower heads, leaves	Paralysis, epilepsy, skin disease, toothache	Tirkey, 2004
19.	*Acorus calamus* L.	Rhizome	Stomachache, anthelmintic, epilepsy, tuberculosis	Sharma and Kumar, 2007; Kala, 2009; Marandi and Britto, 2014
20	*Actiniopteris radiata* (Sw.) Link	Whole plant	Antifertility	Jain et al., 2009
21.	*Adansonia digitata* L.	Fruits pulp	Stomachache	Alawa and Ray, 2012
22.	*Adiantum capillus-veneris* L.	Leaves	Menorrhagia, snake bite, headache, expel worms	Saini, 2008
23.	*Adiantum incisum* Forssk.	Leaves, Rhizome	Jaundice, cold, cough, increase sexual desire in men	Saini, 2008
24.	*Adiantum lunulatum* Burm. f. [Syn.: *A. philippense* L.]	Leaves, plant	Blood dysentery, leprosy, fever, pimples, asthma, bronchitis	Maheshwari et al., 1986; Singh et al., 2005; Saini, 2008

S. No.	Name of the plant species	Plant parts used	Ethnomedicinal uses	References
25.	*Aegle marmelos* (L.) Correa	Fruits, roots, bark, leaves	Dysentery, diarrhea, dyspepsia, stomach disorders, fever, jaundice, diabetes, astringent, piles, bowel complaints, piles, diabetes	Sahu et al., 1983; Maheshwari et al., 1986; Kumar and Rao, 2001; Kumar et al., 2003; Kala, 2009; Jain et al., 2010; Pnghal et al., 2010; Alawa and Ray, 2012; Jain et al., 2012; Singh et al., 2012; Tomar et al., 2012; Anand et al., 2013; Quarnar and Bera, 2014; Sahu et al., 2014; Singh et al., 2014; Kumar et al., 2015; Maity et al., 2015
26.	*Aerva javanica* (Burm.f.) Juss. ex Schult.	Seeds	Headache, rheumatism	Jain et al., 2009
27.	*Aerva lanata* (L.) Juss. ex Schult.	Whole plant, roots, leaves	Ear ache, kidney stones, jaundice, pneumonia, typhoid, urinary problems	Katewa and Galav, 2005; Prachi et al., 2009; Shukla et al., 2010; Singh et al., 2012; Singh et al., 2012
28.	*Agave americana* L.	Roots	Menstrual disorder	Jain et al., 2005a
29.	*Ageratum conyzoides* (L.) L.	Whole plant, leaves	Diarrhea, wounds, leprosy, stomach disorders	Sahu et al., 1983; Upadhyay and Singh, 2005; Shukla et al., 2010; Singh et al., 2012
30.	*Ailanthus excelsa* Roxb.	Leaves, stem bark	Wounds, typhoid, spermatorrhea, febrifuge, laxative, hepatitis, bronchitis, skin disease, joint pain	Jain and Patole, 2001; Jadhav, 2006b; Kumar and Chauhan, 2006; Alawa and Ray, 2012; Kumar et al., 2015; Pandey et al., 2015
31.	*Alangium salvifolium* (L.f.) Wang.	Bark	Jaundice	Kumar and Goel, 2008

TABLE 5.1 (Continued)

S. No.	Name of the plant species	Plant parts used	Ethnomedicinal uses	References
32.	*Albizia lebbeck* (L.) Benth.	Seeds, bark, leaves	Diarrhea, dysentery, gonorrhea, cataract, antitoxic, mouth ulcers, ulcers, antiallergic, cough, snake bite, night blindness, anti-inflammatory	Sahu, 1982; Sahu et al., 1983; Jain and Patole, 2001; Jain et al., 2009; Anand et al., 2013; Sahu et al., 2014; Kumar et al., 2015
33.	*Albizia odoratissima* (L.f.) Benth.	Bark, leaves	Leprosy, cough	Mishra, 2008
34.	*Alhagi maurorum* Medic.	Flower, plant	Bleeding piles, diuretic, laxative	Katewa and Galav, 2005
35.	*Allium cepa* L.	Bulb	Pyorrhea, snake bite, scorpion sting, sunstroke vomiting, impotency, aphrodisiac, tuberculosis	Pandey and Verma, 2002; Shekhawat and Batra, 2006; Shukla et al., 2010; Malviya et al., 2011; Tomar et al., 2012
36.	*Allium sativum* L.	Flakes	Scorpion sting, dog bite, typhoid, jaundice, impotency, aphrodisiac, asthma, cough, respiratory problems	Shekhawat and Batra, 2006; Jadhav, 2008a; Malviya et al., 2011; Singh et al., 2012
37.	*Aloe vera* (L.) Burm.f. [Syn.: *A. barbadensis* Mill.]	Whole plant, leaves, gel, roots, latex	Gonorrhea, skin eruptions, burns, liver disease, mad dog bite, epilepsy, hepatoprotective, anti-inflammatory, eruptions, pimples, blemishes, piles, cough, rheumatic pain, eczema, ulcer, aphrodisiac, stomach pain, constipation, menstrual disorders, hair loss, malaria,	Sahu, 1982; Maheshwari et al., 1986; Jain and Patole, 2001; Katewa et al., 2003; Shekhawat and Batra, 2006; Jadhav, 2006a; Pnghal et al., 2010; Jain et al., 2011; Malviya et al., 2011; Sharma and Kumar, 2011; Alawa and Ray, 2012; Jain et al., 2012; Singh et al., 2012; Sahu et al., 2014; Singh et al., 2014; Kumar et al., 2015
38.	*Alpinia galanga* (L.) Willd.	Rhizome	Aphrodisiac	Malviya et al., 2011
39.	*Astonia scholaris* (L.) R.Br.	Bark	Febrifuge, skin diseases, purgative, chest pain, paralysis, thoracic pain	Maliya, 2007; Marandi and Britto, 2014; Kumar et al., 2015; Maity et al., 2015

S. No.	Name of the plant species	Plant parts used	Ethnomedicinal uses	References
40.	*Alternanthera pungens* Kunth	Whole plant	Gonorrhea	Jain et al., 2009; Shukla et al., 2010
41.	*Alternanthera sessilis* (L.) R.Br. ex DC.	Leaves, roots, inflorescence	Liver disorders, promote conception, headache	Bondya et al., 2006; Jadhav, 2006a; Marandi and Britto, 2014
42.	*Alysicarpus hamosus* Edgew.	Leaves	Eye sight	Tirkey, 2006
43.	*Alysicarpus vaginalis* DC.	Leaves, roots	Eye sight, irregular menses	Tirkey, 2006; Jain et al., 2009
44.	*Amaranthus caudatus* L.	Leaves	Constipation	Katewa and Galav, 2005
45.	*Amaranthus spinosus* L.	Leaves, roots, shoots, whole plant, inflorescence	Spermatorrhea, laxative, eczema, boils, burns, iron tonic, boils, burns, snake bite, skin diseases, leucorrhoea	Maliya, 2004; Upadhyay and Singh, 2005; Bandhyophyay and Mukherjee, 2006; Bondya et al., 2006; Mishra, 2008; Shukla et al., 2010; Singh et al., 2012; Kumar et al., 2015
46.	*Amaranthus viridis* L.	Whole plant	Piles	Singh et al., 2012
47.	*Ammannia baccifera* L.	Whole plant, inflorescence	Fever, skin itching	Katewa and Galav, 2005
48.	*Amomum subulatum* Roxb.	Seeds	Cholera	Khanna, 2002
49.	*Amorphophallus peoniifolius* (Dennst.) Nicolson [Syn.: *A. campanulatus* Bl. ex Decne]	Rhizome	Laxative, digestive disorders, piles, snake bite	Maheshwari et al., 1986; Alawa and Ray, 2012; Kumar et al., 2015
50.	*Ampelocissus latifolia* (Roxb.) Planch. [Syn.: *Vitis latifolia* Roxb.]	Stem, leaves, roots, plant	Typhoid, dental troubles, dysentery, dental troubles, swelling body part	Jadhav, 2008a; Shukla et al., 2010; Quamar and Bera, 2014; Mishra, 2008

TABLE 5.1 (Continued)

S. No.	Name of the plant species	Plant parts used	Ethnomedicinal uses	References
51.	*Ampelopteris prolifera* (Retz.) Copel.	Leaves	Skin diseases, intestinal worms	Saini, 2008
52.	*Anacardium occidentale* L.	Seed oil	Antiseptic	Mairh et al., 2010
53.	*Anacyclus pyrethrum* (L.) Lag.	Flower, roots, seeds	Bleeding gums, toothache, headache	Marandi and Britto, 2014; Sahu et al., 2014
54.	*Andrographis echioides* (L.) Nees	Whole plant	Stomachache, dysentery, fever	Saini, 2008; Marandi and Britto, 2014
55.	*Andrographis paniculata* (Burm.f.) Nees	Root, whole plant, leaves	Fever, malaria, amoebic dysentery, diabetes, intestinal worms, leucorrhoea, fever, skin diseases, anti snake venom	Sahu et al., 1983; Jain and Patole, 2001; Tiwari and Yadav, 2003; Tirkey, 2004; Shukla et al., 2010; Jain et al., 2011; Alawa and Ray, 2012; Jain et al., 2012; Mondal and Rahman, 2012; Singh et al., 2012; Soni et al., 2012; Shrivastava and Kanungo, 2013; Sahu et al., 2014; Singh et al., 2014; Kumar et al., 2015
56.	*Anethum graveolens* L.	Seeds	Kidney stones	Katewa et al., 2003
57.	*Anisomeles indica* (L.) Kuntze	Leaves, whole plant	Pneumonia, nervous disorders, fever	Maliya, 2007; Singh et al., 2012
58.	*Annona reticulata* L.	Seeds, leaves, fruits	Intestinal worms, diabetes, tumor, cancer, indigestion	Sahu et al., 2014
59.	*Annona squamosa* L.	Bark, root, seeds, leaves, fruits	Diarrhea, dysentery, abortifacient, toothache and bleeding in gum, anthelmintic, antiseptic, wound, cancer	Sahu et al., 1983; Jain et al., 2011; Alawa and Ray, 2012; Jain et al., 2012; Mondal and Rahman, 2012; Soni et al., 2012; Sahu et al., 2014

S. No.	Name of the plant species	Plant parts used	Ethnomedicinal uses	References
60.	Anogeissus latifolia (Roxb. ex DC.) Wall. ex Guill. & Perr. (Figure 5.2B)	Bark, gum	Diarrhea, diabetes, liver complaint, wounds, spermatorrhea, for strength	Jadhav, 2006b; Shukla et al., 2010; Alawa and Ray, 2012; Anand et al., 2013; Shrivastava and Kanungo, 2013; Quamar and Bera, 2014
61.	Anogeissus pendula Edgew.	Bark	Gastric disorder	Jain et al., 2009
62.	Antidesma acidum Retz. [Syn.: A. diandrum (Roxb.) B.Heyne ex Roth]	Leaves, roots	Dysentery	Sahu et al., 1983
63.	Argemone mexicana L.	Seeds, roots, plant sap, leaves, latex, whole plant	Gonorrhea, dropsy, Skin disease, eczema, itching, guinea worms, dropsy, jaundice, typhoid, scorpion sting, ringworm, eye diseases, wound, purgative, wounds	Sahu, 1982; Jain and Patole, 2001; Kumar et al., 2003; Maliya, 2004; Upadhyay and Singh, 2005; Jadhav, 2006b; Kumar and Chauhan, 2006; Jain et al., 2009; Shukla et al., 2010; Alawa and Ray, 2012; Singh et al., 2012; Sahu et al., 2014
64.	Argyreia nervosa (Burm.f.) Boj. [Syn.: A. speciosa (L.f.) Sweet] Figure 5.2C	Leaves, roots	Tumor, joint pain, gout, anemia, tonsillitis	Jain et al., 2005a; Jain et al., 2012; Soni et al., 2012; Tirkey, 2004
65.	Argyreia strigosa (Roth) Sant. & Patel	Whole plant	Leucorrhoea, menorrhagia	Jain et al., 2005a
66.	Arisaema tortuosum (Wall.) Schott.	Tubers	Bone fracture, dog bite, liver complaints, stomachache	Katewa et al., 2003; Jain et al., 2005a
67.	Aristolochia bracteolata Lam.	Leaves	Snake bite	Katewa et al., 2003

TABLE 5.1 (Continued)

S. No.	Name of the plant species	Plant parts used	Ethnomedicinal uses	References
68.	*Aristolochia indica* L.	Roots, leaves, whole plant	Fever, leucoderma, antitoxic, jaundice, aphrodisiac, dandruff, antipoison	Tiwari and Yadav, 2003; Saini, 2008; Shukla et al., 2010; Malviya et al., 2011; Singh et al., 2012; Kumar et al., 2015
69.	*Artemisia nilagirica* (C. B.Clarke) Pamp.	Leaves	Irregular menstruation	Bandhyophyay and Mukherjee, 2006
70.	*Artemisia vulgaris* L.	Leaves, roots	Fever, Parkinson's disease, hysteria	Singh et al., 2014
71.	*Artocarpus heterophyllus* Lamk.	Leaves, fruits, roots	Fever, boils, skin diseases, wounds, diabetes, asthma, diarrhea, toothache	Menghani et al., 2010; Sahu et al., 2014; Kumar et al., 2015
72.	*Asparagus racemosus* Willd.	Roots	Typhoid, diuretic, anti-inflammatory, lactation, internal pain, tonic, tumors, diabetes, urination problems in children, leucorrhoea, fever, rheumatoid arthritis, dysentery, laxative	Sahu et al., 1983; Katewa and Galav, 2005; Rai, 2006; Jadhav, 2006b, 2008; Kala, 2009; Jain et al., 2010; Mairh et al., 2010; Shukla et al., 2010; Alawa and Ray, 2012; Mondal and Rahman, 2012; Shrivastava and Kanungo, 2013; Singh et al., 2014; Kumar et al., 2015
73.	*Asphodelus tenuifolius* Cav.	Leaves	Kidney stone, swellings	Katewa and Galav, 2005
74.	*Athyrium falcatum* Bedd.	Leaves	Jaundice, diseases of spleen, diuretic	Saini, 2008
75.	*Averrhoa carambola* L.	Fruits	Fever, antiscorbutic	Upadhyay and Singh, 2005

S. No.	Name of the plant species	Plant parts used	Ethnomedicinal uses	References
76.	*Azadirachta indica* A. Juss.	Leaves, bark, gum, seeds, seed oil, fruits	Small pox, snake bite, scorpion sting, skin diseases, ulcers, leprosy, wounds, malaria, eczema, miscarriages, viral hepatitis, boils, inflammation of eyes, headache, diabetes, fever, cough, rheumatism, constipation, diabetes, toothache, headache, antiseptic, constipation, wound, tuberculosis	Khanna, 2002; Kumar et al., 2003; Shekhawat and Batra, 2006; Bondya et al., 2006; Kumar and Chauhan, 2006a,b; Jadhav, 2006a; Rai, 2006; Menghani et al., 2010; Shukla et al., 2010; Sharma and Kumar, 2011; Alawa and Ray, 2012; Singh et al., 2012; Tomar et al., 2012; Anand et al., 2013; Shrivastava and Kanungo, 2013; Quamar and Bera, 2014; Sahu et al., 2014; Singh et al., 2014; Kumar et al., 2015; Maity et al., 2015
77.	*Bacopa monnieri* (L.) Wettst.	Whole plant, leaves, roots	Fever, malarial fever, asthma, snake bite, blood pressure	Rai, 2006; Shukla et al., 2010; Singh et al., 2014
78.	*Baccharoides anthelmintica* (L.) Moench [Syn.: *Centrantherum anthelminticum* (L.) Gamble, *C. anthelminticum* (L.) Kuntze]	Seeds	Intestinal worms	Jain et al., 2012; Soni et al., 2012
79.	*Balanites aegyptiaca* (L.) Delile)	Fruits, bark	Pneumonia in children, tuberculosis, whooping cough, skin disease, swellings on face	Maheshwari et al., 1986; Jain et al., 2010; Shukla et al., 2010; Alawa and Ray, 2012
80.	*Bambusa bambos* (L.) Voss [Syn.: *B. arundinacea* (Retz.) Willd.]	Newly growing plant, whole plant	For abortion, tuberculosis, wound healing, bronchitis, leprosy	Kumar et al., 2003; Sahu et al., 2014
81.	*Barleria cristata* L.	Stem, young plants	Gummosis, tooth disorders, cough, fever	Pandey and Verma, 2002; Singh et al., 2012; Singh et al., 2014

TABLE 5.1 (Continued)

S. No.	Name of the plant species	Plant parts used	Ethnomedicinal uses	References
82.	*Barleria prionitis* L. Figure 5.2D	Leaves, root, whole plant	Cataract, toothache, laxative, jaundice, hemorrhoids, skin disease, urinary disorder, pyorrhea, fever, pain, poison, cough, toothache, wounds, boils	Katewa and Galav, 2005; Jadhav, 2006b; Saini, 2008; Jain et al., 2009; Shukla et al., 2010; Jain et al., 2012; Sahu et al., 2014; Kumar et al., 2015; Pandey et al., 2015
83.	*Barringtonia acutangula* (L.) Gaertn.	Fruits	Snake bite	Kumar et al., 2015
84.	*Basella alba* L. [Syn.: *B. rubra* L.]	Leaves, fruit, root	Constipation, skin allergy, gonorrhea, conjunctivitis, intestinal disorders	Bondya et al., 2006; Saini, 2008; Sahu et al., 2014
85.	*Bauhinia acuminata* L.	Root bark	Inflammation of liver, vermifuge	Shukla et al., 2010
86.	*Bauhinia purpurea* L.	Roots, stem bark, leaves	Cold, cough, fever, wounds, diarrhea, jaundice	Bondya et al., 2006; Rai; 2006; Anand et al., 2013
87.	*Bauhinia racemosa* Lam.	Stem bark	Conjunctivitis	Jadhav, 2008a
88.	*Bauhinia vahlii* Wight & Arn.	Roots	Facial paralysis	Tirkey, 2004
89.	*Bauhinia variegata* L.	Twig, flower buds, leaves, root, flowers, stem bark	Syphilis, scrofula, ulcers, leprosy, pyorrhea, bleeding gums, mouth ulcer, toothache, laxative, dysentery, diarrhea, skin diseases, tonic, blood purifier, leucorrhoea, piles, diabetes, obesity, asthma, intestinal worms, anthelmintic, chest pain	Sahu, 1982; Pandey and Verma, 2002; Kumar et al., 2003; Tiwari and Yadav, 2003; Shukla et al., 2010; Singh et al., 2012; Anand et al., 2013; Quamar and Bera, 2014; Sahu et al., 2014; Singh et al., 2014
90.	*Benincasa hispida* (Thunb.) Cogn.	Fruits	Jaundice	Kumar et al., 2015
91.	*Berberis asiatica* Roxb. ex DC.	Roots	Wounds, inflammations	Jain and Patole, 2001

S. No.	Name of the plant species	Plant parts used	Ethnomedicinal uses	References
92.	*Bergia suffruticosa* (Del.) Fenzl	Whole plant	Bone fracture, swellings, scorpion sting	Katewa and Galav, 2005
93.	*Bidens biternata* (Lour.) Merr. & Sherff.	Leaves	Ulcers, eye, ear complaints	Jain et al., 2009
94.	*Blepharispermum subsessile* DC.	Roots	Arthritis	Hemadri and Rao, 1989
95.	*Blumea lacera* (Burm.f.) DC.	Leaves	Earache fever	Shukla et al., 2010
96.	*Boerhavia chinensis* (L.) Rottb. [Syn.: *Commicarpus chinensis* (L.) Heim.]	Roots	Spermatorrhea, menorrhagia, leucorrhoea	Kumar and Chauhan, 2006
97.	*Boerhavia diffusa* L.	Whole plant, roots, leaves	Gonorrhea, gall bladder stone, typhoid, diuretic, leucorrhoea, snake bite, kidney stones, jaundice, eye disease, dysentery, urinary disorders, dropsy, rheumatism, gas troubles, elephantiasis	Sahu, 1982; Bondya et al., 2006; Kumar and Chauhan, 2006; Jadhav, 2008a; Jain et al., 2009; Prachi et al., 2009; Shukla et al., 2010; Alawa and Ray, 2012; Jain et al., 2012; Singh et al., 2012; Soni et al., 2012; Kumar et al., 2015
98.	*Boerhavia procumbens* Banks ex Roxb.	Roots	Eye tonic, scorpion sting	Katewa and Galav, 2005
99.	*Bombax ceiba* L. [Syn.: *B. malabaricum* DC.]	Flower, petals, roots, thorns, latex, bark, leaves, tender flowers	Acidity, urinary problems, pimples, spermatorrhea, diarrhea, dysentery, anemia, leucorrhoea, rheumatic pain, menometrorrhagia, urinary troubles, menorrhea, tonic, stimulant, paralysis, impotency	Katewa et al., 2003; Kumar et al., 2003; Maliya, 2004; Bandhyophyay and Mukherjee, 2006; Jadhav, 2006b, 2011; Sharma and Kumar, 2007; Alawa and Ray, 2012; Jain et al., 2012; Singh et al., 2012; Soni et al., 2012; Anand et al., 2013; Sahu et al., 2014

TABLE 5.1 (Continued)

S. No.	Name of the plant species	Plant parts used	Ethnomedicinal uses	References
100.	*Borassus flabellifer* L.	Toddy, petiole	Stomach disorders, diarrhea	Saini, 1996; Mondal and Rahman, 2012
101.	*Boswellia serrata* Roxb. ex Colebr.	Stem bark, gum, leaves	Wounds, toothache, arthritis, eye infection	Kumar and Goel, 2008; Kala, 2009; Jain et al., 2010; Anand et al., 2013
102.	*Brassica rapa* L. [Syn.: *B. campestris* L.]	Seed oil	Mosquito bite, sciatica, toothache, antiseptic	Pandey and Verma, 2002; Kumar et al., 2003; Shekhawat and Batra, 2006; Kumar and Chauhan, 2006
103.	*Bridelia retusa* (L.) A.Juss.	Bark	Rheumatism, diabetes	Quamar and Bera, 2014
104.	*Bryophyllum pinnatum* (Lamk.) Oken [Syn.: *Kalanchoe pinnata* (Lam.) Pers.]	Leaves	Wounds, kidney stones	Jain and Patole, 2001; Prachi et al., 2009; Jadhav, 2011
105.	*Buchanania cochinchinensis* (Lour.) M. R.Almeida [Syn.: *B. lanzan* Spreng.]	Seeds, gum, fruits, bark, seed oil, leaves	Cure antifertility, diarrhea, cough, skin diseases, wounds, glandular swelling of neck, cardiotonic	Saini, 2008; Kala, 2009; Shukla et al., 2010; Jain et al., 2012; Anand et al., 2013
106.	*Bulbophyllum leopardinum* (Wall.) Lindl. ex Wall.	Fruits	Sunstroke, diabetes	Tirkey, 2004
107.	*Butea monosperma* (Lam.) Taub.	Seeds, roots, leaves, flower, bark, gum, stem	Tooth and gum disorders, menstrual pain, tonic, bone fracture, rheumatic pain, eczema, snakebite, scorpion sting, hepatitis, sunstroke, skin diseases, diarrhea, body ache, weakness, leucorrhoea, worm infestation, diabetes, leucoderma, male impotency, cuts, wounds, tuberculosis, inflammation	Pandey and Verma, 2002; Kumar et al., 2003; Tirkey, 2006; Bandhyophyay and Mukherjee, 2006; Jadhav, 2006a, b; Saini, 2008; Jain et al., 2009; Kala, 2009; Shukla et al., 2010; Alawa and Ray, 2012; Singh et al., 2012; Anand et al., 2013; Quamar and Bera, 2014; Singh et al., 2014; Kumar et al., 2015; Pandey et al., 2015;

S. No.	Name of the plant species	Plant parts used	Ethnomedicinal uses	References
108.	*Butea superba* Roxb.	Roots, seeds	Herpes, easy delivery, skin diseases, flatulence, roundworm, arthritis,	Tirkey, 2006; Quamar and Bera, 2014
109.	*Bytneria herbacea* Roxb.	Leaves, roots, twig	Leucorrhoea, swollen legs, wounds, leprosy	Mairh et al., 2010; Jain et al., 2012; Tomar et al., 2012
110.	*Caesalpinia bonduc* (L.) Roxb.	Leaves, seeds	Malarial fever, bleeding in ladies, stomach trouble	Katewa et al., 2003; Jadhav, 2006b; Alawa and Ray, 2012
111.	*Caesulia axillaris* Roxb.	Roots	Mouth sore	Shukla et al., 2010
112.	*Cajanus cajan* (L.) Mill.	Leaves	Cholera, mouth ulcer, corneal opacity	Khanna, 2002; Kumar et al., 2003; Pandey and Verma, 2002; Singh et al., 2012;
113.	*Cajanus scarabaeoides* (L.) Thours [*Atylosia scarabaeoides* (L.) Benth.]	Roots, whole plant	Throat pain, spermatorrhea	Jain et al., 2010; Malviya et al., 2011
114.	*Calligonum polygonoides* L.	Plant	Typhoid	Katewa and Galav, 2005
115.	*Calotropis gigantea* (L.) Dryand.	White flower, latex, leaves	Mental disorders, hepatitis, syphilitic affection, toothache, inflammation, body pain	Katewa et al., 2003; Shukla et al., 2010; Jain et al., 2012; Soni et al., 2012; Kumar et al., 2015; Maity et al., 2015
116.	*Calotropis procera* (Ait.) Dryand.	Roots, leaves, latex, twig, flowers, gynostegium	Pyorrhea, sprain, scorpion sting, wounds, toothache, bleeding gums, rheumatism, asthma, bronchitis, jaundice, hepatitis, syphilitic affection, male sterility, body ache, gonorrhea, leprosy, backache, swelling, malarial fever, wormicidal	Pandey and Verma, 2002; Kumar et al., 2003; Upadhyay and Singh, 2005; Shekhawat and Batra, 2006; Kumar and Chauhan, 2006; Jadhav, 2006b; Maliya, 2007; Mishra, 2008; Jain et al., 2009; Shukla et al., 2010; Jain et al., 2011; Sharma and Kumar, 2011; Singh et al., 2014; Kumar et al., 2015

TABLE 5.1 (Continued)

S. No.	Name of the plant species	Plant parts used	Ethnomedicinal uses	References
117.	*Canavalia cathartica* Thouars [Syn.: *C. virosa* Wight & Arn.]	Seeds	Snake bite	Tomar et al., 2012
118.	*Cannabis sativa* L.	Leaves, flower	Gonorrhea, tonic, pain reliever, piles, wound, ear problem	Sahu, 1982; Kumar et al., 2003; Singh et al., 2012; Singh et al., 2014
119.	*Canscora alata* (Roth) Wall. [Syn.: *C. decussata* (Roxb.) Schult. & Schult.f.]	Whole plant	Stomach trouble, blood purifier	Soni et al., 2012; Jain et al., 2012
120.	*Capparis decidua* (Forssk.) Edgew.	Bark	Purgative, hepatitis	Kumar et al., 2015
121.	*Capsicum annuum* L. [Syn.: *C. frutescens* L.]	Fruits	Scorpion sting, skin itches	Shekhawat and Batra, 2006; Kumar et al., 2015
122.	*Cardiospermum halicacabum* L.	Leaves, seeds	Wounds, swelling	Jain et al., 2009, 2011
123.	*Careya arborea* Roxb. Figure 5.2E	Roots	Snake bite	Jain et al., 2012; Soni et al., 2012
124.	*Carica papaya* L.	Latex from fruits, fruits, seeds	For abortion, dental caries, toothache, scorpion sting, digestant, toothache, anthelmintic, laxative, tonic, aphrodisiac, snake bite	Pandey and Verma, 2002; Kumar et al., 2003; Shekhawat and Batra, 2006; Shukla et al., 2010; Singh et al., 2012; Tomar et al., 2012
125.	*Carissa carandas* L.	Root bark, fruits	Diabetic ulcer, appetizer	Jain et al., 2009; Shukla et al., 2010
126.	*Cascabela thevetia* (L.) Lippold [Syn.: *Thevetia peruviana* (Pers.) K.Schum.]	Latex	Toothache	Pandey and Verma, 2002; Shukla et al., 2010

S. No.	Name of the plant species	Plant parts used	Ethnomedicinal uses	References
127.	*Casearia tomentosa* Roxb. [Syn.: *C. elliptica* Willd.] Figure 5.2F	Roots	Stomach disorder	Jain et al., 2012; Soni et al., 2012
128.	*Cassia fistula* L.	Stem bark, roots, leaves, fruits, flowers, seeds	Skin diseases, diabetes, cardiac disorders, ringworm, hepatitis, purgative, fever, hypertension, burns, rheumatism, snake bite	Jain and Patole, 2001; Kumar et al., 2003; Mishra, 2008; Shukla et al., 2010; Alawa and Ray, 2012; Jain et al., 2012; Soni et al., 2012; Singh et al., 2012; Tomar et al., 2012; Anand et al., 2013; Shrivastava and Kanungo, 2013; Quamar and Bera, 2014; Kumar et al., 2015; Maity et al., 2015
129.	*Cassine glauca* (Rottb.) Kuntze [Syn.: *Elaeodendron glaucum* (Rottb.) Pers.]	Roots	Snake bite	Kala, 2009
130.	*Catharanthus roseus* L.	Root bark, roots	Diabetes, septic wounds, blood dysentery	Menghani et al., 2010; Singh et al., 2012; Singh et al., 2014
131.	*Catunaregam spinosa* (Thunb.) Thirveng.	Fruit	Jaundice	Jain et al., 2012
132.	*Cayratia trifolia* (L.) Domin [Syn.: *Vitis carnosa* (Lam.) Wall.]	Roots	Anemia, body ache	Jain et al., 2009; Kala, 2009
133.	*Ceiba pentandra* (L.) Gaertn. (Malvaceae)	Roots, stem bark	Diarrhea, dysentery, toothache	Quamar and Bera, 2014

TABLE 5.1 (Continued)

S. No.	Name of the plant species	Plant parts used	Ethnomedicinal uses	References
134.	*Celastrus paniculatus* Willd.	Seeds, bark, seed oil	Rheumatic pain, arthritis, increase memory, body swelling, tuberculosis, leprosy, bodyache, stomachache, skin disease, leucorrhoea	Jain and Patole, 2001; Katewa et al., 2003; Tirkey, 2004; Shukla et al., 2010; Alawa and Ray, 2012; Jain et al., 2012; Soni et al., 2012; Quamar and Bera, 2014
135.	*Celosia argentea* L.	Seeds	Urinary diseases, kidney stone, diarrhea, dysentery, burning sensation during urination, ovarian and uterus diseases	Katewa et al., 2003; Bondya et al., 2006; Jain et al., 2009; Shukla et al., 2010; Jain et al., 2011
136.	*Centella asiatica* (L.) Urban.	Leaves, whole plant	Dysentery, paramnesia, brain tonic, insanity, hysteria, spermatorrhea, jaundice, yellow urine	Jain and Patole, 2001; Upadhyay and Singh, 2005; Bondya et al., 2006; Sharma and Kumar, 2007; Alawa and Ray, 2012; Tomar et al., 2012
137.	*Centipeda minima* (L.) A.Br. & Asch.	Whole plant	Toothache	Shukla et al., 2010
138.	*Ceratopteris thalictroides* (L.) Brongn. [Syn.: *C. siliquosa* (L.) Copel.]	Leaves	Cuts and wounds, stomach disorders	Saini, 2008
139.	*Cereus pterogonus* Lam.	Leaves	Muscle pain	Mondal and Rahman, 2012
140.	*Ceropegia candelabrum* L. [Syn.: *C. tuberosa* Roxb.]	Tubers, leaves	Kidney stones	Katewa et al., 2003
141.	*Chamaecrista absus* (L.) H. S.Irwin & Barneby [Syn.: *Cassia absus* L.]	Seeds	Ophthalmic, skin troubles	Shukla et al., 2010
142.	*Cheilanthes farinosa* (Forssk.) Kaulf.	Rhizome, leaves	Urine problems, epilepsy	Singh et al., 2005

S. No.	Name of the plant species	Plant parts used	Ethnomedicinal uses	References
143.	*Cheilocostus speciosus* (J.Koenig) C. D.Specht [Syn.: *Costus speciosus* (Koen.) Sm.] Figure 5.3A	Rhizome	Asthma, sexual debility, abdominal pain, boils, fever, burning sensation in eyes	Tirkey, 2004; Jain et al., 2005a; Alawa and Ray, 2012; Jain et al., 2012; Soni et al., 2012
144.	*Cheilosoria tenuifolia* (Burm. f.) Trevis. [Syn.: *C. tenuifolia* (Burm.f.) Sw.]	Roots	Wounds	Singh et al., 2005
145.	*Chenopodium album* L.	Whole plant, leaves	Intestinal worms, laxative, urinary troubles, colic, piles, cough	Katewa and Galav, 2005; Bondya et al., 2006
146.	*Chlorophytum arundinaceum* Bak.	Roots	Stomachache, asthma, aphrodisiac, diarrhea, menstrual disorders	Tirkey, 2004; Alawa and Ray, 2012
147.	*Chlorophytum borivilianum* Sant. & Fernand.	Tubers	Rheumatism	Katewa et al., 2003
148.	*Chlorophytum tuberosum* (Roxb.) Baker	Tubers, leaves	Skin disease, tonic, diabetes	Kala, 2009; Singh et al., 2014
149.	*Chrozophora rottleri* (Geiss.) A.Juss. ex Spreng.	Leaves	Sun burn, sun stroke	Katewa et al., 2003
150.	*Chrysopogon aciculatus* (Retz.) Trin.	Rhizome	Stomachache, gastrointestinal disorder	Mitra and Mukherjee, 2005a
151.	*Chrysopogon zizanioides* (L.) Roberty [Syn.: *Vetiveria zizanioides* (L.) Nash]	Fruits, roots	For conception, menstrual disorders, dysuria, dyspepsia, headache	Maliya, 2004; Mitra and Mukherjee, 2005a; Shukla et al., 2010
152.	*Cicer arietinum* L.	Seeds	Spider bite, jaundice	Shekhawat and Batra, 2006; Jadhav, 2006b, 2011

A. *Cheilocostus speciosus* (J.Koenig) C.D.Specht | B. *Crateva nurvala* Buch.-Ham.

C. *Curculigo orchioides* Gaertn. | D. *Dendrophthoe falcata* (L.f.) Etting.

E. *Dioscorea bulbifera* L. | F. *Echinops echinatus* Roxb.

FIGURE 5.3

TABLE 5.1 (Continued)

S. No.	Name of the plant species	Plant parts used	Ethnomedicinal uses	References
153.	*Cinnamomum tamala* (Buch.-Ham.) Nees & Eberm.	Leaves	Cough, dysmenorrhea	Bandhyophyay and Mukherjee, 2006; Mondal and Rahman, 2012
154.	*Cissampelos pareira* L.	Rhizome, roots	Diuretic, purgative, diarrhea, hepatitis, antitoxic, fever, cold, bodyache, stomach pain, loose motion, malarial fever, dysmenorrhea	Sahu et al., 1983; Jain and Patole, 2001; Bandhyophyay and Mukherjee, 2006; Mairh et al., 2010; Shukla et al., 2010; Quamar and Bera, 2014; Kumar et al., 2015
155.	*Cissus quadrangularis* L. [Syn.: *Vitis quadrangularis* (L.) Wall.]	Stems	Bone fracture	Tirkey, 2004; Tiwari and Yadav, 2003; Jadhav, 2011; Jain et al., 2011; Alawa and Ray, 2012; Mondal and Rahman, 2012
156.	*Citrullus colocynthis* (L.) Schrad.	Fruits, roots, whole plant	Constipation, rheumatism, antidiabetic	Katewa and Galav, 2005; Jain et al., 2009; Menghani et al., 2010
157.	*Citrus aurantifolia* (Christm. & Panz.) Swingle	Leaves	Reddish eyes	Khanna, 2002
158.	*Citrus aurantium* L.	Fruit	Indigestion, vomiting	Singh et al., 2012
159.	*Citrus medica* L.	Fruits, leaves	Refrigerant, digestive, wasp/bee bite, scorpion sting, headache	Katewa et al., 2003; Kumar et al., 2003; Shekhawat and Batra, 2006
160.	*Clematis gouriana* Roxb. ex DC.	Leaves	Skin disease	Jain et al., 2012; Soni et al., 2012
161.	*Clematis heynei* M. A.Rau & al. [Syn.: *C. triloba* Heyne ex Roth]	Leaves	Ringworm	Jain and Patole, 2001
162.	*Cleome gynandra* L.	Shoot, leaves, seeds	Skin disease, earache, piles, headache	Tiwari and Yadav, 2003; Katewa and Galav, 2005; Kumar and Chauhan, 2005; Shukla et al., 2010

TABLE 5.1 (Continued)

S. No.	Name of the plant species	Plant parts used	Ethnomedicinal uses	References
163.	*Cleome monophylla* L.	Leaves	Urinary infection	Bondya et al., 2006
164.	*Cleome viscosa* L.	Seeds, leaves, whole plant	Bleeding piles, boils, wounds, diabetes	Katewa and Galav, 2005; Jain et al., 2009; Menghani et al., 2010; Shukla et al., 2010
165.	*Clerodendrum indicum* (L.) Kuntze	Roots	Bodyache, burns, ricket	Tirkey, 2004
166.	*Clerodendrum phlomidis* L.f. [Syn.: *C. multiflorum* (Burm.f.) Kuntze]	Flowers, stem, leaves	Miscarriages, fever, earache, gummosis, toothache	Pandey and Verma, 2002; Jadhav, 2006b; Shukla et al., 2010
167.	*Clitoria ternatea* L.	Root bark, seeds, roots	Diuretic, purgative, snake bite, dropsy	Tirkey, 2004; Shukla et al., 2010; Jain et al., 2012; Soni et al., 2012
168.	*Coccinia grandis* (L.) Voigt [*C. cordifolia* Cogn.]	Leaves, roots	Diabetes, jaundice, tuberculosis	Jain et al., 2009; Shukla et al., 2010; Singh et al., 2012; Singh et al., 2014
169.	*Cocculus hirsutus* (L.) W.Theob. [Syn.: *C. hirsutus* (L.) Diels]	Roots, leaves	Rheumatism, venereal diseases, gonorrhea, skin diseases, eczema, joint pain, eyes cooling, leucorrhoea	Katewa and Galav, 2005; Jain et al., 2009; Shukla et al., 2010; Jain et al., 2011; Pandey et al., 2015
170.	*Cocculus pendulus* (Forst. & Forst.) Diels	Leaves, roots	Skin disease, night blindness, cataract, bone fracture, leucorrhoea	Jain et al., 2005a; Katewa and Galav, 2005
171.	*Cocos nucifera* L.	Coir	Bleeding gums	Pandey and Verma, 2002
172.	*Coix lacryma-jobi* L.	Seeds	Dysentery, galactagogue	Mitra and Mukherjee, 2005a
173.	*Colebrookea oppositifolia* Sm.	Leaves	Cuts, wounds	Shukla et al., 2010
174.	*Colocasia esculenta* (L.) Schott.	Leaves, tubers	Abscess, leprosy	Khanna, 2002; Bondya et al., 2006

S. No.	Name of the plant species	Plant parts used	Ethnomedicinal uses	References
175.	*Commelina benghalensis* L.	Leaves	Scabies, wounds	Bondya et al., 2006; Singh et al., 2012
176.	*Commiphora wightii* (Arn.) Bhandari	Plant juice, oleoresin	Antidiabetic	Menghani et al., 2010
177.	*Convolvulus arvensis* L.	Roots	Purgative	Shukla et al., 2010
178.	*Convolvulus microphyllus* Sieb. ex Spreng.	Whole plant	Bleeding, insomnia	Kumar and Chauhan, 2006; Jain et al., 2012; Soni et al., 2012
179.	*Convolvulus prostratus* Forssk. [Syn.: *C. pluricaulis* Choisy]	Whole plant	High blood pressure, brain tonic, memory tonic	Singh et al., 2012; Pandey et al., 2015
180.	*Corbichonia decumbens* (Forssk.) Jacq. ex Exell.	Leaves	Kidney stones, gonorrhea	Katewa and Galav, 2005
181.	*Corchorus aestuans* L.	Seeds, whole plant	Chest congestion, diarrhea	Jain et al., 2009; Shukla et al., 2010
182.	*Corchorus capsularis* L.	Leaves, seeds	Dysentery, puerperal fever, stomachic	Kumar et al., 2003; Bandhyophyay and Mukherjee, 2006
183.	*Corchorus olitorius* L.	Seeds	Hair fall, lice	Shukla et al., 2010
184.	*Cordia dichotoma* Forst. f.	Roots, fruits, leaves, bark	Flatulence, urinary disorder, lungs disease, spleen's disease, influenza, body pain	Tirkey, 2004; Mishra, 2008; Shukla et al., 2010; Quamar and Bera, 2014
185.	*Cordia macleodii* Hook.f. & Thoms.	Leaves, fruits	Fever, dysentery	Mairh et al., 2010
186.	*Cordia sinensis* Lam. [Syn.: *C. gharaf* (Forssk.) Ehr. & Asch.]	Stem, bark	Toothache	Pandey and Verma, 2002

TABLE 5.1 (Continued)

S. No.	Name of the plant species	Plant parts used	Ethnomedicinal uses	References
187.	*Coriandrum sativum* L.	Fruits, twig	Vomiting, toothache	Khanna, 2002; Pandey and Verma, 2002
188.	*Crateva nurvala* Buch.-Ham. Figure 5.3B	Bark	Kidney stones, urinary infection, liver disease	Prachi et al., 2009; Pandey et al., 2015
189.	*Crateva religios a* G.Forst. [Syn.: *C. magna* (Lour.) DC.]	Leaves, bark	Swellings, bladder stone	Kumar and Chauhan, 2006, Sharma and Kumar, 2007
190.	*Crinum asiaticum* L.	Leaves	Chest pain	Maliya, 2007
191.	*Crinum viviparum* (Lam.) R.Ansari & V. J.Nair [Syn.: *C. defixum* Ker-Gawl.]	Leaves	Earache	Kumar et al., 2003
192.	*Crotalaria incana* L.	Roots	Goiter	Tirkey, 2006
193.	*Crotalaria juncea* L.	Leaves, root, flowers, seeds	Jaundice, paralysis, intestinal worms, obesity	Tirkey, 2006; Alawa and Ray, 2012
194.	*Crotalaria medicaginea* Lam.	Leaves	Leucorrhoea	Tirkey, 2006
195.	*Crotalaria nana* Burm.f.	Root	Diarrhea	Jain and Patole, 2001
196.	*Crotalaria orixensis* Willd.	Root	Tuberculosis	Tirkey, 2006
197.	*Crotalaria prostrata* Willd.	Leaves	Eye ailments	Tirkey, 2006
198.	*Croton bonplandianum* Baill.	Leaves	Diarrhea	Kumar et al., 2003
199.	*Croton persimilis* Müll. Arg. [Syn.: *C. oblongifolius* Roxb.]	Plant sap, roots	Eczema, dysentery	Mairh et al., 2010; Alawa and Ray, 2012

S. No.	Name of the plant species	Plant parts used	Ethnomedicinal uses	References
200.	*Cryptolepis dubia* (Burm.f.) M. R.Almeida [Syn.: *C. buchanani* Roem. & Schult.]	Leaves, roots	Eczema, stomach pain	Mairh et al., 2010
201.	*Cullen corylifolium* (L.) Medik. [Syn.: *Psoralea corylifolia* L.]	Seed	Leucoderma	Sharma and Kumar, 2007
202.	*Cuminum cyminum* L.	Fruits	Diarrhea, dyspepsia	Sahu, 1982; Sahu et al., 1983
203.	*Curculigo orchioides* Gaertn. Figure 5.3C	Roots	For lactation, strength, vigor, jaundice, dysentery, leucorrhoea, lunacy, diabetes, jaundice, wound	Sahu, 1982; Shukla et al., 2010; Alawa and Ray, 2012; Jain et al., 2012; Tomar et al., 2012; Singh et al., 2014; Kumar et al., 2015
204.	*Curcuma angustifolia* Roxb.	Rhizome	Wounds, fever	Kala, 2009; Alawa and Ray, 2012
205.	*Curcuma longa* L. [Syn.: *C. domestica* Val.]	Rhizome	Chest pain, jaundice, internal injuries, cough, cold, bone fracture	Khanna, 2002; Kumar et al., 20C3; Tomar et al., 2012; Shukla et al., 2010; Maity et al., 2015
206.	*Cuscuta reflexa* Roxb.	Whole plant	Puerperal fever, cold, cough, swellings, headache, fever, polio, joint pain, for abortion, dandruff, heart problem	Hembrom, 1991; Jain and Patole, 2001; Kumar et al., 2003; Bandhyophyay and Mukherjee, 2006; Kumar and Chauhan, 2006; Rai, 2006; Mishra, 2008; Shukla et al., 2010; Jadhav, 2011; Singh et al., 2012; Singh et al., 2014
207.	*Cyanthillium cinereum* (L.) H. Rob. [Syn.: *Vernonia cinerea* (L.) Less.]	Whole plant, roots, leaves	Leucorrhoea, fever, malarial fever	Bandhyophyay and Mukherjee, 2006; Shukla et al., 2010; Singh et al., 2012; Singh et al., 2014

TABLE 5.1 (Continued)

S. No.	Name of the plant species	Plant parts used	Ethnomedicinal uses	References
208.	Cymbopogon citratus (DC.)Stapf	Leaves	Cholera	Singh et al., 2012
209.	Cymbopogon jwarancusa (Jones) Schult. [Syn.: Andropogon jwarancusa Jones]	Leaves, roots	Fever, diuretic	Kumar et al., 2015
210.	Cymbopogon martini (Roxb.) Wats.	Leaves	Headache	Mitra and Mukherjee, 2005a
211.	Cynanchum viminale (L.) L. [Syn.: Sarcostemma viminale (L.) R.Br.]	Whole plant	Digestive disorders, bone fracture	Katewa and Galav, 2005
212.	Cynodon dactylon (L.) Pers.	Roots, whole plant	Gleet, cuts, wounds, ascites, liver disorder, kidney stones, bleeding nose	Mitra and Mukherjee, 2005a; Upadhyay and Singh, 2005; Kala, 2009; Prachi et al., 2009; Singh et al., 2012; Kumar et al., 2015
213.	Cyperus rotundus L.	Leaves, roots	Snake bite, ascariasis, amoebic dysentery, digestive problem, menstrual complaints	Mishra, 2008; Shukla et al., 2010; Mondal and Rahman, 2012; Singh et al., 2012; Kumar et al., 2015
214.	Dalbergia sissoo Roxb.	Bark, heart wood, leaves	Body pain, liver disorders, jaundice, gonorrhea, bleeding piles, diarrhea, chronic fever, diuretic, leucorrhoea	Kumar et al., 2003; Tirkey, 2006; Shukla et al., 2010; Singh et al., 2012; Anand et al., 2013; Quamar and Bera, 2014; Kumar et al., 2015
215.	Datura innoxia Mill.	Leaves, seeds	Skin diseases, sedative, stomachache, bronchial asthma	Kumar et al., 2003; Singh et al., 2012; Singh et al., 2014

S. No.	Name of the plant species	Plant parts used	Ethnomedicinal uses	References
216.	*Datura metel* L.	Roots, leaves, seeds	Dog bite, mosquito bite, typhoid, to check abortion, toothache, insanity, cerebral complications	Pandey and Verma, 2002; Shekhawat and Batra, 2006; Jadhav, 2008a; Shukla et al., 2010; Sharma and Kumar, 2011; Maity et al., 2015
217.	*Datura stramonium* L.	Leaves	Asthma	Sharma and Kumar, 2007
218.	*Daucus carota* L.	Leaves, tubers	Abortion, antifertility, kidney stones	Sahu, 1982; Prachi et al., 2009
219.	*Dendrocalamus strictus* (Roxb.) Nees	Leaves	Sore throat	Jadhav, 2006, 2008
220.	*Dendrophthoe falcata* (L.f.) Etting. Figure 5.3D.	Stem, leaves	Menstrual troubles, contraceptive, skin diseases	Katewa et al., 2003; Mairh et al., 2010
221.	*Desmodium gangeticum* (L.) DC.	Roots, whole plant, stem bark	Snake bite, diuretic, vomiting, diarrhea, asthma, tuberculosis, goiter	Jain et al., 2005a; Tirkey, 2006; Singh et al., 2012; Singh et al., 2014; Kumar et al., 2015
222.	*Desmodium heterocarpon* (L.) DC.	Root	Backache	Tirkey, 2006
223.	*Desmodium velutinum* (Willd.) DC. [Syn.: *D. latifolium* (Willd.) DC.]	Stem bark	Painful testicles	Tirkey, 2006
224.	*Desmodium oojeinense* (Roxb.) H.Ohasi [Syn.: *Ougeinia dalbergioides* Benth., *O. oojeinensis* (Roxb.) Hochr.]	Stem bark, roots	Dysentery, tumor	Tirkey, 2006; Jain et al., 2012
225.	*Desmodium triflorum* (L.) DC.	Whole plant	Toothache, fever	Tirkey, 2006; Shukla et al., 2010

TABLE 5.1 (Continued)

S. No.	Name of the plant species	Plant parts used	Ethnomedicinal uses	References
226.	*Desmostachya bipinnata* (L.) Stapf	Roots	Gummosis, toothache, stop bleeding	Pandey and Verma, 2002; Kumar et al., 2015
227.	*Dicanthium foveolatum* (Del.) Roberty	Inflorescence	Kidney stone	Jain et al., 2005a
228.	*Dicliptera paniculata* (Forssk.) I.Darbysh. [Syn.: *Peristrophe paniculata* (Forssk.) Brummit]	Whole plant, roots	Bone fracture, sprains, fever, snake bite	Shukla et al., 2010; Singh et al., 2012
229.	*Dicoma tomentosa* Cass.	Roots, branches	Pyorrhea	Katewa and Galav, 2005
230.	*Dillenia pentagyna* Roxb.	Leaves, roots, whole plant, stem bark	Leprosy, high blood pressure, wounds, fistula, pneumonia, skin diseases, blood dysentery	Tirkey, 2004; Alawa and Ray, 2012; Jain et al., 2012; Soni et al., 2012; Quamar and Bera, 2014
231.	*Dioscorea bulbifera* L. Figure 5.3E	Tubers	Piles, ulcers	Mishra, 2008
232.	*Dioscorea hispida* Dennst.	Tubers	Tuberculosis, asthma	Jain and Patole, 2001; Jain et al., 2005a
233.	*Dioscorea pentaphylla* L.	Tubers	Asthma, bronchitis	Jain et al., 2005a
234.	*Diospyros melanoxylon* Roxb.	Flower, leaves, roots, fruits	Urinary diseases, cough, snake bite, scorpion sting, leucorrhoea, cracked soles, dysentery	Mishra, 2008; Kala, 2009; Jain et al., 2010; Jain et al., 2011; Anand et al., 2013; Quamar and Bera, 2014
235.	*Diospyros malabarica* (Desr.) Kostel. [Syn.: *D. peregrina* (Gaertn.) Gürke]	Fruits, bark	Rheumatism, ulcers	Kumar et al., 2015
236.	*Diplocyclos palmatus* (L.) C.Jeffrey	Seeds	Male and Female sterility	Alawa and Ray, 2012

S. No.	Name of the plant species	Plant parts used	Ethnomedicinal uses	References
237.	*Dregea volubilis* (L.f.) Benth. ex Hook.f. [Syn.: *Wattakaka volubilis* (L.f.) Stapf]	Whole plant	Rheumatism	Jain et al., 2011
238.	*Drimia indica* (Roxb.) Jessop [Syn.: *Urginea indica* (Roxb.) Kunth]	Bulb, leaves	Dysmenorrhea, sunstroke, ulcer	Tiwari and Yadav, 2003; Bandhyophyay and Mukherjee, 2006; Alawa and Ray, 2012
239.	*Dryopteris cochleata* (D.Don) C.Chr.	Rhizome, stem	Antidote, epilepsy, leprosy, cuts, wounds, ulcers	Singh et al., 2005
240.	*Echinops echinatus* Roxb. Figure 5.3F	Roots, whole plant	Pneumonia, typhoid, fever, indigestion, headache	Jadhav, 2006, 2008; Shukla et al., 2010; Jain et al., 2011
241.	*Eclipta prostrata* L. [Syn.: *E. alba* (L.) Hassk.]	Leaves, whole plant	Catarrh, dropsy, elephantiasis, fever, dysentery, anemia, jaundice, laxative, cuts, wounds, hair dye, dysmenorrhea, skin diseases, dandruff, liver disorder, spleen enlargement	Kumar et al., 2003; Upadhyay and Singh, 2005; Bandhyophyay and Mukherjee, 2006; Bondya et al., 2006; Rai, 2006; Shukla et al., 2010; Jain et al., 2012; Singh et al., 2012; Soni et al., 2012; Singh et al., 2014; Kumar et al., 2015
242.	*Ehretia laevis* Roxb.	Stem bark	Dysentery	Jain et al., 2012; Soni et al., 2012
243.	*Elephantopus scaber* L.	Roots, leaves, whole plant	Cornea ulcer, pimples, laxative, gummosis, toothache, dysentery, malarial fever, indigestion	Jain and Patole, 2001; Khanna, 2002; Pandey and Verma, 2002; Kumar and Rao, 2002; Maliya, 2004; Shukla et al., 2010; Jain et al., 2012; Soni et al., 2012; Tomar et al., 2012
244.	*Eleusine indica* (L.) Gaertn.	Roots	Snake bite	Mitra and Mukherjee, 2005a

TABLE 5.1 (Continued)

S. No.	Name of the plant species	Plant parts used	Ethnomedicinal uses	References
245.	*Elytraria acaulis* (L.f.) Lind.	Roots	Expulsion of guinea worm, asthma, cholera	Jain et al., 2005a; Jain et al., 2010
246.	*Ensete superbum* (Roxb.) Cheesman	Seeds	Dog bite	Alawa and Ray, 2012
247.	*Enicostema axillare* (Lam.) A.Raynal	Leaves	Malarial fever	Jain et al., 2011
248.	*Equisetum arvense* L.	Plant	Liver diseases	Saini, 2008
249.	*Equisetum ramosissimum* Desf. subsp. *debile* (Vauch.) Hauke	Whole plant	Bone fracture, backache, muscular pain	Singh et al., 2005
250.	*Erycibe paniculata* Roxb.	Young leaves	Night blindness	Tirkey, 2004
251.	*Erythrina stricta* Roxb.	Stem bark	Menometrorrhagia	Bandhyophyay and Mukherjee, 2006
252.	*Eulophia herbacea* Lindl.	Rhizome	Leucorrhoea	Jain and Patole, 2001
253.	*Eulophia ochreata* Lindl.	Tubers	Leukemia	Jain et al., 2005a
254.	*Euphorbia fusiformis* Buch.-Ham. ex D.Don Figure 5.4A	Roots	Poor lactation	Mondal and Rahman, 2012
255.	*Euphorbia hirta* L.	Latex, whole plant, leaves	Gonorrhea, warts, leucoderma, laxative, asthma, cough, bronchitis, wound, leucorrhoea, constipation, stomach problems in children	Sahu, 1982; Katewa and Galav, 2005; Upadhyay and Singh, 2005; Jadhav, 2006, 2011; Maliya, 2007; Shukla et al., 2010; Singh et al., 2012
256.	*Euphorbia neriifolia* L. [Syn.: *E. ligularia* Roxb.]	Leaves, latex	Toothache, cold, fever, whooping cough, asthma, dropsy, colic, jaundice, leprosy, cough, swelling	Pandey and Verma, 2002; Maliya, 2004; Katewa and Galav, 2005; Alawa and Ray, 2012; Tomar et al., 2012

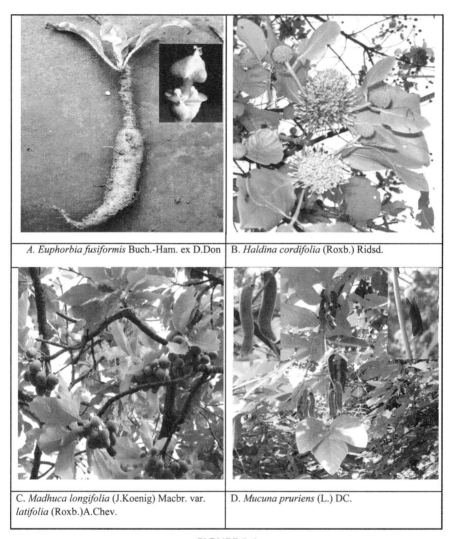

| A. *Euphorbia fusiformis* Buch.-Ham. ex D.Don | B. *Haldina cordifolia* (Roxb.) Ridsd. |
| C. *Madhuca longifolia* (J.Koenig) Macbr. var. *latifolia* (Roxb.)A.Chev. | D. *Mucuna pruriens* (L.) DC. |

FIGURE 5.4

TABLE 5.1 (Continued)

S. No.	Name of the plant species	Plant parts used	Ethnomedicinal uses	References
257.	*Euphorbia thymifolia* L.	Whole plant	Dysentery, spermatorrhea	Shukla et al., 2010
258.	*Evolvulus alsinoides* (L.) L.	Plant, leaves, roots	Febrifuge, asthma, bronchitis, for hair growth, urinary problem, sexual debility	Upadhyay and Singh, 2005; Shukla et al., 2010; Singh et al., 2012; Singh et al., 2014
259.	*Feronia limonia* Swingle	Fruit	Diarrhea, dysentery	Sahu et al., 1983; Kumar et al., 2003
260.	*Ficus benghalensis* L.	Latex, bark, adventitious roots, leaves, fruit	Toothache, dysentery, diarrhea, piles, tooth decay, rheumatism, skin diseases, loss of hair, diabetes, bronchitis, urinary disorder	Pandey and Verma, 2002; Kala, 2009; Jain et al., 2010; Menghani et al., 2010; Singh et al., 2012; Singh et al., 2014
261.	*Ficus palmata* Forssk.	Latex	Leucorrhoea	Shukla et al., 2010
262.	*Ficus racemosa* L. [Syn.: *F. glomerata* Roxb.]	Fruits, root, bark	Diabetes, hepatitis, bleeding	Shukla et al., 2010; Singh et al., 2012; Quamar and Bera, 2014; Kumar et al., 2015
263.	*Ficus religiosa* L.	Stem bark, leaves, root, stem	Throat infection, Gonorrhea, hair fall, snakebite, throat infection, cough, cold, gums	Sahu, 1982; Khanna, 2002; Kumar et al., 2003; Jadhav, 2006, 2008; Shukla et al., 2010; Singh et al., 2012; Quamar and Bera, 2014
264.	*Ficus virens* Dryand.	Bark, leaves	Ulcer, leucorrhoea, dysentery	Khanna, 2002; Shukla et al., 2010; Maity et al., 2015
265.	*Firmiana simplex* (L.) W.Wight [Syn.: *Sterculia urens* Roxb.]	Stem bark, twigs, roots	Skin disease, toothache, sciatica, constipation, liver disease, skin disorders, burn of urinary tract	Tirkey, 2004; Mairh et al., 2010; Quamar and Bera, 2014
266.	*Flacourtia indica* (Burm.f.) Merr.	Bark, leaves, fruits	Dysentery, conjunctivitis, liver problems	Sahu et al., 1983; Shukla et al., 2010; Quamar and Bera, 2014

S. No.	Name of the plant species	Plant parts used	Ethnomedicinal uses	References
267.	*Flemingia macrophylla* (Willd.) Merr. [Syn.: *F. congesta* Roxb.]	Root	Arthritis, epilepsy	Tirkey, 2006
268.	*Flemingia strobilifera* (L.) W. T.Aiton [Syn.: *F. bracteata* (Roxb.) Wight]	Root	Facial paralysis, for easy delivery	Jain and Patole, 2001; Tirkey, 2C06
269.	*Flemingia wightiana* Wight & Arn.	Stem twigs, leaves	Toothache, jaundice	Tirkey, 2004
270.	*Garuga pinnata* Roxb.	Bark	Diabetes	Shukla et al., 2010
271.	*Glinus lotoides* L.	Plant	Indigestion, boils, wounds	Katewa and Galav, 2005
272.	*Gloriosa superba* L.	Tubers, roots, rhizome	Leprosy, snake bite, scorpion sting, wounds, asthma, abortifacient, headache, easy delivery	Jain and Patole, 2001; Katewa et al., 2003; Tiwari and Yadav, 2003; Jain et al., 2005a; Jain et al, 2012; Soni et al, 2012; Tomar et al., 2012
273.	*Glycyrrhiza glabra* L.	Root	Epilepsy	Tomar et al., 2012
274.	*Gmelina arborea* Roxb.	Bark	Laxative, diuretic	Kumar et al., 2015
275.	*Gomphrena celosioides* Mart.	Whole plant	Urolithiasis	Prachi et al., 2009
276.	*Gossypium herbaceum* L.	Epicalyx	Regulation of menstrual cycle	Jadhav, 2006
277.	*Gossypium hirsutum* L. (Malvaceae)	Twigs, stem bark	Asthma, arthritis	Tirkey, 2004
278.	*Grewia hirsuta* Vahl	Terminal twig, roots	Stomachache, leucorrhoea, swollen testicles	Khanna, 2002; Maliya, 2004; Ja:n et al., 2012; Soni et al., 2012;
279.	*Grewia serrulata* DC.	Fruit	Fever	Maliya, 2007

TABLE 5.1 (Continued)

S. No.	Name of the plant species	Plant parts used	Ethnomedicinal uses	References
280.	*Grewia tiliaefolia* Vahl	Leaves	Diabetes	Maliya, 2007
281.	*Gymnema sylvestre* (Retz.) R.Br. ex Sm.	Leaves, roots	Diabetes, cough, rheumatism, ulcers, piles, cataract, liver tonic	Jain and Patole, 2001; Tirkey, 2004; Shrivastava and Kanungo, 2013; Singh et al., 2014
282.	*Gymnosporia emarginata* (Willd.) Thwaites [Syn.: *Maytenus emarginata* (Willd.) Ding Hou]	Leaves	Cold and cough	Jain et al., 2010
283.	*Habenaria plantaginea* Lindl.	Roots	Piles, wounds	Tirkey, 2004
284.	*Haldina cordifolia* (Roxb.) Ridsd. [Syn.: *Adina cordifolia* (Roxb.) Hook.f. ex Brandis] Figure 5.4B	Leaves, stem bark, flower buds	Jaundice, malarial fever, stomach disorders, body pain, inflammation, wounds, ulcers, miscarriages	Jain and Patole, 2001; Jadhav, 2006b, 2011; Jain et al., 2012; Soni et al., 2012; Quamar and Bera, 2014
285.	*Helicteres isora* L.	Fruits, roots, seeds	Indigestion, stomachache, diabetes, snake bite, greeping, for abortion, dysentery, gripping of bowels	Katewa et al., 2003; Tirkey, 2004; Mishra, 2008; Jain et al., 2010; Shukla et al., 2010; Jadhav, 2011
286.	*Helminthostachys zeylanica* (L.) Hoook.f.	Leaves, rhizome	Mouth ulcer, sore throat, constipation, impotency, cough, diarrhea	Pandey and Verma, 2002; Saini, 2008
287.	*Hemidesmus indicus* (L.) R.Br. ex Schult.	Stem, leaves, roots	Toothache, fever, rheumatism, cough, diarrhea, skin diseases, snake bite, epilepsy, toothache, blood pressure, diabetes, cough	Pandey and Verma, 2002; Jain et al., 2005a; Upadhyay and Singh, 2005; Mishra, 2008; Kumar and Goel, 2008; Mairh et al., 2010; Shukla et al., 2010; Jain et al., 2012; Singh et al., 2012; Soni et al., 2012; Singh et al., 2014

S. No.	Name of the plant species	Plant parts used	Ethnomedicinal uses	References
288.	*Hemigraphis hirta* (Vahl) T.Anders.	Whole plant	Urinary disorders	Maliya, 2007
289.	*Hibiscus mutabilis* L.	Petals	Jaundice	Maity et al., 2015
290.	*Hibiscus rosa-sinensis* L.	Petals, buds, leaves, flower, stem bark	Bronchitis, boils, whooping cough, urinogenital problems in males, sprains, kidney troubles, meno-metorhagia, for abortion, aphrodisiac	Kumar et al., 2003; Maliya, 2004; Bandhyophyay and Mukherjee, 2006; Shukla et al., 2010; Jadhav, 2006, 2011; Malviya et al., 2011; Singh et al., 2012; Kumar et al., 2015
291.	*Hippochaete debilis* (Roxb. ex Vaucher) Ching [Syn.: *Equisetum debile* Roxb. ex Vaucher]	Whole plant	Kidney stones	Prachi et al., 2009
292.	*Holarrhena pubescens* (Wall. ex G.Don [Syn.: *H.antidysenterica* Wall.]	Roots, stem bark, seeds, flowers, fruits	Dysentery, malarial fever, laxative, jaundice, digestive troubles, joint pain	Sahu et al., 1983; Tirkey, 2004; Sharma and Kumar, 2007; Kala, 2009; Jain et al., 2010; Jain et al., 2012; Soni et al., 2012; Kumar et al., 2015; Maity et al., 2015
293.	*Holoptelea integrifolia* (Roxb.) Planch.	Bark, leaves	Fever, ringworm, inflammation	Kumar and Chauhan, 2006; Shukla et al., 2010; Anand et al., 2013; Quamar and Bera, 2014
294.	*Hygrophila auriculata* (Schumach.) Haine	Whole plant, leaves, seeds	Venereal diseases, diuretic, swellings, leucorrhoea, aphrodisiac, spermatorrhea	Sahu, 1982; Shukla et al., 2010; Malviya et al., 2011; Kumar et al., 2015;
295.	*Hyptis suaveolens* (L.) Poit.	Stem, seeds, leaves	Gummosis, toothache, soothing agent, pneumonia in children	Pandey and Verma, 2002; Jain et al., 2010; Shukla et al., 2010
296.	*Ichnocarpus frutescens* (L.) W. T.Aiton (	Leaves	Malarial fever, piles	Jain et al., 2012; Soni et al., 2012

TABLE 5.1 (Continued)

S. No.	Name of the plant species	Plant parts used	Ethnomedicinal uses	References
297.	*Impatiens balsamina* L.	Leaves	Boils, wounds, swelling	Katewa and Galav, 2005
298.	*Indigofera astragalina* DC.	Leaves	Diarrhea	Shukla et al., 2010
299.	*Indigofera hirsuta* L.	Whole plant	Body pain	Tirkey, 2006
300.	*Indigofera tinctoria* L.	Whole plant	Abscess, purgative, diuretic, leucorrhoea	Khanna, 2002; Singh et al., 2012; Singh et al., 2014; Kumar et al., 2015
301.	*Ipomoea aquatica* Forssk.	Whole plant, leaves	Itching, dog bite, digestive problem	Khanna, 2002; Shukla et al., 2010; Singh et al., 2014;
302.	*Ipomoea carnea* Jacq.	Leaves	Boils	Khanna, 2002; Shukla et al., 2010
303.	*Ipomoea eriocarpa* R.Br.	Whole plant	Rheumatism, ulcers epilepsy, leprosy	Shukla et al., 2010
304.	*Ipomoea fistulosa* Mart. ex Choisy	Leaves	Swellings	Kumar et al., 2003
305.	*Ipomoea hispida* (Vahl) Roem. & Schult.	Corm	Jaundice, stomach pains	Jain and Patole, 2001
306.	*Ipomoea pes-tigridis* L.	Roots, stem latex	Snake bite, heal cracks	Shukla et al., 2010; Singh et al., 2012
307.	*Ixora coccinea* L.	Flowers	Eczema	Singh et al., 2012; Singh et al., 2014
308.	*Jatropha curcas* L.	Twig, seeds, leaf latex	Pyorrhea, cholera, dysentery, stomach disorder, aphrodisiac, wound	Maliya, 2007; Mairh et al., 2010; Malviya et al., 2011; Singh et al., 2012; Singh et al., 2014
309.	*Jatropha gossypiifolia* L.	Twig	Toothache, bleeding gums	Pandey and Verma, 2002; Shukla et al., 2010

S. No.	Name of the plant species	Plant parts used	Ethnomedicinal uses	References
310.	*Justicia adhatoda* L. [Syn.: *Adhatoda vasica* Nees, *A. zeylanica* Medik.]	Leaves, Stem, roots, flowers	Cough, bronchitis, cold, asthma, headache, bodyache, jaundice, wounds, malaria, dysentery, nose bleeding, skin diseases, fever	Saini, 1996; Kumar et al., 2003; Tirkey, 2004; Katewa and Galav, 2005; Sharma and Kumar, 2007; Mishra, 2008: Jain et al., 2009; Mairh et al., 2010; Shukla et al., 2010; Alawa and Ray, 2012; Mondal and Rahman, 2012; Singh et al., 2012; Tomar et al., 2012; Singh et al., 2014; Kumar et al., 2015; Pandey et al., 2015
311.	*Justicia gendarussa* Burm.f.	Leaves	Menometrorrhagia	Bandhyophyay and Mukherjee, 2006
312.	*Lagerstroemia parviflora* Roxb.	Bark, roots	Laxative, cough, cold	Anand et al., 2013; Maity et al., 2015
313.	*Lannea coromandelica* (Houtt.) Merr.	Bark, gum	Cuts, wounds, injuries, diarrhea, dysentery	Jain and Patole, 2001; Shukla et al., 2010; Jain et al., 2012; Soni et al., 2012; Anand et al., 2013;
314.	*Lantana camara* L. var. *aculeata* (L.) Moldenke	Fruit, flower, roots, stems, leaf	Bleeding gums, decaying teeth, snake bite, stomach pain	Kumar et al., 2003; Pandey and Verma, 2002; Upadhyay and Singh, 2005; Shukla et al., 2010; Singh et al., 2012
315.	*Laphangium luteoalbum* (L.) Tzvelev [Syn.: *Gnaphalium luteo-album* L.]	Plant, whole plant	Constipation, body pain, insomnia, lactation	Sahu, 1982; Mairh et al., 2010
316.	*Launaea asplenifolia* (Willd.) Hook.f.	Leaves, stem	Bodyache	Maliya, 2007
317.	*Launaea procumbens* (Roxb.) Ram. & Raj.	Whole plant, leaves	Fever, piles, antidote for poisoning	Kumar et al., 2003; Kumar and Chauhan, 2006; Singh et al., 2012

TABLE 5.1　(Continued)

S. No.	Name of the plant species	Plant parts used	Ethnomedicinal uses	References
318.	*Lawsonia inermis* L.	Leaves, seeds, roots	Jaundice, fever, burning micturition, dysentery, inflammation	Kumar et al., 2003; Katewa and Galav, 2005; Sharma and Kumar, 2007; Tomar et al., 2012
319.	*Leea asiatica* (L.) Ridsdale [Syn.: *L. crispa* L.]	Tuberous roots	Hydrocele	Tomar et al., 2012
320.	*Leea macrophylla* Roxb. ex Hornem.	Roots	Sexual debility in men, chest pain	Jain et al., 2005a; Jain et al., 2012; Soni et al., 2012
321.	*Leonotis nepetifolia* (L.) R.Br.	Flower	Skin disease, paralysis, eczema, itching	Tarafder and Raichoudhuri, 1981; Maity et al., 2015
322.	*Lepidagathis incurva* Buch.Ham. ex D.Don	Leaves	Cough	Shukla et al., 2010
323.	*Leucas aspera* (Willd.) Link	Leaves	Eye pain, cough, cold	Tiwari and Yadav, 2003; Singh et al., 2012; Singh et al., 2014
324.	*Leucas cephalotes* (Koen. ex Roth) Spreng.	Whole plant, leaves	Stomachache, mouth ulcer, gummosis, toothache, headache	Khanna, 2002; Pandey and Verma, 2002; Shukla et al., 2010
325.	*Leucas lanata* Benth.	Leaves	Pain	Jain and Patole, 2001
326.	*Leucas urticifolia* (Vahl) Sm.	Flower, leaves	Cold, cough, fever	Katewa and Galav, 2005;
327.	*Limnophila rugosa* (Roth) Merr.	Plant	Stomach disorders	Kumar and Goel, 2008
328.	*Limonia acidissima* Groff	Leaves, fruits	Leucorrhoea, diarrhea, dysentery	Bandhyophyay and Mukherjee, 2006; Shukla et al., 2010
329.	*Linum usitatissimum* L.	Seeds	Lice	Shekhawat and Batra, 2006

S. No.	Name of the plant species	Plant parts used	Ethnomedicinal uses	References
330.	*Lippia alba* (Mill.) N. E.Br. ex Britton & P.Wilson [Syn.: *L. geminata* Kunth]	Leaves	Rabies	Kumar and Goel, 2008
331.	*Litsea glutinosa* (Lour.) Rob. (Syn.: *L. chinensis* Lam.)	Stem bark, bark	Blood urine, leucorrhoea, bone fracture, sprain, vomiting	Tiwari and Yadav, 2003; Shukla et al., 2010; Jain et al., 2012; Soni et al., 2012
332.	*Litsea monopetala* (Roxb.) Pers.	Bark, leaves	Abscess, bruise	Mairh et al., 2010
333.	*Ludwigia octovalvis* (Jacq.) P. H.Raven [Syn.: *Jussiaea suffruticosa* L.]	Roots, leaves	Dysentery, infantile diarrhea, puerperal fever	Tirkey, 2004; Bandhyophyay and Mukherjee, 2006
334.	*Ludwigia perennis* L. [Syn.: *L. parviflora* Roxb.]	Plant	Eczema, skin disease	Jain and Patole, 2001
335.	*Luffa echinata* Roxb.	Fruits	Dog bite	Kumar and Chauhan, 2006
336.	*Lygodium flexuosum* (L.) Sw.	Leaves	Epilepsy, jaundice, skin diseases, diarrhea, sprains, dysentery, scabies, carbuncles, burn wounds, rheumatism	Tirkey, 2004; Singh et al., 2005; Saini, 2008; Tomar et al., 2012
337.	*Madhuca longifolia* (J.Koenig) Macbr. var. *latifolia* (Roxb.) Chevalier [Syn.: *M. indica* Gmel.] Figure 5.4C	Stem bark, roots, leaves, flower, twig fruits	Intestinal worms, bleeding gums, ulcers, diabetes, hydrocele, burns, for strength, diabetes, bone fracture, rheumatism, body pain, pyorrhea, snake bite, stomach disorder	Kala, 2009; Quamar and Bera, 2014; Shukla et al., 2010; Jain et al., 2012; Mondal and Rahman, 2012; Anand et al., 2013; Shrivastava and Kanungo, 2013; Singh et al., 2014
338.	*Magnolia champaca* (L.) Baill. ex Pierre [Syn.: *Michelia champaca* L.]	Roots	Urination problem in children	Mondal and Rahman, 2012

TABLE 5.1 (Continued)

S. No.	Name of the plant species	Plant parts used	Ethnomedicinal uses	References
339.	*Mallotus philippensis* (Lam.) Muell.-Arg.	Stem bark, fruits	Jaundice, skin diseases, ringworm, blisters in ears	Tirkey, 2004; Shukla et al., 2010; Jain et al., 2012; Soni et al., 2012; Anand et al., 2013
340.	*Mallotus roxburghianus* Lam.	Plant, seeds	Tape worms, ulcers	Mairh et al., 2010
341.	*Malvastrum coromandelianum* (L.) Garcke	Seeds	Premature ejaculation	Khanna, 2002; Shukla et al., 2010
342.	*Mangifera indica* L.	Unripe fruits, latex, bark, seeds	Antifertility, spider bite, cholera, pyorrhea, jaundice, stomach disorders, bleeding piles, skin diseases	Maity and Manna, 2000; Pandey and Verma, 2002; Kumar et al., 2003; Shekhawat and Batra, 2006; Shukla et al., 2010; Singh et al., 2012; Tomar et al., 2012; Quamar and Bera, 2014; Maity et al., 2015
343.	*Marsilea minuta* L.	Sporocarps, leaves, whole plant	Throat inflammation, insomnia, eye diseases, toothache, burning sensation while urinating	Khanna, 2002; Bondya et al., 2006; Saini, 2008; Tomar et al, 2012
344.	*Martynia annua* L.	Roots, leaves	Swellings, boils, rheumatism, worms scorpion sting,	Katewa and Galav, 2005; Jadhav, 2011; Shukla et al., 2010
345.	*Medicago sativa* L.	Leaves	Night blindness	Jadhav, 2006
346.	*Melia azedarach* L.	Leaves	Chest pain	Maliya, 2007
347.	*Mentha arvensis* L.	Leaves	Mouth ulcer, toothache	Pandey and Verma, 2002; Kumar et al., 2003
348.	*Merremia emarginata* (Burm.f.) Hall.f.	Whole plant, leaves	Wounds, boils, pain killer, snake bite, for hair growth	Jain et al., 2010, 2011; Shukla et al., 2010; Pandey et al., 2015

S. No.	Name of the plant species	Plant parts used	Ethnomedicinal uses	References
349.	*Mimosa hamata* Willd.	Seeds, leaves	Sexual weakness in males, wounds, ulcer	Katewa and Galav, 2005
350.	*Mimosa pudica* L.	Whole plant, root, leaves	Scorpion sting, measles, piles, wound healing	Kumar et al., 2003; Singh et al., 2012; Singh et al., 2014
351.	*Mitragyna parvifolia* (Roxb.) Korth.	Leaves, stem bark	Wounds, dropsy	Shukla et al., 2010; Jain et al., 2012; Soni et al., 2012; Anand et al., 2013; Quamar and Bera, 2014
352.	*Mollugo pentaphylla* L.	Whole plant	Promote menstrual discharge	Upadhyay and Singh, 2005
353.	*Momordica charantia* L.	Leaves, fruits	Mouth ulcer, piles, diabetes, constipation	Pandey and Verma, 2002; Kumar and Chauhan, 2006; Tomar et al., 2012; Shrivastava and Kanungo, 2013;
354.	*Moringa oleifera* Lam.	Leaves, bark, fruits, flower buds	Heart ailments, rheumatic pain, high blood pressure, hypotension, night blindness, stomach pain	Bondya et al., 2006; Kumar and Chauhan, 2006; Maliya, 2007; Shukla et al., 2010; Singh et al., 2012; Tomar et al., 2012; Quamar and Bera, 2014
355.	*Morus alba* L.	Leaves	Goiter	Maliya, 2004; Maliya, 2007
356.	*Mucuna imbricata* DC.	Roots	Madness, hysteria	Tirkey, 2004
357.	*Mucuna pruriens* (L.) DC. [Syn.: *M. prurita* Hook.] Figure 5.4D	Seeds, roots	Scorpion sting, asthma, facial paralysis, leprosy, swollen legs, mental disorder	Jain and Patole, 2001; Katewa and Galav, 2005; Tirkey, 2006; Singh et al., 2012
358.	*Murdannia edulis* (Stokes) Faden	Root	Aphrodisiac, spermatorrhea	Malviya et al., 2011
359.	*Murraya koenigii* Spreng.	Leaves	Diarrhea, dysentery, fever	Sahu et al., 1983; Maliya, 2007
360.	*Murraya paniculata* (L.) Jack	Leaves, root bark	Cough, hysteria, rheumatism	Shukla et al., 2010

TABLE 5.1 (Continued)

S. No.	Name of the plant species	Plant parts used	Ethnomedicinal uses	References
361.	*Musa balbisiana* Colla	Roots	Kidney and urinary tract complaints	Prachi et al., 2009
362.	*Musa paradisiaca* L. [Syn.: *M. sapientum* L.]	Fruits, stem	Jaundice, leucorrhoea, diarrhea, cough, pyorrhea	Khanna, 2002; Jadhav, 2011; Shukla et al., 2010; Mondal and Rahman, 2012; Kumar and Chauhan, 2006
363.	*Nelumbo nucifera* Gaertn.	Rhizome, fruits	Leucorrhoea, mental weakness, vomiting in children	Khanna, 2002; Bandhyophyay and Mukherjee, 2006; Jadhav, 2006; Shukla et al., 2010
364.	*Neolamarckia cadamba* (Roxb.) Bosser [Syn.: *Anthocephalus cadamba* (Roxb.) Miq., *A.chinensis* (Lam.) A.Rich.]	Leaves, Stem bark	Stomach pain, wounds, fever, rheumatism	Anand et al., 2013; Shukla et al., 2010
365.	*Nepeta hindostana* (Heyne ex Roth) Haines	Leaves	Mouth ulcer	Pandey and Verma, 2002
366.	*Neptunia oleracea* Lour.	Roots	Blood dysentery	Jadhav, 2011
367.	*Nerium oleander* L. [Syn.: *N. indicum* Mill.]	Seeds, roots	Wounds, inflammation	Shukla et al., 2010; Singh et al., 2012
368.	*Nervilia concolor* (Blume) Schltr. [Syn.: *N. aragoana* Goud.	Leaves	Wound	Maliya, 2007
369.	*Nicotiana tabacum* L.	Leaf midrib, leaves, flower	Pyorrhea, asthma	Pandey and Verma, 2002; Singh et al., 2012; Singh et al., 2014
370.	*Nigella sativa* L.	Seed	Pyorrhea	Pandey and Verma, 2002

S. No.	Name of the plant species	Plant parts used	Ethnomedicinal uses	References
371.	Nyctanthes arbor-tristis L.	Leaves, seeds, stem bark	Diuretic, fever, rheumatism, diabetes, purgative, sciatica, ringworm, baldness, internal injuries, hypertension	Kumar et al., 2003; Tirkey, 2004; Jadhav, 2011; Shukla et al., 2010; Jain et al., 2012; Soni et al., 2012; Tomar et al., 2012; Maity et al., 2015
372.	Nymphaea nouchali Burm.f.	Rhizome, roots	Piles, dysentery, dyspepsia, urination problem in children	Katewa et al., 2003; Shukla et al., 2010; Mondal and Rahman, 2012
373.	Nymphaea rubra Roxb. ex Andrews	Flower	Meno-metarrhagia	Bandhyophyay and Mukherjee, 2006
374.	Ochna pumila Buch.- Ham. ex D.Don	Roots	Urinary troubles, weak eyesight	Tirkey, 2004
375.	Ochna squarrosa L.	Roots	Tuberculosis chest pain	Tirkey, 2004
376.	Ocimum americanum L.	Seeds	Skin diseases	Katewa and Galav, 2005
377.	Ocimum basilicum L.	Leaves	Mouth ulcer, pus in ear, malaria, cough, abdominal pain	Pandey and Verma, 2002; Kumar et al., 2003; Jadhav, 2006; Singh et al., 2012
378.	Ocimum gratissimum L.	Plant, leaves	Rheumatism, paralysis, toothache, seminal weakness	Upadhyay and Singh, 2005; Malviya et al., 2011
379.	Ocimum tenuiflorum L. [Syn.: O.sanctum L.]	Leaves, seeds	Lice, prevent conception, cold, cough, toothache, spermatorrhea, earache, fever, epilepsy	Kumar et al., 2003; Shekhawat and Batra, 2006; Kumar and Chauhan, 2006; Rai, 2006; Jadhav, 2006; Shukla et al., 2010; Singh et al., 2012; Tomar et al., 2012; Singh et al., 2014
380.	Operculina turpethum (L.) Manso	Leaves	Itching, ringworm	Shukla et al., 2010
381.	Ophioglossum reticulatum L.	Whole plant	Stomach disorders, heart diseases, cuts	Saini, 2008

TABLE 5.1　(Continued)

S. No.	Name of the plant species	Plant parts used	Ethnomedicinal uses	References
382.	*Oroxylum indicum* (L.) Kurz Figure 5.5A	Leaves, roots, mature fruits, stem bark	Diarrhea, asthma, rheumatism, chicken pox, small pox	Shukla et al., 2010; Jain et al., 2012; Maity et al., 2015
383.	*Oryza sativa* L.	Seed	Mouth ulcer	Pandey and Verma, 2002
384.	*Oxalis corniculata* L.	Whole plant, leaves	Headache, fever, dysentery, skin and eye diseases, physical weakness, malaria, stomachic, refrigerant	Khanna, 2002; Kumar et al., 2003; Upadhyay and Singh, 2005; Bondya et al., 2006; Shukla et al., 2010; Singh et al., 2012; Singh et al., 2014; Pandey et al., 2015
385.	*Paedaria foetida* L. [Syn.: *P. scandens* Lour.]	Leaves, plant	Piles, indigestion, gastric problems, sores in skin, dysentery	Tomar et al., 2012; Kumar and Goel, 2008; Mondal and Rahman, 2012
386.	*Pandanus tectorius* Sol. ex Park.	Roots	Prevent conception	Jadhav, 2006
387.	*Papaver somniferum* L.	Plant	Allergic diseases	Pandey et al., 2015
388.	*Pedalium murex* L.	Plant, fruits	Spermatorrhea, complaints of urinary system	Katewa and Galav, 2005; Kumar and Chauhan, 2006
389.	*Pennisetum glaucum* (L.) R.Br. [Syn.: *P. typhoides* (Burm.f.) Stapf & Hubb]	Grain	Lice	Shekhawat and Batra, 2006
390.	*Pergularia daemia* (Forssk.) Chiov.	Leaves	Uterine tonic	Sahu, 1982
391.	*Persicaria barbata* (L.) H.Hara [Syn.: *Polygonum barbatum* L.]	Whole plant	Fever	Shukla et al., 2010

A. *Oroxylum indicum* (L.) Kurz

B. *Premna herbacea* Roxb.

C. *Pueraria tuberosa* (Willd.) DC.

D. *Soymida febrifuga* (Roxb.) A. Juss

E. *Strychnos nux-vomica* L.

F. *Tribulus terrestris* L.

FIGURE 5.5

TABLE 5.1 (Continued)

S. No.	Name of the plant species	Plant parts used	Ethnomedicinal uses	References
392.	*Phoenix sylvestris* (L.) Roxb.	Root	Toothache	Pandey and Verma, 2002
393.	*Phyla nodiflora* (L.) E.Greene	Whole plant	Injury	Shukla et al., 2010
394.	*Phyllanthus amarus* Schumm. & Thonn.	Whole plant, leaves	Jaundice, leucorrhoea	Jain et al., 2011; Tomar et al., 2012
395.	*Phyllanthus emblica* L. [Syn.: *Emblica officinalis* Gaertn.]	Fruits, bark, leaves	Inflammation of eyes, eye disease, gonorrhea, diabetes, dysentery, laxative, hepatitis, cold, blood cancer, cough, anemia, stomach trouble, leucorrhoea, aphrodisiac, heart disorders, diarrhea; dyslipidemia	Upadhyay and Singh, 2005; Malviya et al., 2011; Singh et al., 2012; Tomar et al., 2012; Anand et al., 2013; Singh et al., 2014; Kumar et al., 2015; Maity et al., 2015
396.	*Phyllanthus fraternus* Webster	Plant, seed	Gonorrhea, genitourinary diseases, menorrhagia	Sahu, 1982; Sharma and Kumar, 2007
397.	*Phyllanthus niruri* L.	Whole plant	Jaundice, diuretic	Singh et al., 2012; Singh et al., 2014
398.	*Phyllanthus reticulatus* Poir. [Syn.: *Kirganelia reticulata* (Poir.) Baill.]	Leaves	Cooling agent	Kumar and Chauhan, 2006
399.	*Physalis minima* L.	Leaves, fruits	Dropsy, colic complaints, constipation, leucorrhoea	Katewa and Galav, 2005; Shukla et al., 2010
400.	*Piper betle* L.	Leaves	Antipyretic, carminative	Pandey et al., 2015
401.	*Piper longum* L.	Leaves	Cough	Mondal and Rahman, 2012
402.	*Piper nigrum* L.	Fruits	Urination problem in children	Mondal and Rahman, 2012

S. No.	Name of the plant species	Plant parts used	Ethnomedicinal uses	References
403.	*Pithecellobium dulce* (Roxb.) Benth.	Bark	Fever	Shukla et al., 2010
404.	*Plantago ovata* Forssk. [Syn.: *P. ispaghula* Roxb. ex Fleming]	Roots	Diarrhea, dysentery	Sharma and Kumar, 2007
405.	*Pleopeltis macrocarpa* (Bory ex Willd.) Kaulf. [Syn.: *Pleopeltis lanceolata* Kaulf.]	Leaves	Cold, cough, headache, cuts, wounds	Saini, 2008
406.	*Pluchea lanceolata* (DC.) Clarke	Leaves	Joint pain	Pandey et al., 2015
407.	*Plumbago zeylanica* L.	Roots, whole plant, leaves	Stomach disorder, body ache, rheumatic pain, abdominal pain, cuts, wounds, headache, skin diseases, eczema, scabies, ringworm, leucoderma	Jain and Patole, 2001; Tirkey, 2004; Sharma and Kumar, 2007; Shukla et al., 2010; Jain et al., 2011; Jain et al., 2012; Singh et al., 2012; Singh et al., 2014
408.	*Polygonum plebeium* R.Br.	Whole plant	Bowel complaints, pneumonia, eczema	Katewa and Galav, 2005; Bondya et al., 2006
409.	*Porana paniculata* Roxb.	Roots	Wounds, abortion	Tirkey, 2004
410.	*Portulaca oleracea* L.	Whole plant, leaves	Scorpion sting, toothache, scurvy, liver diseases, refrigerant	Katewa and Galav, 2005; Shekhawat and Batra, 2006; Bondya et al., 2006; Singh et al., 2012
411.	*Premna herbacea* Roxb. Figure 5.5B	Roots	Tuberculosis, joint pains	Tirkey, 2004

TABLE 5.1 (Continued)

S. No.	Name of the plant species	Plant parts used	Ethnomedicinal uses	References
412.	*Premna mollissima* Roth [Syn.: *P. latifolia* var. *mucronata* (Roxb.) Clarke]	Bark	Skin diseases	Kumar and Goel, 2008
413.	*Prosopis cineraria* (L.) Druce [Syn.: *P. spicigera* L.]	Leaves, seeds, stem	Skin diseases, rheumatism	Sharma and Kumar, 2011; Singh et al., 2012; Singh et al., 2014
414.	*Psidium guajava* L.	Leaves	Cholera, diarrhea, vomiting, toothache	Pandey and Verma, 2002; Upadhyay and Singh, 2005
415.	*Pteris longifolia* L.	Rhizome, leaves	Scrophula, tuberculosis, diarrhea, dysentery	Saini, 2008
416.	*Pterocarpus marsupium* Roxb.	Gum, bark, stem, leaves	Fever, diabetes, skin diseases	Rai, 2006; Anand et al., 2013; Shrivastava and Kanungo, 2013
417.	*Pterospermum acerifolium* (L.) Willd.	Calyx, flower	Glandular swelling, menstrual disorders, sunstroke	Sikarwar and Kaushik, 1992; Tirkey, 2004
418.	*Pueraria tuberosa* (Willd.) DC. Figure 5.5C	Tubers	To increase memory, chest pain, body swelling	Tirkey, 2004; Jain et al., 2005a; Tirkey, 2006; Jain et al., 2012
419.	*Punica granatum* L.	Leaves, fruit rind	Dysentery, pyorrhea, for improving memory	Sahu et al., 1983; Pandey and Verma, 2002; Singh et al., 2012
420.	*Putranjiva roxburghii* Wall. [Syn.: *Drypetes roxburghii* (Wall.) Hurusawa]	Fruits	Skin allergy	Khanna, 2002; Shukla et al., 2010
421.	*Ranunculus sceleratus* L.	Whole plant	Asthma, pneumonia, rheumatism	Kaur and Vashistha, 2014
422.	*Raphanus sativus* L.	Roots, juice, leaves, seeds	Bee/wasp bite, scorpion sting, jaundice, bladder stones	Shekhawat and Batra, 2006; Sharma and Kumar, 2007; Shukla et al., 2010

S. No.	Name of the plant species	Plant parts used	Ethnomedicinal uses	References
423.	*Rauvolfia serpentina* Benth. ex Kurz.	Roots, leaves	Snake bite, uterine contraction, labor pain, blood pressure	Tirkey, 2004; Singh et al., 2012; Singh et al., 2014
424.	*Ricinus communis* L.	Leaves, seed oil, seeds, roots	Migraine, dandruff, pneumonia fever, rheumatism, skin diseases, epilepsy, kidney stones, piles, intestinal worms, scorpion sting	Jadhav, 2006; Kala, 2009; Prachi et al., 2009; Shukla et al., 2010; Jain et al., 2011; Singh et al., 2012; Singh et al., 2014
425.	*Rosa damascena* Mill.	Flowers	Stomach problem	Singh et al., 2012
426.	*Rotheca serrata* (L.) Steane & Mabb. [Syn.: *Clerodendrum serratum* (L.) Moon]	Roots, leaves	Swollen legs, snake bite, malarial fever, bronchitis, bodyache	Tirkey, 2004; Kumar and Goel, 2008; Shukla et al., 2010
427.	*Rubia cordifolia* L.	Root	Bone fracture	Sharma and Kumar, 2007
428.	*Rubus lagerbergii* Lind. [Syn.: *R. maritimus* (L. ex F.Aresch.) Elmq.]	Seeds	Spermatorrhea	Khanna and Kumar, 2000
429.	*Rumex dentatus* L.	Whole plant	Sunstroke	Shukla et al., 2010; Singh et al., 2012
430.	*Rungia pectinata* (L.) Nees	Leaves	Small pox	Upadhyay and Singh, 2005
431.	*Ruta graveolens* L.	Leaves	Carminative	Kumar, 2012
432.	*Saccharum officinarum* L.	"Gur" from stem juice (Jaggery)	Mad dog bite, jaundice	Shekhawat and Batra, 2006; Singh et al., 2012
433.	*Salvinia natans* (L.) All.	Plants	Ringworm, eczema, acidity	Saini, 2008
434.	*Sansevieria hyacinthoides* (L.) Druce	Leaves	Snake bite	Jadhav, 2011
435.	*Saraca asoca* (Roxb.) de Wild.	Bark, flower	Uterine diseases, miscarriage	Kumar et al., 2003; Bandhyophyay and Mukherjee, 2006

TABLE 5.1 (Continued)

S. No.	Name of the plant species	Plant parts used	Ethnomedicinal uses	References
436.	*Schleichera oleosa* (Lour.) Merr. [Syn.: *S. oleosa* (Lour.) Oken]	Flower	Hair tonic	Anand et al., 2013
437.	*Scoparia dulcis* L.	Whole plant, roots, leaves	Toothache, diarrhea, menorrhagia	Upadhyay and Singh, 2005; Shukla et al., 2010
438.	*Selaginella bryopteris* (L.) Baker	Leaves	Venereal diseases, stomachache	Singh et al., 2005; Shukla et al., 2010
439.	*Selaginella ciliaris* (Retz.) Spring.	Whole plant	Venereal diseases, kidney stones	Saini, 2008
440.	*Semecarpus anacardium* L.f.	Seed oil, fruits	Rheumatism, cough, piles, indigestion	Shukla et al., 2010; Jain et al., 2012; Singh et al., 2012; Anand et al., 2013; Quamar and Bera, 2014; Singh et al., 2014
441.	*Senna alata* (L.) Roxb. [Syn.: *Cassia alata* L.]	Leaves	Ringworm, scabies, laxative	Jain et al., 2009; Kumar et al., 2015
442.	*Senna obtusifolia* (L.) H. S.Irwin & Barneby [Syn.: *Cassia obtusifolia* L.]	Leaves, seeds	Foul ulcers, itchy eruptions	Upadhyay and Singh, 2005
443.	*Senna occidentalis* (L.) Link [Syn.: *Cassia occidentalis* L.]	Fruit, leaves, roots, seeds	Skin diseases, snake bite, cough, laxative, menometrorrhagia, mad dog bite	Khanna, 2002; Mishra, 2008; Shukla et al., 2010; Kumar et al., 2015; Bandhyophyay and Mukherjee, 2006
444.	*Senna sophera* (L.) Roxb. [Syn.: *Cassia sophera* L.]	Bark, seeds, leaves	Diabetes	Shukla et al., 2010
445.	*Senna sulfurea* (Collad.) H. S.Irwin & Barneby [Syn.: *Cassia glauca* Lamk.]	Wood	Jaundice	Alawa and Ray, 2012

S. No.	Name of the plant species	Plant parts used	Ethnomedicinal uses	References
446.	*Senna tora* (L.) Roxb. [Syn.: *Cassia tora* L.]	Whole plant, leaves, seeds, flower, root	Ringworm, asthma, gonorrhea, cuts, wounds, rheumatic pain, skin diseases, digestive problems, laxative, joint pain, jaundice	Katewa et al., 2003; Bondya et al., 2006; Mishra, 2008; Jain et al., 2009; Shukla et al., 2010; Kumar et al., 2015; Jadhav, 2011; Singh et al., 2012
447.	*Senna uniflora* (Mill.) H. S.Irwin & Barneby [Syn.: *Cassia sericea* Sw.]	Flower	Jaundice	Jadhav, 2011
448.	*Sesamum indicum* L. [*S. orientale* L.]	Seeds	Loose teeth, urinary troubles, tonic, diuretic	Pandey and Verma, 2002; Kumar et al., 2003; Shukla et al., 2010
449.	*Sesbania sesban* (L.) Merr. [Syn.: *S. aegyptiaca* Pers.]	Seeds, roots, stem	For abortion, antifertility, pneumonia in children	Tirkey, 2006; Mishra, 2008
450.	*Sesbania grandiflora* (L.) Pers.	Flower, roots	Dysmenorrhea, cold and cough, diarrhea, dysentery	Bandhyophyay and Mukherjee, 2006; Bondya et al., 2006
451.	*Shorea robusta* Roxb. ex Gaertn.	Leaves, gum, bark	Skin diseases, stomach pain, dysentery, bone fracture, wound	Jain and Patole, 2001; Shukla et al., 2010; Mondal and Rahman, 2012; Quamar and Bera, 2014; Singh et al., 2014
452.	*Sida acuta* Burm.f.	Leaves, whole plant	Intestinal worms, snake bite, bone fracture, impotency	Kumar et al., 2003; Upadhyay and Singh, 2005; Shukla et al., 2010; Malviya et al., 2011; Maity et al., 2015
453.	*Sida cordata* (Burm.f.) Borss.	Leaves, roots	Cuts, aphrodisiac	Kumar et al., 2003; Shukla et al., 2010
454.	*Sida cordifolia* L.	Root bark, roots, whole plant	Gonorrhea, asthma, leucorrhoea, spermatorrhoes, Leg pain, aphrodisiac, micturition, fever, spermatorrhea	Sahu, 1982; Jain and Patole, 2001; Upadhyay and Singh, 2005; Shukla et al., 2010; Malviya et al., 2011; Singh et al., 2014

TABLE 5.1 (Continued)

S. No.	Name of the plant species	Plant parts used	Ethnomedicinal uses	References
455.	*Sida rhombifolia* L.	Leaves, whole plant	Ulcer, tuberculosis, spermatorrhea	Maliya, 2004; Shukla et al., 2010
456.	*Sida ovata* Forssk.	Seeds	Lumbago	Katewa and Galav, 2005
457.	*Smilax ovalifoia* Roxb. ex D.Don	Roots	Piles	Jha and Verma, 1996
458.	*Smilax perfoliata* Lour. [Syn.: *S. prolifera* Roxb)	Roots, leaves, stem	Sunstroke, burning urination, wounds	Tirkey, 2004
459.	*Smilax zeylanica* L.	Stem, roots	Toothache, boils, piles, leucorrhoea, dysentery	Pandey and Verma, 2002; Mairh et al., 2010; Shukla et al., 2010; Quamar and Bera, 2014; Maity et al., 2015
460.	*Solanum americanum* Mill. [Syn.: *S. nigrum* L.]	Juice, whole plant, leaves	Liver enlargement, diarrhea, fever, jaundice, liver and skin diseases	Kumar et al., 2003; Maliya, 2007; Upadhyay and Singh, 2005; Shukla et al., 2010; Singh et al., 2012; Singh et al., 2014
461.	*Solanum erianthum* D. Don	Leaves	Jaundice	Maliya, 2007
462.	*Solanum ferox* L. [Syn.: *S. indicum* L.]	Fruit	Migraine, cough, cold	Maliya, 2007
463.	*Solanum surattense* Burm.f.	Leaves, roots, fruits	Skin diseases, piles kidney stones	Katewa and Galav, 2005; Prachi et al., 2009
464.	*Solanum virginianum* L. [Syn.: *S. xanthocarpum* Schrad. & Wendl.]	Roots, whole plant, fruits, seeds, leaves	Gonorrhea, cold, cough, pneumonic fever, toothache, bleeding gum, mouth ulcer, asthma, snake bite	Sahu, 1982; Pandey and Verma, 2002; Kumar et al., 2003; Upadhyay and Singh, 2005; Jain et al., 2010; Shukla et al., 2010; Jain et al., 2011; Mondal and Rahman, 2012; Tomar et al., 2012; Singh et al., 2014
465.	*Sonchus asper* (L.) Hill	Plant	Wounds, boils, swellings	Katewa and Galav, 2005

S. No.	Name of the plant species	Plant parts used	Ethnomedicinal uses	References
466.	*Sonchus oleraceus* (L.) L.	Leaves, plant	Stomach pain, liver disease	Jain and Patole, 2001; Katewa and Galav, 2005
467.	*Soymida febrifuga* (Roxb.) A. Juss. Figure 5.5D	Bark	Muscular pain, bone fracture, for blood clotting	Kala, 2009; Jain et al., 2011; Mondal and Rahman, 2012
468.	*Spermacoce articularis* L.f. [Syn.: *Borreria articularis* (L.f.) F. N. Williams]	Whole plant, leaves	Toothache, conjunctivitis, stomach pain	Jain and Patole, 2001; Pandey and Verma, 2002; Jain et al., 2009
469.	*Spermacoce mauritiana* Gideon	Leaves	Mastitis	Bandhyophyay and Mukherjee, 2006
470.	*Sphaeranthus indicus* L.	Young twigs, fruits, leaves, whole plant, seeds	Earache, aphrodisiac, stomach disorders, earache, indigestion, piles, dysentery, impotency	Khanna, 2002; Katewa and Galav, 2005; Kumar and Goel, 2008; Jain et al., 2010; Shukla et al., 2010; Malviya et al., 2011; Pandey et al., 2015
471.	*Spilanthes calva* DC.	Flower heads	Tooth ache	Upadhyay and Singh, 2005
472.	*Spondias pinnata* (L.f.) Kurz.	Bark	Diarrhea	Mondal and Rahman, 2012
473.	*Stephania japonica* (Thunb.) Miers	Leaves	Dysmenorrhea	Bandhyophyay and Mukherjee, 2006
474.	*Streblus asper* Lour.	Twigs, bark, roots, latex	Toothache, dental carries, snake bite, scorpion sting, stomach ailments, puerperal fever	Pandey and Verma, 2002; Maliya, 2004; Bandhyophyay and Mukherjee, 2006; Mairh et al., 2010
475.	*Strychnos nux-vomica* L. Figure 5.5E	Bark, whole plant	Epilepsy, spermatorrhea	Shukla et al., 2010; Malviya et al., 2011

TABLE 5.1 (Continued)

S. No.	Name of the plant species	Plant parts used	Ethnomedicinal uses	References
476.	*Syzygium aromaticum* Merr. & Perr. [Syn.: *Eugenia caryophyllata* Thunb.]	Flower buds	Sore throat, tooth ache, bleeding in gum	Jadhav, 2006; Mondal and Rahman, 2012
477.	*Syzygium cumini* (L.) Skeels	Bark, fruits, leaves, seeds, gum	Diarrhea, diabetes, scorpion sting, leucorrhoea, spermatorrhea, body pain, diabetes, stomach problems, indigestion	Upadhyay and Singh, 2005; Shekhawat and Batra, 2006; Kumar and Chauhan, 2006; Menghani et al., 2010; Shukla et al., 2010; Jain et al., 2011; Jain et al., 2005a; Singh et al., 2012; Tomar et al., 2012; Shrivastava and Kanungo, 2013; Quamar and Bera, 2014; Singh et al., 2014
478.	*Tabernaemontana divaricata* (L.) R.Br.	Roots	Toothache, scorpion sting, to treat enhanced heart beat	Khanna, 2002; Pandey and Verma, 2002; Shukla et al., 2010
479.	*Tagetes erecta* L.	Leaves	Earache, bee/wasp bite	Kumar et al., 2003; Shekhawat and Batra, 2006
480.	*Tamarindus indica* L.	Seeds, leaves, fruits, bark	Scorpion sting, dysentery, tooth ache, bleeding gum, laxative, fever, gastric pain, dandruff	Shekhawat and Batra, 2006; Bondya et al., 2006; Sharma and Kumar, 2011; Mondal and Rahman, 2012; Singh et al., 2012
481.	*Tamarix dioica* Roxb. ex Roth	Leaves	Liver disorder, digestive, laxative	Khanna and Kumar, 2000; Kumar et al., 2003
482.	*Tecomella undulata* (Sm.) Seem.	Roots, bark	Leucorrhoea, eczema, eruptions	Katewa and Galav, 2005
483.	*Tectaria gemmifera* (Fée) Alston [Syn.: *T. coadunata* C. Chr.]	Rhizome, leaves	Anthelmintic, stomach pain, respiratory disorders, septic ulcers	Singh et al., 2005; Saini, 2008

S. No.	Name of the plant species	Plant parts used	Ethnomedicinal uses	References
484.	*Tectona grandis* L.f.	Wood	Laxative, piles, dysentery, skin diseases	Singh et al., 2012; Quamar and Bera, 2014
485.	*Tephrosia purpurea* (L.) Pers.	Whole plant, roots, root bark	Gonorrhea, fever, diabetes, liver disorder, dysentery, laxative, urinary disorders, impotency dyspepsia, diarrhea, rheumatism, stomach pain, anthelmintic	Sahu, 1982; Tirkey, 2004; Katewa and Galav, 2005; Tirkey, 2006; Mishra, 2008; Shukla et al., 2010; Jain et al., 2012; Singh et al., 2012
486.	*Terminalia alata* Heyne ex Roth	Bark	Diarrhea	Quamar and Bera, 2014
487.	*Terminalia arjuna* (Roxb. ex DC.) Wight & Arn. [Syn.: *Terminalia cuneata* Roth]	Fruits, roots, leaves, bark	Colic pain, cough, fever, headache, earache, spermatorrhea, heart disease, bone fracture, dysentery, high blood pressure, cardiac disorder	Khanna, 2002; Kumar et al., 2003; Rai, 2006; Sharma and Kumar, 2007; Mondal and Rahman, 2012; Anand et al., 2013; Quamar and Bera, 2014; Singh et al., 2014; Pandey et al., 2015
488.	*Terminalia bellirica* (Gaertn.) Roxb.	Fruits	Laxative, stomach trouble, menstrual disorder, cough, cold	Shukla et al., 2010; Anand et al , 2013; Quamar and Bera, 2014; Singh et al., 2014
489.	*Terminalia chebula* Retz.	Fruits	Astringent, laxative, bleeding and ulceration of gums, diabetes, constipation, cough, cold, asthma	Jain and Patole, 2001; Mairh et al., 2010; Shukla et al., 2010; Tomar et al., 2012; Anand et al., 2013; Shrivastava and Kanungo, 2013; Quamar and Bera, 2014; Maity et al., 2015
490.	*Thespesia populnea* Sol. ex Corr.	Stem bark, young branches	Neuritis, mad dog bite, promote conception	Tirkey, 2004; Jadhav, 2006
491.	*Tiliacora racemosa* Colebr. [Syn.: *T. acuminata* (Lamk.) Miers]	Stem	Constipation	Maliya, 2004

TABLE 5.1 (Continued)

S. No.	Name of the plant species	Plant parts used	Ethnomedicinal uses	References
492.	*Tinospora cordifolia* (Willd.) Miers ex Hook. f. &Thoms.	Stem juice, stem, leaves, roots	Typhoid, psoriasis, diabetes, headache, puerperal fever, aphrodisiac, cough, cold, for smooth delivery	Kumar et al., 2003; Bandhyophyay and Mukherjee, 2006; Shukla et al., 2010; Jadhav, 2011; Malviya et al., 2011; Singh et al., 2012; Shrivastava and Kamungo, 2013; Singh et al., 2014; Maity et al., 2015
493.	*Tinospora sinensis* (Lour.) Merr.	Leaves	Mosquito bite	Shekhawat and Batra, 2006
494.	*Toona ciliata* Roem.	Bark, stem	Mouth ulcer, tonsillitis, dysentery	Pandey and Verma, 2002; Shukla et al., 2010; Quamar and Bera, 2014
495.	*Trachyspermum ammi* (L.) Spreng.	Seeds, plant, fruits	Regulate menstrual period, eye diseases, gastric diseases, digestive problems, toothache	Pandey and Verma, 2002; Jadhav, 2006; Mondal and Rahman, 2012; Pandey et al., 2015
496.	*Trewia nudiflora* L.	Fruits, seeds	Whooping cough	Maliya, 2007; Shukla et al., 2010
497.	*Trianthema portulacastrum* L.	Leaves	Jaundice, kidney stones, leucorrhoea	Prachi et al., 2009; Shukla et al., 2010
498.	*Trianthema triquetra* Rottler & Willd.	Plant	Rheumatism	Katewa and Galav, 2005
499.	*Tribulus terrestris* L. Figure 5.5F	Fruits, whole plant	Urine disorders, kidney stones, aphrodisiac	Prachi et al., 2009; Shukla et al., 2010; Malviya et al., 2011; Singh et al., 2012; Singh et al., 2014
500.	*Trichodesma indicum* (L.) R.Br.	Roots	Snake bite	Pandey et al., 2015
501.	*Trichosanthes dioica* Roxb.	Fruit	Spermatorrhea	Sahu, 1982

S. No.	Name of the plant species	Plant parts used	Ethnomedicinal uses	References
502.	*Trichosanthes tricuspidata* Lour.	Fruits	Rheumatism	Maliya, 2004
503.	*Tridax procumbens* (L.) L.	Leaves	Cuts, wounds, piles, leucorrhoea, eczema	Mishra, 2008; Shukla et al., 2010; Jain et al., 2011; Singh et al., 2012; Singh et al., 2014
504.	*Trigonella foenum-graecum* L.	Seeds	Diabetes	Menghani et al., 2010
505.	*Tripodanthus acutifolius* (Ruiz & Pav.) Tiegh. [Syn.: *Loranthus ligustrinus* Wall.]	Stem	Polio	Tirkey, 2004
506.	*Triticum aestivum* L.	Seeds	Dysuria, dog bite	Mitra and Mukherjee, 2005a; Shekhawat and Batra, 2006
507.	*Triumfetta rhomboidea* Jacq.	Roots	Boils, inflamed eyelids	Upadhyay and Singh, 2005; Shukla et al., 2010
508.	*Tylophora indica* (Burm.f.) Merr.	Whole plant	Asthma	Tirkey, 2004
509.	*Typha domingensis* Pers. [Syn.: *T. angustata* Bory & Chaub.]	Inflorescence	Wound healing	Katewa et al., 2003
510.	*Vanda tessellata* (Roxb.) Hook. ex G.Don [Syn.: *V. roxburghii* R.Br.] Figure 5.6A	Plant, roots, leaves, whole plant	Rheumatism, arthritis, earache, bone fracture, paralysis, fever	Tiwari and Yadav, 2003; Tirkey, 2004; Jain et al., 2005a; Upadhyay and Singh, 2005; Shukla et al., 2010; Jain et al., 2012; Singh et al., 2012
511.	*Ventilago denticulata* Willd. Figure 5.6B	Stem	Eye inflammation	Shukla et al., 2010; Quamar and Bera, 2014

TABLE 5.1 (Continued)

S. No.	Name of the plant species	Plant parts used	Ethnomedicinal uses	References
512.	*Ventilago maderaspatana* Gaertn.	Bark	Infecundity (female)	Tomar et al., 2012
513.	*Vernonia amygdalina* Del.	Leaves	Diabetes, parasitic worms, malaria, constipation, fever, cough	Kumar and Verma, 2011
514.	*Vigna unguiculata* (L.) Walp. [Syn.: *Dolichos biflorus* L.]	Seeds	Diuretic, jaundice	Kumar et al., 2015
515.	*Vitex negundo* L.	Roots, leaves, seed oil, bark	Menstrual disorders, headache, cataract, sprains, bodyache, muscular pain, old wound	Katewa and Galav, 2005; Mishra, 2008; Kumar and Goel, 2008; Mairh et al., 2010; Jain et al., 2011; Pandey et al., 2015
516.	*Vitex peduncularis* Wall. ex Schauer	Stem bark	Cough, cold, body ache, fever	Tirkey, 2004; Kumar and Goel, 2008
517.	*Volkameria inermis* L. [Syn.: *Clerodendrum inerme* (L.) Gaertn.]	Leaves	Psoriasis, skin affections	Singh et al., 2012
518.	*Waltheria indica* L. [Syn.: *W. americana* L.]	Roots	Leucorrhoea, spermatorrhea	Katewa and Galav, 2005
519.	*Withania somnifera* (L.) Dunal	Leaves, roots, whole plant	Male sterility, for conception, fever, skin diseases, leucorrhoea, tonic, aphrodisiac, tuberculosis, rheumatic pain	Tirkey, 2004; Katewa and Galav, 2005; Shukla et al., 2010; Jain et al., 2011; Malviya et al., 2011; Singh et al., 2014
520.	*Wrightia tinctoria* (Roxb.) R.Br.	Seeds	Aphrodisiac, diabetes	Malviya et al., 2011; Shrivastava and Kanungo, 2013;

S. No.	Name of the plant species	Plant parts used	Ethnomedicinal uses	References
521.	*Xanthium strumarium* L. [Syn.: *X. indicum* Koen.]	Fruits, leaves, root	Eczema, scabies, migraine, small pox, conjunctivitis, anthelmintic	Katewa and Galav, 2005; Jadhav, 2008a; Shukla et al., 2010; Singh et al., 2012; Singh et al., 2014
522.	*Zaleya govindia* (Buch.-Ham. ex G.Don) Nair	Roots	Syphilis, regularize menstruation	Katewa and Galav, 2005
523.	*Zea mays* L.	Flower, styles	Skin disease, kidney stones	Mitra and Mukherjee, 2005a; Prachi et al., 2009
524.	*Zingiber officinale* Rosc.	Rhizome	Cough, digestive problem, bronchitis, throat infection	Kumar et al., 2003; Mondal and Rahman, 2012
525.	*Ziziphus jujuba* Mill. [Syn.: *Z. mauritiana* Lam.]	Leaves, fruits	Nose bleeding, cold, cough, cool drink	Kumar and Chauhan, 2006; Jain et al., 2010; Shukla et al., 2010
526.	*Ziziphus nummularia* (Burm.f.) Wt. & Arn.	Root bark, bark, leaves, fruit, root	Pyorrhea, blood purifier, dysentery, scorpion sting, anthelmintic	Kumar and Chauhan, 2006; Jain et al., 2011; Singh et al., 2012; Anand et al., 2013; Singh et al., 2014
527.	*Ziziphus oenoplia* (L.) Mill.	Ripe fruits	Dysentery, burning sensation during urination	Shukla et al., 2010; Quamar and Bera, 2014
528.	*Ziziphus rugosa* Lamk.	Whole plant	Bodyache	Kala, 2009

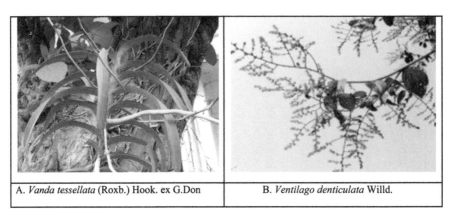

| A. *Vanda tessellata* (Roxb.) Hook. ex G.Don | B. *Ventilago denticulata* Willd. |

FIGURE 5.6

5.3 ETHNOMEDICINAL PLANT RESOURCES AND THEIR UTILIZATION

Perusal of such a vast literature revealed that altogether 528 plant species have been used by different ethnic and traditional communities of these two regions of India for treatment of a wide range of diseases and ailments like urino-genital problems, male sterility, hematuria, gout, rheumatism, tuberculosis, back pain, skin diseases, nervous disorders, baldness, etc. The enlisted 528 species are distributed over 112 plant families. Ten dominant families in respect of their maximum number of recorded species are Leguminosae (74 species), Malvaceae (30 species), Compositae (29 species), Lamiaceae (24 species), Apocynaceae (21 species), Convolvulaceae (17 species), Euphorbiaceae (16 species), Poaceae (16 species), Acanthaceae (12 species) and Solanaceae (12 species). It has been found that different plant parts have been used as effective ingredients for preparation of ethnomedicine in the area. In 245 cases, leaves were used in ethnomedicine preparation followed by underground parts (206 cases), whole plant (118 cases), bark (87 cases), seeds (86 cases), fruits (75 cases), flower and floral parts (51 cases), etc. (Figure 5.7).

The Indo-Gangetic region encompasses most parts of the northern and eastern India. From this region many research articles on ethnomedicine have been published till date. From the state of Uttar Pradesh, Dixit and Pandey (1984) have recorded the uses of 14 locally available plants used by the natives of Jhansi and Lalitpur sections of Bundelkhand. During

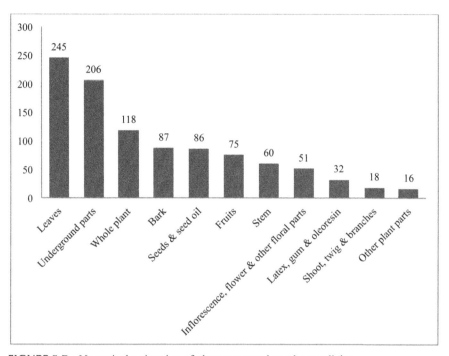

FIGURE 5.7 Numerical estimation of plant parts used as ethnomedicine.

ethnobotanical survey of Jaunsar-Bawar, an area inhabited by hilly tribals in Uttar Pradesh, it was observed that about 100 plants are being used by the local Jaunsari tribe for the treatment of various ailments (Jain and Puri, 1984). An ethnobotanical exploration study by Singh et al. (2002) presents the folk medicinal uses of 125 plants by tribes of Sonbhadra district in Uttar Pradesh. Maliya (2004) has highlighted the uses of 16 ethnomedicinal plants which are traditionally utilized by the Tharu tribals and other rural inhabitants of Bahraich district. In the year 2005, Upadhyay and Singh have documented ethnomedicinal uses of 30 plant species from Tikri forest of Gonda district. From Muzaffarnagar district in Uttar Pradesh, Prachi et al. (2009) have recorded 15 plant species which are used as urolithiatic agents. Singh and Singh (2009) made an ethnobotanical study of medicinal plants in Chandauli district in Uttar Pradesh and reported uses of 40 medicinal plants. Prakash (2011) has explored traditional knowledge of some threatened and potential ethnomedicinal plants among the tribals of Uttar Pradesh. Singh and Dubey (2012) reported use of 143 medicinal plants in Sonbhadra district of Uttar Pradesh. Anand et al. (2013) have enlisted ethnobotanical uses of 30 tree species found in Sonbhadra district. On the other hand, Rahul (2013a) carried out an ethnobotanical survey in Taindol village of Jhansi district and reported 57 medicinal plants. In the same year, he (Rahul, 2013b) made an ethnobotanical study on medicinal shrubs used by the people in Lakhmanpura region of Bundelkhand, Uttar Pradesh. Kumar and Akhtar (2013) reported ethnomedicinal uses of 14 species of Solanaceae from Eastern Uttar Pradesh. Kumar et al. (2015) described biodiversity and indigenous uses of medicinal plants in Chandra Prabha Wildlife sanctuary in Uttar Pradesh. Other reports on ethnomedicinal plants of Uttar Pradesh include Maheshwari et al. (1981, 1986), Saxena and Vyas (1981), Maheshwari and Singh (1984), Singh and Maheshwari (1985, 1989), Singh et al. (1994, 1998, 2007, 2010), Singh and Prakash (1994, 1996), Ranjan (1996), Singh (1996), Khanna and Kumar (2000), Prakash and Singh (2000), Kumar (2004), Nigam and Kumar (2005), Pandey and Kumar (2006), Singh and Narain (2009), etc.

A review work has been carried out on 21 ethnobotanical plants from the state of Punjab (Kaur, 2015). Yadav and Bhandoria (2012) have carried out ethnobotanical exploration in Mahendergarh district of Haryana and reported 56 species of medicinal plants which are used for treating more than 60 diseases. Kaur and Vashistha (2014) documented ethnomedicinal uses of 71 plant species from Karnal district in Haryana. Singh et al. (2015) carried out an ethnobotanical survey of common medicinal plants in Bhiwani in Haryana and reported medicinal uses of 60 plant species. Singh (2016) reported ethnomedicinal uses of 30 plant species from Bhiwani district in Haryana. Singh et al. (2016) carried out ethnobotanical study related to child care, with emphasis to commonly available herbs in various parts of Haryana state and reported use of 36 flowering plant species. Ethnomedicinal uses of 66 plant species of Haryana state was given by Gitika and Kumar (2016).

Many ethnomedicinal survey works have been carried out in different parts of Rajasthan. About 610 species of medicinal plants have been used by 4.2 million population of tribals of Rajasthan (Singh and Pandey, 1998). Joshi (1995) brought out a book on ethnobotany of primitive tribes in Rajasthan. Katewa et al. (2001) gave an account of 24 ethnomedicinal grasses of Rajasthan traditionally used for treating various ailments. Deora and Jhala (2002) have made a detail ethno-medico-botanical study of Kotra-Udaipur and reported the uses of 43 plant species from this area. Katewa et al. (2003) have recorded some medicinal and food plants used by the ethnic people reside in and around Aravalli hills. An ethnobotanical survey of tribal area of southern Rajasthan was carried out by Jain et al. (2004) for documentation of some tribal herbal medicines and reported the use of 53 plant species for curing sexual diseases and for family planning. A floristic survey of ethnomedicinal plants occurring in the tribal area of Rajasthan was conducted by Katewa et al. (2004) to assess the potentiality of plant resources for modern treatments and recorded uses of 61 plant species. Jain et al. (2005a) have recorded phytotherapeutic claims of 20 plant species used by the tribals of southern Rajasthan. Jain et al. (2005b) carried out a study in Sitamata Wildlife sanctuary in Rajasthan and documented the medicinal utility of herbs belonging to 243 genera. In the year 2006, Kumar and Chauhan carried out an ethnomedicinal survey in Keoladeo National Park and documented a considerable number of medicinal plants. For the treatment of animal and insect bite, 23 plant species were recorded as household remedies by Shekhawat and Batra (2006) from Bundi district. In 2007, Jain et al. recorded ethnomedicinal uses of biodiversity from Tadgarh-Raoli Wildlife sanctuary. Choudhary et al. (2008) have made a review on ethnobotanical studies in Rajasthan. Ethnomedicinal uses of *Acacia jacquemontii*, a tree of Thar desert was given by Choudhary et al. (2009). Sharma and Khandelwal (2010) described traditional uses of 36 plant species used by the tribals in Dang regions in Rajasthan as cooling agents during summer. Upadhyay et al. (2010) conducted ethnobotanical study in eastern Rajasthan and reported the use of 213 plant species for treating various ailments. Other reports on ethnomedicinal plants of Rajasthan include Katewa and Guria (1997), Katewa and Sharma (1998), Meena et al. (2003, 2013), Seema and Kumar (2005), Sharma and Kumar (2006, 2011), Meena and Yadav (2007, 2010a, b), Katewa (2009), Meena (2011), Meena and Kumar (2012), Mishra et al. (2014).

In the state of Bihar, many ethnobotanical surveys have been carried out by a number of researchers (Hembrum, 1991; Jain, 1996; Jha and Verma, 1996; Kumar and Pandey, 1990; Sahoo and Mudgal, 1993; Sriwastwa and Verma, 1981; Tarafdar and Raichaudhuri, 1981, 1990; Varma and Pandey, 1990). From an extensive study in 1998, Upadhyay et al. have documented 54 plants from 35 families which are used for the treatment of skin diseases. Ethnomedicinal uses of *Vernonia amygdalina* have thoroughly been discussed by Kumar and Verma in 2011. Recently ethnobotanical importance of 6 wild tree species of Bihar have explored by Jha and Yogesh (2015).

Jharkhand is a state of forest and tribal cultures and many ethnobotanical studies have been made in this state of India. Bondya and Sharma (2004) have recorded 11 ethnomedicinal plants of antidiabetic property from

Bharagora Block of this state. In 2007, Chandra et al., have inventoried 28 medicinal plants used in twelve ethnomedicinal formulations which are used by the traditional herbal practitioners of Ranchi. Tomar et al. (2012) have reported 58 herbal formulations from the state of Jharkhand which are used in 39 different disorders. Marandi and Britto (2014) have carried out an ethnobotanical survey in the tribal areas of Latehar district and reported the traditional uses of 67 plant species. Ethnobotanical knowledge on 38 plants was documented by Maity et al. (2015) from different tribal communities of Patratu, Hazaribagh district in Jharkhand.

On the other hand from West Bengal, Banerjee (2000) gave ethnobotanical uses of a few plant species from Birbhum district. Mitra and Mukherjee (2005) have reported 27 ethnobotanical usages of 16 grasses from West Dinajpur district. Bandyopadhyay and Mukherjee (2006) have enlisted 28 herbal prescriptions used by the ethnic communities of Koch Bihar district. Ethnomedicinal uses of plants for gynecological, urinogenital and other related problems by the tribal people of Birbhum and Murshidabad districts of West Bengal was given by Ghosh et al. (2011). An ethnobotanical survey have been carried out by Mondal and Rahaman (2012) in some selected areas of Birbhum district of West Bengal and Dumka district of Jharkhand to collect information on the uses of 28 ethnomedicinal plants which are used by the tribal people of these area in 10 different ethnomedicinal preparations for curing 10 types of diseases and ailments. Practices in ethnomedicine by the economically challenged people of Ayodhya hills in Purulia district of West Bengal was given by Chanda et al. (2012). Other reports on ethnomedicinal plants of West Bengal include Maity and Manna (2000), Mitra and Mukherjee (2005b, 2009, 2010), Mandal and Mukherjee (2008), Mahanta and Pal (2012).

Ethnobotanical exploration in India was first started in Central India and till a good number of research projects are going on in the different areas of this part. In the earlier phase of ethnobotanical research in India, S. K. Jain and his associates have done extensive survey in the tribal areas of Central India (Jain, 1963, 1965, 1975, 1991). An ethnobotanical survey was conducted by Maheshwari et al. (1986) in the tribal blocks of Jhabua district in Madhya Pradesh. Bhalla et al. (1992, 1996) have described ethnomedicinal uses of *Indigofera* and various members of Asteraceae from Bundelkhand region of Madhya Pradesh. Jain (1992) gave an account on ethnomedicinal plants of Sahariya tribals of Madhya Pradesh. Indigenous medicinal knowledge of hill Korwas of this state has been documented by Bajpai and

Mitra (1997). Singh et al. (2005) have reported 8 ethnomedicinal pterido-phytes from Amarkantak. Bondya et al. (2006) have enlisted ethnomedicinal uses of 21 leafy vegetables from the Achanakmar Amarkantak Biosphere Reserve. Jadhav (2006a) have documented the ethnomedicinal uses of 22 plant species from Maalgamdi in Ujjain district. Plant sources used for the treatment of various types of fever by Bhil tribe of Ratlam District was dis-cussed by Jadhav (2006c). Dwivedi et al. (2006) reported the herbal prepa-rations used for gastrointestinal diseases and disorders by tribal and rural people of Satna district in Madhya Pradesh. Rai (2006) has recorded the use of some plants for curing different types of fever including malaria from the Baiga, Bhariya and Gond tribes of Madhya Pradesh. Ethno medicinal plants used by Bhil tribes of Matrunda in Ratlam district were recorded by Jadhav (2007). Ethnomedicinal plants remedies for snake bite and scorpion sting among the Bhil tribes of Ratlam district was given by Jadhav (2008b). In the year 2008, Saini has documented traditional uses of 17 species of pte-ridophytes among Baiga tribes of Amarkantak in Anuppur district. Mishra (2008) has reported ethnomedicinal uses of 31 angiospermic species utilized by the Korku tribal community from Khandwa district. Ethnomedicinal plants used for dental troubles and antipyretic agents among the tribes of Ratlam district were documented by Jadhav (2009, 2010). Dahare and Jain (2010) have reported ethnomedicinal uses of 47 plant species used by Korku and Gond tribes of Multai Tehsil. Shukla et al. (2010) have carried out an ethnobotanical study and reported medicinal uses of 166 plant species in Rewa district of this state. Wagh and Jain (2010a,b) have carried out eth-nobotanical survey among Bheel and Bhilala tribes of Jhabua district and reported 15 and 20 ethnomedicinal plants, respectively which are used for the treatment of various ailments and diseases like, mouth ulcer, constipa-tion, earache, headache, asthma, ringworm, jaundice, snake bite, etc. Nath and Khatri (2010b, 2011) have made extensive ethnomedicinal survey in different districts of Madhya Pradesh and recorded more than 100 plant spe-cies from this region. In the year 2011, Jadhav has made an ethnobotanical survey in Ratlam district with emphasis on doctrine of signatures. Malaviya et al. (2011) have made a review on indigenous herbal remedies used in this state for improving sexual performance and treating problems associated with sexuality. Jadhav and Rawat (2011) recorded ethnomedicinal plants used in the treatment of various ailments by Bhilala tribe of Alirajpur dis-trict. In the year 2012, Alawa and Ray have reported ethnomedicinal uses of 86 plant species for the treatment of 35 diseases from Dhar district. Soni et al. (2012) have enlisted the ethnic uses of 34 plant species from Dindori district. Pathan et al. (2012) have documented some ethnomedicinal plants for treating jaundice in Rajgarh district. In another study from Madhya Pradesh, Bharti (2013) has reported 80 ethnomedicinal plants for treating 34 types of diseases from Shadol district. In another study, Rahul (2013c) has carried out an ethnobotanical survey at Orchha Wildlife Sanctuary region and reported the uses of 65 plant species. Sandya and Sandya (2013) have recorded the use of 25 ethnomedicinal plant species from Baiga tribal villages. From the study of Thakur et al. in 2014, it has been found that tribals of Alirajpur have used 15 plant species as ethnomedicine. For the treatment of joint diseases, Wagh and Jain (2014) have reported 35 ethnomedicinal plant species used by the tribal people of Jhabua district of Madhya Pradesh. Recently, Alawa et al. (2016) have carried out an ethnobotanical survey and reported the folklore claims of 24 plant species used by Bhil tribes of Dhar district. Other reports on ethnomedicinal uses of Madhya Pradesh include Shrivastava (1985), Rai and Pandey (1987), Rai (1987, 1988, 1993), Rai and Ojha (1989), Maheshwari et al. (1990), Oomachan et al. (1990), Prasad et al. (1990), Shah and Singh (1990), Dwivedi and Pandey (1992), Sikarwar and Maheshwari (1992), Rai and Nonhare (1992), Khan (1993), Lal (1993), Prasad and Pandey (1993), Maheshwari (1996), Rai et al. (1996), Jain (1998a,b), Kumar (1999), Rai et al. (2000, 2004a,b,c), Dwivedi (2003), Rai and Nath (2005a), Nath and Khatri (2010a), Kumar (2012), etc.

From the state of Chhattisgarh many studies have so far been carried out to explore the ethnomedicinal plant resources of this state of India. Tirkey et al. (2006) gave an exhaustive review on ethnobotanical research in Chhattisgarh. Tirkey (2004) has reported 50 ethnomedicinal plants used by the tribals of Chhattisgarh. Ethnomedicinal uses of 30 plant species used by the tribal people of Jashpur district of this state was explored in 2007 by Ekka and Dixit. Kala (2009) has carried out a study on native uses of ethnobotanical species in the south Surguja district and enlisted 36 plant species. Ekka (2011a) has presented a historical overview of ethnobotanical works of Chhattisgarh state considering 315 publications on ethnobotany of Chhattisgarh and closely bordering topics. In the same year Ekka (2011b) documented some ethnomedicine for anti-fertility commonly used by the tribals in Chhattisgarh. In another work, he has reported the uses of 17 ethnomedicinal plants from this state (Ekka, 2011c). Ekka and Ekka (2013b) have enlisted 27 ethnomedicinal plant species used by Birhor tribes of Chhattisgarh. Traditional knowledge of 64 ethnomedicinal plant species have been enlisted which were used by the tribal and rural people in Korea district for the treatment of some common diseases (Kushwaha et al., 2013). In 2014, a detail of ethnomedicinal plant knowledge was studied by Chatterjee from Surguja region. Kujur and Ahirwar (2015) have documented the folkloric claims on 34 ethnomedicinal plants used by various tribes of Jashpur district. Pandey et al. (2015) have reported the use of 29 medicinal plants used by tribal and folk people of Chhattisgarh. An ethnobotanical study was conducted by Singh and Bharti (2015) in different areas of Raigarh district of Chhattisgarh. Very recently traditional knowledge on ethnomedicinal plants used by Kamar tribe and Oraon tribe has been documented from the state of Chhattisgarh (Ekka, 2016; Ekka et al., 2016). Other publications on ethnomedicinal plants of Chhattisgarh include Rai et al. (2003, 2005b), Ekka (2012a,b, 2013), Ekka and Ekka (2013), Masih et al. (2013).

5.4 CONCLUSION

It is understood that the Indo-Gangetic region and Central India, the two agro-climatic zones of India are quite rich in its ethnomedicinal knowledge and related plant resources. Present review enlisted 528 ethnomedicinal plants which will help in future to prepare a vast inventory of ethnomedicinal plants and traditional knowledge of these two agro-climatic zones of India and it will be an asset for further scientific exploitation in the area of conservation and bioprospecting of plant resources. This list of ethnomedicinal plants is not an exhaustive one, but it represents a good portion of rich ethnomedicinal plant resources of this entire area of India. It is expected that further intensive review of the literature from these two zones of India will certainly add more number of medicinal species to this inventory of 528 ethnomedicinal plants and its related traditional knowledge.

Now, one striking point is to be noted here that out of 244 research articles reviewed, no one has been analyzed with ethnomedicinal tools to measure the importance and reliability of the ethnomedicinal remedies or medicinal plants documented. The lack of quantitative analysis of the data documented from these two zones of India, makes them quite unsuitable for bioprospecting of natural products. As bioprospecting scientists now-a-days favor the ethnomedicinal claims or information which are statistically analyzed with the help of suitable quantitative ethnobotanical indices.

Statistically analyzed data will equally be helpful also in selecting the prioritized medicinal plants for their conservation which are found important to the ethnic communities and rare in that area. Though, a few papers have recently been published from the states of Uttar Pradesh and West Bengal where ethnobotanical data have been analyzed through statistical tools. Finally, this huge ethnobotanical data documented so far from this area of India should be compiled in respect of medicinal plant species, their disease curing ability, user's ethnic community, etc. and analyzed through proper statistical tools and software, which then be an immense help in bioprospecting and medicinal plant resource management of that area.

KEYWORDS

- **Central India**
- **ethnobotany**
- **ethnomedicinal plants**
- **Indo-Gangetic region**
- **utilization**

REFERENCES

Alawa, K. S., & Ray, S. (2012). Ethnomedicinal plants used by tribals of Dhar district, Madhya Pradesh, India. *CIB Tech J. Pharmaceut. Sci., 1,* 7–15.

Alawa, K. S., Ray, S., & Dubey, A. (2016). Folk lore claims of some ethnomedicinal plants used by Bhil tribes of Dhar district, Madhya Pradesh. *Bioscience Discovery, 7*(1), 60–62.

Anand, R. K., Singh, M. P., Dwivedi, S. V., Ram, S., & Khare, N. (2013). Ethnobotanical study of trees found in district Sonbhadra, Uttar Pradesh. *Technofame, 2*(1), 1–5.

Bajpai, H. R., & Mitra, M. (1997). Indigenous medical practices of hill Korwas of Madhya Pradesh. *J. Human Ecol., 9*(3), 295–298.

Bandyopadyay, S., & Mukherjee, S. K. (2006). Traditional medicine used by the ethnic communities of Koch Bihar district (West Bengal, India). *J. Trop. Med. Plants, 7*(2), 303–312.

Banerjee, A. (2000). Ethnobotany of a few plant species in the eroded soil of Birbhum, West Bengal. In: Maheshwari, J. K. (ed.), Ethnobotany and Medicinal plants of Indian Subcontinent. Scientific Publishers, Jodhpur, India, pp. 527–530.

Bhalla, S., Patel, J. R., & Bhalla, N. P. (1992). Ethnomedicinal studies of genus *Indigofera* from Bundelkhand region of M. P. *J. Econ. Taxon. Bot. Addl. Ser., 10,* 221–332.

Bhalla, S., Patel, J. R., & Bhalla, N. P. (1996). Ethnomedicinal observations on some Astera-ceae of Bundelkhand region, Madhya Pradesh. *J. Econ. Taxon. Bot. Addl. Ser. 12*, 175–178.

Bharti, V. K. (2013). An ethnobotanical study of medicinal plants in Shadol district of Mad-hya Pradesh, India. *Intern. J. Scientific Res., 4*(10), 1501–1505.

Bodding, P. O. (1925). Studies in Santhal medicines and connected folklore: Part I, Santhals and disease. Mem. *Asiatic Soc. Bengal, 10*(1), 1–132.

Bodding, P. O. (1927). Santal medicine. The Book Trust, Calcutta.

Bondya, S. L., Khanna, K. K., & Singh, K. P. (2006). Ethnomedicinal uses of leafy vegeta-bles from the tribal folk-lore of Achanakmar-Amarkantak Biosphere reserve (Madhya Pradesh and Chhattisgarh). *Ethnobotany, 18*, 145–148.

Bondya, S. L., & Sharma, H. P. (2004). Ethnobotanical studies on plants used in diabetes (Madhumeha) under the Baharagora block of Jharkhand. *Ethnobotany, 16*, 139–140.

Bressers, J. (1951). The Botany of Ranchi district, Bihar. Catholic Press, Ranchi.

Chanda, S., Sen, S., & Pal, D. C. (2012). Practices in ethnomedicine by the economically challenged people of Ayodhya hills, Purulia district, West Bengal. In: Maiti, G. G., & Mukherjee, S. K. (eds.), Multidisciplinary approaches in Angiosperm Systematics. University of Kalyani, Kalyani, India. pp. 633–636.

Chandra, R., Mahato, M., Mondal, S. C., Kumar, K., & Kumar, J. (2007). Ethnomedicinal for-mulations used by traditional herbal practioners of Ranchi, Jharkhand. *Indian J. Trad. Knowl., 6*(4), 599–601.

Chatterjee, A. K. (2014). Study of ethno-medicinal plants among the tribals of Surguja region (C. G.). *Intern. J. Advanced Computer Theory and Engineering, 3*(2), 56–60.

Choudhary, K., Singh, M., & Pillai, U. (2008). Ethnobotanical survey of Rajasthan- An update. *American-Eurasian J. Bot., 1*(2), 38–45.

Choudhary, K., Singh, M., & Shekhawat, N. S. (2009). Ethnobotany of *Acacia jacquemontii* Benth. – An uncharted tree of Thar desert, Rajasthan, India. *Ethnobotanical leaflets, 13*, 668–678.

Dahare, D. K., & Jain, A. K. (2010). Ethnobotanical studies on plant resources of Tahsil Multai, district Betul, Madhya Pradesh, India. Ethnobotanical Leaflets, 14: 694–705.

Deora, G. S., & Jhala, G. P. S. (2002). Ethno-medico-botanical diversity of Kotra, Udaipur. In: Dadhich, L. K., & Sharma, A. P. (eds.), Biodiversity: Strategies for conservation. A. P. H. Publishing Corporation, New Delhi, India. pp. 317–329.

Dixit, R. D., & Pandey, H. C. (1984). Plants used as folk medicine in Jhansi and Lalitpur sec-tions of Bundelkhand, U. P. *Intern. J. Crude Drug Res. 22*, 48–51.

Dutta, B. K., & Dutta, P. K. (2005). Potential of ethnobotanical studies in North East India: An overview. *Indian J. Trad. Knowl., 4*(1),7–14.

Dwivedi, S. N. (2003). Ethnobotanical studies and conservational strategies of wild and natu-ral resources of Rewa district, Madhya Pradesh. *J. Econ. Taxon. Bot. 27*(1), 233–244.

Dwivedi, S. N., Dwivedi, S., & Patel, P. C. (2006). Medicinal plants used by the tribal and rural people of Satna district, Madhya Pradesh for the treatment of gastrointestinal dis-eases and disorders. *Nat. Prod. Radiance 5*(1), 60–63.

Dwivedi, S. N., & Pandey, A. (1992). Ethnobotanical studies on wild and indigenous species of Vindhya Plateau, Herbaceous Flora. *J. Econ. Taxon. Bot. Addl. Ser., 10*, 143–150.

Ekka, A. (2011a). A Historical overview of ethnobotanical literature of Chhattisgarh (India): A graphic review and future directions. *Res. J. Sci. Tech., 3*(4), 220–224.

Ekka, A. (2011b). Medicinal plants used by local inhabitants in Jashpur District, Chhattis-garh, India. Proc. of Climate change and its effect on biodiversity, pp. 32–36.

Ekka, A. (2011c). Folklore claims of some medicinal plants used by tribal community of Chhattisgarh, India. *Res. J. Biol. 1*(1), 16–20.

Ekka, A. (2012a). Some rare and less known plants used by Gond and Korwa tribes in their healthcare from Chhattisgarh. In: Paul, V. I. (ed.), Biodiversity: Issues, Impacts, Remediation and Significances. VL Media Solutions, New Delhi, pp. 381–387.

Ekka, A. (2012b). Some traditional medicine for anti-fertility used by the tribals in Chhattisgarh, India. *Intern. J. Biol. Pharm. & Allied Sci. 1*(2), 108–112.

Ekka, A. (2013). Some rare plants used by Hill-Korwa in their healthcare from Chhattisgarh. *Intern. J. Life Sci. Biotech. & Pharma Res. 2*(1), 198–203.

Ekka, A. (2016). Traditional medicament used by Kamar tribes of Chhattisgarh, India. *Intern. J. Interdisciplinary Res. 2*(2), 508–515.

Ekka, A., & Ekka, N. S. (2013a). Traditional plants used for snakebite by Oraon tribe of Jashpur district, Chhattisgarh. *Intern. J. Advanced Res. Management and Social Sci., 2(6)*, 1–9.

Ekka, A., & Ekka, N. S. (2013b). Traditional health care in Birhor tribes of Chhattisgarh. *Intern. J. Interdisciplinary Res., 3(6),* 476–483.

Ekka, M. K., Prasad, H., & Tiwari, P. (2016). Traditional uses of medicinal plants practiced by the Oraon tribe of Jashpur district (C. G.), India. *IOSR J. Environ. Sci. Toxicol. & Food Tech., 1*(1), 60–64.

Ekka, N. R., & Dixit, V. K. (2007). Ethno-pharmacognostical studies of medicinal plants of Jashpur district (Chhattisgarh). *Intern. J. Green Pharm., 1*(1), 2–4.

Ghosh, A., Rahman, C. H., & Mandal, S. (2011). Observations on the ethnomedicinal uses of plants for gynecological, urinogenital and other related problems by the tribal people of Birbhum and Murshidabad districts, West Bengal, India. *J. Econ. Taxon. Bot., 35*(1), 17–26.

Gitika & Kumar, M. (2016). Ethnobotanical study of some medicinal plants of Haryana, India. *World J. Pharm. & Pharmaceut. Sci. 5*(8), 1717–1736.

Hembrum, P. P. (1991). Tribal medicine in Chotanagpur and Santal Parganas of Bihar, India. Ethnobotany, 3: 97–99.

Jadhav, D. (2006a). Ethno-medicinal survey of Maalgamdi in Ujjain district, Madhya Pradesh, India. *Ethnobotany, 18,* 157–159.

Jadhav, D. (2006b). Ethnomedicinal plants used by Bhil tribes of Bibdod, Madhya Pradesh. *Indian J. Trad. Knowl. 5*(2), 263–267.

Jadhav, D. (2006c). Plant sources used for the treatment of types of fevers by Bhil tribe of Ratlam District, Madhya Pradesh. *J. Econ. Taxon. Bot. 30*(4), 909–911.

Jadhav, D. (2007). Ethno medicinal plants used by Bhil tribes of Matrunda, District Ratlam, Madhya Pradesh, India. *Bull. Bot. Surv. India, 49,* 203–206.

Jadhav, D. (2008a). Amulets and other plant wearings believed to be contact therapy among tribals of Ratlam district (MP), India. *Ethnobotany, 20,* 144–146.

Jadhav, D. (2008b). Ethnomedicinal plants remedies for snake bite and scorpion sting among the Bhil tribes of Ratlam District (Madhya Pradesh). *J. Non-Timber Forest Prod., 15*(2), 127–128.

Jadhav, D. (2009). Ethnomedicinal plants used for dental troubles by the Tribes of Ratlam District, Madhya Pradesh. *J. Econ. Taxon. Bot. 33*(Supp.), 59–62.

Jadhav, D. (2010). Ethnomedicinal plants used as antipyretic agents among the Bhil tribes of Ratlam District, Madhya Pradesh. *Indian Forester, 136*(6), 843–846.

Jadhav, D. (2011). Ethnomedicinal survey of tribal inhabited localities of Ratlam district (MP) in the light of 'doctrine of signature'. *Ethnobotany, 23,* 121–124.

Jadhav, D., & Rawat, S. S. (2011). Ethnomedicinal plants used in the treatment of various ailments by Bhilala Tribe of Alirajpur District (MP). *J. Econ. Taxon. Bot. 35*(4), 654–657.

Jain, A., Katewa, S. S., Choudhary, B. L., & Galav, P. K. (2004). Folk Herbal medicines used in birth control and sexual diseases by tribals of Southern Rajasthan, India. *J. Ethnopharmacol., 90*(1), 171–177.

Jain, A., Katewa, S. S., & Galav, P. K. (2005a). Some phytotherapeutic claims by tribals of southern Rajasthan. *Indian J. Trad. Knowl., 4*(3), 291–297.

Jain, A., Katewa, S. S., Galav, P. K., & Nag, A. (2007). Unrecorded ethnomedicinal uses of biodiversity from Tadgarh-Raoli wildlife sanctuary, Rajasthan, India. *Acta Botanica Yunnanica, 29*(3), 337–344.

Jain, A., Katewa, S. S., Galav, P. K., & Sharma, P. (2005b). Medicinal plant diversity from the Sitamata wild life sanctuary. Chittorgarh district India. *J. Ethnopharmacol.*, 102(3), 543–557.

Jain, A. K. (1992). Ethnobotanical studies of Sahariya tribals of M. P. with special reference to medicinal plants. *J. Econ. Taxon. Bot., 16*, 227–232.

Jain, A. K., & Patole, S. N. (2001). Less known medicinal plants among some tribal and rural communities of Pachmarhi forest (M. P.). *Ethnobotany, 13*, 96–100.

Jain, A. K., Vairale, M. G., & Singh, R. (2010). Folklore claims on some medicinal plants used by Bheel tribe of Guna district Madhya Pradesh. *Indian J. Trad. Knowl., 9* (1), 105–107.

Jain, A. K., Wagh, V. V., & Kadel, C. (2011). Some ethnomedicinal plant species of Jhabua district, Madhya Pradesh. *Indian J. Trad. Knowl., 10*, 538- 540.

Jain, S. C., Jain, R., & Singh, R. (2009). Ethnobotanical survey of Sariska and Siliserh regions from Alwar district of Rajasthan, India. *Ethnobotanical Leaflets, 13*, 171–188.

Jain, S. K. (1963). Observation on Ethnobotany of tribals of M. P. *Vanyajati, 11*(4), 177–187.

Jain, S. K. (1965). Medicinal plant-lore of tribals of Bastar. *Bull. Bot. Surv. India, 8*, 237–251.

Jain, S. K. (1975). Ethnobotany of Central India Tribals. *J. Indian Bot. Soc. (Abstract), 1*(6), 63.

Jain, S. K. (1991). Observations on ethnobotany of the tribals of Central India. In: Jain, S. K. (ed.), Glimpses of Indian Ethnobotany. Oxford and IBH, New Delhi, pp. 192–198.

Jain, S. K., & Borthakur, S. K. (1980). Ethnobotany of the Mikirs of India. *Economic Bot., 34*(3), 264–272.

Jain, S. P. (1996). Ethno-medico-botanical survey of Chaibasa Singbhum district, Bihar. *J. Econ. Taxon. Bot. Addl. Series, 12*, 403–407.

Jain, S. P., Gupta, N., Saini, S., & Prakash, A. (2012). Ethno-medico-botanical survey of Chindwara district, Madhya Pradesh. In: Maiti, G. G., & Mukherjee, S. K. (eds.), Multidisciplinary approaches in Angiosperm Systematics. University of Kalyani, Kalyani, India. pp. 617–624.

Jain, S. P., & Puri, H. S. (1984). Ethnomedicinal plants of Jaunsar–Bawar hills, Uttar Pradesh, India. *J. Ethnopharmacol., 12*(2), 213–222.

Jha, A. K., & Yogesh, G. (2015). Some wild trees of Bihar and their ethnobotanical study. *IOSR J. Res. & Method in Education, 5*(6), 74–76.

Jha, R. R., & Verma, S. K. (1996). Ethnobotany of Sauria Paharias of Santhal Pargana, Bihar: 1. Medicinal plants. *Ethnobotany, 8*, 31–35.

Joshi, P. (1995). Ethnobotany of primitive tribes in Rajasthan. Printwell, Jaipur.

Kala, C. P. (2009). Aboriginal uses and management of ethnobotanical species in deciduous forests of Chhattisgarh state in India. *J. Ethnobiol. Ethnomed., 5*, 20 Doi: 10.1186/1746-4269-5-20.

Katewa, S. S. (2009). Indigenous people and forests: Perspectives of an ethnobotanical study from Rajasthan (India). In: Ramawat, K. G. (ed.), Herbal Drugs: Ethnomedicine to Modern Medicine. Springer, Berlin, pp. 33–56.

Katewa, S. S., Chaudhary, B. L., & Jain, A. (2004). Folk herbal medicines from tribal area of Rajasthan, India. *J. Ethnoharmacol., 92*(1), 41–46.

Katewa, S. S., Chaudhary, B. L., Jain, A., & Galav, P. K. (2003). Traditional uses of plant biodiversity from Aravalli hills of Rajasthan. *Indian J. Trad. Knowl., 2*(1), 27–39.

Katewa, S. S., & Galav, P. K. (2005). Traditional herbal medicines from Shekhawati region of Rajasthan. *Indian J. Trad. Knowl., 4*(3), 237–245.

Katewa, S. S., & Guria, B. D. (1997). Ethnomedicinal observation on certain wild plants from Southern Aravalili hills of Rajasthan. *Vasundharaa, 2,* 85–88.

Katewa, S. S., Guria, B. D., & Jain, A. (2001). Ethnomedicinal and obnoxious grasses of Rajasthan, India. *J. Ethnopharmacol., 76*(3), 293–297.

Katewa, S. S., & Sharma, R. (1998). Ethnomedicinal observation from certain watershed areas of Rajasthan. *Ethnobotany, 10,* 46–49.

Kaur, R. (2015). Ethnobotanical studies of some of the traditionally important medicinal plants of Punjab (India). *Intern. J. Current Res. Acad. Rev., 3*(5), 262–271.

Kaur, R., & Vashistha, B. D. (2014). Ethnobotanical studies on Karnal district, Haryana, India. *Intern. Res. J. Biol. Sci., 3*(8), 46–55.

Khan, S. S. (1993). Ethnomedicinal studies on plants of Bhopal district of M. P. PhD thesis, Burkatullah University, Bhopal, India.

Khanna, K. K. (2002). Unreported ethnomedicinal uses of plants from the tribal and rural folklore of Gonda district, Uttar Pradesh. *Ethnobotany, 14,* 52–56.

Khanna, K. K., & Kumar, R. (2000). Ethnomedicinal plants used by the Gujjar tribe of Saharanpur district, UP. *Ethnobotany, 12,* 17–22.

Kujur, M. K., & Ahirwar, R. K. (2015). Folklore claims on some ethnomedicinal plants used by various tribes of district Jashpur, Chhattisgarh, India. *Intern. J. Curr. Microbiol. & Appl. Sci., 4*(9), 860–867.

Kumar, A. (2004). Ethnobotanical aspects of pharmacological flora used by Tharu tribes in Terai belt of North-Eastern Uttar Pradesh. PhD thesis, Dr. R. M. L. Avadh University, Faizabad, India.

Kumar, A. (2012). Ethnomedicinal investigation of some plant used by the tribals of Bori Wildlife Sanctuary in Hoshangabad district (M. P.). *Ethnobotany, 24,* 123–125.

Kumar, A., & Pandey, A. K. (1990). Ethnobotanical notes on certain medicinal plants used by tribals of Bihar. *Ecology, Environment and Conservation. 4,* 65.

Kumar, A., Tewari, D. D., & Pandey, Y. N. (2003). Ethnophytotherapeutics among Tharus of Beerpur Semra forest range of Balrampur district, U. P. *J. Econ. Taxon. Bot., 27*(4), 839–844.

Kumar, M. S., Ankit, S., Gautam, D. N. S., & Kumar, S. N. (2015). Biodiversity and indigenous uses of medicinal plant in the Chandraprabha wildlife sanctuary, Chandauli district of Uttar Pradesh. *Intern. J. Biodiversity.* doi: 10.1155/2015./394307.

Kumar, S., & Chauhan, A. K. S. (2006). Less-known medicinal plant species in Keoladeo National Park, Bharatpur, Rajasthan. Ethnobotany, 18, 153–156.

Kumar, S., & Goel, A. K. (2008). A purview of odoriferous plants found around Itchagarh hills in Ormanjhi block of Jharkhand. *Ethnobotany, 20,* 135–137.

Kumar, S., & Verma, S. K. (2011). Ethnobotanical investigations on *Vernonia amygdalina* (Asteraceae) in Bihar. *Ethnobotany, 23,* 135–137.

Kumar, V. (1999). Some indigenous tools of Surguja district, Madhya Pradesh, India. *Ethno-botany, 11,* 135–137.

Kumar, V., & Akhtar, M. (2013). Ethnomedicinal solanaceous plants of Eastern Uttar Pradesh. *Indian J. Life Sci., 2*(2), 95–98.

Kumar, V., & Jain, S. K. (1998a). A contribution to Ethnobotany of Surguja district in Madhya Pradesh, India. *Ethnobotany, 10,* 89–96.

Kumar, V., & Jain, S. K. (1998b). Some less known ethnomedicine among the tribals of Sur-guja district in Madhya Pradesh. *J. Non-Timber Forest Products, 6,* 110–113.

Kumar, V., & Rao, R. R. (2002). Medicinal uses of *Elphantopus scaber* by some tribals in Central India. *J. Trop. Med. Plants, 3*(2), 219–221.

Kushwaha, K., Tripathi, R. K., & Dwivedi, S. N. (2013). Medicinal plants used in the treat-ment of some common diseases by the tribal and rural people in Korea district of Chhat-tisgarh. *Intern. J. Pharm. Life Sci., 4*(10), 3023–3027.

Lal, B. (1993). Ethnobotany of Baigas of Madhya Pradesh: a preliminary report. *Arunachal Forest News, 11*(10), 17–20.

Mahanta, A. K., & Pal, D. C. (2012). Some ethnomedicinal uses of forest weeds among San-tal tribes in Paschim Medinipur district, W. B., India. In: Maiti, G. G., & Mukherjee, S. K. (eds.), Multidisciplinary approaches in Angiosperm Systematics. University of Kalyani, Kalyani, India. pp. 652–654.

Maheshwari, J. K. (1996). Ethnobotanical documentation of primitive tribes of Madhya Pradesh. *J. Econ. Taxon. Bot. Addl. Series 12,* 206–213.

Maheshwari, J. K., Kalakoti, B. S., & Lal, B. (1986). Ethnomedicine of Bhil tribe of Jhabua District, Madhya Pradesh. *Ancient Sci. Life, 4,* 255–261.

Maheshwari, J. K., Painuli, R. M., & Dwivedi, R. P. (1990). Note on ethnobotany of the Oraon and Korwa tribes of Madhya Pradesh. In: Jain, S. K. (ed.), Contribution to eth-nobotany of India. Scientific Publishers, Jodhpur, pp. 75–90.

Maheshwari, J. K., & Singh, J. P. (1984). Contribution to the ethnobotany of Bhoxa tribe of Bijnor and Pauri, Garhwal district, U. P. *J. Econ. Taxon. Bot., 5,* 251–259.

Maheshwari, J. K., Singh, K. K., & Saha, S. (1981). The ethnobotany of Tharus of Kheri dis-trict, Uttar Pradesh. Economic Botany Information Service, NBRI, Lucknow, pp. 1–48.

Maheshwari, J. K., Singh, K. K., & Saha, S. (1986). Ethnobotany of tribals of Mirzapur dis-trict, Uttar Pradesh. Economic Botany, Information Service, NBRI, Lucknow, pp. 1–38

Mairh, A. K., Mishra, P. K., Kumar, J., & Mairh, A. (2010). Ethnomedicinal formulations used by traditional herbal practitioners of Ranchi, Jharkhand. *Indian J. Trad. Knowl., 9*(3), 467–470.

Maity, A., & Manna, C. K. (2000). Some ethnomedicines used by the Santhal of Baghmundi-Ajodhya hills regions of Purulia district, West Bengal in controlling fertility. *Ethno-botany, 12,* 72–76.

Maity, D., Dey, S. K., Chatterjee, S., & Maiti, G. G. (2015). Ethnobotany and environmen-tal management by the tribal communities of Patratu, Hazaribagh district, Jharkhand. *Exploratory Animal and Med. Res., 5*(1), 44–61.

Maliya, S. D. (2004). Some new or less known folk medicines of district Bahraich, Uttar Pradesh, India. *Ethnobotany, 16,* 113–115.

Maliya, S. D. (2007). Traditional fruit and leaf therapy among Tharus and indigenous people of district Bahraich, India. *Ethnobotany, 19,* 131–133.

Malviya, N., Jain, S., Gupta, V. B., & Vyas, S. (2011). Indigenous herbal remedies used by tribals of Madhya Pradesh for improving their sexual performance and problem associ-ated with sexuality. *Intern. J. Res. Ayurveda & Pharm. 2*(2), 399–402.

Mandal, S. K., & Mukherjee, A. (2008). Medicinal uses of plants as revealed from tribal communities in Purulia District, West Bengal. In: Patil, D. A. (ed.), Herbal Cures Traditional Approach. Aavishkar Publishers, Distributors, Jaipur, India. pp. 295–301.

Mao, A. A., Hynniewta, T. H., & Sanjappa, M. (2009). Plant wealth of Northeast India with reference to ethnobotany. *Indian J. Trad. Knowl., 8(*1), 96–113.

Marandi, R. R., & Britto, S. J. (2014). Ethnomedicinal plants used by the Oraon tribals of Latehar district of Jharkhand, India. *Asian J. Pharm. Res. 4*(3), 126–133.

Masih, V., Sahu, P. K., & Singh, M. (2013). Observation on ethno-medicinal herbs of Dantewada, Chhattisgarh, India. *Intern. J. Drug Discovery and Herbal Res. 3*, 644–648.

Meena, K. L. (2011). Ethnobotany of Garasia Tribe, Rajasthan, India. LAP LAMBERT Academic Publishing, Saarbrücken, Germany.

Meena, K. L., Ahir, P. C., & Dhaka, V. (2013). Ethnomedicinal survey of medicinal plants for sexual debility and birth control by tribals of Southern Rajasthan, India. *Photon, 118*, 238–244.

Meena, R., & Kumar, A. (2012). Ethnobotanical survey of medicinal plants from Baran District of Rajasthan, India. *Photon, 117*, 199–203.

Meena, S. L., Sharma, K. C., & Gopalan, R. (2003). Ethnomedicinal plants of Karauli district, Rajasthan. In: Singh, V., & Jain, A. P. (eds.), Ethnobotany and Medicinal plants of India and Nepal. Vol. 1. Scientific Publishers, Jodhpur, India. pp. 177–180.

Meena, K. L., & Yadav, B. L. (2007). Some ethnomedicinal plants of Rajasthan. In: Trivedi, P. C. (ed.), Ethnomedicinal Plants of India. Aavishkar Publishers and Distributors, Jaipur, India. pp. 33–44.

Meena, K. L., & Yadav, B. L. (2010a). Some ethnomedicinal plants of southern Rajasthan. *Indian J. Trad. Knowl. 9*(1), 169–172.

Meena, K. L., & Yadav, B. L. (2010b). Some traditional etnomedicinal plants of southern Rajasthan. *Indian J. Trad. Knowl., 9*(3), 471–474.

Menghani, E., Pareek, A., Negi, R. S., & Ojha, C. K. (2010). Antidiabetic potentials of various ethno-medicinal plants of Rajasthan. *Ethnomedicinal Leaflets, 14*, 578–583.

Mishra, L., Dixit, Y., & Singh, M. (2014). Studies on ethnomedicinal plants of Shekhawati region, Rajasthan, having hypoglycemic properties. *Indian J. Fundam. & Appl. Life Sci., 4*(2), 62–66.

Mishra, S. (2008). Ethnomedicinal studies of Korku tribe, with Gotra names derived from plant names, from Khandwa district in Madhya Pradesh. *Ethnobotany, 20*, 122–127.

Mitra, S., & Mukherjee, S. K. (2005a). Ethnobotanical uses of grasses by the tribals of West Dinajpur district, West Bengal. *Indian J. Trad. Knowl. 4*(4), 396–402.

Mitra, S., & Mukherjee, S. K. (2005b). Root and rhizome drugs used by the tribals of West Dinajpur in Bengal. *J. Trop. Med. Plants, 6(*2), 301–315.

Mitra, S., & Mukherjee, S. K. (2009). Some abortifacient plants used by the tribal people of West Bengal. *Nat. Prod. Radiance, 8*(2), 167–171.

Mitra, S., & Mukherjee, S. K. (2010). Ethnomedicinal uses of some wild plants of North Bengal plains for gastro-intestinal problems. *Indian J. Trad. Knowl., 9*(4), 705–712.

Mondal, S., & Rahman, C. H. (2012). Medicinal plants used by the tribal people of Birbhum district of West Bengal and Dumka district of Jharkhand in India. *Indian J. Trad. Knowl., 11*(4), 674–679.

Nath, V., & Khatri, P. K. (2010a). Documentation of traditional knowledge on ethnomedicinal information from traditional herbal healers in Jabalpur and Seoni district, Madhya Pradesh. *J. Trop. Forestry, 26*(3), 26–42.

Nath, V., & Khatri, P. K. (2010b). Traditional knowledge on ethno-medicinal uses prevailing in tribal pockets of Chindwara and Betula districts, Madhya Pradesh, India. *African J. Pharm. & Pharmacol, 4*(9), 662–670.

Nath, V., & Khatri, P. K. (2011). Traditional knowledge on ethno-medicinal uses prevailing in tribal pockets of Harda and Raisen districts, Madhya Pradesh. *Indian Forester, 137*(9), 1071–1076.

Nigam, G., & Kumar, V. (2005). Some ethno–medicinal plants of Jhansi District. *Flora and Fauna, 11*(1), 91–93.

Oommachan, M., Bajaj, A., & Masih, S. K. (1990). Ethnobotanical observations at Pachmarhi. *J. Trop. Forestry, 6*(2), 157–161.

Pandey, B., Pandey, P., & Paikara, D. (2015). Some important medicinal plants used by tribal people of Chhattisgarh. *Indian J. Life Sci. 5*(1), 67–69.

Pandey, H. P., & Verma, B. K. (2002). Plants in oral healthcare among the aborigins of Gonda and Balrampur Regions, U. P., India. *Ethnobotany 14*, 81–86.

Pandey, R. S., & Kumar, A. (2006). An ethnobotanical study in the Vindhyan region, Uttar Pradesh. *Indian J. Forestry, 29*(4), 389–394.

Panghal, M., Arya, V., Yadav, S., Kumar, S., & Yadav, J. P. (2010). Indigenous knowledge of medicinal plants used by Saperas community of Khetawas, Jhajjar District, Haryana, India. *J. Ethnobiol. Ethnomed., 6*, 4. doi: 10.1186/1746-4269-6-4.

Pathan, F. M., Shah, K. W., & Sanghi, S. B. (2012). Medicinal plants used by tribals of Rajgarh district of Madhya Pradesh for jaundice. *RGC Res. Journ., 1*(3), 1–5.

Prachi, Chauhan, N., Kumar, D., & Kasana, M. S. (2009). Medicinal plants of Muzaffarnagar district used in treatment of urinary tract and kidney stones. *Indian J. Trad. Knowl., 8*(2), 191–195.

Prakash, A. (2011). Uses of some threatened and potential ethnomedicinal plants among the tribals of Uttar Pradesh and Uttarakhand in India. In: Proc. National conference on Forest Biodiversity: Earth Living resource. Uttar Pradesh Biodiversity Board, pp. 93–99.

Prakash, A., & Singh, K. K. (2000). Observations on some high valued ethnomedicinal plants among the tribals of Uttar Pradesh. *J. Med. & Aromatic Plant Sci., 23*, 519–522.

Prasad, R., & Pandey, R. K. (1993). Ethno-medico-botanical studies on indigenous medicinal plants of Lamni and Achanakmar forests of Bilaspur district of M. P. *J. Trop. Forestry, 9*, 27–40.

Prasad, R., Pandey, R. K., & Bhattacharya, P. (1990). Socio-economic and ethno-medico–botanical studies of Patalkot region. A case study of Bhariya tribes. In: Proc. National Seminar on Medicinal and Aromatic plants. SFRI, Jabalpur: pp. 46–59.

Pushpangadan, P. (2002). Biodiversity and emerging benefit sharing arrangements- Challanges and opportunities for India. *Proc. National Acad. Sci., India, Section B: Biol. Sci., 68*(3), 297–314.

Quamar, M. F.& Bera, S. K. (2014). Ethno-medico-botanical studies of plant resources of Hoshangabad district of Madhya Pradesh, India. Retrospect and Prospect. *J. Plant Sci. Res., 1*(1), 101–105.

Rahaman, C. H. (2015). Tribal medicine: Traditional therapeutic uses of plants among the tribes of Birbhum district, West Bengal. In: Hussain, S. E., & Saha, M. (eds.), India's indigenous medical systems- A cross disciplinary approach. Primus Books, Delhi, India, pp. 267–298.

Rahaman, C. H., & Karmakar, S. (2015). Ethnomedicine of Santal tribe living around Susunia hill of Bankura district, West Bengal, India: The quantitative approach. *J. Appl. Pharmaceut. Sci., 5*(2), 127–136.

Rahaman, C. H., & Pradhan, B. (2011). A study on the ethnomedicinal uses of plants by the tribal people of Birbhum district, West Bengal. *J. Econ. Taxon. Bot., 35*(3), 529–534.

Rahul, J. (2013a). An ethnobotanical study of medicinal plants in Taidol village, District Jhansi, region of Bundelkhand, Uttar Pradesh, India. *J. Med. Plants Studies, 1*(5), 59–71.

Rahul, J. (2013b). Ethnobotanical study of medicinal shrubs used by people in Lakhmanpura region of Bundelkhand, Uttar Pradesh, India. *Intern. J. Sci. & Nature, 4*(2), 362–364.

Rahul, J. (2013c). An ethnomedicinal survey of Orchha Wildlife Sanctuary region of Tikamgarh district, Madhya Pradesh, *India. J. Bot. Res., 4*(1), 31–34.

Rai, B. K., Ayachi, S. S., & Rai, A. (1996). A note on ethno-medicines from Central India. *J. Econ. Taxon. Bot. Addl. Series, 12,* 186–191.

Rai, M. K. (1987). Ethno-medical studies of Patalkot and Tamia (Distt. Chindwaraa), M. P. – Plants used as tonic. *Ancient Sci. Life, 7*(2), 119–121.

Rai, M. K. (1988). Ethnomedicinal survey of Patalkot and Tamiya (District Chhindwara) – I: Plants used against pores and skin ailments and liver problems. *J. Econ. Taxon. Bot., 12*(2), 337–339.

Rai, M. K. (1993). A note on ethnomedicinal studies of Chhindwara plants used in snake and scorpion bite. *Aryavaidyan, 7*(1), 48–50.

Rai, M. K., & Nonhare, B. P. (1992). Ethnomedicinal studies of Bichhua (Distt. Chhindwara) Madhya Pradesh. *Indian Medicine, 4*(3), 7–10.

Rai, M. K., & Ojha, G. C. (1989). Ethnomedicinal studies of Chhindwara district, Madhya Pradesh. Plants used in stomach disorders. *Indian Medicine, 1*(2), 1–5.

Rai, M. K., & Pandey, A. K. (1997). Folk medicines of Gond tribe of Seoni District, M.P. India. *J. Non-Timber Forest Prod., 4,* 61–69.

Rai, M. K., Pandey, A. K., & Acharya, D. (2000). Ethno-medicinal plants used by Gond tribe of Bhanadehi district, Chhindwara Madhya Pradesh. *J. Non-Timber Forest Prod., 7,* 237–241.

Rai, R. (2006). Studies on indigenous herbal remedies in cure of fever by tribals of Madhya Pradesh. In: Proceeding of National Symposium on Tribal Health, October 19 20, organized by Regional medical research Center for tribals, Jabalpur, pp. 177–182.

Rai, R., & Nath, R. (2005a). Use of medicinal plants by traditional herbal healers in Central India. *Indian Forester, 131*(3), 463–468.

Rai, R., & Nath, V. (2005b). Some lesser known oral herbal contraceptives in folk claims as anti-fertility and fertility induced plants in Bastar region of Chhattisgarh. *J. Natural Remedies, 5*(2), 153–159.

Rai, R., Nath, V., & Shukla, P. K. (2002). Ethno-medicinal studies on Bhariya Tribes in Satpura plateau of Madhya Pradesh. *New Agriculturist, 13,* 109–114.

Rai, R., Nath, V., & Shukla, P. K. (2003). Ethnobiogy of Hill Korwa Tribes Chhattisgarh. *J. Trop. Forestry, 19,* 35–46.

Rai, R., Nath, V., & Shukla, P. K. (2004a). Ethnobotanical studies in Patalkot Valley in Chhindawara district of Madhya Pradesh. *J. Trop. Forestry, 20*(2) 38–50.

Rai, R., Nath, V., & Shukla, P. K. (2004b). Characteristics and ethnobotanical studies on primitive tribes of Madhya Pradesh. In: Govil, J. N. (ed.), Recent progress in medicinal plants. Chapter Ethno- medicine and Pharmacognosy. Research Book Centre, New Delhi, India, 8 (37), 543– 552.

Rai, R., Nath, V., & Shukla, P. K. (2004c). Ethnobiological studies on Bhariya tribes of Madhya Pradesh. *J. Trop. Forestry, 20*(1), 150–160.

Ranjan, V. (1996). Some Ethnomedicinal plants of Lalitpur District, Uttar Pradesh, India. In: Jain, S. K. (ed.), Ethnobiology in Human Welfare. Deep Publications, New Delhi, pp. 149–150.

Sahoo, A. K., & Mudgal, V. (1993). Ethnobotany of South Chotanagpur (Bihar). *Nelumbo, 35,* 40–59.

Sahu, P. K., Masih, V., Gupta, S., Sen, D., & Tiwari, A. (2014). Ethnomedicinal plants used in the healthcare systems of tribes of Dantewada, Chhattisgarh India. *American J. Plant Sci., 5,* 1632–1643.

Sahu, T. R. (1982). An ethnobotanical study of Madhya Pradesh. 1: Plants used against various disorders among tribal women. *Ancient Sci. Life, 1*(3), 178–181.

Sahu, T. R., Sahu, I., & Dakwale, R. N. (1983). Further contributions towards the ethnobotany of Madhya Pradesh 2: Plants used against diarrhea and dysentery. *Ancient Sci. Life, 2*(3), 169–170.

Saini, D. C. (2008). Traditional uses of pteridophytes among Baiga tribes of Amarkantak, Anuppur district, M. P. *Ethnobotany, 20,* 65–69.

Saini, V. K. (1996). Plants in the welfare of tribal women and children in certain areas of Central India. In: Jain, S. K. (ed.), Ethnobiology in Human Welfare. Deep Publications, New Delhi, India. pp. 140–144.

Sandya, G. S., & Sandya, K. (2013). Ethnomedicinal plants used by Baiga tribes in Mandla district, Madhya Pradesh (India). *Intern. J. Scientific Res., 4*(2), 2017–2020.

Saxena, H. O. (1986). Observation on the ethnobotany of Madhya Pradesh. *Bull. Bot. Surv. India, 28,* 149–156.

Saxena, H. O., & Dutta, P. K. (1975). Studies on the ethnobotany of Orissa. *Bull. Bot. Surv. India, 17,* 124–131.

Saxena, A. P., & Vyas, K. M. (1981). Ethnobotanical records on infectious diseases from tribals of Banda district. *J. Econ. Taxon. Bot. 2,* 191–194.

Seema & Kumar, A. (2005). Study of some traditional medicinal plants used by tribal peoples of Rajasthan in human ailments. *Intern. J. Mendel, 22,* 47–48.

Shah, N. C., & Singh, S. C. (1990). Hitherto unreported phytotherapeutical makes use of from tribal pockets of Madhya Pradesh (India). *Ethnobotany, 2,* 91–95.

Sharma, H., & Kumar, A. (2011). Ethnobotanical studies on medicinal plants of Rajasthan (India): A review. *J. Med. Plants Res., 5,* 1107–1112.

Sharma, L., & Khandelwal, S. (2010). Traditional uses of plants as cooling agents by the Tribal and traditional communities of Dang region in Rajasthan, India. *Ethnobotanical Leaflets, 14,* 218–224.

Sharma, L. K., & Kumar, A. (2006). Ethnobotanical and phytochemical studies on some selected medicinal plants of Rajasthan. *Indian J. Environ. Sci., 10,* 51–53.

Sharma, L. K., & Kumar, A. (2007). Traditional medicinal practices of Rajasthan. *Indian J. Trad. Knowl., 6,* 531–533.

Sharma M., & Kumar, A. (2011). Ethnobotanical and pharmacognostical studies of some medicinal plants: Tribal medicines for health care and improving quality of life. Lambert Academic Publishers. Germany.

Sharma, S., & Thokchom, R. (2014). A review on endangered medicinal plants of India and their conservation. *J. Crop & Weed, 10*(2), 205–218.

Shekhawat, D., & Batra, A. (2006). Ethnobotany of some household remedies used against animal and insect bit in Bundi district, Rajasthan. *Ethnobotany, 18,* 131–134.

Shrivastava, R. K. (1985). Herbal remedies used by the Bhils of Madhya Pradesh. Oriental Med, Kyoto, Japan. pp. 389–395.

Shrivastava, S., & Kanungo, V. K. (2013). Ethnobotanical survey of Surguja district with special reference to plants used by Uraon tribe in treatment of Diabetes. *Intern. J. Herbal Med., 1*(3), 127–130.

Shukla, A. N., Srivastava, S., & Rawat, A. K. S. (2010). An ethnobotanical study of medicinal plants of Rewa district, Madhya Pradesh. *Indian J. Trad. Knowl., 9*(1), 191–202.

Sikarwar, R. L. S., & Kaushik, J. P. (1992). Some less known medicinal uses of trees among the Sahariyas of Morena district, MP, India. *Ethnobotany, 4,* 71–74.

Sikarwar, R. L. S., & Maheshwari, J. K. (1992). Some unrecorded ethnobotaanical plants from Amarkantak plateau of M. P. *Bull. Tribal Res. Dev. Inst. 20,* 19–22.

Singh, A. (2016). An ethnobotanical study of medicinal plants in Bhiwani district of Haryana, India. *J. Med. Plants Studies 4*(2), 212–215.

Singh, A., & Dubey, N. K. (2012). An ethnobotanical study of medicinal plants in Sonebhadra district of Uttar Pradesh, India with reference to their infection by foliar fungi. *J. Med. Plants Res., 6*(14), 2727–2746.

Singh, A., Singh, G. S., & Singh, P. K. (2012). Medico-ethnobotanical inventory of Reukoot forest division of district Sonbhadra, Uttar Pradesh, India. *Indian J. Nat. Prod. Resources, 3*(3), 448–457.

Singh, A., Singh, P., Singh, G., & Pandey, A. K. (2014). Plants used in primary health practices in Vindhya region of Eastern Uttar Pradesh, India. *Intern. J. Herbal Med., 2*(2), 31–37.

Singh, A., & Singh, P. K. (2009). An ethnobotanical study of medicinal plants in Chandauli district of Uttar Pradesh, India. *J. Ethnopharmacol., 121,* 324–329.

Singh, A., Tak, H. S., Singh, L., Kumar, A., & Kumar, S. (2015). Ethnobotanical survey of common medicinal plants in Bhiwani, Haryana, India. *World J. Pharmaceut. Sci., 3*(3), 492–499.

Singh, A. K., Raghubanshi, A. S., & Singh, J. S. (2002). Medical ethnobotany of the tribals of Sonaghati of Sonbhadra district, Uttar Pradesh, India. *J. Ethnopharmacol. 81*(1), 31–41.

Singh, J., Singh, N., Satpal, Sharma, K., & Singh, B. (2016). Observations on plant formulations for pediatric use in Haryana, India. *J. Global Biosciences 5*(2), 3656–3664.

Singh, K. K., Kalakoti, B. S., & Prakash, A. (1994). Traditional phytotherapy in the health care of Gond tribals of Sonbhadra district, Uttar Pradesh. *J. Bombay Nat. Hist. Soc. 91*(3), 386–390.

Singh, K. K., & Prakash, A. (1994). Indigenous phytotherapy among the Gond tribe of Uttar Prades, India. *Ethnobotany, 6,* 37–41.

Singh, K. K., & Prakash, A. (1996). Observations on ethnobotany of Kol tribe of Varanasi district, Uttar Pradesh, India. *J. Econ. Taxon. Bot. Addl. Series, 12,* 133–137.

Singh, K. K., & Maheshwari, J. K. (1985). Forest in the life and economy of the tribals of Varanasi district, U.P., *J. Econ. Taxon. Bot., 6,* 109–116.

Singh, K. K., & Maheshwari, J. K. (1989). Traditional herbal remedies among the Tharus of Bahraich district, U.P., India. *Ethnobotany, 1,* 51–56.

Singh, K. K., Saha, S., & Maheshwari, J. K. (1998). Ethnobotanical use of some fern amongst the tribal area of Uttar Pradesh. *Indian fern Journal, 6,* 66–67.

Singh, P. K., Kumar, V., Tiwari, R. K., Sharma, A., Rao, C. V., & Singh. R. H. (2010). Medico-ethnobotany of 'Chatara' block of district of Sonebhadra, Uttar Pradesh, India. *Adv. Biol. Res., 4*(1), 65–80.

Singh, P. K., Singh, R. H., & Kumar, V. (2007). Medicinal Plants used by Gond tribe of 'Dudhi' District Sonebhadra, Uttar Pradesh, India. *Flora and Fauna, 13*(1), 50–54.

Singh, R. K. (1996). Ethnobotanical observations on Sonbhadra district of southern Uttar Pradesh, India: Utilization and conservation. In: Jain, S. K. (ed.), Ethnobiology in Human Welfare. Deep Publications, New Delhi, India, pp. 145–148.

Singh, S., Dixit, R. D., & Sahu, T. R. (2005). Ethnomedicinal uses of Pteridophytes of Amarkantak, Madhya Pradesh. *Indian J. Trad. Knowl., 4*(4), 392–395.

Singh, U., & Bharti, A. K. (2015). Ethnobotanical study of plants of Raigarh area, Chhattisgarh, India. *Intern. Res. J. Biol. Sci., 4*(6), 36–43.

Singh, U., & Narain, S. (2009). Ethno-botanical wealth of Mirzapur district, U. P. *Indian Forester, 135*(1), 185–197.

Singh, V., & Pandey, R. P. (1980). Medicinal plant lore of the tribals of East Rajasthan. *J. Econ. Taxon. Bot. 1,* 137–147.

Singh, V., & Pandey, R. P. (1998). Ethnobotany of Rajasthan, India. Scientific Publishers, Jodhpur.

Singh, V. P., & Jadhav, D. (2011). Ethnobotany of Bhil tribe– A case study among the Bhils of Ratlam district (Madhya Pradesh). Scientific Publishers, Jodhpur.

Soni, V., Prakash, A., & Nema, M. (2012). Study on ethno-medico-botany of some plans of Dindori district of Madhya Pradesh, India. *Intern. J. Pharm. Life Sci., 3*(8), 1926–1929.

Sriwastwa, D. K., & Verma, S. K. (1981). An ethnobtanical study of Santal Parganas, Bihar. *Indian Forester, 107*(1), 30–41.

Tarafder, C. R., & Raichoudhuri, H. N. (1981). Less known medicinal uses of plants among the tribals of Hazaribagh district, Bihar. In Jain, S. K. (ed.), Glimpses of Indian Ethnobotany, Oxford and IBH, New Delhi. pp. 208–217.

Tarafdar, C. R., & Raichaudhuri, H. N. (1990). Less known medicinal uses of plants among the tribals of Hazaribagh district of Bihar. In: Jain, S. K. (ed.), Contributions to Indian Ethnobotany of India. Scientific Publishers, Jodhpur. pp. 91–100.

Thakur, A., Naqvi, S. M. A., Aske, D. K., & Sainkhediya, J. (2014). Study of some ethnomedicinal plants used by tribals of Alirajpur, Madhya Pradesh, India. *Res. J. Agri. & Forestry Sci., 2*(4), 9–12.

Tirkey, A. (2004). Some ethnobotanical plant species of Chhattisgarh State. *Ethnobotany,* 16: 118–124.

Tirkey, A. (2006). Some ethnomedicinal plants of family Fabaceae of Chhattisgarh state. *Indian J. Trad. Knowl. 5*(4), 551–553.

Tirkey, A. (2007). Notable ethnomedicinally important plant species of Chhattisgarh. In: Trivedi, P. C. (ed.), Ethnomedicinal plants of India. Aavishkar Publisher and Distributors, Jaipur, Rajasthan, India. pp. 188–233.

Tirkey, A. (2008). Ethnobotany in Chhattisgarh (India): A Graphic review and future Directions. In: Patil, D. A. (ed.), Herbal Cures: Traditional Approaches. Aavishkar Publisher and Distributors, Jaipur, India, India. pp. 340–347.

Tirkey, A. (2009). Some rare and less known ethnomedicinal plants used by tribal community from Chhattisgarh. In: Trivedi, P. C. (ed.), Indigenous ethnomedicinal plants. Pointer Publishers, Jaipur, India, pp. 185–189.

Tirkey, A., Kumar, V., Sikarwar, R. L. S., & Jain, S. K. (2006). Ethnobotanical research in Chhattisgarh– A conspectus. *Ethnobotany, 18,* 67–76

Tiwari, A. P., Joshi, B., & Ansari, A. A. (2012). Medicinal uses of some weeds of Uttar Pradesh, India. *Researcher, 4*(7), 67–72.

Tiwari, D. K., & Yadav, A. (2003). Ethnobotanical investigation of some medicinal plants availed by Gond tribe of Naoradehi Wild Life Sanctuary, Madhya Pradesh. *Anthropologist, 5*(3), 201–202.

Tomar, J. B., Bishnoi, S. K., & Saini, K. K. (2012). Healing the tribal way: Ethnomedicinal formulations used by the tribes of Jharkhand, India. *Intern. J. Med. Aromatic Plants, 2*(1), 97–105.

Upadhyay, B., Parveen, Dhaker, A. K., & Kumar, A. (2010). Ethnomedicinal and ethnopharmaco-statistical studies of Eastern Rajasthan, India. *J. Ethnopharmacol. 129,* 64–86.

Upadhyay, O. P., Kumar, K., & Tiwar, R. K. (1998). Ethnobotanical study of skin treatment uses of medicinal plants of Bihar. *Pharmaceut. Biol., 36*(3), 167–172.

Upadhyay, P. B., Roy, S., & Kumar, A. (2007). Traditional uses of medicinal plants among thee rural communities of Churu district in the Thar desert, India. *J. Ethnopharmacol. 113,* 387–399.

Upadhyay, R., & Singh, J. (2005). Ethnomedicinal uses of plants from Tikri forest of Gonda District (U.P.). *Ethnobotany, 17,* 167–170.

Varghese, S. V. D. E. (1996). Applied ethnobotany- A case study on the Kharias of Central India. Deep Publications, New Delhi, India.

Varma, S. K., & Pandey, A. K. (1990). Ethnobotanical notes on certain medicinal plants used by tribals of Bihar. *J. Econ. Taxon. Bot. 14*(2), 329–333.

Varma, S. K., Pandey, A. K., & Sriwastawa, D. K. (1999). Ethnobotany of Santhal Pargana. Narendra Publishing House, New Delhi, India.

Verma, P., Khan, A. A., & Singh, K. K. (1995). Traditional phytotherapy among the Baiga tribe of Shahdol district of Madhya Pradesh, India. *Ethnobotany, 7,* 69–73.

Wagh, V. V., & Jain, A. K. (2010a). Ethnomedicinal observations among the Bheel and Bhilala tribe of Jhabua District, Madhya Pradesh, India. *Ethnobotanical Leaflets 14,* 715–720.

Wagh, V. V., & Jain, A. K. (2010b). Traditional herbal remedies among Bheel and Bhilala tribes of Jhabua district, Madhya Pradesh. *Intern. J. Biol. Tech. 1*(2), 20–24.

Wagh, V. V., & Jain, A. K. (2014). Herbal remedies used by the tribal people of Jhabua district, Madhya Pradesh for the treatment of joint diseases. *Intern. J. Phytother., 4*(2), 63–66.

Yadav, S. S., & Bhandoria, M. S. (2013). Ethnobotanical exploration in Mahendergarh district of Haryana (India). *J. Med. Plants Res. 7*(18), 1263–1271.

CHAPTER 6

AN OVERVIEW OF ETHNOVETERINARY MEDICINES OF THE INDO-GANGETIC REGION

R. L. S. SIKARWAR

Arogyadham (J.R.D. Tata Foundation for Research in Ayurveda and Yoga Sciences), Deendayal Research Institute, Chitrakoot, Dist. Satna (M.P.) – 485334, India, E-mail: rlssikarwar@rediffmail.com, sikarwarrls@gmail.com

CONTENTS

ABSTRACT

The Indo-Gangetic region consists mainly with Madhya Pradesh, Chhattisgarh, Rajasthan, Haryana, Punjab Plains, Delhi, Uttar Pradesh, Jharkhand, Bihar and Plains of West Bengal of Indian States. The review of literature on ethnoveterinary medicines reveals that there are about 449 species of plants belonging to 325 genera and 108 families employed by the tribal and rural communities of Indo-Gangetic region for the treatment of more than 200 type's of ailments, diseases and disorders of their pets/ domesticated animals. The information regarding the uses of plants for veterinary purpose is transmitted from one generation to another orally. This indigenous knowledge and practice of tribal and rural people is based on locally available herbs and are very effective to cure diseases and disorders of animals. Therefore, all these herbs should be screened scientifically in order to investigate newer sources of ethnoveterinary drugs and medicines of herbal origin.

6.1 INTRODUCTION

Millions of people around the world have an intimate relationship with their livestock. Many people depend on their livestock. Animals provide them with food, clothing, labor, fertilizers and cash/money and act as a store of wealth and a medium of exchange. Animals are a vital part of culture and in many societies and are regarded as equal as humans. To keep animals healthy, traditional healing practices have been applied for centuries and have been passed down orally from generation to generation. Before the introduction of western medicine, all livestock keepers relied on these traditional practices only. According to the WHO reference, at the moment, at least 80% of the people in developing countries depend largely on these practices for the control and treatment of various diseases that affect both animals and humans. These traditional healing practices of animal's health are called 'ethnoveterinary medicine'.

Ethnoveterinary medicine consists of local people's knowledge, skills, methods, practices and beliefs pertaining to animal health and production (MacCorkle, 1986). Ethnoveterinary medicine often provides cheaper options than comparable western drugs and the products are locally available and easily accessible. In the face of these and other factors there is increasing interest in the field of ethnoveterinary research and development (Masika et al., 2000). Livestock raisers and healers throughout the world use traditional veterinary practices to prevent and treat common animal ailments and diseases.

In India Ethnoveterinary practices were in vogue since ancient times. In ancient India the Vedic literature particularly Atharvaveda is a repository of traditional medicine including prescriptions for the treatment of animal diseases. Scriptures such as Skand Purana, Devi Purana, Matsya Purana, Agni purana, Garuda purana and Lingapurana and books written by Charaka, Susruta and Shalihotra documented treatment of animal diseases using medicinal plants. The history of traditional veterinary science dates back to the period of Mahabharata. During the battle of Mahabharata thousands of animals were wounded/killed and also suffered from various diseases which were then treated with medicinal plants. Prince Nakul and Prince Sahadev were the physicians of horses and cows, respectively. Indian medicinal treatises like Charaka, Susruta and Harita Samhits contain references of care of animals. The greatest and most revered teacher of veterinary science was Shalihtra, the father of veterinary science in India (Raikwar and Maurya, 2015).

The Indo-Gangetic region consists mainly with Madhya Pradesh, Chhattisgarh, Rajasthan, Haryana, Punjab Plains, Delhi, Uttar Pradesh, Jharkhand, Bihar and Plains of West Bengal of Indian states and occupied the core zone of the country. The area is very rich in cultural as well as biological diversity. The forest of Indo-Gangetic region consists mainly of Tropical semi-evergreen forests, Tropical moist deciduous forests, Tropical dry deciduous forests and Tropical thorn forests. Indo-Gangetic region enjoys the widespread Indian monsoon climate with maximum rain falling between the ends of June to September. The average annual rainfall varies from 700 mm to 2000 mm. The maximum temperature recorded highest as 47°C in May and June and minimum sometimes reaching as low as 1°C in December and January.

The various tribal communities inhabit in several states of Indo-Gangetic region like Madhya Pradesh, Chhattisgarh, Jharkhand, West Bengal, Bihar, southern parts of Uttar Pradesh and Rajasthan. The dominant tribes are Abujhmaria, Baiga, Bhil, Bhilala, Gond, Kol, Korku, Korwa, Oraon, Sahariya Munda, Oraon, Santhal, Meena, Garasia, Dhodia, etc. They inhabit in and around the forest areas and mainly dependent on forest resources for fulfillment of their routine requirements such as for food, medicine, fodder, fiber, hunting and fishing, household and agricultural equipments, etc. Although they mainly depend upon forest resources for their livelihood but apart from local and rural people they also raise domestic animals such as cows, buffaloes, oxen, goats, sheep, hen, dogs, pigs, etc. for milk, agriculture and commercial purposes.

As the modern medicine is either not available or not affordable for the poor tribals therefore, they use traditional veterinary practices to prevent and treat common animal ailments and diseases with the help of locally available medicinal herbs.

The ethnoveterinary medicine is easily available, cheap and effective, cures many diseases and causes no side effects. Besides, no complicated technology is required for its preparation and application. The tribal people have been using this traditional system for a very long time. They have acquired this valuable knowledge from centuries of experience and trial and error methods and this knowledge has passed down from one generation to next generation.

But it has been observed that the younger generation of the tribal communities neither has aptitude to acquire the knowledge nor the experience to practices of ethnoveterinary medicine effectively. Therefore, it is necessary that before this valuable traditional knowledge is lost forever it must be properly documented from old and experienced tribal medicine men.

6.2   REVIEW OF LITERATURE

A survey of literature reveals that hundreds of research papers have so far been published on different aspects of ethnomedicine from different districts of Central India especially on human medicines. But, in Ethnoveterinary medicine, very little work has so far been done. Sikarwar et al. (1994) explored different districts of Central India and collected 31 species under 28 genera and 26 families as used by aboriginals to treat common ailments of livestock. Sikarwar (1996) carried out Ethnoveterinary study in Morena district of Madhya Pradesh and illustrated 35 plant species which have been used by tribal and rural communities. Sikarwar and Kumar (2005) again conducted ethnoveterinary study in Central Indian states (M.P. & C.G.) and listed 35 plant species belonging to 22 families as used for the treatment of ailments and diseases of domestic animals. Kadel and Jain (2006) presented 43 species of flowering plants used in 20 animal diseases by the ethnic tribes of Jhabua district. Shukla et al. (2007) gave an account of 17 wild species of plants belonging to 17 genera and 14 families, as practiced by tribal communities of Achanakmar-Amarkantak Biosphere Reserves of Madhya Pradesh and Chhattisgarh. Jadhav (2009) explored 21 plant species used as veterinary medicine in Ratlam district of Madhya Pradesh. Satya and Solanki (2009) reported 72 plant species which are used as veterinary medicine from West Nimar of Madhya Pradesh. Dwivedi et al. (2009) gave an account of 32-plant species ethnoveterinary importance form Vindhya, Malva and Nimar regions. Tripathi and Singh (2010) presented 27 plant species which have been used by the villagers of Chitrakoot region of Satna district for the treatment of their domestic animals. Nigam and Sharma (2010) described 46 ethnoveterinary plants from Jhansi district of U.P. which is surrounded three sides from M.P. Dar et al. (2011) also illustrated ethnoveterinary value of some plant species utilized by rural people of Jhansi district, Bundelkhand region. Chouhan and Ray (2012) carried out ethnoveterinary survey of Alirajpur district of M.P. and presented 30 plant species under 30 genera and 24 families as practiced by Bhil, Bhilala, Barela and Pateleya tribes. Shrivastava et al. (2012) explored Gwalior district and observed 23 plant species which have been used in 21 diseases of goats (*Cypris communis*). Jatav et al. (2014) conducted study in Shivpuri district and reported 32 plant species belonging to 30 genera and 19 families as ethnoveterinary medicines. Verma (2014) carried out study on ethnoveterinary medicinal plants in Tikamgarh district of Bundelkhand, Central India and a total of 41 plant species in 39 genera and 25 families used for the treatment of 36 diseases. Garima and Richhariya (2015) reported 23 plants species of 20 families as ethnoveterinary medicine used by tribal and rural communities of Chitrakoot, Satna district of Madhya Pradesh. Sanghi (2014) collected 17 species of plants belonging to 15 genera and 11 families from Tendukheda of Narsinghpur district of Madhya Pradesh. Patil and Deshmukh (2015) gathered from ethnic people specially Gond and Korku in the tribal pockets of Betul district of Madhya Pradesh and a total 25 species belonging to 25 genera, representing 19 families as employed for 14 types of animal diseases.

Ekka (2015) carried out ethnoveterinary study on Oraon tribals on northeast, Chhattisgarh and recorded 37 species of ethnoveterinary plants belonging to 30 families and 35 genera.

Jha et al. (1991) described 23 plant species under 23 genera and 18 families utilized as veterinary medicine in Darbhanga district of North Bihar. Mishra et al. (1996) also published ethnoveterinary medicinal plants from the same district. Singh (1987) enumerated 13 plant species as veterinary medicines from Chhotanagpur. 20 species under 19 genera and 16 families are used as veterinary by tribals of Bengal, Orissa and Bihar as reported by Pal (1980).

Some plants used as veterinary medicines by Bhils of Rajasthan was given by Sebastian and Bhandari (1984). Sebastian (1984) explored forest area of Rajasthan and reported that 27 plant species are used in ethnoveterinary medicine and galactagogue. Katewa and Chaudhary (2000) made an eth-noveterinary survey of plants of Rajasamand district in Rajasthan. Deora et al. (2004) reported ethnoveterinary uses of 20 plants from northwest part of Udaipur district. Kumar et al. (2004) recorded ethnoveterinary medicines from 128 villages of 4 districts of Indian Arid zone relates to 39 species belonging to 25 families. Galav et al. (2007) explored Mount Abu region of Rajasthan and described 44 formulations used to treat 20 diseases of domestic animals from 33 plant species belonging to 26 families. Nag et al. (2009) reported 24 plants species of veterinary importance under 30 meth-ods of treating 16 types of veterinary health problems from Kotda tehsil of Udaipur district of Rajasthan. Ahir et al. (2012) described some tradi-tional ethnoveterinary medicinal plants of Kumbhalgarh wildlife sanctuary in Rajasthan. A study of Takhar (2004) reveals ethnoveterinary medicinal uses of 37 plant species belonging to 25 families of southern Rajasthan. Galav et al. (2013) also carried out ethnoveterinary survey of tribal area of Rajasthan, which yielded veterinary uses of 59 plant species belonging to 55 genera of Angiosperms. Meena (2014) carried out ethnoveterinary study in Pratapgarh district of Rajasthan and listed 24 plant species under 24 gen-era and 23 families as ethnoveterinary medicine. A review article entitled "Ethnoveterinary practices in Rajasthan, India-A Review" was published by Yadav et al. (2012). Singh et al. (2014) described the local knowledge and its associated skills, practices, beliefs and social practices pertaining to health care and healthful husbandry of sheep by Raikas, a nomadic livestock rear-ing group from Marwar region of Rajasthan. Yadav et al. (2015) under taken study on one hundred and twenty tribal livestock owners from eight villages of two tehsils to document different ethnoveterinary practices. Gupta et al. (2015) recorded ethnoveterinary remedies of camel diseases of Rajasthan and listed uses of 31 plant materials and 13 household materials.

A medicobotanical study in relation to veterinary medicine of Shahjahanpur district, Uttar Pradesh was carried out by Sharma (1996). In Uttar Pradesh, Pandey et al. (1999) recorded 30 ethnoveterinary plants, com-monly used for the treatment of domestic animals by the aboriginal people of Gonda region. Prajapathi and Verma (2004) described ethnoveterinary plants of Mahoba district in Uttar Pradesh. Singh et al. (2011) recorded ethnoveter-inary healthcare practices in Marihan sub-division of district Mirzapur, Uttar Pradesh. Kumar and Bharri (2012a, 2012b) enumerated folk veterinary med-icines from Jalaun (57 plant species have been found to be used against 21 ailments of livestock in the form of 27 medicinal formulations) and Sitapur districts (57 medicinal formulas), respectively. Kumar and Bharati (2013) conducted field study in Uttar Pradesh and collected 83 medicinal plants used to treat 36 livestock ailments. Kumar et al. (2011) documented phytotherapeutic knowledge and health care management practices among the Tharu tribal community of Uttar Pradesh. In that study 59 phytotherapeutic practices using 48 plant species were documented for management of 18 types of healthcare problems of domesticated animals. Kumar and Kumar (2013) gave the details of 83 medicinal plants used to treat 36 livestock ailments in Uttar Pradesh. Mangal (2015) studied Gorakhpur district and a total 62 plant species have been found to be used for ethnoveterinary preparations for treating 37 diseases and ailments of animals.

In Haryana, Tosham block of Bhiwani district, a study on veterinary medicines was carried out by Yadav et al. (2014) and reported 54 medicinal plant species belonging to 37 families.

Mukherjee and Namhata (1988) described herbal veterinary medicine as prescribed by the tribals of Bankura district. Mandal and Chauhan (2000) presented the results of a survey conduced in Bankura district of West Bengal on the management procedures for the treatment of digestive disorders, wounds, fractures, respiratory diseases, throat swelling, intestinal helminthes, jaundice, ectoparasites, yoke gall, wounds in the vagina and foot and mouth disease. Herbal veterinary medicine from the tribal areas of Midnapore and Bankura district in West Bengal was given by Ghosh (2003). Bandhyopadhyay and Mukherjee (2005) recorded 25 ethnoveterinary prescriptions in which 23 plant species have been used. Mitra and Mukherjee (2007) enumerated 23 ethnoveterinary medicinal plants used by the four major tribal communities of the Uttar and Dakshin Dinajpur districts to treat the ailments of Cattle, to promote better lactation and also to improve the quality of meat, egg, etc. which are being traditionally used. Das and Tripathi (2009) documented ethnoveterinary practices and socio-cultural values associated with animal husbandry in rural Sundarbans in West Bengal. Rahaman et al. (2009) carried out studies on ethnoveterinary medicinal plants used by the tribals of Birbhum district in West Bengal. Dey and De (2010) enumerated 25 plant species used by the aboriginals of Purulia district of West Bengal. Mandal and Rahman (2014) recorded 40 plants which are used in preparation of 35 types of ethnoveterinary remedies for curing 22 types of diseases and ailments. Saha et al. (2014) recorded 70 phytotherapeutic practices involving 60 plants which were used to treat 34 types of diseases and disorders of livestock among the tribal community of Malda district of West

Bengal. Mandal and Rahman (2016) recorded the use of 25 medicinal plant species by the tribals of Birbhum district in West Bengal.

The aforesaid literature reveals that the good work on ethnoveterinary medicine have so far been carried out in Madhya Pradesh and Rajasthan but no significant work on this aspect has been done from Chhattisgarh, Delhi, Haryana and Punjab states which is rich in cultural and biological diversity. Except Sikarwar et al. (1994) and Sikarwar (1996) all research papers published on ethnoveterinary medicine after Jain (1999) are the major contribution on this aspect from Central India.

6.3   ENUMERATION

The plants used in veterinary medicine are enumerated alphabetically by botanical names, followed by family (in parenthesis), local names of different districts (district name written in parenthesis) are given in Table 6.1.

6.4   DISCUSSION

The data given above reveals that there are 449 plant species belonging to 325 genera and 108 families are used by tribal and rural people of Indo-Gangetic region for the treatment of more than 200 different types of ailments, diseases and disorders.

The Fabaceae is the largest family and contributes 35 species in Ethnoveterinary medicines. This is followed by Euphorbiaceae (24), Poaceae (22), Asteraceae (19), Lamiaceae and Solanaceae (16 each), Caesalpiniaceae (15) Cucurbitaceae (12), Malvaceae (11), Apiaceae and Mimosaceae (10 each), Convolvulaceae, Asclepiadaceae, Amaranthaceae and Zingiberaceae (9 each), Apocynaceae Liliaceae and Araceae (8 each), Rubiaceae, Anacardiaceae and Brassicaceae (7 each), Verbenaceae and Tiliaceae (6 each), Rutaceae, Arecaceae and Acanthaceae (5 each), Menispermaceae, Combretaceae, Sapindaceae, Meliaceae and Moraceae (4 each), Myrsinaceae, Lythraceae, Orchidaceae, Aristolochiaceae, Cleomaceae, Capparaceae, Loranthaceae, Vitaceae, Leeaceae, Bignoniaceae and Rosaceae (3 each), Bombacaceae, Oxalidaceae, Orobanchaceae, Crassulaceae, Ranunculaceae, Polygonaceae, Commelinaceae, Sapotaceae, Papaveraceae, Boraginaceae, Burseraceae, Theaceae, Gentianaceae, Zygophyllaceae, Rhamnaceae, Sterculiaceae, Amarhyllidaceae, Annonaceae

TABLE 6.1 Ethnoveterinary Plants of Indo-Gangetic Plains and Central India

S. No.	Name of the Plant	Part(s) used	Diseases	References
1	*Abrus precatorius* L. (Fabaceae) L.N.: Gongchi (Gwalior), Chirmu (Shivpuri), Gumchi, Ratti (Narsinghpur), Chirmi, Charmoi (Udaipur, Sirohi, Arid zone) Gunj (Jaspur and Surguja), Gunchi (Betul) Safed Chirmi (Southern Rajasthan) (Figure 6.1A)	Seeds + flour of *Pennisetum typhoides*; Leaves, root; Leaf paste; Root paste; Leaf paste; Ash of leaves of *Abrus precatorius* + *Calotropis procera*; Crushed roots, seeds; Seed paste; Seed paste	Rhinitis; acute colic, ephemeral fever, skin allergy; Swelling; Tumor in mastitis; Swelling; Cleanse stomach after delivery; Cough, cold and pneumonia, constipation; Expulsing of placenta	Shrivastava et al., 2012; Jatav et al., 2014; Sanghi, 2014; Nag et al., 2009; Ekka, 2015; Sebastian, 1984; Patil and Deshmukh, 2015; Takhar, 2004; Kumar et al., 2004
2	*Abutilon indicum* (L.) Sweet. (Malvaceae) L.N.: Kanghi (Amarkantak, Chitrakoot, Shivpuri, Jaspur and Surguja, Rajasthan), Petari (Betul), Kakai (Sitapur, Uttar Pradesh)	Leaf powder; Leaves; Leaf paste + butter milk; Leaf extract + Milk; Leaf decoction + Whey; New leaves of *Syzygium cumini* + leaves of *Mangifera indica* + *Acacia nilotica* ssp. *indica* + *Euphorbia reticulata* + *Aegle marmelos* + *Abutilon indicum* + common salt; Paste of seeds and leaves; Crushed leaves	Diarrhea; Arthritis; Dysentery; Diarrhea; Hematuria; Diarrhea; Hematuria; Constipation, Hematuria; Blood in urine	Shukla et al., 2007; Tripathi and Singh, 2010; Jatav et al., 2014; Ekka, 2015; Patil and Deshmukh, 2015; Kumar and Bharati, 2012b; Galav et al., 2013; Kumar and Bharati, 2013
3	*Acacia catechu* (*L.f.*) Brandis (Mimosaceae) L.N.: Katha (Chitrakoot, Sitapur), Kattha (Bundelkhand, Gorakhpur), Khair (Mount Abu, Jalaun), Katha (Darbhanga)	Stem bark paste; Wood powder; Young leaves and flowers; Latex of *Acacia catechu* + Leaves of *Cassia occidentalis*; Katha + Lime + Alum + Vaseline; Bark decoction of *Acacia catechu* + *Acacia nilotica* ssp. *indica* + *Ficus glomerata* + *Syzygium cumini* + *Butea monosperma*; Paste of kattha + *Areca catechu*; Crushed gum + Leaves of *Cassia occidentalis*	Wound; Growing of papillae; Easy delivery; Placenta protrudes outside before delivery; Wounds; Foot and mouth disease; Prolepse; Expulsion of placenta	Tripathi and Singh, 2010; Mishra et al., 2010; Galav et al., 2007; Kumar and Bharti, 2012a; Mishra et al., 1996; Kumar and Bharti, 2012b; Mangal, 2015; Kumar and Bharati, 2013

FIGURE 6.1

TABLE 6.1 (Continued)

S. No.	Name of the Plant	Part(s) used	Diseases	References
4	*Acacia leucophloea* (Roxb.) Willd. (Mimosaceae) L.N.: Ramja (Morena), Rijua (Udaipur, Mount Abu)	Stem bark paste; Root decoction; Pods, root extract	Dislocated bones; Bone fracture; Ouster induction, Bone fracture	Sikarwar, 1996; Nag et al., 2009; Galav et al., 2007
5	*Acacia nilotica* (L.) Del. ssp *indica* (Benth.) Brenan (Mimosaceae) L.N.: Babool (Morena, Gwalior, Narsinghpur, Gonda, Sitapur, Rajasthan), Bawal (Udaipur), Deshi babool (Southern Rajasthan, Arid zone), Kikar (Bhiwani), Babul (Rajasthan)	Spines decoction; Leaves + Leaves of Nimbu (*Citrus limon*) + baking soda; Leaves and Bark powder; Flower paste, bark extract; Bark decoction; Tender pods; Bark decoction of *Acacia nilotica ssp. indica + Ficus glomerata + Syzygium cumini + Butea monosperma + Acacia catechu*, stem bark decoction + Alum, stem bark decoction of *Acacia nilotica ssp. indica + Butea monosperma*, leaf extract + common salt; Thorns decoction; Paste of seeds, Decoction of dried bark; Twigs, fruits, leaf powder + Cow butter + Sugar; Crushed leaves + Leaves *Syzygium cumini*, *Mangifera indica* and *Aegle marmelos*, leaf extract, crushed gum + Criolite + rhizome of *Zingiber officinale*, ash of stem bark + vinegar of *Saccharum officinarum*; Seed paste, Decoction of dried bark powder, bark infusion: Bark infusion	Colic pain; Bloat (Tympanitis); Maggot wounds; Jaundice, dysentery; Foot and mouth disease; Increase lactation; Foot and mouth disease, conjunctivitis; Removal of placenta after delivery; Conceive, Indigestion, gas problems; Diarrhea, kill stomach worms, easy delivery; Blood in urine, conjunctivitis, Expulsion of placenta, Wound, injury; Facilitate conceiving, Indigestion, gastric problem, cooling; Sun stroke/heat stroke	Sikarwar, 1996; Shrivastava et al., 2012; Sanghi, 2014; Verma, 2014; Nag et al., 2009; Pandey et al., 1999; Kumar and Bharati, 2012b; Galav et al., 2013; Takhar, 2004; Yadav et al., 2014; Kumar and Bharati, 2013; Kumar et al., 2004; Gupta et al., 2015

TABLE 6.1 (Continued)

S. No.	Name of the Plant	Part(s) used	Diseases	References
6	*Acalypha indica* L. (Euphorbiaceae) L.N.: Dudhiya (Jaspur and Surguja), Kuppi (Darbhana)	Leaf paste + common salt; Leaf paste + Juice of *Citrus limon*	Wounds; Scabies	Ekka, 2015; Mishra et al., 1996
7	*Acanthospermum hispidum* DC. (Asteraceae) L.N.: Gokharu (West Nimar)	Leaves	General tonic	Satya and Solanki, 2009
8	*Achyranthes aspera* L. (Amaranthaceae) L.N.: Addajhara (Morena), Latjira (Shivpuri), Chirchiri (Vindhyan region), Andhijhada (West Nimar), Chirchiri (North Bihar), Chirchitha (Jaspur and Surguja), Tatjeera (Sitapur), Modo kanto, Adhijhara (Rajasthan)	Root paste + *Ferula asafoetida* + paste of *Calotropis procera* leaves; Leaf juice + saffron; Roasted seeds; Leaves; Aerial parts decoction; Leaf juice + saffron; Crushed root, root paste + black pepper; Root decoction + rhizome of *Curcuma longa*, whole plant decoction, Leaf juice	Bronchitis; Watering in eyes; Bronchitis; Appetizer; Diuretic; Watering in eyes; Placenta and umbilical chord not came out; Fever, Stomachache; Easy removal of placenta; Opacity of cornea	Sikarwar, 1996; Jatav et al., 2014; Dwivedi et al., 2009; Satya and Solanki, 2009; Jha et al., 1991; Ekka, 2015; Kumar and Bharati, 2012b; Galav et al., 2013
9	*Acorus calamus* L. (Araceae) L.N.: Bach (Chitrakoot)	Leaf paste, rhizome	Wounds	Gautam and Richhariya, 2015

S. No.	Name of the Plant	Part(s) used	Diseases	References
10	*Adiantum capillus-veneris* L. (Adiantaceae) L.N.: Hansraj (Amarkantak)	Plant paste with mustard oil	Skin diseases	Shukla et al., 2007
11	*Aegle marmelos* (L.) Correa (Rutaceae) L.N.: Bel (Jhabua, Chitrakoot, Shivpuri, Tikamgarh), Bael (Narsinghpur) Bel pathar (Bhiwani), Bail (Uttar Pradesh), Bilpatra (Banswara)	Fruit paste; Fruit pulp; Fruit; Leaf paste + turmeric, fruits; Leaf paste + seed oil of *Ricinus communis*; Fruit paste + Dried zinger; Crushed leaves + Leaves of *Abutilon indicum*; Effected tail pierce of thorn and dipped into boiled Soyabean or Mustard oil	Abortion; Injury, Constipation; Flatulence, gastric problem; wounds, gastric problem; Sun burn; Dysentery and Diarrhea; Blood in urine; Tail gangrene	Kadel and Jain, 2006; Tripathi and Singh, 2010; Jatav et al., 2014; Sanghi, 2014; Verma, 2014; Yadav et al., 2014; Kumar and Bharati, 2013; Yadav et al., 2015
12	*Aerva lanata* (L.) Juss. ex Schult. (Amaranthaceae) L.N.: Chhotibui (Rajasthan)	Crushed roots	Antidote in snake bite	Galav et al., 2013; Gupta et al., 2015
13	*Aerva persica* (Burm.f.) Merrill (Amaranthaceae) L.N.: Safed bui (Rajasthan)	Flower decoction, root decoction, whole plant extract, Inflorescence decoction	Digestive disorder, dysuria, foot and mouth disease, Eating soil, foot and mouth disease	Galav et al., 2013
14	*Aerva pseudotomentosa* Blatt. and Hall. (Amaranthaceae) L.N.: Safed bui (Southern Rajasthan, Arid zone)	Crushed inflorescence; warmed inflorescence	Swelling, Inflammations	Takhar, 2004; Kumar et al., 2004

TABLE 6.1 (Continued)

S. No.	Name of the Plant	Part(s) used	Diseases	References
15	*Ageratum conyzoides* (L.) L. (Asteraceae) L.N.: Kubbi, Khar (Bilaspur), Gangarigera (Baster, Amarkantak)	Whole plant paste, Leaf juice; Bark paste	Healing of wounds, Cut and wounds; Ailing animals	Sikarwar and Kumar, 2005; Shukla et al., 2007
16	*Ailanthus excela* Roxb. (Simaroubaceae) L.N.: Lohagal (West Nimar), Aduso (Udaipur, Sirohi), Adua (Udaipur), Maharukh (Betul)	Crushed bark; Bark paste, warmed leaves; Leaf paste; Bark paste + Goat milk; Leaf decoction	Appetizer; Ailing animals, swellings; Control ticks and lice, nose wound; Remove maggots from wound	Satya and Solanki, 2009; Sebastian, 1984; Deora et al., 2004; Patil and Deshmukh, 2015
17	*Alangium salvifolium* (L. f.) Wangerin (Alangiaceae) L.N.: Akol (West Nimar), Ankol (Rajasthan) (Figure 6.1B)	Plant powder; Root paste + Butter milk, leaf paste	Antidote against poisonous herbs; Antidote to dog bite, malarial fever, enlargement of liver	Satya and Solanki, 2009; Galav et al., 2013
18	*Albizia lebbeck* (L.) Benth. (Mimosaceae) L.N.: Siras (Bhiwani)	Leaf juice	Eye problems	Yadav et al., 2014
19	*Aleurites moluccana* (L.) Willd. (Euphorbiaceae) L.N.: Latjeera (Marwar)	Root paste; Root	Diarrhea, Internal parasites	Dudi and Meena, 2015

S. No.	Name of the Plant	Part(s) used	Diseases	References
20	*Allium cepa* L. (Liliaceae) L.N.: Piyaj (Jhabua, Jhansi, Tikamgarh, Darbhanga, Uttar Pradesh), Kanda (Alirajpur, West Nimar), Kando (Ratlam), Onion (Gorakhpur, Banswara)	Fruit paste, Bulb pieces; Bulb paste; Bulb paste + mustard oil; Bulb paste + Jaggery; Bulb extract; Bulb paste + leaf ash of *Musa paradisiaca* + mustard oil; Bulb juice; Onion paste + Ghee + Jaggery; Paste of onion + leaves of *Cannabis sativa* + Turmeric powder + Mustard oil, Paste of Onion + Alum + *Cannabis sativa* + Turmeric powder; Bulb crushed + Flower buds of *Syzygium aromaticum* + Oil of *Brassica napus*, bulb crushed + gum of *Acacia nilotica* + fruits of *Trachyspermum ammi*; Bulb paste + *Curcuma longa* powder + Common salt + Mustard oil; Paste of Pyaj + *Trachyspermum ammi* + *Coriandrum sativum*; Bulb + Mustard oil fed	Indigestion, Bad taste of mouth cavity (Dandki disease); Remove maggots from wounds; Hoof diseases; Nasal Secretion; Expel the insect from eyes; Ectoparasites; Cough, cold and fever; Dysentery; Swelling in shoulder, fracture; Mastitis, anorexia; Wound; Fever; Anemia	Kadel and Jain, 2006; Chouhan and Ray, 2014; Jadhav, 2009; Satya and Solanki, 2009; Nigam and Sharma, 2010; Verma, 2014; Mishra et al., 1996; Mangal, 2015; Kumar and Bharati, 2013; Yadav et al., 2015; Dudi and Meena, 2015; Gupta et al., 2015
21	*Allium sativum* L. (Liliaceae) L.N.: Lasun (Jhabua, Chitrakoot), Lahsun (Gwalior, Shivpuri, Darbhanga, Gonda, Sitapur, Rajasthan, Bhiwani), Lehson (Jhansi)	Bulbs; Bulb juice; Bulbs + mustard oil, Bulbs + mustard oil + ash of cow dung cake + curd; Bulb paste + bees wax, milk + cooking oil + Bulb paste; Seeds + bulb paste of *Allium cepa*; Bulb juice; *Triticum aestivum* flour + garlic pods + *Brassica campestris* oil; Green plants; Cloves boiled in milk; Bulbs boiled in *Brassica juncea* oil; Bulb paste + 2 eggs of hen + milk; Garlic paste + Elaichi + Jaggery; Bulb paste; Bulb + Mustard oil fed	Brain disease, Earache; Cough and Cold; Indigestion; Diarrhea, food poisoning; Injuries, Snake bite; Bronchitis; Cough, cold and fever; Dysentery; Cough; Cough and flatulence; Foot and mouth disease; Impaction and lumbago; Cold and Fever; External parasites; Anemia	Kadel and Jain, 2006; Tripathi and Singh, 2010; Mishra et al., 2010; Shrivastava et al., 2012; Jatav et al., 2014; Nigam and Sharma, 2010; Mishra et al., 1996; Pandey et al., 1999; Kumar and Bharati, 2012b; Galav et al., 2013; Yadav et al., 2014; Gupta et al., 2015

TABLE 6.1 (Continued)

S. No.	Name of the Plant	Part(s) used	Diseases	References
22	*Alocasia indica* Schott. (Araceae) L.N.: Kanda (Darbhanga), Maan (Gonda), Maanmoor (Sitapur), Mein (Gorakhpur)	Decoction of tuber + *Trigonella corniculata*; Leaves; Roots crushed + *Cinnamomum camphora*; Leaves	Respiratory trouble; Loss of estrus; Regulate estrus cycle	Mishra et al., 1996; Pandey et al., 1999; Kumar and Bharati, 2012b; Mangal, 2015
23	*Alocasia macrorrhiza* (L.) G. Don (Araceae) L.N.: Arbi (Uttar Pradesh)	Rhizome crushed with camphor, Rhizome paste	Prepare for mating, furunculous, boils	Kumar and Bharati, 2013
24	*Aloe vera* L. (Liliaceae) L.N.: Gwarpatha (Jhabua), Gheekuwar (Chitrakoot), Gubarpata (Shivpuri), Gheekumari (Vindhyan region), Gheegwar (West Nimar), Ghikumari (Jhansi), Ghritkumari (Jaspur and Surguja)	Leaf pulp; Leaf pulp + curd; Leaf paste; Leaf mucilage; Leaves + leaves of *Jatropha curcas*; Leaf pulp + sore milk + water; Leaf pulp	Mastitis (Thanela disease); Injuries; Unconsciousness (Drooping head); Mastitis; Swelling of udder; Burns; Drooping head; Burned part	Kadel and Jain, 2006; Tripathi and Singh, 2010; Jatav et al., 2014; Dwivedi et al., 2009; Satya and Solanki, 2009; Nigam and Sharma, 2010; Ekka, 2015; Gupta et al., 2015
25	*Amaranthus spinosus* L. (Amaranthaceae) L.N.: Kantamarish (Amarkantak), Cholai (Alirajpur)	Whole plant, Plant decoction; Whole plant paste	Lactation, Delivery complaints; Wounds	Shukla et al., 2007; Chouhan and Ray, 2014

S. No.	Name of the Plant	Part(s) used	Diseases	References
26	*Amaranthus tricolor* L. (Amaranthaceae) L.N.: Genhari	Aerial part decoction	Galactagogue	Jha et al., 1991
27	*Amaranthus viridis* L. (Amaranthaceae) L.N.: Chaulai (Jhansi)	Seeds + water	Tympany	Nigam and Sharma, 2010
28	*Ammannia baccifera* L. (Lythraceae) L.N.: Akasia (Ratlam)	Plant extract	Induce fertility	Jadhav, 2009
29	*Amorphophallus peoniifolius* (Dennst) Nicolson (Araceae) L.N.: Bhabdi (Jhabua), Jangli Suran, Bhahna Kand (Chhindwara)	Corm decoction and paste	Body pain	Sikarwar and Kumar, 2005
30	*Ampelocissus latifolia* (Roxb.) Planch (Vitaceae) L.N.: Emlosa (Tikamgarh), Eamlaua (Panna), Emlaura (Morena), Dokarbel (West Nimar)	Root paste; Root paste + jaggery; Root paste	As a tonic, lactation; Blood dysentery	Sikarwar and Kumar, 2005; Sikarwar, 1996; Satya and Solanki, 2009

TABLE 6.1 (Continued)

S. No.	Name of the Plant	Part(s) used	Diseases	References
31	*Andrographis paniculata* (Burm.*f.*) Wall. (Acanthaceae) L.N.: Kalmegh (Shivpuri), Bhuineem (Jaspur and Surguja), Karsuyya (Gonda)	Whole plant decoction	Fever and cough; Fever, cough, loss of appetite, liver disorder and physical debility	Jatav et al., 2014; Ekka, 2015; Pandey et al., 1999
32	*Anethum graveolens* L. (Apiaceae) L.N.: Soya (Darbhanga)	Decoction of seeds + Seeds of *Trachyspermum copticum*	Crack in palate	Mishra et al., 1996
33	*Anisomeles indica* (L.) Kuntze (Lamiaceae) L.N.: Kuschor (Jhabua), Bonrmal (Bilaspur), Phulmajri (Mount Abu, Udaipur), Ghabro (Rajasthan)	Leaf decoction; Animal pass under the branch (Myth), Stem extract; Whole plant infusion, root paste; Whole plant decoction	Body inflammation; Foot and mouth disease, Indigestion, bone fracture, uterus dislocation, weakness after delivery; Flatulence, leucorrhoea	Sikarwar and Kumar, 2005; Galav et al., 2007; Deora et al., 2004; Galav et al., 2013
34	*Annona reticulata* L. (Annonaceae) L.N.: Custard apple Marwar)	Leaf paste	Ectoparasites	Dudi and Meena, 2015

S. No.	Name of the Plant	Part(s) used	Diseases	References
35	*Annona squamosa* L. (Annonaceae) L.N.: Sitaphal (Bastar, Jhabua, Alirajpur, Rajasthan), Sarifa (Amarkantak), Sitafal (Vindhyan region, Udaipur), Seetaphal (West Nimar), Sharifa (Jhansi, Darbhanga), Mondal (Chhotanagpur)	Leaf paste, Juice of unripe fruits, Leaf decoction, Seed paste; Leaf paste; Seed powder; Leaf juice + Asafoetida Leaf paste + lime; Leaf paste; Green leaves; Leaves; Leaf paste	Cut and wounds, Worms in stomach, Lice, house flies, mosquitoes and snails, Bone fracture; Wounds; Ectoparasites; Kill maggots; Foot disease; Foot and mouth disease; Ectoparasites; Wounds	Sikarwar and Kumar, 2005; Kadel and Jain, 2006; Shukla et al., 2007; Chouhan and Ray, 2014; Dwivedi et al., 2009; Satya and Solanki, 2009; Nigam and Sharma, 2010; Nag et al., 2009; Mishra et al., 1996; Singh, 1987; Gupta et al., 2015
36	*Anogeissus latifolia* (Roxb.) Wall. ex Bedd. (Combretaceae) L.N.: Dhavada (Mount Abu), Dhawada (Rajasthan)	Bark extract; Bark decoction	Fever	Galav et al., 2007, 2013
37	*Apium graveolens* L. (Apiaceae) L.N.: Ajmod (Sitapur)	Seeds + fruits of *Melia azadirach* + butter milk + dried unripe fruits of *Mangifera indica* + seeds of *Trachyspermum ammi* + *Ferula asafoetida* + gum of *Acacia nilotica* ssp. *indica* + *Allium sativum* + *Allium cepa* crushed together; Fruits crushed + Fruits of *Trachyspermum ammi* + *Ferula asafoetida* + un ripe fruits of *Mangifera indica*	Anorexia	Kumar and Bharati, 2012b; Kumar and Bharati, 2013
38	*Arachis hypogaea* L. (Fabaceae) L.N.: Mungphali (Jhansi)	Seed oil + salt	Twitching	Nigam and Sharma, 2010

TABLE 6.1 (Continued)

S. No.	Name of the Plant	Part(s) used	Diseases	References
39	*Ardisia solanacea* Roxb. (Myrsinaceae) Amarkantak	Tender shoot paste + bamboo leaves + sugar + hen's egg	Bone fracture	Shukla et al., 2007
40	*Argemone mexicana* L. (Papaveraceae) L.N.: (Amarkantak, Tikamgarh), Peeli Kateeli (Sitapur), Satyanashi (Bhiwani) (Figure 6.1C)	Root paste; Leaf and fruit juice; Fruits decoction; Plant decoction, plant fed; Tisane	Eczema; Foot infection, rheumatism; Fever; Constipation, removal of retained placenta; Fever	Shukla et al., 2007; Verma, 2014; Kumar and Bharati, 2012b; Yadav et al., 2014; Kumar and Bharati, 2013
41	*Argyria nervosa* (Burm.f.) Bojer. (Convolvulaceae) L.N.: Marang harlu (Singhbhum)	Leaf paste	Safe delivery	Pal (1980)
42	*Arisaema tortuosum* (Wall.) Schott. (Araceae) L.N.: Dhei (Panna), Safed telia kand (Shahdol), Jhattawan (Betul), Samp bhutta (Bilaspur), Suran (Mount Abu)	Tuber paste; Tuber paste + salt + chili paste; Tuber decoction	As tonic; Throat swellings; Flatulence and other gastric problem	Sikarwar and Kumar, 2005; Sikarwar et al., 1994; Galav et al., 2007
43	*Aristolochia bracteata* Retz. (Aristolochiaceae) L.N.: Kidamari (Udaipur)	Leaf juice	Wounds, tympanitis	Deora et al., 2004

S. No.	Name of the Plant	Part(s) used	Diseases	References
44	*Aristolochia bracteolata* Lam. (Aristolochiaceae) L.N.: Girdhan (West Nimar), Acchho (Jaspur and Surguja)	Leaf juice: Leaf heated in Sesame oil	Wounds; Skin infection and wounds	Satya and Solanki, 2009; Ekka, 2015
45	*Aristolochia indica* L. (Aristolochiaceae) L.N.: Acchho (Jaspur and Surguja), Ishar-lori (Midnapur), Gorisal (Rajasthan)	Leaf paste + pepper; Root paste + 7 Black pepper; Root powder + Wheat chapatti	Insect bite; Snake bite; Fever	Ekka, 2015; Pal, 1980; Galav et al., 2013
46	*Artemisia nilagirica* (C. B. Clarke.) Pamp. (Asteraceae) L.N.: Karanj (Marwar)	Decoction of leaf, root, leaf powder	Internal parasite	Dudi and Meena, 2015
47	*Artemisia parviflora* Roxb. (Asteraceae) L.N.: Dhatura (Marwar)	Leaves	Ectoparasites	Dudi and Meena, 2015
48	*Asparagus adscendens* Roxb. (Liliaceae) L.N.: Satavari (Narsinghpur)	Root paste	Lactation	Sanghi, 2014

TABLE 6.1 (Continued)

S. No.	Name of the Plant	Part(s) used	Diseases	References
49	*Asparagus racemosus* Willd. (Liliaceae) L.N.: Satawar (Jhabua, Jalaun), Satavari (Chitrakoot), Shatavari (West Nimar, Tikamgarh, Betul), Naharkanta (Udaipur), Sahasmoosali, Undhkant (Udaipur, Sirohi), Satavar (Bhiwani)	Root powder; Root paste + sugar; Root powder or whole plant; Root powder; Stem paste; 110 g roots + 250 g fruits of *Mesua ferea* + 22 g rhizome powder of *Zingiber officinale* + 125 g fruits of *Trachyspermum ammi*, 100 g roots + 25 g Fruits of *Trigonella foenum–graecum* + 25 g fruits of *Foeniculum vulgare* + 250 g fruits of *Terminalia chebula* + 50 g fruits of *Terminalia bellirica* + 25 g fruits of *Piper nigrum* + 25 g flower buds of *Syzygium aromaticum* + 25 g fruits of *Elettaria cardamomum* + 250 g mustard oil + 250 g bulbs of *Allium cepa* + 1 kg Jaggery; Tubers; Crushed roots; Plant fed; Crushed roots + Fruits of *Trigonella foenum–graecum* + *Foeniculum vulgare* + *Terminalia chebula* + *Terminalia bellirica* + *Piper nigrum* + *Elettaria cardamomum* + Flower buds of *Eugenia caryophyllus* + Bulb of *Allium cepa* + Jaggery	Increase lactation; Arthritis; Broken horn; Mastitis; Stopped mulching; Enhance milk yield; Heat production; Mastitis	Kadel and Jain, 2006; Tripathi and Singh, 2010; Satya and Solanki, 2009; Verma, 2014; Nag et al., 2009; Kumar and Bharti, 2012a; Sebastian, 1984; Patil and Deshmukh, 2015; Yadav et al., 2014; Kumar and Bharati, 2013

S. No.	Name of the Plant	Part(s) used	Diseases	References
50	*Azadirachta indica* A. Juss. (Meliaceae) L.N.: Neem (Jhabua, Amarkantak, Chitrakoot, Gwalior, Vindhyan region, West Nimar, Jhansi, Narsinghpur, Tikamgarh, Jalaun, Jaspur and Surguja, Darbhanga, Udaipur, Sirohi, Sitapur, Rajasthan, Gorakhpur, Southern Rajasthan, Bhiwani, Banswara, Marwar, Arid zone)	Leaf decoction; Bark decoction; Leaf paste + Turmeric + Ajwain seeds powder + black salt, Leaf paste + salt; Leaf infusion; Leaf paste; Leaf paste, Leaf paste + lime, Leaf paste + *Citrus limon* juice, Leaves + salt, Leaf paste; Leaves; Leaf poultice, Leaves; Seed oil; Leaf decoction + salt; Leaf paste + caster oil; Bark paste + bark paste of *Acacia nilotica*; 25 g leaves + 50 leaves of *Trachyspermum ammi* + black salt + 100 g Jaggery + 200 ml water; 100 ml leaf extract + 100 ml root extract of *Withania somnifera* + 100 g ash of *Opuntia elatior* thorns; Seed oil; Leaf decoction, green leaves, Leaves steam, leaf decoction; Leaves; Oil; Leaves, neem oil, Alum + Jaggery + leaves, boiled leaves, Leaves + Jaggery (Gur), leaf paste; Ash of leaves + Ghee; Tender twigs and leaves, Leaf paste + Hairs + Mustard oil; Seed oil; Ash of leaves + Clarified butter; Leaves; Leaf paste, burning of leaves of Neem + *Citrus medica* + *Calotropis procera* near goat flock, leaf paste; Ash of leaves + Butter oil; Leaf extract + Butter milk + Rock salt; Seed oil, Leaves crushed in mustard oil	Foot and Mouth disease, Ecto-parasites (lice and bugs on skin), Skin diseases (Khori disease), Earache, Wounds; Intestinal worms; Constipation, External parasites, Immunity; Indigestion; Wounds; Bovicolasis, Injury, Scabies, Volvulus, Wounds; Antipyretic, thrust, nausea, vomiting, skin disease, ulcer; Cut, Injury, Intestinal worms, skin diseases; Foot and Mouth disease, wounds; Remove insect from the eyes; Insect bites; Wounds; Worm infection; Diphtheria; Wound; Skin disease, stomatitis, prolepses of uterus; Glands in throat; Antidote to snake bite; Conjunctivitis; Leucorrhoea; Round worm, tapworm, Brucellosis/premature abortion, Foot 7 mouth disease; Fracture; Calf does not suck milk from mothers teat; Swelling, Inflammations; Stomach ache; Fractured horns; Conjunctivitis; Wound; Internal parasites, ectoparasites, skin disease; Boils; Internal parasites, external parasites	Kadel and Jain, 2006; Shukla et al., 2007; Tripathi and Singh, 2010; Misrra et al., 2010; Chouhan and Ray, 2014; Shrivastava et al., 2012; Jatav et al., 2014; Dwivedi et al., 2009; Satya and Solanki, 2009; Nigam and Sharma, 2010; Sanghi, 2014; Verma, 2014; Kumar and Bharati, 2012a; Ekka, 2015; Mishra et al., 1996; Sebastian, 1984; Kumar and Bharati, 2012b; Galav et al., 2013; Mangal, 2015; Takhahr, 2004; Yadav et al., 2014; Yadav et al., 2014; Kumar and Bharati, 2013; Yadav et al., 2015; Dudi and Meena, 2015; Kumar et al., 2004; Gupta et al., 2015

TABLE 6.1 (Continued)

S. No.	Name of the Plant	Part(s) used	Diseases	References
51	*Azanza lampas* (Cav.) Alef. (Malvaceae) L.N.: Jangli bhindi (Raigarh)	Root paste	Paralysis	Sikarwar et al., 1994
52	*Balanites aegyptiaca* (L.) Del. (=*B. roxburghii* Planch.) (Simaroubaceae) L.N.: Hingot (Morena, Mount Abu), Hingota (Jhansi, Tikamgarh, Southern Rajasthan, Arid zone)	Fruit stone paste + Chili, Leaf paste; Seed paste + water; Stem bark powder, bark paste; Seed paste; Endosperm paste	Constipation, Eye conjunctivitis; Neck inflammation; Increasing milk and removal of intestinal worms, snake bite; Expulsion of placenta	Sikarwar, 1996; Nigam and Sharma, 2010; Galav et al. 2007; Takhar, 2004; Kumar et al., 2004
53	*Bambusa arundinacea* (Retz.) Willd. (Poaceae) L.N.: Bans (Jhansi, Tikamgarh), Baas (Jalaun, Gorakhpur)	Dried leaves + lukewarm water; Leaves, rhizome and leaf paste; Leaves + leaves of *Basella alba* + straw of *Oryza sativa*; Leaves, Leaf decoction + Jaggery; Paste of kernel of Bamboo shoots + Kapoor + Kunain; Ash of soft sprout of Bamboo + Alum powder; Decoction of leaves, flowers	Retard placenta; Easy delivery, diarrhea; Placenta and umbilical cord not expelled; Placenta did not fall; Foot and mouth disease; Navel infection; Expulsion of placenta	Nigam and Sharma, 2010; Verma, 2014; Kumar and Bharti, 2012a; Mangal, 2015; Yadav et al., 2015
54	*Bambusa bambos* (L.) Voss (Poaceae) L.N.: Bas (Uttar Pradesh)	Nodal roots crushed + Pinch of quinine + Camphor	Foot and mouth disease	Kumar and Bharati, 2013
55	*Bambusa spinosa* Roxb. (Poaceae) L.N.: Bans (Darbhanga)	Leaves	Blood dysentery	Mishra et al., 1996
56	*Barleria prionitis* L. (Acanthaceae) L.N.: Vajradanti (Udaipur)	Leaf paste	Skin diseases of camel and dogs.	Nag et al., 2009

S. No.	Name of the Plant	Part(s) used	Diseases	References
57	*Basella alba* L. (Basellaceae) L.N.: Poi (Gorakhpur)	Leaf paste; Leaves crushed + Leaves of *Bambusa bambos*	Placenta did not fall; Retention of placenta and umbilical cord after delivery	Mangal, 2015; Kumar and Bharati, 2013
58	*Bauhinia racemosa* Lam. (Caesalpiniaceae) L.N.: Astu (Sehore, Dewas), Koinar (Amarkantak)	Root decoction, Leaf paste	Stop abortion, Dysentery	Sikarwar and Kumar, 2005; Shukla et al., 2007
59	*Bauhinia variegata* L. (Caesalpiniaceae) L.N.: Kachnal (Surguja), Kachnar (Jhansi, Gonda)	Root paste; Root paste + black cow urine; Bark decoction	Expel placenta after delivery; Blindness; Foot and Mouth disease	Sikarwar et al., 1994; Nigam and Sharma, 2010; Pandey et al., 1999
60	*Begonia picta* Sm. (Begoniaceae) L.N.: Bahari (Mandla)	Leaf paste	Kill lice of body	Sikarwar et al., 1994
61	*Biophytum sensitivum* (L.) DC. (Oxalidaceae) L.N.: Rajarani (Dhar)	Leaf paste	Cuts	Sikarwar et al., 1994
62	*Boerhavia diffusa* L. (Nyctaginaceae) L.N.: Punarnava (Chitrakoot, Bhiwani), Patharchatta (Gwalior), Pattharchatta (West Nimar), Gadahpurain (North Bihar), Pattharchatta (Gorakhpur), Punerva (Uttar Pradesh)	Root powder; Leaves + Bark paste of *Moringa oleifera* + bulbs of *Allium sativum*; Root juice; Aerial part decoction; Leaf paste + salt; Whole plant fed; Whole plant pounded + Fruits of *Terminalia chebula* + Oil of *Ricinus communis*	Black quarter, Liver diseases; Rheumatism; Diarrhea and dysentery; Diuretic, cardiac stimulant; Mastitis; Removal of retained placenta; Expulsion of placenta	Gautam and Richhariya, 2015; Shrivastava et al., 2012; Satya and Solanki, 2009; Jha et al., 1991; Mangal, 2015; Yadav et al., 2014; Kumar and Bharati, 2013

TABLE 6.1 (Continued)

S. No.	Name of the Plant	Part(s) used	Diseases	References
63	**Bombax ceiba** L. (= Salmalia malabaricum (DC.) Schott and Endl.) (Bombacaceae) L.N.: Semra (Morena), Semal (Jhabua, Satna, Vindhyan region, Mount Abu, Gonda, Sitapur, Gorakhpur), Semel (Darbhanga), Katsawar (Betul), Bamboo (Banswara)	Stem bark paste + turmeric powder; Leaf paste, Flower juice; Bone fracture; Stem bark; Flowers + Stem bark of Alangium salvifolium; Fruit cotton + Ghee; Bark paste + Jaggery; Stem bark paste; Seeds paste of Trigonella foenum-graecum + Black salt + Cuminum cyminum + Ferula asafoetida + leaves of Bombax ceiba; Bark decoction, foment of boiled bark; Crushed leaves + Stems of Bombax ceiba; Flowers and Leaves fed, sticks + Mud	Dislocated bones; Wounds, Prolapse of uterus; Flatulence, Indigestion; Easy removal of placenta; Indigestion and flatulence; Hematuria, excess flatulence; Dysentery; Flatulence, bleeding; Dysentery; Flatulence, dysentery; Fracture, Sprain; Retention of placenta, removal of retained placenta; Fracture	Sikarwar, 1996; Kadel and Jain, 2006; Gautam and Richhariya, 2015; Dwivedi et al., 2009; Galav et al. 2007; Mishra et al., 1996; Pandey et al., 1999; Patil and Deshmukh, 2015; Kumar and Bharati, 2012b; Mangal, 2015; Yadav et al., 2015
64	**Borassus flabbelifer** L (Arecaceae) L.N.: Palm (Darbhanga)	Ash of inflorescence + Jaggery	Post delivery clearance of uterus	Mishra et al., 1996
65	**Boswellia serrata** Roxb. (Burseraceae) L.N.: Salar (Rajasthan) (Figure 6.1D)	Bark decoction	Arthritis, indigestion, windiness, flatulence	Galav et al., 2013; Gupta et al., 2015
66	**Brassica campestris** L. (Brassiccaceae) L.N.: Sarson (Jhabua, Gwalior, Mount Abu, Darbhanga, Gonda), Sarso (Arid zone)	Seed oil, Seed paste; Ash of cycle tire + mustard oil; 500 ml oil + a pinch of Ferula asafoetida Seed oil + seeds of Piper nigrum; Warmed mustard oil, mustard oil + egg, Oil, oil + salt; Oil + Asafoetida + Garlic	Skin disease, Mastitis (Thanela disease), Brain disease, Earache, Falling of tail; Wounds; Foot rot; Indigestion and flatulence; Cough, cold and fever; Hemorrhagic septicemia, cough, Etching; Diarrhea, foot and mouth disease; Gastric problem	Kadel and Jain, 2006; Shrivastava et al., 2012; Galav et al., 2007; Mishra et al., 1996; Mangal, 2015; Dudi and Meena, 2015; Kumar et al., 2004

S. No.	Name of the Plant	Part(s) used	Diseases	References
67	*Brassica juncea* (L.) Czern. and Coss. (Brassicaceae) L.N.: Rai (Mount Abu), Raai (Uttar Pradesh)	Bulb of *Allium cepa* + *A. sativum* + seeds of *B. juncea* + butter milk; 250 g Seed oil + 250 g Jaggery + 50 g fruits of *Trigonella foenum-graecum*; Seed oil; Lukewarm oil	Diarrhea; Placenta and umbilical cord not expelled; Watering of eyes, Headache; Infection in tail; External parasites	Galav et al., 2007; Kumar and Bharti, 2012a; Pandey et al., 1999; Kumar and Bharati, 2013; Gupta et al., 2015
68	*Brassica napus* L. (Brassicaceae) L.N.: Sarsoo (Jalaun), Sarson (Southern Rajasthan),	Seed oil; Seed oil + seed oil of *Sesamum indicum* + *Ferula asafoetida* + Garlic paste; Seed oil + Jaggery + *Trigonella foenum-graecum*, seed oil	Scabies; Gas problem; Retention of placenta and umbilical cord after delivery, infection in tail	Kumar and Bharti, 2012a; Takhar, 2004; Kumar and Bharati, 2013
69	*Brassica rapa* L. ssp. *campestris* (L.) Clapham (Brassicaceae) L.N.: Sarso (Alirajpur), Sarsu (Ratlam)	Seed oil + Paste of bulb of *Allium cepa* L.; Mustard oil + *Curcuma longa*	Wounds; Indigestion (Afra disease)	Chouhan and Ray, 2014; Jadhav, 2009
70	*Bridelia retusa* (L.) Spreng. (Euphorbiaceae) L.N.: Palati (Amarkantak)	Stem bark paste	Prevent abortion	Shukla et al., 2007
71	*Bryophyllum daigremontianum* (Raym., Hamet and Perrier) A. Berger (Crassulaceae) L.N.: Patherchat (Bhiwani)	Leaves fed	Urinary problem	Yadav et al., 2014
72	*Buchanania lanzan* Spreng. (=*B. latifolia* Roxb.) (Anacardiaceae) L.N.: Char (Chitrakoot), Chironji (Gwalior, Jhansi), Charoli (Betul)	Bark extract; Kernels; Root bark powder + cow milk; Gum resin	Washing wounds; Hyperthermia; Backbone fracture; Bone fracture	Gautam and Richhariya, 2015; Shrivastava et al., 2012; Nigam and Sharma, 2010; Patil and Deshmukh, 2015

TABLE 6.1 (Continued)

S. No.	Name of the Plant	Part(s) used	Diseases	References
73	*Bunium persicum* (Boiss.) B. Fedtsch. (Apiaceae) L.N.: Kala jeera (Sitapur, Uttar Pradesh)	Sodium bicarbonate + Tatric acid + seeds of *Bunium persicum* + seeds of *Trigonella foenum-graecum* + Juice of *Citrus aurantifolia* + Sulamani namak + Black salt; Fruits + Fruits of *Trigonella foenum-graecum* + Fruit juice of *Citrus aurantiifolia* + Black salt + Sodium bicarbonate + Tatric acid + Sulamni namak	Flatulence	Kumar and Bharati, 2012b; Kumar and Bharati, 2013
74	*Butea monosperma* (Lam.) Taub. (Fabaceae) L.N.: Chhiula (Tikamgarh), Chhola (Raigarh, Seoni), Tesu (Chhindwara), Palash (Jhabua, Chitrakoot, West Nimar), Dhak (Bundelkhand, Tikamgarh, Sitapur), Khakara (Mount Abu), Palas (Betul), Khankra (Pratapgarh), Taysoou (Uttar Pradesh) (Figure 6.1E)	Bark decoction; Bark paste; Seed paste + turmeric; Warmed leaves; Seed paste; Seed powder + salt + water; Seed oil + seed oil of Mango; Flower decoction; Bark powder + animal urine, bark extract; Stem bark, root decoction; Tender pods; Bark decoction of *Butea monosperma* + *Acacia nilotica* ssp. *indica* + *Ficus glomerata* + *Syzygium cumini* + *Acacia catechu*; Flower powder, bark; Flower paste + Straw of *Triticum aestivum* + Mustard oil, Flower decoction	Swelling; Dysentery; Foot and mouth disease (Khurpaka); Expel intestinal worms; Deworming; Skin inflammation; Dysurea, paralysis; Foot and mouth disease, gastroenteritis; Hematuria; Foot and mouth disease; Cut and wounds, weakness; Stop urination, Flatulence	Sikarwar and Kumar, 2005; Kadel and Jain, 2006; Tripathi and Singh, 2010; Mishra et al., 2010; Gautam and Richhariya, 2015; Nigam and Sharma, 2010; Satya and Solanki, 2009; Verma, 2014; Galav et al., 2007; Patil and Deshmukh, 2015; Kumar and Bharati, 2012b; Meena, 2014; Kumar and Bharati, 2013

S. No.	Name of the Plant	Part(s) used	Diseases	References
75	*Caesalpinia bonduc* (L.) Roxb (Caesalpiniaceae) L.N.: Gataran (Bastar)	Seed paste	Kill worms	Sikarwar and Kumar, 2005
76	*Caesalpinia crista* L. emend Dandy and Excel (Caesalpiniaceae) L.N.: Kali gather (West Nimar)	Seeds + leaves of *Vitex negundo*, seed powder	Fever, as an anthelmintic	Satya and Solanki, 2009
77	*Caesalpinia decapetala* (Roth) Alston (Caesalpiniaceae) L.N.: Kather, Karongsi (Udaipur, Sirohi)	Leaves	Mouth sores	Sebastian, 1984
78	*Cajanus cajan* (L.) Huth. (Fabaceae) L.N.: Arhar (Gwalior, Darbhanga, Sitapur)	Leaf paste; Green pods paste + cold water; Decoction of Pulse + Root of *Withania somnifera*	Wounds; Dysentery; Blood dysentery	Shrivastava et al., 2012; Mishra et al., 1996; Kumar and Bharati, 2012b
79	*Cajanus scarabaeoides* (L.) du Petit-Thou (=*Atylosia scarabaeoides* Benth.) (Fabaceae) L.N.: Gonj (Morena), Bankulthi (Amarkantak), Gaisani (Chhotanagpur)	Leaf paste, Whole plant paste; Plant decoction + Seeds of *Dolichos biflorus*	Diarrhea; Estrus; Diarrhea; Dysentery	Sikarwar, 1996; Shukla et al., 2007; Mishra et al., 2010; Singh, 1987
80	*Calligonum comosum* L.' Her.(Polygonaceae) L.N.: Fog (Bhiwani)	Plant fed	Urinary problems	Yadav et al., 2014

TABLE 6.1 (Continued)

S. No.	Name of the Plant	Part(s) used	Diseases	References
81	**Calligonum polygonoides** L. (Polygonaceae) L.N.: Phog (Rajasthan, Southern Rajasthan, Arid zone)	Whole plant extract, whole plant decoction; Phyllods paste; Leaf paste	Constipation, urinary problems; Stop water fall from nose; Initiate curding process	Galav et al., 2013; Takhar, 2004; Kumar et al., 2004
82	**Calotropis gigantea** (L) R.Br. ex Ait. (Asclepiadaceae) L.N.: Safed Aak (Baster, Bundelkhand, Shivpuri), Safed Akaua (Morena), Safed Ankuru (Ratlam), Safed madar (Vindhyan region), Akaun (Darbhanga), Madar (Jaspur and Surguja), Akban (Chhotanagpur), Maddar (Uttar Pradesh)	Leaf and root paste; Burnt root powder + sesame oil; Warm leaves; Root; Latex; Flower buds, Burnt root; Flower buds + *Ricinus communis* leaves + salt; Root paste + Pepper + Garlic; Leaves + Mustard oil; Heated leaves; Root paste + Fruits of *Piper nigrum* + Oil of *Brassica napus*, leaf paste + *Brassica napus* oil	Wounds; Foot and mouth disease; Running nose; Boils, cuts, injury, wounds, blisters; Diarrhea, Dysentery, Wounds; Dysentery; Fever; Swelling inflammation; Swelling; Fever, shivering	Sikarwar and Kumar, 2005; Sikarwar, 1996; Mishra et al., 2010; Jatav et al., 2014; Jadhav, 2009; Dwivedi et al., 2009; Mishra et al., 1996; Ekka, 2015; Singh, 1987; Kumar and Bharati, 2013

S. No.	Name of the Plant	Part(s) used	Diseases	References
83	*Calotropis procera* (Ait.) R.Br. (Asclepiadaceae) L.N.: Madar (Jhabua, Malva region, Sitapur), Akwan (Chitrakoot), Akda (Alirajpur), Aak (Gwalior, Shivpuri, Narsinghpur, Gorakhpur, Southern Rajasthan, Bhiwani, Banswara, Arid zone), Akkaua (Jhansi, Tikamgarh), Akdo (Mount Abu), Maddar (Jalaun), Akra (Pratapgarh, Rajasthan), Mandar (Gorakhpur), Aakda (Rajasthan)	Leaf juice, Latex; Leaves, Leaves warmed with mustard oil; Leaf paste; Latex; Root powder + milk; Leaves; Leaves; Latex + seed oil of *Arachis hypogaea* + red lead; Warmed leaves; Flower paste + jaggery, latex; Latex; 50 g Sindur (Red oxide of lead) + 200 ml mustard oil + 1 leaves of *Calotropis procera*; Root paste + Black pepper + mustard oil; Smoke of stem; Latex + Hydrochloric acid + latex of *Mangifera indica*; Latex, Stem + Geru + Sindoor; Latex, Gynostegium + Butter milk, fresh bark, ash of stem + ghee, flower decoction; Tail is dipped into latex; leaves fed, dried leaves fed; Leaf paste + *Brassica napus* oil; Ash of stem + Desi ghee; Frothed solution; Latex, Gynostegium + Butter milk, fresh bark tied as bandage, stem ash + ghee, Flower decoction: Root paste	Earache, Conjunctivitis, Falling of tail; Swelling, Indigestion; Bone fracture; Skin diseases; Urine retention; Diarrhea and Dysentery; Stomachache; Tumor; Swelling; Easy delivery, snake bite; Skin infection; Shivering; Fever; Khurpaka; Scorpion bite; Burst of tail; Wobble (Kampana); Hasten suppuration, Cracking of teats; Infestation of worms in the stomach; Cracking of teats, wounds, gastro-intestinal worms; Removal of retained placenta after delivery, kill intestinal worms, increase milk quantity; Shivering; Wound; Diarrhea; Suppuration, cracking of teats, infestation of worms in stomach, cure cracking of teats, wound healing, gastro-intestinal parasitic worms; External parasites	Kadel and Jain, 2006; Tripathi and Singh, 2010; Mishra et al., 2010; Chouhan and Ray, 2014; Shrivastava et al., 2012; Jatav et al., 2014; Dwivedi et al., 2009; Nigam and Sharma, 2010; Sanghi, 2014; Verma, 2014; Galav et al., 2007; Kumar and Bharti, 2012a; Meena, 2014; Galav et al., 2013; Mangal, 2015; Takhar, 2004; Yadav et al., 2014; Kumar and Bharati, 2013; Yadav et al., 2015; Dudi and Meena, 2015; Kumar et al., 2004; Gupta et al., 2015
84	*Camellia sinensis* (L.) O. Kuntze (Theaceae) L.N.: Chai (Gonda)	Leaf decoction	Eye trouble	Pandey et al., 1999

TABLE 6.1 (Continued)

S. No.	Name of the Plant	Part(s) used	Diseases	References
85	*Canavalia ensiformis* (L.) DC. (Fabaceae) L.N.: Barseem (Darbhanga)	Barseem leaves + Boiled seeds of *Dolichos biflorus* L.	Increase lactation	Mishra et al., 1996
86	*Cannabis sativa* L. (Cannabinaceae) L.N.: Bhang (Jhansi, Marwar)	Leaf powder + whey + water; Paste of dried leaves	Loose motion; Diarrhea	Nigam and Sharma, 2010; Dudi and Meena, 2015
87	*Capparis decidua* (Forssk) Edgew. (Capparaceae) L.N.: Kareel (Morena), Karil (West Nimar), Ker (Rajasthan), Kair (Southern Rajasthan), Keir (Aridzone)	Root paste; Whole plant paste; Syrup made of shoots of *Cissus quadrangularis* + ash of stem of *Capparis decidua* + roots of *Ziziphus jujuba* + Jaggery + milk; Powder of charcoal of root + *Sesamum orientale* oil, Extract of fresh stem, fruit fed	Conjunctivitis; Inflamed shoulders; Bone fracture, wound, gastric problem	Sikarwar, 1996; Satya and Solanki, 2009; Galav et al., 2013; Takhar, 2004; Kumar et al., 2004
88	*Capparis zeylanica* L. (Capparaceae) L.N.: Waghata (Betul) (Figure 6.1F)	Leaves paste + Edible oil	Bone fracture	Patil and Deshmukh, 2015
89	*Capsicum annuum* L. (Solanaceae) L. N. Shimla mirch (Shivpuri), Mirch (Jhansi, Southern Rajasthan), Mirchi (Pratapgarh), Lal mirach (Arid zone)	Fruits; Fruit paste + seeds of *Allium sativum* + seeds of *Piper nigrum* + seeds of *Cuminum cyminum* + alum + water; Paste of red chilies; Dried fruit powder	Cattle castrated; Dulness; Khurpaka; Dog bite	Jatav et al., 2014; Nigam and Sharma, 2010; Meena, 2014; Takhar, 2004; Kumar et al., 2004

S. No.	Name of the Plant	Part(s) used	Diseases	References
90	*Capsicum frutescens* L. (Solanaceae) L.N.: Lal mirch (Jhansi, Bhiwani)	Fruit powder + rock salt + Jaggery; Fruits with food material	Lunacy; Heat production	Nigam and Sharma, 2010; Yadav et al., 2014
91	*Cardiospermum halicacabum* L. (Sapindaceae) L.N.: Nagad Gomchi (Morena), Ramano, Kalicharmoi (Udaipur, Sirohi)	Leaf paste + whey; Root paste	Filariasis, Diarrhea; Remove lice and other parasitic insects	Sikarwar, 1996; Sebastian, 1984
92	*Careya arborea* (Roxb) (Lecythidaceae) L. N. Kumbhi, (Jhabua, Jabalpur, Jaspur and Surguja), Kumodi, Kumbhi (Bastar) (Figure 6.1G)	Leaf paste, Stem bark paste; Ripe friuts	Wounds, Diarrhea; Dysentery	Sikarwar and Kumar, 2005; Sikarwar et al., 1994; Ekka, 2015
93	*Carissa carandas* L. (Apocynaceae) L.N.: Karonda (Jhansi, West Nimar)	Rootpaste + Coconut oil; Root paste	Maggots of wounds	Nigam and Sharma, 2010; Satya and Solanki, 2009
94	*Carissa spinarum* L. (Apocynaceae) L.N.: Jangli karonda (Raigarh), Gadasur (Singhbhum)	Root paste; Root paste + *Madhuca longifolia* flower liquor	Wounds; Removing worm of sore	Sikarwar et al., 1994; Pal, 1980
95	*Carum carvi* L. (Apiaceae) L.N.: Ajwain (Vindhyan region)	Seed powder	Flatulence, Indigestion	Dwivedi et al., 2009

TABLE 6.1 (Continued)

S. No.	Name of the Plant	Part(s) used	Diseases	References
96	*Cassia angustifolia* Vahl (Caesalpiniaceae) L.N.: Senna (Southern Rajasthan), Sonamukhi (Aridzone)	Pos and leaves	Gastric problem	Takhar, 2004; Kumar et al., 2004
97	*Cassia fistula* L. (Caesalpiniaceae) L.N.: Barron (Morena), Amaltas (Amarkantak, Bundelkhand, Shivpuri, Narsinghpur, Tikamgarh, Betul, Udaipur, Sirohi, Rajasthan), Garmalo (Ratlam), Germala (West Nimar), Bahawa (Mount Abu), Bandarlouri (Jaspur and Surguja), Karmalo (Udaipur, Sirohi), Murju-Baha (Chhotanagpur), Banderlatia (Midnapur)	Fruit paste, Stem bark paste with cold water; Stem bark powder; Seed paste; Leaves, fruits; Fruit extract; Fruit pulp + water, fruit decoction; Leaf paste, bark paste + pepper; Pod paste + wheat bread, Leaf paste + mustard oil, leaves and ripe pode paste; Pods; Root paste; Bark powder + water; Root paste + + Black pepper (*Piper nigrum*) + Leaf juice of *Artocarpus heterophyllus*; Flower decoction; Leaves + resin of *Shorea rousta* are burnt; Bark/pods boiled in water fed after cooling, pods fed	Anthrax, Diarrhea; Dysentery; Insect bite, Swelling, rheumatism, facial paralysis; Indigestion (Afra disease); Gas and acidity, Cold, throat infection; Rheumatism, fever; Indigestion, Improve appetite, as purgative, severe constipation; Flatulence; Snake bite; Diarrhea; Swelling; Cough and cold; Antidote to snake bite; Constipation	Sikarwar, 1996; Sikarwar et al., 1994; Shukla et al., 2007; Mishra et al., 2010; Jatav et al., 2014; Jadhav, 2009; Satya and Solanki, 2009; Sanghi, 2014; Verma, 2014; Galav et al., 2007; Ekka, 2015; Sebastian, 1984; Singh, 1987; Patil and Deshmukh, 2015; Pal, 1980; Gupta et al., 2015
98	*Cassia pumila* Lam. (Caesalpiniaceae) L.N.: Aabala (Udaipur, Srohi)	Whole plant	Galactagogue	Sebastian, 1984
99	*Catharanthus roseus* (L.) G.Don (Apocynaceae) L.N.: Sadabhar (Alirajpur)	Whole plant extract	Wounds	Chouhan and Ray, 2014

S. No.	Name of the Plant	Part(s) used	Diseases	References
100	*Catunaregum spinosa* (Thunb.) Thirv. (Rubiaceae) L.N.: Kharedi (Rajasthan)	Whole plant + Jaggery + rhizome powder of *Curcuma longa*	Diarrhea	Galav et al., 2013
101	*Cayratia trifolia* (L.) Domin. (Vitaceae) L.N.: Ramcharan (Morena), Khersiliya (Dhar), Choti gurbel (Ratlam), Tifankand (Betul)	Root paste + turmeric paste; Root extract; Tuber paste	Foot and mouth disease, Throat swelling; Dog bite; Yoke sore	Sikarwar, 1996 Sikarwar et al., 1994; Jadhav, 2009; Patil and Deshmukh, 2015
102	*Ceiba pentandra* (L.) Gaertn. (Bombacaceae) L.N.: Semal (Alirajpur), Simal (Marwar)	Stem bark paste; Bark paste	Wounds; Internal parasites	Chouhan and Ray, 2014; Dudi and Meena, 2015
103	*Celosia argentea* L. (Amaranthaceae) L.N.: Gorkha (Rajasthan)	Root juice + Fruit or leaf extract of *Tamarindus indica*; Root juice + Fruits or leaves of *Tamarindus indica*	Food poisoning	Galav et al., 2013; Gupta et al., 2015
104	*Centella asiatica* (L.) Urban (Apiaceae) L.N.: Kharaiti (Uttar Pradesh)	Whole plant + Butter milk, Whole plant decoction	Diarrhea due to sun-stroke, Expulsion of placenta during pregnancy	Kumar and Bharati, 2013
105	*Chenopodium album* L. (Chenopodiaceae) L.N.: Bathua (Satna)	Leaf powder	Sore and wounds	Gautam and Richhariya, 2015
106	*Chlorophytum tuberosum* (Roxb.) Baker (Liliaceae) L.N.: Musli (Jhabua)	Leaves	Brain disease	Kadel and Jain, 2006

TABLE 6.1 (Continued)

S. No.	Name of the Plant	Part(s) used	Diseases	References
107	*Chloroxylon swietenia* DC. (Meliaceae) L.N.: Bharhi, (Surguja)	Leaf paste	Swelling	Sikarwar, and Kumar, 2005
108	*Chrozophora rottleri* (Geis.) Juss. ex Spreng. (Euphorbiaceae) L.N.: Kala Dhatura (Morena, Vindhyan region)	Leaf paste + whey; Leaf paste	Kill wound worms; Maggots of wounds	Sikarwar, 1996; Dwivedi et al., 2009
109	*Cicer arietinum* L. (Fabaceae) L.N.: Gram (Gwalior), Chana (Darbhanga, Sitapur, Bhiwani)	Gram flour + butter milk; Chana + *Hordeum vulgare* porridge; Imbibed seeds + Jaggery, Seeds power + water; Soaked seeds; Paste of soaked seeds	Dysentery; Blood dysentery; Flatulence, Diarrhea in winter season; Increase the milk; Diarrhea	Shrivastava et al., 2012; Mishra et al., 1996; Kumar and Bharati, 2012b; Yadav et al., 2014; Dudi and Meena, 2015
110	*Cinnamomum camphora* (L.) Ness and Eberm. (Lauraceae) L.N.: Kapoor (Darbhanga, Gonda, Sitapur, Gorakhpur)	Kapoor + Sugar + Lime + Napthalene; Kapoor; Kapoor + crushed roots of *Alocasia indica Triticum aestivum* bread; Kapoor + Mustard oil	Wounds; Ectoparasites; Regulate estrus cycle; Redness and mucous eyes	Mishra et al., 1996; Pandey et al., 1999; Kumar and Bharati, 2012b; Mangal, 2015
111	*Cissampelos pareira* L. (Menispermaceae) L.N.: Kalipar (Mount Abu)	Root extract	Snake bite and Dog bite	Galav et al., 2007

S. No.	Name of the Plant	Part(s) used	Diseases	References
112	*Cissus quadrangularis* L. (Vitaceae) L.N.: Harjod (Chitrakoot, Shivpuri), Arhand (Ratlam), Harjor (Vindhyan region, Jaspur and Surguja), Harjori (West Nimar), Hadjor (Narsinghpur), Hadjor (Mount Abu), Hadjod (Betul)	Leaf paste; Stem paste + Amarbel; Leaf decoction + pepper + garlic; Stem extract + wheat bread; Tender stem juice, stem paste; Leaf and stem paste + turmeric + sesame oil; Whole plant extract; Leaf decoction + Pepper + Garlic; Stem paste	Bone fracture; Fever; Fracture; Bone fracture; Diarrhea, Bone fracture; Bone fracture; Fever; Bone fracture	Tripathi and Singh, 2010; Gautam and Richhariya, 2015; Jatav et al., 2014; Jadhav, 2009; Dwivedi et al., 2009; Satya and Solanki, 2009; Sanghi, 2014; Galav et al., 2007; Ekka, 2015; Patil and Deshmukh, 2015
113	*Cistanche tubulosa* (Schremk) Wight. (Orobanchaceae) L.N.: Lauki mula (Southern Rajasthan, Arid zone)	Paste of whole plant	Bovine Viral Mammatus	Takhar, 2004; Kumar et al., 2004
114	*Citrullus colocynthis* (L.) Schr. (Cucurbitaceae) L.N.: Dimbo (Jaspur and Surguja), Gad tumba (Rajasthan), Tumba (Southern Rajasthan), Gadumba (Bhiwani), Tumba (Marwar, Arid zone) (Figure 6.1H)	Root decoction; Root decoction, Root juice + honey + mustard oil; Paste of roasted seeds, warmed fruits, fruits + salt, fruits; Fruits + Whole plant of *Solanum surratense*, Fruits fed; Fruit powder; Paste of roasted seeds; Warmed fruits, fruit is kept under burning coal and morning mixed with salt and given, fruit fed; *Citrullus colocynthis* + Salt + Water, fruit pulp and root decoction	Cough; Constipation, easy opening of uterus during delivery; Wounds, tonic, skin irritation, stomach disorders, increase lactation; Dysentery, Improving digestion; Bloat; Wounds to kill germs: Tonic, cure skin irritation, stomach disorder, increase milk production, dissuade from eating sand; Surra/Trypanosomiasis, constipation	Ekka, 2015; Galav et al., 2013; Takhar, 2004; Yadav et al., 2014; Dudi and Meena, 2015; Kumar et al., 2004; Gupta et al., 2015

TABLE 6.1 (Continued)

S. No.	Name of the Plant	Part(s) used	Diseases	References
115	*Citrus aurantifolia* (Christm.) Swing (Rutaceae) L.N.: Neebu (Sitapur)	Juice + Sodium bicarbonate + Tatric acid + seeds of *Bunium persicum* + seeds of *Trigonella foenum-graecum* + Sulamani namak + Black salt	Flatulence	Kumar and Bharati, 2012b
116	*Citrus limon* (L.) Burm. *f.* (Rutaceae) L.N.: Nimbu (Gwalior, Jhansi)	Leaves + leaves of Babool (*Acacia nilotica* (L.) Del. Ssp. *indica* (Benth.) Hill. + baking soda, Fruit juice + Leaf paste of *Azadirachta indica*; Fruit juice	Bloat (Tympanitis), Scabies; Blindness	Shrivastava et al., 2012; Nigam and Sharma, 2010
117	*Citrus medica* L. (Rutaceae) L.N.: Nimbu (Rajasthan), Bara Nimbu (Marwar)	Fruit juice + Sugar; Paste of leaves	Induce lactation; Ectoparasites	Galav et al., 2013; Dudi and Meena, 2015
118	*Clematis triloba* Heyne ex Roth (Ranunculaceae) L.N.: Ran-mogra (Betul)	Leaf paste	Throat swelling	Patil and Deshmukh, 2015
119	*Cleome gynandra* L. (Cleomaceae) L.N.: Safed Hul-hul (Morena, Rajasthan); Hulhul (Chitrakoot, Vindhyan region), Hurhur (North Bihar), Hurhura (Sitapur)	Leaf paste; Seed paste; Aerial part decoction; Plant paste; Leaf paste; Whole plant extract	Wounds; Skin diseases; Anthelmintic, Vermifuge, Anti-rheumatic; Maggots in wounds; Eczema; Maggots	Sikarwar, 1996; Dwivedi et al. 2009; Jha et al., 1991; Kumar and Bharati, 2012b; Galav et al., 2013; Kumar and Bharati, 2013
120	*Cleome simplicifolia* (Camp.) Hook*f.* and Thoms. (Cleomaceae) L.N.: Hurera (West Nimar)	Whole plant ash + sesame oil, seed paste	Wounds, arthritis	Satya and Solanki, 2009

S. No.	Name of the Plant	Part(s) used	Diseases	References
121	*Cleome viscosa* L. (Cleomaceae) L.N.: Pivli Tilwan (Betul), Hulhul (Rajasthan)	Leaf decoction, seed powder + water; Seed paste + water	Killing maggots of sores, Epilepsy; Diarrhea, Fever	Patil and Deshmukh, 2015; Galav et al., 2013
122	*Cleistanthus collinus* Benth.ex Hook.*f.* (Euphorbiaceae) L.N.: Kama, (Baster),	Bark decoction, Fruit paste	Wounds	Sikarwar and Kumar, 2005
123	*Clerodendrum indicum* (L.) Kuntze (Verbanaceae) L.N.: Bhati (Gorakhpur)	Leaf decoction	Etching	Mangal, 2015
124	*Clerodendrum infortunatum* L. (Verbenaceae) L.N.: Bhantwas (Gonad)	Leaf decoction	Ectoparasites	Pandey et al., 1999
125	*Clerodendrum multiflorum* Baker (=*C. phlomidis* L.) (Verbenaceae) L.N.: Arni (West Nimar, Udaipur), Piri chamgar (Singhbhum)	Plant juice; Leaf juice; Decoction of plant; fresh plant paste, leaf juice	Kill lice, itching; Hooves rot, maggots of wounds; Dyspepsia, diarrhea, as anthelmintic	Satya and Solanki, 2009; Deora et al., 2004; Pal, 1980
126	*Clitoria ternatea* L. (Fabaceae) L.N.: Gokarni (Betul)	Root powder	Scorpion sting	Patil and Deshmukh, 2015

TABLE 6.1 (Continued)

S. No.	Name of the Plant	Part(s) used	Diseases	References
127	*Coccinia grandis* (L.) Voigt. (Cucurbitaceae) L.N.: Jangle Kundri (Jaspur and Surguja)	Leaf extract + Ghee	Wounds	Ekka, 2015
128	*Coccinia indica* Wt. and Arn. (Cucurbitaceae) L.N.: Tilkor (Darbhanga)	Leaf juice	Cough, cold and fever	Mishra et al., 1996
129	*Cocculus hirsutus* (L.) Diels (Menispermaceae) L.N.: Chirenta (Morena), Ma baba na baklo (Udaipur, Sirohi)	Leaf paste; Plant	Diarrhea; Lactagogue	Sikarwar, 1996; Sebastian, 1984
130	*Cocculus pendulus* (Forst.) Diels (Menispermaceae) L.N.: Pilwan (Rajasthan)	Ash of stem + Cow milk	Mastitis	Galav et al., 2013
131	*Cocos nucifera* L. (Arecaceae) L.N.: Nariyal (Alirajpur, Gorakhpur)	Seed oil; Coconut oil + Caustic soda	Wounds; Mastitis	Chouhan and Ray, 2014; Mangal, 2015
132	*Coleus amboinicus* Lour. (Lamiaceae) L.N.: Konchi, (Mandla)	Root paste	Blood dysentery	Sikarwar and Kumar, 2005
133	*Coleus forskohlii* Briq. (=*C. barbatus* Benth.) (Lamiaceae) L.N.: Bayada (Sitapur)	Roots + *Ferula asafoetida* + seeds of *Sorghum vulgare* + water	Anorexia	Kumar and Bharati, 2012b

S. No.	Name of the Plant	Part(s) used	Diseases	References
134	*Colocasia esculenta* (L.) Schott (Araceae) L.N.: Saru (Amarkantak), Jangli Arbi (West Nimar), Arvi (Marwar)	Crushed tubers; Corm paste; Dried leaves frothed solution leaf paste	Increasing lactation; Wounds; Diarrhea	Shukla et al., 2007; Satya and Solanki, 2009; Dudi and Meena, 2015
135	*Corallocapus epigaeus* (Rottl. and Willd) Hook. *f.* (Cucurbitaceae) L.N.: Mirchia kand (West Nimar, Rajasthan)	Bulb paste + mustard oil; Tuber paste + whole plant of *Tinospora cordifolia*	Abdominal digestion; Tonsillitis	Satya and Solanki, 2009; Galav et al., 2013
136	*Commelina suffruticosa* Blume (Commelinaceae) L.N.: Nagel gera (Baster, Morena)	Whole plant paste	Wounds to kill worms	Sikarwar and Kumar, 2005; Sikarwar, 1996.
137	*Convolvulus arvensis* L. (Convolvulaceae) L.N.: Galobi (Ratlam)	Whole plant	Lactation	Jadhav, 2009
138	*Corchorus capsularis* L. (Tiliaceae) L.N.: Chaunch Bhaji (Baster), Patsun (Sitapur)	Seed paste; Dried flowers, fruit paste; Crushed dried flower	Kill wound worms; Prolongation of expulsion of placenta, Expel placenta	Sikarwar and Kumar, 2005; Kumar and Bharati, 2012b; Kumar and Bharati, 2013
139	*Corchorus depressus* (L.) Stocks (Tiliaceae) L.N.: Chamkas (Southern Rajasthan)	Leaves fed	Stomachache	Takhar, 2004; Kumar et al., 2004

TABLE 6.1 (Continued)

S. No.	Name of the Plant	Part(s) used	Diseases	References
140	**Cordia dichotoma** Forst.f. (Ehretiaceae) L.N.: Goonda (Rajasthan), Leshwa (Bhiwani)	Leaf juice + Honey: Warmed leaves are tied	Foot and mouth disease; Cracked nipples	Galav et al., 2013; Yadav et al., 2014
141	**Coriandrum sativum** L. (Apiaceae) L.N.: Dhania (Chitrakoot, Vindhyan region, Jhansi, Tikamgarh, Jaspur and Surguja, Darbhanga, Gonda)	Fruit paste + black salt; Leaf paste; Seed powder + leaves of *Lawsonia inermis* + water; Seed powder + leaf paste of *Lawsonia inermis*; Fruit powder; Leaves; leaf paste + chips of soap: Fruit powder + Jaggery	Indigestion; Mastitis; Loose motion; Loose motion; Facilitate conception; Increase lactation; Mastitis; Omitting of conception	Tripathi and Singh, 2010; Dwivedi et al., 2009; Nigam and Sharma, 2010; Verma, 2014; Ekka, 2015; Mishra et al., 1996; Pandey et al., 1999
142	**Costus speciosus** Sm. (Costaceae)L.N.: Kandua (Jhabua), Nalguj (Betul) (Figure 6.2I)	Root paste; Root stock	Fever; Rheumatic pain	Sikarwar and Kumar, 2005; Patil and Deshmukh, 2015
143	**Crinum latifolium** L. (Amaryllidaceae) L.N.: Barapungania (Jhabua)	The juice of bulbous root	Fever	Sikarwar and Kumar, 2005
144	**Crinum pratense** Herb (Amaryllidaceae) L.N.: Jalsatawar (Baster)	Root stock paste	Wounds	Sikarwar and Kumar, 2005

FIGURE 6.2

TABLE 6.1 (Continued)

S. No.	Name of the Plant	Part(s) used	Diseases	References
145	*Crotalaria burhia* Buch.-Ham. (Fabaceae) L.N.: Sinio (Pratapgarh, Southern Rajasthan, Arid zone), Jhunda (Rajasthan)	Root decoction; Whole plant decoction; Root decoction	Expel retained placenta; Urinary problem, easy removal of placenta; Expel retained placenta	Meena, 2014; Galav et al., 2013; Takhar, 2004; Kumar et al., 2004
146	*Crotalaria juncea* L. (Fabaceae) L.N.: San (Morena, Chitrakoot), Sanai (Darbhanga)	Leaf paste; Seeds	Wounds; Cut and wounds; Post delivery clearance of uterus	Sikarwar, 1996; Tripathi and Singh, 2010; Mishra et al., 1996
147	*Crotalaria linifolia* L (Fabaceae) L.N.: Korsa (Chhotanagpur)	Plat + Common salt	Remove the maggots of wounds	Singh, 1987
148	*Crotalaria spectabilis* Roth (Fabaceae) L.N.: Sakesing, Xar-shunka, Haradi jiling (Midnapur)	Seeds, fresh plants	Reduce toxic effects	Pal, 1980
149	*Cucumis melo* L. (Cucurbitaceae) L.N.: Phoot kachari (West Nimar), Kachra (Pratapgarh, Southern Rajasthan), Kanchra (Arid zone)	Fruit; Fruit paste + Butter; Fruit powder + Ghee	Abdominal distention, Appetizer, dysentery, abdominal disorders; Dysentery and gastric problems; Dysentery, gastric problems	Satya and Solanki, 2009; Meena, 2014; Takhar, 2004; Kumar et al., 2004

S. No.	Name of the Plant	Part(s) used	Diseases	References
150	*Cucumis callosus* (Rottler) Cogn. (Cucurbitaceae) L.N.: Kachri (Bhiwani, Rajasthan)	Crushed fruits + Whey; Fruits crushed + Sugar + Oil + Water, Fruits + *Trachyspermum ammi*	Stomachache; Diarrhea; Constipation	Yadav et al., 2014; Gupta et al., 2015
151	*Cuminum cyminum* L. (Apiaceae) L.N.: Jeera (Jhabua, Gwalior, Vindhyan region, Sitapur)	Seed powder; seeds + Jaggery; Fruit powder + Seed powder of *Trigonella foenum-graecum* + Fruit powder of *Trachyspermum ammi* + Seed powder of *Cassia tora* + fruit powder of *Piper nigrum* + *Ferula asafoetida* Seeds; Seeds paste of *Trigonella foenum-graecum* + Black salt + *Cuminum cyminum* + *Ferula asafoetida* + leaves of *Bombax ceiba*	Indigestion; Agalactiae, Constipation; Mastitis; Flatulence, dysentery	Kadel and Jain, 2006; Shrivastava et al., 2012; Dwivedi et al., 2009; Kumar and Bharati, 2012b;
152	*Curculigo orchiodes* Gaertn. (Hypoxidaceae) L.N.: Kali musli (Jhabua, Udaipur), Ran musli (West Nimar), Tar-muli (Chhotanagpur)	Stem paste; Root; Root paste + wheat bread; Tuber juice	Foot and Mouth disease; Maggots of wounds; Foot and Mouth disease; Eye diseases	Kadel and Jain, 2006; Satya and Solanki, 2009; Nag et al., 2009; Singh, 1987
153	*Curcuma angustifolia* Dalz. and Gibs. (Zingiberaceae) L.N.: Sapaini (Jaspur and Surguja)	Rhizome extract + Ghee	Running nose	Ekka, 2015

TABLE 6.1 (Continued)

S. No.	Name of the Plant	Part(s) used	Diseases	References
154	**Curcuma amada** Roxb. (Zingiberaceae) L.N.: Amba haldi (Udaipur), Jangli haldi (Mount Abu), Aama haldi (Sitapur)	Rhizome paste, Dried rhizome powder + 100 g rock salt + 100 g alum + 100 g *Piper longum* + 100 g seeds of *Trachyspermum ammi* + 250 g jaggery + 100 g milk fat; Rhizome + 2–3 bulbs of *Allium cepa* + Jaggery; Decoction of rhizome + *Eichhornia crassipes*	Fever, Easy removal of placenta; Indigestion and flatulence; Hoof wound, inflammation of shoulder	Nag et al., 2009; Galav et al., 2007; Kumar and Bharati, 2012b; Kumar and Bharati, 2013
155	**Curcuma longa** L. (= C. *domestica* Valeton) (Zingiberaceae) L.N.: Haldi (Jhabua, Alirajpur, Jhansi, North Bihar, Darbhanga Jalaun, Pratapgarh, Gorakhpur, Bhiwani, Marwar), Halad (Mount Abu, Gonda)	Rhizome paste, Leaf juice; Rhizome powder; Rhizome powder + rock salt + pure ghee; Rhizome decoction; Powder of rhizome (50 g) + alum (10 g) + jaggery (2 kg); Crushed rhizome; Rhizome powder; Paste of rhizome + *Acacia catechu* (kattha) + *Piper betle*; Rhizome decoction; Rhizome; Rhizome powder + Jaggery; Rhizome + Mango pickle; Paste of Rhizome + *Citrus medica* Paste of Haldi + *Cynodon dactylon* + Salt, leaf solution	Bad taste of mouth cavity (Dandki disease), Conjunctivitis; Wounds; Swelling of teats; Carminative, Anti-rheumatic; Bone fracture; Flatulence; Bad taste or grains in mouth cavity; Crack in palate; Physical debility; Hindura disease; Placenta did not fall; Gastric problem; Ectoparasites, foot and mouth disease, fever	Kadel and Jain, 2006; Chouhan and Ray, 2014; Nigam and Sharma, 2010; Jha et al., 1991; Galav et al., 2007; Kumar and Bharti, 2012a; Kumar and Bharati, 2013; Mishra et al., 1996; Pandey et al., 1999; Meena, 2014; Mangal, 2015; Yadav et al., 2014; Dudi and Meena, 2015
156	**Curcuma pseudomontana** L. (Zingiberaceae) L.N.: Ran Halad (Betul)	Leaf decoction + salt	Tymphany	Patil and Deshmukh, 2015

S. No.	Name of the Plant	Part(s) used	Diseases	References
157	*Cuscuta reflexa* Roxb. (Convolvulaceae) L.N.: Amerbel (Satna, West Nimar, Bhiwani), Amar-bel (Rajasthan)	Stem; Plant paste; Whole plant + leaves of *Datura innoxia* + boiled in mustard oil; Plant decoction	Skin diseases; Uterine prolapsed; Lumbago, rheumatic pain; Poisonous venom	Gautam and Richhariya, 2015; Satya and Solanki, 2009; Galav et al., 2013; Yadav et al., 2014
158	*Cyamopsis tetragonoloba* (L.) Taub. (Fabaceae) L.N.: Gawar (Udaipur, Sirohi), Guar (Pratapgarh, Southern Rajasthan, Bhiwani), Guar phalli (Arid zone)	Plants; Boiled seeds; Boiled seeds + Oil; Fried seeds; Filtrate of boiled seeds water + Caster oil	Enhance lactation; Weakness; Induce heat; Dysentery	Sebastian, 1984: Meena, 2014; Takhar, 2004; Yadav et al., 2014; Kumar et al., 2004
159	*Cyanotis tuberosa* Schultes (Commelinaceae) L.N.: Jereng-Arok (Chhotanagpur)	Tuber paste + Common salt	Fever	Singh, 1987
160	*Cymbopogon martini* (Roxb.) Watson (Poaceae) L.N.: Rohido (Udaipur)	Whole plant fumigation	Fever	Nag et al., 2009
161	*Cymbopogon schoenanthus* (L.) Spreng. (Poaceae) L.N.: Gyaliyo (Udaipur, Sirohi)	Plants	Galactagogue	Sebastian, 1984
162	*Cynodon dactylon* (L.) Pers. (Poaceae) L.N.: Duba (Jhabua, Tikamgarh)	Plant; Aerial plant, leaf juice	Increase lactation; Conjunctivitis	Kadel and Jain, 2006; Verma, 2014

TABLE 6.1 (Continued)

S. No.	Name of the Plant	Part(s) used	Diseases	References
163	*Cyperus rotundus* L. (Cyperaceae) L.N.: Montha (Bundelkhand), Dongli (West Nimar)	Root powder; Plant	Intestinal worms; Tonsillitis, stomach disorders	Mishra et al., 2010; Satya and Solanki, 2009
164	*Dalbergia latifolia* Roxb. (Fabaceae) L.N.: Shisham (Jabalpur, Satna, Jaspur and Surguja, Gonda), Kalashisham (Shivpuri)	Stem bark decoction; Leaf juice; Stem bark paste + Garlic + pepper; Roasted fruits	Hoop diseases; Skin eruptions, indigestion; Lazy in grazing; Diarrhea, liver disorder, sunstroke	Sikarwar et al., 1994; Gautam and Richhariya, 2015; Jatav et al., 2014; Ekka, 2015; Pandey et al., 1999
165	*Dalbergia sissoo* Roxb. ex DC. (Fabaceae) L.N.: Shisham (Jhansi, Tikamgarh, Rajasthan), Tali (Rajasthan)	Leaf paste + water; Leaf juice; Leaf infusion, soaked leaves; Leaf infusion	Blisters and leg sore; Stop bleeding; Sunstroke, dysuria, heat in the body; Diarrhea; Sun stroke/heat stroke	Nigam and Sharma, 2010; Verma, 2014; Galav et al., 2013; Gupta et al., 2015
166	*Datura innoxia* Mill. (=*D. alba* Nees) (Solanaceae) L.N.: Dhatura (Jhabua, Mount Abu), Datura (Darbhanga)	Root paste; Leaf extract; Fruits + Seeds of *Brassica juncea* + Rhizome of *Zingiber officinale*	Fever; Tonsillitis; Strength and warmth	Kadel and Jain, 2006; Galav et al., 2007; Mishra et al., 1996
167	*Datura metel* L. (Solanaceae) L.N.: Datura (Alirajpur), Dhatura (Shivpuri, Jhansi, Tikamgarh, Jaspur and Surguja, Gonda)	Leaf paste; Roasted fruits; Root powder; Ripe fruit paste, Leaf and root paste; Diarrhea, dysentery, cough	Wounds; dysentery and cough; Bleeding; Cold, wound healing; Roasted fruits	Chouhan and Ray, 2014; Jatav et al., 2014; Nigam and Sharma, 2010; Verma, 2014; Ekka, 2015; Pandey et al., 1999

S. No.	Name of the Plant	Part(s) used	Diseases	References
168	*Datura stramonium* L. (Solanaceae) L.N.: Kala dhatura (Chitrakoot), Dhturo (Ratlam); Dhatur (North Bihar), Dhatura (Gorakhpur, Banswara)	Leaf paste with ghee; Seed extract; Leaf decoction; Fruit; 1–2 kg grinded seeds	Kill wound worms; Appetizer; Nerve disease, Anti-rheumatic; Diarrhea; Anestrous	Tripathi and Singh, 2010; Jadhav, 2009; Jha et al., 1991; Mangal, 2015; Yadav et al., 20015
169	*Delonix elata* Gamble (Caesalpiniaceae) L.N.: Handeda (Udaipur)	Stem bark decoction	Dysentery, diarrhea	Deora et al., 2004
170	*Delonix regia* (Bojer ex Hook.) Raf. (Caesalpiniaceae) (Tikamgarh)	Bark extract + black pepper + garlic	Fever	Verma, 2014
171	*Dendrocalamus strictus* Nees (Poaceae) L.N.: Bans (Amarkantak, Shivpuri, Mount Abu, Udaipur, Sirohi, Gonda, Southern Rajasthan, Arid zone)	Leaf decoction; Roasted fruits; Infusion of tender stem; Leaves; Tender culms paste; Decoction of stripes	Dysentery; Dysentery and cough; Fractured bone, expel intestinal worms, As stomach washer; Worm infection; Expulsion of placenta	Shukla et al., 2007; Jatav et al., 2014; Galav et al., 2007; Sebastian, 1984; Pandey et al., 1999; Takihar, 2004; Kumar et al., 2004
172	*Dendrophthoe falcata* (L.f.) Etting (Loranthaceae) L.N.: Dudeli (Pratapgarh)	Leaves ties	Bone fracture	Meena, 2014
173	*Desmodium triflorum* (L.) DC. (Fabaceae) L.N.: Tinpatiya (Satna)	Leaves, Whole plant	Wounds, Galactagogue	Gautam and Richhariya, 2015

TABLE 6.1 (Continued)

S. No.	Name of the Plant	Part(s) used	Diseases	References
174	*Desmostachya bipinnata* (L.) Stapf (Poaceae) L.N.: Kush (Gonda), Dab (Rajasthan)	Crushed Inflorescence; Whole plant decoction;	Lack of estrus; Flatulence	Pandey et al., 1999; Galav et al., 2013
175	*Dichrostachys cinerea* (L.) Wt. and Arn. (Mimosaceae) L.N.: O (Udaipur, Sirohi), Amna (Udaipur), Giya-khair kolai (Pratapgarh) (Figure 6.2I)	Fruit decoction; Stem bark powder; Root bark extract + stem bark extract of *Butea monosperma* + *Ziziphus mauritiana*	Enhance lactation; Urinary complaints; Stomachache	Sebastian, 1984; Deora et al., 2004; Meena, 2014
176	*Digitaria adsendens* (H. B., and K.) Henr. (Poaceae) L.	Seeds + Seeds of *Tribulus terrestris*, *Pedalium murex* + Crushed fruits of *Cucumis melo* + *Citrullus lanatus*	Constipation	Galav et al., 2013
177	*Dioscorea bulbifera* L. (= *D. sativa* Thunb.) (Dioscoreaceae) L.N.: Gethikanda (Surguja), Vidarikand (Udaipur, Marwar)	Tuber paste	Wounds; Mastitis; Diarrhea	Sikarwar et al., 1994; Nag et al., 2009; Dudi and Meena, 2015
178	*Diospyros melanoxylon* Roxb. (Ebenaceae) L.N.: Tendu (Satna)	Stem bark extract, Unripe fruits	Diarrhea, Lactation	Gautam and Richhariya, 2015
179	*Diplocyclos palmatus* (L.) Jeffery (Cucurbitaceae) L.N.: Rowa (Amarkantak), Shivlingi (Satna, Betul)	Fruit paste; unripe fruits, Seeds; Crushed fruits or leaves + Butter milk	Kill pest; Fever, colitis; Ephemeral fever	Shukla et al., 2007; Gautam and Richhariya, 2015; Patil and Deshmukh, 2015

S. No.	Name of the Plant	Part(s) used	Diseases	References
180	**Dolichos biflorus** L. (Fabaceae) L.N.: Kurthi (Darbhanga)	Boile seeds + Leaves of *Canavalia insiformis*	Increase lactation	Mishra et al., 1996
181	**Echinops echinatus** Roxb. (Asteraceae) L.N.: Kalthori (Shivpuri), Untkatara (West Nimar), Undh-katalo, Datura (Udaipur, Sirohi)	Root paste; Chopped plant; Plant	Wounds; Increasing lactation; To make barren animal fertile	Sikarwar and Kumar, 2005; Satya and Solanki, 2009; Sebastian, 1984
182	**Eclipta prostrata** (L.) L. (=*E. alba* (L.) Hassk.) (Asteraceae) L.N.: Bhrangraj (Alirajpur), Ghamira (Vindhyan region), Bhringraj (Narsinghpur), Bhangrayya (Gonda), Hatu-Kesari (Chhotanagpur)	Whole plant poultice; Whole plant; Leaf juice; Leaf paste; Leaf decoction + mustard oil; Leaf juice, leaf + stem paste; Root paste + seed oil of *Carthamus tinctorius* (Kusum)	Cut and wounds; Blindness, Bronchitis; Cut, injury; Swelling; Wounds; Cut, foot and mouth disease, dysentery; ulcer and Wounds	Chouhan and Ray, 2014; Jatav et al., 2014; Dwivedi et al., 2009; Sanghi, 2014; Verma, 2014; Pancey et al., 1999; Singh, 1987
183	**Berberis aristata** DC. (Berberidaceae) L.N.: Kilmodi (Marwar)	Rcoot decoction	Fever	Dudi and Meena, 2015
184	**Eichhornia crassipes** (Mart.) Solms (Pontederiaceae) L.N.: Jalkumbhi (Sitapur)	Decoction of *Eichhornia crassipes* + rhizome of *Curcuma amada*	Wound of hoof	Kumar and Bharati, 2012b

TABLE 6.1 (Continued)

S. No.	Name of the Plant	Part(s) used	Diseases	References
185	*Elephantopus scaber* L. (Asteraceae) L.N.: Kirmar, Jangli bhilwa, (Baster) Chilbil (Raigarh), Manjurjhuti (Amarkantak), Minjurchundi (Jaspur and Surguja)	Root pastes; Leaf paste with sugarcandy; Whole plant paste	Kill worms of wounds; Diarrhea; Dysentery	Sikarwar and Kumar, 2005; Shukla et al., 2007; Ekka, 2015
186	*Elettaria cardamomum* (L.) Maton (Zingiberaceae) L.N.: Elachi (Bhiwani)	Fruits + Jaggery	Cold	Yadav et al., 2014
187	*Eleusine coracana* (L.) Gaertn. (Poaceae)L.N.: Maal (Mount Abu)	Flour bread	Diarrhea, mastitis (locally called "thanella"	Galav et al., 2007
188	*Elytraria acaulis* (L.f.) Lindau (Acanthaceae) L.N.: Dasmul (Udaipur), Marua (Darbhanga)	Dried root decoction; Flour bread	As an antinematode; Dysentery	Nag et al., 2009; Mishra et al., 1996
189	*Embelia ribes* Burm.f. (Myrsinaceae) L.N.: Wawding (Betul)	Leaf decoction + Whey, boiled fruits	Dysentery	Patil and Deshmukh, 2015
190	*Emblica officinalis* Gaertn. (Euphorbiaceae) L.N.: Amla (Chhindwara, Betul), Aonla (Surguja), Amla (Satna)	Bark paste; Fruit powder	Wounds; Abdominal disorder	Sikarwar and Kumar, 2005; Gautam and Richhariya, 2015

S. No.	Name of the Plant	Part(s) used	Diseases	References
191	*Enicostema axillare* (Lam.) A. Raynal (= *E. hyssopifolium* Blume) (Gentianaceae) L.N.: Naio (Jhabua), Nami (Udaipur), Namme (Rajasthan)	Entire plant infusion; Leaves	Worms in stomach; Intestinal worms; Fever	Kadel and Jain, 2006; Deora et al., 2004; Galav et al., 2013
192	*Entada pursaetha* Benth. (Mimosaceae) L.N.: Gar-Karla (Chhotanagpur)	Cotyledons paste + Polyporus fungus	Swelling of neck, Diarrhea	Singh, 1987
193	*Ephedra ciliata* Fisch. and Mey. ex C. A. May. (Gnetaceae) Oontphog (Rajasthan)	Whole plant decoction	Constipation	Galav et al., 2013
194	*Equisetum diffusum* D.Don (Equisetaceae) L.N.: Hadjudwa (Gorakhpur), Harjorwa (Uttar Pradesh)	Leaf paste + Geru + Milk;	Fracture; Joint inflammation	Mangal, 2015; Kumar and Bharati, 2013
195	*Equisetum officinales* D.Don (Equisetaceae)L.N.: Harjorwa (Jalaun)	Stems + Geru (Oker red lumber) crushed together + Goat/cow milk	Arthritis	Kumar and Bharti, 2012a
196	*Eruca vesicaria* (L.) Cav. (Brassicaceae) L.N.: Taramira Bhiwani)	Whole plant fed	Mastitis	Yadav et al., 2014
197	*Erycibe paniculata* Roxb. (Convolvulaceae) L.N.: Kari lata (Chhotanagpur)	Leaf and stem decoction	Sprains	Singh, 1987

TABLE 6.1 (Continued)

S. No.	Name of the Plant	Part(s) used	Diseases	References
198	*Erythrina suberosa* Roxb. (Fabaceae) L.N.: Parabdo (Ratlam)	Stem bark ash + coconut oil	Blisters, Wounds	Jadhav, 2009
199	*Erythrina variegata* L. (= *E. indica* Lam.) (Fabaceae) L.N.: Pangara (West Nimar), Farhad (Darbhanga)	Bark; Piece of wood tied in neck	Anti-inflammatory; Wounds	Satya and Solanki, 2009; Mishra et al., 1996
200	*Eucalyptus camaldulensis* Dehnh. (Myrtaceae) L.N.: Nilgiri (West Niamr)	Oil	Inflammation of foot.	Satya and Solanki, 2009
201	*Eulophia nuda* Lindl. (Orchidaceae) L.N.: Kukadi kand (West Nimar)	Bulb paste + whey, bulb paste + hot water + soda	Acidity, inflammation, constipation, diuretic	Satya and Solanki, 2009
202	*Euphorbia caducifolia* Haines (Euphorbiaceae) Thor (Pratapgarh)	Latex	Mouth disease	Meena, 2014
203	*Euphorbia chamaesyce* L. (Euphorbiaceae) L.N.: Gudni (Morena)	Whole plant paste + Coal powder + whey	Wounds	Sikarwar, 1996
204	*Euphorbia dracunculoides* Lam. (Euphorbiaceae) L.N.: Jainti (Morena, Shivpuri)	Whole plant decoction	Kill lice of body	Sikarwar, 1996; Sikarwar et al., 1994.

S. No.	Name of the Plant	Part(s) used	Diseases	References
205	*Euphorbia fusiformis* Buch.- Ham. ex D. Don (Euphorbiaceae) L.N.: Khagoni (Jhabua), Dudhia kand (West Nimar), Pahadi muli (Udaipur), Khurdun (Gorakhpur)	Root paste; Crushed root + hey; Tuber; Paste of fruit + red chilies + water	Dysentery, fever; Lactagogue; Hemorrhagic septicemia	Sikarwar and Kumar, 2005; Satya and Solanki, 2009; Nag et al., 2009; Mangal, 2015
206	*Euphorbia heterophylla* L.(Euphorbiaceae) L.N.: Dudhia (Mandla)	Whole plant paste	Increase lactation	Sikarwar et al., 1994.
207	*Euphorbia hirta* L. (Euphorbiaceae) L.N.: Dudhi (Jhabua, Alirajpur, Shivpuri, Jaspur and Surguja, Rajasthan), Bangenda (Amarkantak), Sahjana (West Nimar)	Leaf paste; Whole plant decoction with pepper; Latex; Milk; Plant paste; Latex; Whole plant paste; Leaves, Leaf paste + leaf paste of *Cynodon dactylon*	Abortion, Antifertility; Cough and cold; Wounds; Wounds; Increase lactation; Wounds; Diarrhea, Fever	Kadel and Jain, 2006; Shukla et al., 2007; Chouhan and Ray, 2014; Jatav et al.. 2014; Satya and Solanki, 2009; Ekka, 2015; Galav et al, 2013
208	*Euphorbia indica* Lam. (=*E. parviflora* L.) (Euphorbiaceae) L.N.: Chhoti dudhi (West Nimar)	Crushed plant	Dissolve nails, iron pieces	Satya and Solanki, 2009
209	*Euphorbia microphylla* Heye. (Euphorbiaceae) L.N.: Choti dudhee (Udaipur)	Crushed whole plant	Stop bleeding after delivery	Deora et al, 2004

TABLE 6.1 (Continued)

S. No.	Name of the Plant	Part(s) used	Diseases	References
210	*Euphorbia nerifolia* L. (Euphorbiaceae) L.N.: Kanti dudhi (Jhabua), Danda- thor (Mount Abu)	Root paste; Latex, roots fed	Antifertility; Antiseptic, Ouster induction and easy removal of placenta after delivery	Kadel and Jain, 2006; Galav et al., 2007
211	*Euphorbia nivulea* Buch.-Ham. (Euphorbiaceae) L.N.: Thuaar (Dhar)	Latex	Throat swellings	Sikarwar et al. 1994.
212	*Euphorbia tirucalii* L. (Euphorbiaceae) L.N.: Danda thor (Rajasthan)	Latex	Skin diseases like dermatitis, eczema	Galav et al., 2013
213	*Euphorbia thymifolia* L. (Euphorbiaceae) L.N.: Chhoti dudhi, Lal dudhi (West Nimar), Barris-Gora (Chhotabagpur)	Crushed plant + *Asparagus racemosus* + wheat dough; Plant paste + Leaf paste of *Scoparia dulcis*	Lactagogue	Satya and Solanki, 2009; Singh, 1987
214	*Fagonia indica* Burm.f. (Zygophyllaceae) L.N.: Dhamasa (Southern Rajasthan)	Whole plant decoction, Whole plant decoction + Tobacco leaves powder	Cough, kill lice	Takhar, 2004; Kumar et al., 2004
215	*Feronia limonia* (L.) Swingle (=*F. elephantum* Corr.) (Rutaceae) L.N.: Kavit (Jhabua), Kainth (Gwalior, Tikamgarh)	Fruit paste; Leaf paste; Leaf paste + water	Abortion; Wounds; Kill intestinal worms	Kadel and Jain, 2006; Shrivastava et al., 2012; Verma, 2014

S. No.	Name of the Plant	Part(s) used	Diseases	References
216	*Ferula asafoetida* L. (Apiaceae) L.N.: Hing (Jhabua, Sitapur, Rajasthan), Heeng (Gwalior, Bhiwani)	Exudes; Exudes + Fruit powder of *Cuminum cyminum* + Seed powder of *Trigonella foenum-graecum* + Fruit powder of *Trachyspermum ammi* + Seed powder of *Cassia tora* + fruit powder of *Piper nigrum*; Exudes; Seeds paste of *Trigonella foenum-graecum* + *Cuminum cyminum* + *Ferula asafoetida* + leaves of *Bombax ceiba*; Resin, Resin + milk; Resin + *Linum usitatissimum* oil	Indigestion; Constipation, Mastitis; Flatulence, dysentery; Infection of nematode in intestines; flatulence, gas problem	Kadel and Jain, 2006; Shrivastava et al., 2012; Kumar and Bharati, 2012b; Kumar and Bharati, 2013; Galav et al., 2013; Yadav et al., 2014
217	*Ficus benghalensis* L. (Moraceae) L.N.: Bargad (Alirajpur, Narsinghpur, Tikamgarh, Sitapur, Marwar), Vadlo (Ratlam)	Latex; Leaf ash + water; Latex; Root juice; Root decoction; Asafoetida + Ajwain + jaggery; Leaves + Leaves of *Annona squamosa* + Linseed oil	Maggot infested wounds; Indigestion (Afra disease); Maggot wounds; Stomachache; Prolongation of expulsion of placenta; Retention of placenta and umbilical cord; Bloat; Snake and scorpion bite	Chouhan and Ray, 2014; Jadhav, 2009; Sanghi, 2014; Verma, 2014; Kumar and Bharati, 2012b; Kumar and Bharati, 2013; Dudi and Meena, 2015
218	*Ficus hispida* L.f. (Moraceae) L.N.: Ottelawa (Betul), Bhui gular (West Nimar), Damura (Singhbhum)	Root paste; Crushed fruits + warm water; Fruit paste + warm water, boiled ripe fruits, fresh leaves	Weakness; Lactagogue; Baboes, secretion of milk, expulsion of placenta	Sikarwar and Kumar, 2005; Satya and Solanki, 2009; Pal, 1980

TABLE 6.1 (Continued)

S. No.	Name of the Plant	Part(s) used	Diseases	References
219	*Ficus racemosa* L. (=*F. glomerata* Roxb.) (Moraceae) L.N.: Gular (Shivpuri, Jhansi, Narsinghpur, Mount Abu, Sitapur, Rajasthan), Goolar (Darbhanga), Umar (Udaipur, Sirohi, Pratapgarh), Loandra, Dumbari (Midnapur),	Fruits, Leaves; Fruits; Bark decoction + butter milk; Latex; Leaves; Bark decoction + paste of 2 Onion + 21 black pepper + *Madhuca longifolia* liquor; Bark decoction of *Ficus glomerata* + *Acacia nilotica* ssp. *indica* + *Syzygium cumini* + *Butea monosperma* + *Acacia catechu*; Bark paste + stem sap of Banana; Bark paste	Renal problems, dermatitis, diarrhea, ulcer, anthelmintic; Removal of placenta after delivery; Diarrhea; Skin diseases; Blood dysentery; As a galactagogue; Plague; Foot and mouth disease; Diarrhea; Snake bite	Jatav et al., 2014; Nigam and Sharma, 2010; Sanghi, 2014; Galav et al., 2007; Mishra et al., 1996; Sebastian, 1984; Pal, 1980, Kumar and Bharati, 2012b; Meena, 2014; Gupta et al., 2015
220	*Ficus religiosa* L. (Moraceae) L.N.: Peepal (Morena, Chitrakoot, Gwalior, Vindhyan region, Tikamgarh, Darbhanga, Bhiwani), Pipal (Rajasthan)	Stem bark paste; Smashed fruits; Soft leaves + Jaggery; Bark decoction; Leaf juice; Bark decoction, leaf powder; fruits; Leaf extract; Bark decoction	Constipation; inducing fertility; Fracture; Foot and mouth disease; Tonsils; Foot and mouth disease; Bronchitis; To bring cow heat; Dysuria, hematuria; Removal of retained placenta	Sikarwar, 1996; Tripathi and Singh, 2010; Shrivastava et al., 2012; Dwivedi et al., 2009; Verma, 2014; Mishra et al., 1996; Galav et al., 2013; Yadav et al., 2014
221	*Foeniculum vulgare* Mill (Apiaceae) L.N.: Saunf (Bhiwani)	Seed powder; Paste of seeds + *Swertia chirayita* + *Zingiber officinale*	Diarrhea; Bloat	Yadav et al., 2014; Dudi and Meena, 2015
222	*Gardenia gummifera* L. (Rubiaceae) L.N.: Dururing, Dururi (Midnapur)	Paste of leaf buds, powder of dry leaf buds	Repellent of flies from wounds, anthelmintic	Pal, 1980

S. No.	Name of the Plant	Part(s) used	Diseases	References
223	*Gardenia latifolia* Ait. (Rubiaceae) L.N.: Phetra (West Nimar)	Oil	Foot and mouth disease	Satya and Solanki, 2009
224	*Garuga pinnata* Roxb. (Burseraceae) L.N.: Tuj (Jalaun)	Stem bark + Rhizome powder of *Zingiber officinale* + alum + *Swertia chirata*	Scabies	Kumar and Bharti, 2012a
225	*Gentiana kurroo* Royle (Gentianaceae) L.N.: Kutali (Marwar)	Root paste	Fever	Dudi and Meena, 2015
226	*Geranium pratense* L. (Geraniaceae) L.N.: Chalmodi (Marwar)	Leaf decoction	Fever	Dudi and Meena, 2015
227	*Gloriosa superba* L. (Liliaceae) L.N.: Kaliharikand (Chhindwara, West Nimar), Kalihari (Shivpuri, Mount Abu, Udaipur), Kalgari (Rajasthan) (Figure 6.2K)	Root paste; Leaves; Crushed bulb; Tuber paste; Root paste or juice; Tuber paste; Crushed stem bark + Rhizome of *Zingiber officinale* + Stem of *Swertia chirata* + Pinch of potassium alum	Wounds to kill worms; Swelling, chronic ulcer, colic pain; Wounds to kill maggots; Foot and mouth disease; Maggots of wounds; Foot and mouth disease; Itching of skin	Sikarwar et al., 1994; Jatav et al., 2014; Satya and Solanki, 2009; Galav et al., 2007; Deora et al., 2004; Galav et al., 2013; Kumar and Bharati, 2013
228	*Glossocardia bosvalea* (L. f.) DC. (Astereceae) L.N.: Pittpara (Betul),	Leaf powder	Lameness	Sikarwar and Kumar, 2005
229	*Glycine max* Merril. (Fabaceae) L.N.: Soyabean (Marwar)	Frothed solution	Diarrhea, bloat, fever	Dudi and Meena, 2015

TABLE 6.1 (Continued)

S. No.	Name of the Plant	Part(s) used	Diseases	References
230	*Gossypium arboretum* L. (Malvaceae) L.N.: Narma (Gonda)	Root extract	Lack of estrus (Heat)	Pandey et al., 1999
231	*Gossypium hirsutum* L. (Malvaceae) L.N.: Binola (Bhiwani)	Seeds	Increase butter quantity in milk	Yadav et al., 2014
232	*Grewia abutilifolia* Vent. ex Juss. (Tiliaceae) L.N.: Gengchi (Pratapgarh)	Root powder decoction	Bone fracture	Meena, 2014
233	*Grewia hirsuta* Vahl (Tiliaceae) L.N.: Gudsakri (West Nimar)	Root powder + water	Bone fracture	Satya and Solanki, 2009
234	*Grewia tiliaefolia* Vahl (Tiliaceae) L.N.: Farangdee (Udaipur)	Root infusion	Bone fracture	Deora et al., 2004
235	*Gymnema sylvestre* (Retz.) Schultes (Asclepiadaceae) L.N.: Gudmar (Jhansi, Jaspur and Surguja)	Leaf paste; Leaf paste + Pepper + Garlic + Pinch of salt	Cataract; Fever	Nigam and Sharma, 2010; Ekka, 2015
236	*Hedychium coronarium* Koenig (Zingiberaceae) L.N.: Gulbakaoli (Shahdol)	Rhizome paste	Kill intestinal worms	Sikarwar et al., 1994.

S. No.	Name of the Plant	Part(s) used	Diseases	References
237	*Helianthus annuus* L. (Asteraceae) L.N.: Surajmukhi (Jhansi, Bhiwani)	Seed powder + coconut oil; Seed oil	Swelling of testis; Smooth delivery	Nigam and Sharma, 2010; Yadav et al., 2014
238	*Helicteres isora* L. (Sterculiaceae) L.N.: Marorphali (Vindhyan region, Jhansi), Marophalli (Mount Abu)	Dried fruits; Fruit powder + seed powder of *Cuminum cyminum* + whey; Pods paste	Stomachache; Dysentery	Dwivedi et al., 2009; Nigam and Sharma, 2010; Galav et al., 2007
239	*Hemidesmus indicus* (L.) R.Br. (Asclepiadaceae) L.N.: Dudhiya parhi (Jaspur and Surguja) (Figure 6.2L)	Whole plant paste	Wounds	Ekka, 2015
240	*Hibiscus cannabinus* L. (Fabaceae) L.N.: Sun (Uttar Pradesh)	Leaf extract + Flower extract of *Allium sativum* + water	Constipation	Kumar and Bharati, 2013
241	*Hibiscus rosa-sinensis* L. (Malvaceae) L.N.: Gudhal (Jhansi, Tikamgarh)	Bark decoction; Leaves + *Origanum majorana*	Twitching; Snake bite and scorpion sting	Nigam and Sharma, 2010; Verma, 2014; Dudi and Meena, 2015
242	*Hiptage bengalensis* (L.) Kurz. (Malpighiaceae) L.N.: Jal (Udaipur, Sirohi)	Leaves	More Lactation	Sebastian, 1984

TABLE 6.1 (Continued)

S. No.	Name of the Plant	Part(s) used	Diseases	References
243	*Holarrhena pubescens* Wall. ex Don (=*H. antidysenterica* Wall. ex DC.) (Apocynaceae) L.N.: Kuretha (Morena), Koriya (Jaspur and Surguja)	Stem bark paste + Turmeric paste + Alum + Liquor; Stem bark decoction	Anthrax; Dysentery	Sikarwar, 1996; Ekka, 2015
244	*Holoptelea integrifolia* (Roxb.) Planch. (Ulmaceae) L.N.: Ohla (Alirajpur), Bandar bati (Rajasthan)	Leaf paste; Leaf juice; Leaves	Wounds; Ectoparasites; Eczema	Chouhan and Ray, 2014; Verma, 2014; Galav et al., 2013
245	*Hordeum vulgare* L. (Poaceae) L.N.: Jau (Ratlam), Barley (Gorakhpur)	Flour; Boiled Barley	Body weakness; Anorexia	Jadhav, 2009; Mangal, 2015
246	*Hydrilla verticillata* (L.f.) Royle (Hydrocharitaceae) L.N.: Harshringar Paani waala (Uttar Pradesh)	Whole plant crushed + Rhizome of *Curcuma longa*	Prepare for mating	Kumar and Bharati, 2013
247	*Hygrophila schullii* (Buch.-Ham. M.R and S. M. Almeida (Acanthaceae) L.N.: Telmakhana (West Nimar)	Ash of whole plant + oil	Wound of shoulder	Satya and Solanki, 2009

S. No.	Name of the Plant	Part(s) used	Diseases	References
248	*Hyoscymus niger* L. (Solanaceae) L.N.: Khurasani Ajwain (Jhansi, Sitapur)	Seed oil; Seeds + Seeds of *Myristica fragrans*	Remove maggots of wounds; Regulate estrus cycle	Nigam and Sharma, 2010; Kumar and Bharati, 2012b
249	*Hyptis suaveolens* (L.) Poit. (Lamiaceae) L.N.: Madhuban (Durg),	Leaf juice	Eye inflammation, conjunctivitis	Sikarwar and Kumar, 2005; Sikarwar, 1996.
250	*Indigofera cassioides* Rottl. ex DC. (Fabaceae) L.N.: Jhinni Patti (Chhindwara)	Leaf fumes	Bodyache	Sikarwar and Kumar, 2005
251	*Indigofera oblongifolia* Forssk (Fabaceae) L.N.: Jangli neel (West Nimar)	Leaves + leaves of *Momordica charantia* and *Momordica dioica* + salt	Abdominal distention	Satya and Solanki, 2009
252	*Indigofera tinctoria* L.(Fabaceae) L.N.: Neel (Gwalior, West Nimar)	Leaves; Plant powder	Mastitis; Boils	Shrivastava et al., 2012; Satya and Solanki, 2009
253	*Ipomoea carnea* Jacq. (Convolvulaceae) L.N.: Naseda (Alirajpur)	Warmed leaves	Wounds	Chouhan and Ray, 2014
254	*Ipomoea nil* (L.) Roth (Convolvulaceae) L.N.: Farman, Baramansi (Udaipur, Sirohi)	Plants	As health tonic and galactagogue	Sebastian, 1984

TABLE 6.1 (Continued)

S. No.	Name of the Plant	Part(s) used	Diseases	References
255	*Ipomoea turbinata* Lag. (=*Calonyction muricatum* (L.) G.Don) (Convolvulaceae) L.N.: Khotlaiya (Vindhyan region)	Seed powder	Constipation	Dwivedi et al., 2009
256	*Jatropha curcas* L. (Euphorbiaceae) L.N.: Ratanjot (Jhabua, Jhansi, Mount Abu), Adusa (Udaipur)	Latex; Milky sap; Root paste; Latex, root powder	Wounds; Yolk sore; As a purgative; Eye diseases, bronchial disease	Kadel and Jain, 2006; Nigam and Sharma, 2010; Nag et al., 2009; Galav et al., 2007
257	*Jatropha gossypifolia* L. (Euphorbiaceae) L.N.: Lal-gab-jhara (Chhotanagpur)	Plant paste	Wounds	Singh, 1987
258	*Justicia adhatoda* L. (=*Adhatoda vasica* Medic.) (Acanthaceae) L.N.: Adusa (Morena, Chitrakoot, Alirajpur, Shivpuri, Vindhyan region, West Nimar, Tikamgarh, Udaipur, Mount Abu, Jaspur and Surguja); Bakas (North Bihar), Rusa (Gonda), Adhoosh (Gorakhpur)	Leaf paste + jaggery; Leaf paste; Leaf poultice; Leaf decoction; Leaf powder; Crushed leaves; Leaf juice + bark juice of *Syzygium cuminii*; Leaf and Stem decoction; Bark paste + Bark paste of *Bombax ceiba*; Leaf extract + 1 pod of *Cassia fistula* + 5 leaves of *Holarrhena antidysenterica*, Leaf decoction; Stem + bark decoction; Decoction of leaves + leaves of *Ficus religiosa*	Bronchitis; Cough; Healing wounds; Fever; Bronchitis; Expel intestinal worms; Diarrhea and dysentery; Expectorant; Diarrhea; Throat infection, Fever and cough, ectoparasites and skin diseases; Fever; Fever, cough, ectoparasites; Milk fever	Sikarwar, 1996; Sikarwar et al., 1994; Tripathi and Singh, 2010; Chouhan and Ray, 2014; Jatav et al., 2014; Dwivedi et al., 2009; Satya and Solanki, 2009; Verma, 2014; Jha et al., 1991; Nag et al., 2009; Galav et al., 2007; Ekka, 2015; Pandey et el., 1999; Mangal, 2015

S. No.	Name of the Plant	Part(s) used	Diseases	References
259	*Kalanchoe pinnata* (Lam.) Pers. (Crassulaceae) L.N.: Goja (Marwar)	Paste	Ectoparasites	Dudi and Meena, 2015
260	*Lallemantia royleana* Benth. (Lamiaceae), L.N.: Tutmalanga (Darbhanga)	Tutmalanga + seeds *of Lens culinaris* + pulp of *Aloe vera* + *Acacia catechu* (Kattha) + flower of *Thevetia neriifolia* + turpentine oil	Swelling of foot	Mishra et al., 1996
261	*Lantana camara* L. (Verbenaceae) L.N.: Kirmich (Alirajpur), Phulari (Shivpuri), Lailumri (West Nimar)	Leaf juice: Leaves, flowers; Leaves paste	Wounds; Skin itches, wounds, scabies; Cut and wounds	Chouhan and Ray, 2014; Jatav et al., 2014; Satya and Solanki, 2009
262	*Lasia aculeata Lour.* (Araceae) L.N.: Janumsaru (Midnapur)	Paste of petioles + 7 tuber of Cyprus rotundas + 21 black pepper	Throat diseases	Pal, 1980
263	*Launaea procumbens* (Roxb.) Ram. and Raj. (Asteraceae) L.N.: Kadvi galobi (Ratlam)	Plant	Increasing lactation	Jadhav, 2009
264	*Lawsonia inermis* L. (Lythraceae) L.N.: Mehadi (Morena, Vindhyan region), Mehendi (Jhansi), Mehandi (West Nimar), Henna (Bhiwani, Marwar)	Leaf paste; Leaf decoction; Leaf paste; Leaf powder + water; Crushed leaves; Leaf powder; Paste	Foot and mouth diseases; Wounds; Foot and mouth disease; Hematuria; Acidity, diarrhea, Stomach disorder; Maintain pregnancy; Skin disease	Sikarwar, 1996; Chouhan and Ray, 2014; Dwivedi et al., 2009; Nigam and Sharma, 2010; Satya and Solanki, 2009; Yadav et al., 2014

TABLE 6.1 (Continued)

S. No.	Name of the Plant	Part(s) used	Diseases	References
265	*Lecanthus peduncularis* (Royle) Weed. (Urticaceae) L.N.: Bicchu (Satna)	Fruit paste, Leaf poultice	Eczema, Ring worm, Sores	Gautam and Richhariya, 2015
266	*Leea macrophylla* Roxb. (Leeaceae) L.N.: Antharun kand (Betul)	Root paste	Carbuncles	Patil and Deshmukh, 2015
267	*Leea robusta* Roxb. (Leeaceae) L.N.: Horim, Hatkan (Jalpaiguri)	Root extract + Ginger, root extract + 7 black pepper + common salt	Diarrhea and dysentery, fever	Pal (1980)
268	*Lens culinaris* Medic (Fabaceae) L.N.: Masoor (Darbhanga)	Boiled seeds + Gur (Jaggery) + *Musa sapientum* milk; Masur + *Lallemantia royleana* + pulp of *Aloe vera* + *Acacia catechu* (Kattha) + flower of *Thevetia neriifolia* + turpentine oil; Cooked seeds; Cooked seeds + Seeds of *Trigonella foenum-graecum*	Increase lactation; Swelling of foot; Prepare for mating, Difficulty breathing, chewing and swelling, loss of appetite, fever	Mishra et al., 1996; Mishra et al., 1996; Kumar and Bharati, 2013
269	*Leonurus japonicas* Houtt (Lamiaceae) L.N.: Bar (North Bihar)	Stem, leaf and flower decoction	Vulnerary, febrifuge	Jha et al., 1991
270	*Lepidium sativum* L. (Brassicaceae) L.N.: Chansur (Vindhyan region), Chamsur (Marwar)	Seed poultice; Roasted powder	Mastitis; Bloat	Dwivedi et al., 2009; Dudi and Meena, 2015

S. No.	Name of the Plant	Part(s) used	Diseases	References
271	*Leptadenia pyrotechnica* (Forssk) Decne (Asclepiadaceae) L.N.: Kneep (Udaipur), Kheemp (Rajasthan, Southern Rajasthan)	Whole plant infusion; Tender shoot pieces; Decoction of stem pieces + seeds of *Trachyspermum ammi* + *Sesamum orientale* oil; stem decoction; Filtrate of stem water	Uterus prolapse; Ouster induction; Expulsion of placenta; smooth movement of joints; Flatulence	Deora et al., 2004; Galav et al., 2013; Takhar, 2004; Kumar et al., 2004
272	*Leptadenia reticulata* (Retz.) Wt. and Arn. (Asclepiadaceae) L.N.: Devdali (West Nimar)	Plant juice, Plant paste	Worm infestation, Wounds	Satya and Solanki, 2009
273	*Leucas aspera* (Willd.) Link (= *L. plukenetii* (Roth) Spreng.) (Lamiaceae) L.N.: Gumma (Satna, Jalaun), Dargal (Rajasthan), Gum (Gorakhpur)	Leaf decoction; Whole plant + rhizome of *Piper longum*; Leaf decoction; Leaf juice + Ghee, plant decoction; Crushed leaves	Cough, cold, Respiratory diseases; Anorexia; Fever; Indigestion; Cough, Fever	Gautam and Richhariya, 2015; Kumar and Bharti, 2012a; Galav et al., 2013; Mangal, 2015; Kumar and Bharati, 2013
274	*Leucas cephalotes* Spreng. (Lamiaceae) L.N.: Gumma (Vindhyan region, Darbhanga); Guma (North Bihar), Khumbi (Rajasthan)	Leaf juice; Whole plant decoction; Plant juice; Whole plant	Flatulence, indigestion; Anthelmintic, Vermifuge; Snake bite; As tonic	Dwivedi et al., 2009; Jha et al., 1991; Mishra et al., 1996; Galav et al., 2013
275	*Leucas lavandulifolia* Sm. (Lamiaceae) L.N.: Guma (Gonda)	Crushed leaves + 21 black pepper	Loss of appetite, cuts	Pandey et al., 1999

TABLE 6.1 (Continued)

S. No.	Name of the Plant	Part(s) used	Diseases	References
276	*Levisticum officinale* W. D. J. Koch. (Apiaceae) L.N.: Ajwain (Uttar Pradesh)	Fruits crushed + Potassium alum	Flatulence, anorexia	Kumar and Bharati, 2013
277	*Lindernia hyssopioides* (L.) Haines (Scrophulariaceae) L.N.: O (Udaipur, Sirohi)	Leaf paste	External insect parasites	Sebastian, 1984
278	*Linum usitatissimum* L. (Linaceae) L.N.: Alsi (Jhansi, West Nimar, Jalaun, Sitapur), Teesi (Darbhanga)	Seed oil + gun powder + lime water; Grinded seeds + wheat flour; 3 kg *Cicer aritinum* flour + 10 g Ganja (*Cannabis sativa*) + Teesi oil; 500 g oil cake of *Linum usitatissimum* + 100 g Jaggery; Seeds + gum of *Sterculia urens*	Burns; General health tonic; Dysentery; Decrease in milk secretion; Little and frequent urine	Nigam and Sharma, 2010; Satya and Solanki, 2009; Mishra et al., 1996; Kumar and Bharti, 2012a; Kumar and Bharati, 2012b; Kumar and Bharati, 2013
279	*Lobelia alsinoides* Lam. (Lobeliaceae) L.N.: Jangli Tambakhu (West Nimar)	Leaves	Foot and mouth disease	Satya and Solanki, 2009
280	*Lolium temulentum* L. (Poaceae) L.N.: Bisa gash (Midnapur)	Plant given as fodder but before that Fruit extract of *Tamarindus indica* is given	Convulsion and Paralysis	Pal (1980)
281	*Loranthus ampulaceus* Roxb. (Loranthaceae) L.N.: Banda (Gonda)	Bark and Leaves decoction	Swelling of shoulders	Pandey et al., 1999

S. No.	Name of the Plant	Part(s) used	Diseases	References
282	*Luffa acutengula* (L.) Roxb.(Cucurbitaceae) L.N.: Kadvi taroi (Ratlam), Torai (Gorakhpur)	Seed extract + water; Leaf paste	Appetizer; Mastitis	Jadhav, 2009; Mangal, 2015
283	*Luffa cylindrica* (L.) Roem. (Cucurbitaceae) L.N.: Turai, Phatkuli (Narsinghpur), Jangli turri (Mount Abu), Phalla (Chotanagpur), Turai (Sitapur, Uttar Pradesh)	Leaf paste; Fruit decoction + seeds of *Trachyspermum ammi* + salt; Smoke of the dry monocarp of the fruits; Fruit juice	Insect bites; Flatulence; Dyspepsia; Anuresis; Stop urination	Sanghi, 2014; Galav et al., 2007; Singh, 1987; Kumar and Bharati, 2012b; Kumar and Bharati, 2013
284	*Lycium barbatum* L. (Solanaceae) L.N.: Mureli (Southern Rajasthan), Murali (Arid zone)	Root decoction; A snuff of powder roots	Expulsion of placenta, Respiratory problems	Takhar, 2004; Kumar et al., 2004
285	*Lycopercon esculentum* Mill. (Solanaceae) L.N.: Tamatar (Sitapur, Uttar Pradesh)	Ripen fruits + Common salt	Worm infestation	Kumar and Bharati, 2012b; Kumar and Bharati, 2013
286	*Macrotyloma uniflorum* (Lam.) Verd. (Fabaceae) L.N.: Kulthi (West Nimar), Kurthi (North Bihar)	Seeds + pulp of fruits of *Aegle marmelos*; Aerial part decoction	Lactation; Diuretic	Satya and Solanki, 2009; Jha et al., 1991

TABLE 6.1 (Continued)

S. No.	Name of the Plant	Part(s) used	Diseases	References
287	*Madhuca longifolia* (Koen.) Macbr. (=*M. indica* J. F. Gmel.) (Sapotaceae) L.N.: Mahua (Jhabua, West Nimar, Tikamgarh, Sitapur, Banswara), Moha (Betul), Mori (Pratapgarh)	Flower juice; Flower paste; Dried flowers; Boiled flowers; Bark paste + bark paste of *Soymida febrifuga*; Flower paste + Jaggery + water; Leaf decoction, Fruit paste; Bark decoction; Seed oil of *Brassica campestris* is applied on leaves; Alcohol of Mahua	Cough and Cold; Swelling due to injury; Malnutrition; Wounds; Sprain; Fever; Blood diarrhea, diphtheria; Maggots in hoof; Swelling; Prolapse	Kadel and Jain, 2006; Tripathi and Singh, 2010; Gautam and Richhariya, 2015; Chouhan and Ray, 2014; Satya and Solanki, 2009; Verma, 2014; Patil and Deshmukh, 2015; Kumar and Bharati, 2012b; Kumar and Bharati, 2013; Meena, 2014; Yadav et al., 20015
288	*Maerua arenaria* (DC.) Hook.f. and Thoms. (Capparaceae) L.N.: Jethivela (Udaipur)	Crushed roots + Butter milk	Tympanitis, throat swelling	Deora et al., 2004
289	*Mangifera indica* L. (Anacardiaceae) L.N.: Am (Morena, Jhabua, West Nimar, Tikamgarh), Ambo (Ratlam), Aam (Gonda, Uttar Pradesh)	Pickled fruits; Flower paste + turmeric powder + Onion paste; Three years pickled fruits; Pickled fruit paste + *Curcuma longa* + mustard oil; Bark paste + lime; Fruit paste + wheat bread; Bark decoction + Barley + Jaggery; New leaves of *Syzygium cumini* + leaves of *Mangifera indica* + *Acacia nilotica* ssp. *indica* + *Euphorbia reticulata* + *Aegle marmelos* + *Abutilon indicum* + common salt; Crushed kernels decoction + Lal fitkari	Constipation; Swelling due to injury; Indigestion; Indigestion (Afra disease); Diarrhea and dysentery; Indigestion; Lactation problem and physical debility; Hematuria; Foot and mouth disease	Sikarwar, 1996; Tripathi and Singh, 2010; Kadel and Jain, 2006; Jadhav, 2009; Satya and Solanki, 2009; Verma, 2014; Pandey et al., 1999; Kumar and Bharati, 2012b; Kumar and Bharati, 2013

S. No.	Name of the Plant	Part(s) used	Diseases	References
290	*Maytenus emarginatus* (Ruiz and Pav.) Loes (Celastraceae) L.N.: Kangera (Bhiwani)	Burnt ash of leaves + Mustard oil	Cracked nipples	Yadav et al., 2014
291	*Melia azedarach* L. (Meliaceae) L.N.: Bakain (Darbhanga, Gonda, Sitapur), Bakayan (Rajasthan)	Leaf decoction; Leaves; Fruits + butter milk + dried unripe fruits *of Mangifera indica* + seeds of *Trachyspermum ammi* + *Ferula asafoetida* + gum of *Acacia nilotica ssp. indica* + seeds of *Apium graveolens* + *Allium sativum* + *Allium cepa* crushed together; Leaves	Anthelmintic, vermifuge, diuretic; Loss of appetite, weakness, decreased milk secretion; Anorexia; Intestinal parasites	Jha et al., 1991; Pandey et al., 1999: Kumar and Bharati, 2012b; Galav et al., 2013
292	*Mentha arvensis* L. (Lamiaceae) L.N.: Podina (Tikamgarh)	Leaf paste + leaf paste of *Centella asiatica*	Fever	Verma, 2014
293	*Mesua ferrea* L. (Calophyllaceae) L.N.: Mangrail (Uttar Pradesh)	Crushed fruits + roots of *Asparagus racemosus* + Rhizome of *Zingiber officinale* + *Trachyspermum ammi*	Mastitis	Kumar and Bharati, 2013
294	*Milletia extensa* (Benth.) Baker (Fabaceae) L.N.: Golhar (Morena)	Root paste, Leaf paste	Wounds to kill worms, Kill lice of body	Sikarwar, 1996.
295	*Mimosa pudica* L. (Mimosaceae) L.N.: Lajwanti (Jhabua, Shivpuri, Jhansi, Jalaun), Chuimui (Jaspur and Surguja), Chui-mui (Uttar Pradesh)	Leaf extract; Leaf paste + Pepper + garlic + onion + saffron; Leaf paste + chapattis; 50 g seeds + 500 ml unboiled milk; Leaf paste + Pepper + Garlic + Onion + Saffron; Crushed leaves + Mustard oil, seeds + raw milk	Prolapse of uterus; Fever; Maggot wounds; Diphtheria; Fever; Expulsion of placenta, difficulty breathing, chewing, swelling, loss of appetite, fever	Kadel and Jain, 2006; Jatav et al., 2014; Nigam and Sharma, 2010; Kumar and Bharti, 2012a; Ekka, 2015; Kumar and Bharati, 2013

TABLE 6.1　(Continued)

S. No.	Name of the Plant	Part(s) used	Diseases	References
296	*Mimusops elengi* L. (Sapotaceae) L.N.: Bakul, Mollshri (Narsinghpur)	Bark decoction	Urinary infection	Sanghi, 2014
297	*Mitragyna parviflora* (Roxb.) Korth (Rubiaceae) L.N.: Kaim (Morena)	Bark decoction	Filariasis	Sikarwar, 1996
298	*Momordica charantia* L. (Cucurbitaceae) L.N.: Karela (West Nimar, Gonda)	Leaf paste; Leaf extract	Thorny growth on tongue; Lack of estrus	Satya and Solanki, 2009; Pandey et al., 1999
299	*Momordica dioica* Roxb. ex Willd. (Cucurbitaceae) L.N.: Kankodi (Udaipur, Sirohi) (Figure 6.2M)	Plants	As a blood purifier, injuries	Sebastian, 1984
300	*Moringa oleifera* Lam. (Moringaceae) L.N.: Sahajan (Seoni), Saunjana (Gwalior, Tikamgarh), Surajano (Ratlam), Muniga (North Bihar), Sohjan (Darbhanga)	Bark decoction; Bark paste + bulbs of *Allium sativum* + Leaves of *Boerhavia diffusa*: Root extract + water; Leaf paste, pod paste, root juice; Bark decoction; Leaf decoction; Leaves + Cow milk, Leaves + turmeric powder	Influenza; Rheumatism; Appetizer; Diarrhea and dysentery, rheumatism, ulcer; Cardiac stimulant, anti-rheumatic; Swelling due to injury; Bloat; Snake and scorpion bite	Sikarwar and Kumar, 2005; Shrivastava et al., 2012; Jadhav, 2009; Verma, 2014; Jha et al., 1991; Mishra et al., 1996; Dudi and Meena, 2015

S. No.	Name of the Plant	Part(s) used	Diseases	References
301	*Mucuna pruriens* (L.) DC. (Fabaceae) L.N.: Kemuch (Ratlam); Kabachchu (North Bihar), Kabachh (Darbhanga), Khajkurari (Betul), Kenvach (Rajasthan) (Figure 6.2N)	Fruit hairs; Aerial part decoction; Leaves; Tender leaf; Pods + Jaggery	Intestinal worms; Anthelmintic, vermifuge, nervous disease, diuretic; Increase lactation; Lactation; Infertility	Jadhav, 2009; Jha et al., 1991; Mishra et al., 1996; Patil and Deshmukh, 2015; Gupta et al., 2015
302	*Musa paradisiaca* L. (Musaceae) L.N.: Ker (Ratlam), Kela (Jhansi, Tikamgarh, Bhiwani, Uttar Pradesh)	Spate extract; Fruit paste + sugar candy; Leaf and root; Camphor tablet inserted in ripen fruit and given; Put the leaf and gently push inside	Prolapse of uterus; Blisters, hoof sore; Body heat; Mastitis; Expulsion of placenta	Jadhav, 2009; Nigam and Sharma, 2010; Verma, 2014; Yadav et al., 2014; Kumar and Bharati, 2013
303	*Myristica fragrans* Houtt. (Myristicaceae) L.N.: Jaiphal (Sitapur), Jaiful (Uttar Pradesh)	Seeds + Seeds of *Hyoscymus niger* + Black salt, seed + seeds of *Piper nigrum* + roots of *Withania somnifera*	Regulate estrus cycle, Diarrhea in winter season; Prepare for mating	Kumar and Bharati, 2012b; Kumar and Bharati, 2013
304	*Nerium oleander* L. (Apocynaceae) L.N.: Kaner (Alirajpur)	Seed ash + mustard oil	Wounds	Chouhan and Ray, 2014
305	*Nicotiana rustica* L. (Solanaceae) L.N.: Tambaku (Gonda)	Crushed leaves, crushed leaves + Mustard oil	Eye trouble, ectoparasites	Pandey et al., 1999

TABLE 6.1 (Continued)

S. No.	Name of the Plant	Part(s) used	Diseases	References
306	*Nicotiana tabacum* L. (Solanaceae) L.N.: Tambaku (Bundelkhand, Vindhyan region), Tobacco (Darbhanga, Gorakhpur), Tanbaku (Pratapgarh), Tambakhu (Rajasthan)	Leaf fumes; Seeds; Leaf juice; Dried leaves + fruit powder of *Capsicum annuum*; Leaves + Sodium carbonate + Mustard oil; Decoction of roots + Curd	Ticks and lice; Intestinal worms; Ectoparasites; Hematuria; Flatulence; Ticks and mites	Mishra et al., 2010; Gautam and Richhariya, 2015; Dwivedi et al., 2009; Mishra et al., 1996; Meena, 2014; Galav et al., 2013; Mangal, 2015
307	*Nyctanthes arbor-tristis* L. (Oleaceae) L.N.: Harsingar (Alirajpur, Vindhyan region), Harsinghar (North Bihar, Darbhanga) (Figure 6.2O)	Leaf decoction	Maggot infested wounds; Fever; Expectorant, anthelmintic, vermifuge; Fever	Chouhan and Ray, 2014; Dwivedi et al., 2009; Jha et al., 1991; Mishra et al., 1996
308	*Ocimum americanum* L. (Lamiaceae) L.N.: Jangali tulsi (Udaipur), Bapchi (Rajasthan)	Seed (200 g) + 250 g milk fat + 500 g wheat flour + 500 g jaggery; Paste of seeds of *Terminalia catappa* + Jhadis lac + *Foeniculum vulgare* + *Ocimum americanum*	Asthmatic attack; Heat in body, leucorrhoea	Nag et al., 2009; Galav et al., 2013
309	*Ocimum gratissimum* L. (Lamiaceae)	Leaf paste	Removal of ectoparasites	Verma, 2014
310	*Ocimum tenuiflorum* L. (=*O. sanctum* L.) (Lamiaceae) L.N.: Tulsi (Jhabua, Satna, Vindhyan region, Tikamgarh, Jalaun, Rajasthan, Uttar Pradesh)	Leaf extract; Seed powder; Leaf decoction; Leaf extract; Leaf paste; Paste of leaves + Honey + Calf urine; Leaf extract	Brain disease; Maggots and wounds; Cough and cold; Conjunctivitis; Scabies; Cancer; Conjunctivitis	Kadel and Jain, 2006; Dwivedi et al., 2009; Verma, 2014; Kumar and Bharti, 2012a; Galav et al., 2013; Kumar and Bharati, 2013

S. No.	Name of the Plant	Part(s) used	Diseases	References
311	*Opuntia elatior* Mill. (Cactaceae) L.N.: Hatha thoohar (West Nimar), Nagphani (Betul), Nagfani (Uttar Pradesh)	Stem paste; Heated half part of phylloclade; Thorn ash + Root extract of *Withania somnifera* + Leaf extract of *Azadirachta indica*	Bone fracture; Swelling; Difficulty breathing, chewing and swelling, loss of appetite, fever	Satya and Solanki, 2009; Patil and Deshmukh, 2015; Kumar and Bharati, 2013
312	*Origanum majorana* L. (Lamiaceae) L.N.: Marva (Marwar)	Leaves, leaf paste	Internal parasites, foot and mouth disease	Dudi and Meena, 2015
313	*Orobanche aegyptiaca* Pers. (Orobanchaceae) L.N.: Agia (Morena; Chitrakoot), Bhatua ghans (Midnapur)	Whole plant; Plant paste + seeds of *Ocimum sanctum* + red mud, plant decoction + bark decoction of *Shorea robusta*	Increasing lactation; Boils, diarrhea	Sikarwar, 1996; Tripathi and Singh, 2010; Pal, 1980
314	*Oryza sativa* L. (Poaceae) L.N.: Dhan (Jhabua, Jhansi, Tikamgarh, Darbhanga, Gorakhpur), Chawal (Sitapur)	Powder; Bran + seeds of *Gossypium* sp.; Rice grains + black gram + black salt + black pepper; Baked seeds; Boiled rice water; Plant	Prolapse of uterus; Cough; Increase lactation; Dysentery; Diarrhea; Placenta did not fall	Kadel and Jain, 2006; Nigam and Sharma, 2010; Verma, 2014; Mishra et al., 1996; Kumar and Bharati, 2012b; Mangal, 2015
315	*Oxalis corniculata* L. (Oxalidaceae) L.N.: Tinpatia (Chitrakoot), Tulsi (Gorakhpur), Chuka (Uttar Pradesh)	Leaf paste; Leaf juice; Leaf extract + Common salt	Neck swelling; Redness and mucous eyes; Hanging of placenta	Tripathi and Singh, 2010; Mangal, 2015; Kumar and Bharati, 2013

TABLE 6.1 (Continued)

S. No.	Name of the Plant	Part(s) used	Diseases	References
316	*Paederia scandens* (Lour.) Merr. (Rubiaceae) L.N.: Gandhapasarani (North Bihar)	Aerial part decoction	Anthelmintic, vermifuge, diuretic	Jha et al., 1991
317	*Pandanus tectorius* Park. ex du Roi (Pandanaceae) L.N.: Kayvara (Sitapur)	Crushed flower	Regulate estrus cycle, Prepare for mating	Kumar and Bharati, 2012b; Kumar and Bharati, 2013
318	*Papavar somniferum* L. (Papaveraceae) L.N.: Opium (Banswara)	Semisolid paste	Anestrous	Yadav et al., 20015
319	*Parkinsonia aculeata* L. (Caesalpiniaceae) L.N.: Parkinsonia (Southern Rajasthan), Vilayati kikar (Arid zone)	Leaf paste	Constipation	Takhar, 2004; Kumar et al., 2004
320	*Pedalium murex* L. (Pedaliaceae) L.N.: Dakhani gokhru (Rajasthan), Bada gokharu (Southern Rajasthan, Arid zone)	Whole plant; Whole plant fed	Cooling effect; Stomachache	Galav et al., 2013; Takhar, 2004; Kumar et al., 2004
321	*Pennisetum glaucum* (L.) R.Br. (Poaceae) L.N.: Bajra (Bhiwani)	Fried millet	Heat production	Yadav et al., 2014

S. No.	Name of the Plant	Part(s) used	Diseases	References
322	*Pennisetum typhoides* (Burm,/.) Stapf. and Hubb. (Poaceae) L.N.: Bazra (Gwalior), Bajra (Southern Rajasthan, Rajasthan)	Seed flour + Seeds of *Abrus precatorius*; Buffed chapattis of Seeds flour + Moath bean seeds flour; Aqueous paste of inflorescence; Dried Bajra + Ghee	Rhinitis; Mouth disease; Skin itching; Diarrhea	Shrivastava et al., 2012; Takhar, 2004; Kumar et al., 2004; Gupta et al., 2015
323	*Pentanema indicum* (L.) King (Asteraceae) L.N.: Gurumi (Gorakhpur)	Fruit paste + Turmeric	Indigestion	Mangal, 2015
324	*Pergularia daemia* (Forssk) Chiov. (Asclepiadaceae) L.N.: Dudheli (Udaipur, Sirohi)	Latex	Facilitate easy delivery	Sebastian, 1984
325	*Peristrophe paniculata* (Forssk.) Brummitt (Acanthaceae) L.N.: Kuljeera (West Nimar)	Plant ash + ghee + oil or paraffin jelly	Wounds	Satya and Solanki, 2009
326	*Phellorinia inquinans* Berk (Agaricaceae) L.N.: Desi mushroom (Arid zone)	Cut into pieces and dried in shade and fed	Fractured bones	Kumar et al., 2004
327	*Phoenix dactylifera* L. (Arecaceae) L.N.: Khajur (Jhabua)	Root paste	Worms in stomach	Kadel and Jain, 2006
328	*Phoenix humilis* Royle (Arecaceae) L.N.: Chhind (Baster)	Leaf decoction	Wounds	Sikarwar and Kumar, 2005

TABLE 6.1 (Continued)

S. No.	Name of the Plant	Part(s) used	Diseases	References
329	*Phoenix sylvestris* (L.) Roxb. (Arecaceae) L.N.: Khajur (Pratapgarh)	Fresh leaflets	Weakness	Meena, 2014
330	*Phyllanthus amarus* Schum. and Thonn. (Euphorbiaceae) L.N.: Buiawla (Alirajpur)	Leaf paste	Wounds	Chouhan and Ray, 2014
331	*Phyllanthus fraternus* Webst. (Euphorbiaceae) L.N.: Bhui amla (West Nimar), Bhui umbri (Udaipur)	Crushed whole plant Ash of 250 g leaves + 100 ml *Pongamia pinnata* oil	Lactagogue; Septicemia	Satya and Solanki, 2009; Nag et al., 2009
332	*Phyllanthus reticulatus* Poir. (Euphorbiaceae) L.N.: Teekhar (Uttar Pradesh)	Crushed leaves + Leaf buds of *Syzygium cumini* + *Mangifera indica* + Leaves of *Acacia nilotica* + *Aegle marmelos* + *Abutilon indicum* + Water + Salt	Blood in urine	Kumar and Bharati, 2013
333	*Physalis minima* L. (Solanaceae) L.N.: Badi popti (West Nimar)	Crushed seeds + oil cake, Leaf paste	Appetizer, snake bite	Satya and Solanki, 2009
334	*Piper betle* L. (Piperaceae) L.N.: Paan (Darbhanga, Gorakhpur)	Leaves + Seeds of *Trachyspermum copticum* + *Piper nigrum* + Rhizome of *Zinziber officinale* + *Allium sativum*; Leaf paste + Raw Kattha	Indigestion; Broken horn	Mishra et al., 1996; Mangal, 2015
335	*Piper longum* L. (Piperaceae) L.N.: Pipari (North Bihar)	Fruit decoction	Antibiotic, carminative	Jha et al., 1991

S. No.	Name of the Plant	Part(s) used	Diseases	References
336	*Piper nigrum* L (Piperaceae) L.N.: Kalapeepar (Jhabua), Kalimirach (Gwalior, Sitapur, Bhiwani), Marich (Darbhanga)	Seeds; Seed paste; Fruit powder + Seed powder of *Trigonella foenum-graecum* + Fruit powder of *Trachyspermum ammi* + fruit powder of *Cuminum cyminum* + *Ferula asafoetida* Seeds + *Ferula asafoetida* + *Zingiber officinal* + *Curcuma longa*; Seeds + seeds of *Cuminum cyminum* + Green *Capsicum annuum* + *Aloe barbadensis*; Seeds + Mustard oil, Seed powder + mustard oil; Seed powder + Butter; Cow ghee + *Piper nigrum*	Cough and Cold; Constipation; Indigestion; Mastitis; Fever, constipation; Poisonous venom; Snake bite	Kadel and Jain, 2006; Shrivastava et al., 2012; Mishra et al., 1996; Kumar and Bharati, 2012b; Yadav et al., 2014; Gupta et al., 2015
337	*Pithacellobium dulce* (Roxb.) Benth. (Mimosaceae) L.N.: Imerati (Gorakhpur)	Root paste + Alum + Salt	Swelling in shoulder	Mangal, 2015
338	*Pistacia integerrima* Stewart ex Brandia (Anacardiaceae) L.N.: Guruju (Marwar)	Paste of stem	Bloat	Dudi and Meena, 2015
339	*Plantago orbignyana* Steinh. ex Decne (=*P. ovata* Phil.) (Plantaginaceae) L.N.: Isabgol (Uttar Pradesh)	Husk + Seed powder of *Brassica juncea* + Zinger + Rock salt + Turpentine oil + Oil of *Brassica napus*	Mastitis	Kumar and Bharati, 2013

TABLE 6.1 (Continued)

S. No.	Name of the Plant	Part(s) used	Diseases	References
340	**Plectranthus barbatus** Andrews (=Coleus barbatus (Andrews) Benth. (Lamiaceae) L.N.: Bayada (Uttar Pradesh)	Root + Sorghum bicolor + Gum of Ferula asafoetida	Anorexia	Kumar and Bharati, 2013
341	**Plumbago zeylanica** L. (Plumbaginaceae) L.N.: Chitawar (Morena, Shivpuri), Chitrak (Alirajpur, Mount Abu, Rajasthan), Chirayta (Bhiwani) (Figure 6.2P)	Root paste + jaggery; Root paste; Leaf decoction + seeds of Trachyspermum ammi; Seed decoction + Trachyspermum ammi + leaves of Plumbago zeylanica; Twig powder + Meeth soda	Stomatis, Stomach pain; Wounds; Indigestion and flatulence; Flatulence; Improve appetites	Sikarwar, 1996; Sikarwar et al., 1994; Chouhan and Ray, 2014; Galav et al., 2007; Galav et al., 2013; Yadav et al., 2014
342	**Polygonum barbatum** L. (Polygonaceae) L.N.: Mirmiri (Gorakhpur)	Leaf paste	Boils	Mangal, 2015
343	**Pongamia pinnata** (L.) Pierre (Fabaceae) L.N.: Kanji (Satna, Shivpuri), Karanja (Jhansi), Karanj (Narsinghpur), Koranjo (Jaspur and Surguja), Kanja (Gonda)	Leaves, Seed oil; Leaf paste + pepper, Stem bark decoction; Seed oil + phosphorus powder; Leaf paste + pepper; Leaf paste + Pepper, Stem bark decoction Leaf paste + Black pepper, seed powder	Galactagogue; Skin diseases; Fever, dysentery; Ring worm; Fever; Dysentery; Fever, Cough	Gautam and Richhariya, 2015; Jatav et al., 2014; Nigam and Sharma, 2010; Sanghi, 2014; Ekka, 2015; Pandey et al., 1999
344	**Porana paniculata** Roxb. (Convolvulaceae) L.N.: Masbandhi (Jaspur and Surguja),	Tuber paste	Bone fracture;	Ekka, 2015

S. No.	Name of the Plant	Part(s) used	Diseases	References
345	*Portulaca oleracea* L. (Portulacaceae) L.N.: Nunia (Bhiwani)	Whole plant fed	Prevent excessive bleeding after delivery	Yadav et al., 2014
346	*Premna serratifolia* L. (Lamiaceae) L.N.: Arni (Bhiwani)	Leaf juice	Wounds to kill germs	Yadav et al., 2014
347	*Prosopis chilensis* (Molina) Stuntz (*P. juliflora* DC.) (Mimosaceae) L.N.: Vilayati babool (Southern Rajasthan, Arid zone)	Leaf paste	Wounds; Affection of foot pad	Takhar, 2004; Kumar et al., 2004
348	*Prosopis cineraria* (L.) Druce (Mimosaceae) L.N.: Janti (Bhiwani)	Poultice of bark	Pimples and wounds	Yadav et al., 2014
349	*Prunus persica* (L.) Stokes (Rosaceae) L.N.: Aaru (Uttar Pradesh)	Crushed leaves + Salt	Anorexia	Kumar and Bharati, 2013
350	*Psidium guajava* L. (Myrtaceae)	Leaf decoction	Fever	Verma, 2014
351	*Pupalia atropurpurea* Moq. (Amaranthaceae) L.N.: Sihitti (Udaipur	Root paste	Infection in anus	Nag et al., 2009

TABLE 6.1 (Continued)

S. No.	Name of the Plant	Part(s) used	Diseases	References
352	*Pueraria tuberosa* (Roxb. ex Willd.) DC. (Fabaceae) L.N.: Gajua (Jhabua), Patalkumbra (Baster)	Tuber	Increasing secretion of milk	Sikarwar and Kumar, 2005
353	*Radermachera xylocarpa* (Roth) K. Schum. (Bignoniaceae) L.N.: Kharseng (West Nimar)	Crushed roots + water	Gas and acidity	Satya and Solanki, 2009
354	*Raphanus sativus* L. (Brassicaceae) L.N.: Muli (Sitapur, Bhiwani)	One root + seeds of *Trigonella foenum-graecum* + Mustard oil + Bulbs of *Allium sativum*; Plant fed	Otitis; Maintain pregnancy	Kumar and Bharati, 2012b; Kumar and Bharati, 2013Yadav et al., 2014
355	*Ranunculus pulchellus* C. A. Meyer (Ranunculaceae) L.N.: Kush (Shivpuri)	Plant paste	Cuts and wounds	Jatav et al., 2014
356	*Rauvolfia serpentina* Benth. ex Kurz (Apocynaceae) L.N.: Isargat (North Bihar)	Aerial part	Nervous diseases, eye diseases	Jha et al., 1991
357	*Rhus mysorensis* G.Don (Anacardiaceae) L.N.: Dansarae (Rajasthan)	Leaf paste	Allergy, rashes, eczema	Galav et al., 2013

S. No.	Name of the Plant	Part(s) used	Diseases	References
358	*Ricinus communis* L. (Euphorbiaceae) L.N.: Arand (Jhabua, Bhiwani), Andi (Chitrakoot, Narsinghpur, North Bihar, Darbhanga), Arandi (Alirajpur, Narsinghpur, Sitapur, Vindhyan region, Mount Abu, Pratapgarh, Southern Rajasthan, Arid zone)	Ash of leaves; Seed oil; Leaf juice; Seed oil, Leaves; Leaves; Seeds; Leaf decoction; Seed oil; Leaves; Seed oil; Seeds + water; Seed cotyledons; Seed oil; Seed oil; Seeds + Sodium bicarbonate + Water; Seed oil	Wounds; Constipation; Wounds; Constipation, Diarrhea and Dysentery; Intestinal worm; Constipation; Anthelmintic, vermifuge, anti-rheumatic; Indigestion and flatulence; Increase lactation; Constipation; Regulate estrus cycle; Prepare for mating Stomachache; Stomach disorders, throat problem; Gastric problem; Flatulence; Constipation	Kadel and Jain, 2006; Tripathi and Singh, 2010; Chouhan and Ray, 2014; Dwivedi et al., 2009; Sanghi; 2014; Verma, 2014; Jha et al., 1991; Galav et al., 2007; Mishra et al., 1996; Kumar and Bharati, 2012b; Meena, 2014; Takhar, 2004; Kumar et al., 2004 Yadav et al., 2014; Kumar and Bharati, 2013; Gupta et al., 2015
359	*Rivea hypocrateriformis* (Desr.) Choisy (Convolvulaceae) L.N.: Fang (West Nimar)	Crushed leaves	Anti-inflammatory	Satya and Solanki, 2009
360	*Rosa damascena Mill.* (Rosaceae) L.N.: Gulab (Sitapur)	Gulab jal + Alum + Juice of *Citrus aurantifolia*	Conjunctivitis	Kumar and Bharati, 2012b
361	*Rosa cymosa* Tratt. (Rosaceae) L.N.: Gulab (Bhiwani)	Petals, Gulkand fed	Increasing milk	Yadav et al., 2014

TABLE 6.1 (Continued)

S. No.	Name of the Plant	Part(s) used	Diseases	References
362	*Saccharum bengalense* Retz. (Poaceae) L.N.: Sarkanda, Jhunda, Moonj (Bhiwani)	Young leaves fed	Removal of retained placenta	Yadav et al., 2014
363	*Saccharum officinarum* L. (Poaceae) L.N.: Ganna (Jhansi, Sitapur), Santa, Hanta (Pratapgarh)	Leaf; Local rum made; Dried leaves; Whisky, whisky + sugar, jaggery; Chaffed leaves + Grounded seed rice + Seed of *Madhuca longifolia*; Leaves + Luke warm water	Retard placenta; Expel placenta, Cold; Separate placenta after delivery; Cold, worm infection; Retention of placenta; Expulsion of placenta	Nigam and Sharma, 2010; Kumar and Bharati, 2012b; Meena, 2014; Kumar and Bharati, 2013; Yadav et al., 2015
364	*Saccharum spontaneum* L. (Poaceae) L.N.: Kans (Bhiwani)	Plant fed	Heat production	Yadav et al., 2014
365	*Salvadora oleoides* Decne (Salvadoraceae) L.N.: Pelu (Morena)	Leaf paste + Turmeric and Garlic paste	Hemorrhagic septicemia	Sikarwar, 1996.
366	*Salvadora persica* L. (Salvadoraceae) L.N.: Pala (West Nimar), Khara jal (Southern Rajasthan), Jaal (Bhiwani), Pilu (Marwar), Meetha jal (Arid zone)	Burnt leaves + cow urine; Leaf paste; Leaves fed; Paste of leaf buds + *Milletia pinnata*; Leaf ash + Water, Leaf paste applied, Leaf paste fed, Dried fruit soaked in water and fed	Wounds to kill worms; Irritation, remove lice, constipation, gas problem; Increase milk production; Foot and mouth disease; Skin irritation, remove lice and other parasitic insects, constipation and gastric problem, as coolant	Satya and Solanki, 2009; Takhar, 2004; Yadav et al., 2014; Dudi and Meena, 2015; Kumar et al., 2004

S. No.	Name of the Plant	Part(s) used	Diseases	References
367	*Sapindus laurifolius* Vahl (Sapindaceae) L.N.: Ritha (West Nimar)	Fruit powder	Snake bite	Satya and Solanki, 2009
368	*Sapindus mukorossi* Gaertn. (Sapindaceae) L.N.: Ritha (Marwar)	Frothed solution of fruit	Ectoparasites	Dudi and Meena, 2015
369	*Sapindus trifolius* L. (Sapindaceae) L.N.: Ritha (Singhbhum)	Bark decoction + Powder of leaves of *Vitex negundo*, nuts soaked in water	Ulcer, intestinal worms	Pal, 1980
370	*Sarcostemma viminale* (L.) R.Br. (Asclepiadaceae) L.N.: Bhurbel (West Nimar)	Crushed roots + water	Anti-inflammatory	Satya and Solanki, 2009
371	*Sauromatum venosum* (Ait.) Kunth (Araceae) L.N.: Bhasmakanda (Satna)	Tuber	Tonic	Gautam and Richhariya, 2015
372	*Schleichera oleosa* (Lour.) Oken (Sapindaceae) L.N.: Kusum (Raigarh, Chhindwara, Betul), Kusum phal (West Nimar)	Seed oil Fruit powder + sugar	Itching, leg swellings; Wounds to kill maggots	Sikarwar and Kumar, 2005; Satya and Solanki, 2009
373	*Scindapsus officinalis* Schott. (Araceae) L.N.: Gachpipal (Jaspur and Surguja)	Whole plant paste	Bone fracture	Ekka, 2015

TABLE 6.1 (Continued)

S. No.	Name of the Plant	Part(s) used	Diseases	References
374	*Semecarpus anacardium* L.*f.* (Anacardiaceae) L.N.: Bhilawa (West Nimar), Bibba (Betul), Kohbar (Gorakhpur), Bhilama (Banswara)	Crushed seeds; Fruit + Jawar bread (*Hordeum vulgare*); Leaf paste; 1 fruit daily for 4 days	Cold; Mouth diseases; Worms in wound; Anestrous	Satya and Solanki, 2009; Patil and Deshmukh, 2015; Mangal, 2015; Yadav et al., 20015
375	*Senna auriculata* (L.) Roxb. (=*Cassia auriculata* L.) (Caesalpiniaceae) L.N.: Tarbad (West Nimar), Aval (Narsinghpur), Anwal (Udaipur)	Leaf paste + coconut oil, Leaf decoction, Leaves + salt; Leaf paste; Leaf infusion, Leaves + Jaggery	Sprain, gas, acidity, diarrhea; Pox; Hooves, Foot and mouth disease	Satya and Solanki, 2009; Sanghi, 2014; Deora et al., 2004
376	*Senna occidentalis* (L.) Link (=*Cassia occidentalis* L.) (Caesal iniaceae) L.N.: Babai (Jalaun), Malvi punwad (Udaipur), Chippi-kaali (Uttar Pradesh)	30 g leaves + mustard oil; Leaf paste, seed powder; Crushed pods + Oil of *Brassica juncea*	Fever; Eczema, skin diseases, increase milk production; Fever	Kumar and Bharti, 2012a; Deora et al., 2004; Kumar and Bharati, 2013
377	*Senna tora* (L.) Roxb. (=*Cassia tora* L.) (Caesalpiniaceae) L.N.: Puwadiya (Alirajpur), Pamar (Gwalior), Pumar (Shivpuri), Chakoda (Jaspur and Surguja), Puwario (Udaipur, Sirohi)	Seed paste; Seed powder + Seed powder of *Trigonella foenum-graecum* + Fruit powder of *Trachyspermum ammi* + fruit powder of *Piper nigrum* + fruit powder of *Cuminum cyminum* + *Ferula asafoetida* Seed paste; Seeds	Wounds; Constipation; Skin disease; More lactation	Chouhan and Ray, 2014; Shrivastava et al., 2012; Jatav et al., 2014; Ekka, 2015; Sebastian, 1984

S. No.	Name of the Plant	Part(s) used	Diseases	References
378	*Sesamum orientale* L. (=*S. indicum* L.) (Pedaliaceae) L.N.: Til (Mount Abu, Southern Rajasthan)	Seed oil, Seed oil + Turmeric powder; Seed oil; Sesame cake; Oil; Seed oil + Turmeric powder, Seed oil; Seed oil + sugar	Foot and mouth disease; Dysentery, horn wound; Maggot wound; Bloat, ectoparasites; Foot infection, dysentery; Arthritis	Galav et al., 2007; Takhar, 2004; Yadav et al., 2015; Dudi and Meena, 2015; Kumar et al., 2004; Gupta et al., 2015
379	*Shorea robusta* Gaertn.f. (Dipterocarpaceae) L.N.: Sar (Udaipur)	Leaves, root decoction	As an antinimatode, Typhoid and fever	Nag et al., 2009
380	*Sida acuta* Burm.f. (Malvaceae) L.N.: Wala (Alirajpur)	Leaf juice	Wounds	Chouhan and Ray, 2014
381	*Sida cordata* (Burm.f.) Borss. (Malvaceae)	Plant	Galactagogue	Sebastian, 1984
382	*Sida cordifolia* L. (Malvaceae) L.N.: Bariyar (North Bihar), Farhad (Jalaun, Gorakhpur)	Leaf and fruit decoction; Leaves + *Ferula asafoetida* + Camphor (*Cinnamomum camphora*) crushed together; Leaf paste + *Ferula asafoitida* + Kapoor	Diuretic, cardiac, anti-rheumatic, eye diseases; Wound or injury; Foot and mouth disease	Jha et al., 1991; Kumar and Bharti, 2012a; Kumar and Bharati, 2013; Mangal, 2015
383	*Sida ovata* Forssk (Malvaceae) L.N.: Khariti (Rajasthan)	Whole plant	Pain reliever	Galav et al., 2013
384	*Smilax wightii* DC. (Liliaceae) L.N.: Ranpawan (Raigarh)	Leaf paste	Diarrhea	Sikarwar et al., 1994.

TABLE 6.1 (Continued)

S. No.	Name of the Plant	Part(s) used	Diseases	References
385	*Solanum americanum* Mill. (Solanaceae) L.N.: Buiregni (Alirajpur)	Fruit paste + Leaf paste of *Heteropogon contortus*	Wounds	Chouhan and Ray, 2014
386	*Solanum melongena* L. **var.** *esculentum* (Solanaceae) L.N.: Baigan (Sitapur)	Roasted fruits; Baked fruit + Ghee + salt; Fruit paste + Black salt	Inflammation of shoulder; Irregularity in estrus cycle; Swelling in shoulder	Kumar and Bharati, 2012b; Kumar and Bharati, 2013; Mangal, 2015
387	*Solanum nigrum* L. (Solanaceae) L.N.: Lal Gongachi (Dhar) (Figure 6.3Q)	Leaf paste	Body swelling	Sikarwar et al., 1994.
388	*Solanum surattense* Burm. *f.* (Solanaceae) L.N.: Regni Kataiya (North Bihar)	Leaf, flower decoction	Expectorant, carminative	Jha et al., 1991
389	*Sorghum bicolor* (L.) Moench. (Poaceae) L.N.: Juwar	Flour; Seeds + roots of *Coleus barbatus* + *Ferula asafoetida* + water	Dysentery; Anorexia	Jadhav, 2009; Kumar and Bharati, 2012b
390	*Sorghum halepense* (L.) Pers. (Poaceae) L.N.: Bru (Pratapgarh)	Inflorescence or Caryopsis	Diarrhea and weakness	Meena, 2014
391	*Sorghum vulgare* (L.) Pers. (Poaceae) L.N.: Jwar (Vindhyan region), Jawar	Tender leaves; Seed flour + whey	Intestinal worms of infants; Loose motion	Dwivedi et al., 2009; Nigam and Sharma, 2010

FIGURE 6.3

TABLE 6.1 (Continued)

S. No.	Name of the Plant	Part(s) used	Diseases	References
392	*Soymida febrifuga* (Roxb.) A. Juss (Meliaceae) L.N.: Rohan (Raigarh, Dewas, West Nimar), Rohin (Rajasthan), Rohina (Gorkhpur)	Bark paste; Crushed bark; Crushed bark + bark of *Butea monosperma* + Butter milk; Crushed leaves	Diarrhea; Inflammation of foot; Diarrhea; Constipation	Sikarwar and Kumar, 2005; Galav et al., 2013; Mangal, 2015
393	*Spermacoce stricta* L. (Rubiaceae) L.N.: Agio (Rajasthan)	Whole plant decoction	Vulvo-vaginal-uterine-prolapse	Galav et al., 2013
394	*Spondias pinnata* L. (Anacardiaceae) L.N.: Ambirlo (Jaspur and Surguja)	Fresh fruit paste	Dysentery	Ekka, 2015
395	*Sterculia urens* Roxb. (Sterculiaceae) L.N.: Kurlu (Jaspur and Surguja), Kateera (Sitapur) (Figure 6.3R)	Stem bark; Gum + *Linum ustatissimum*	Wounds; Little and frequent urine	Ekka, 2015; Kumar and Bharati, 2012b; Kumar and Kumar, 2013
396	*Steriospermum chelonoides* (L.f.) DC. (Bignoniaceae) L.N.: Padal (West Nimar)	Crushed leaves	Wounds to prevent bleeding	Satya and Solanki, 2009
397	*Styrax benzoin* Dry. (Styracaceae) L.N.: Lobaan (Sitapur)	Resin + Gayru	Expulsion of placenta, Mastitis	Kumar and Bharati, 2012b

S. No.	Name of the Plant	Part(s) used	Diseases	References
398	*Swertia chirayita* (Roxb. ex Flem.) Karst. (Gentianaceae) L.N.: Chiraita (Gonda)	Plant decoction	Fever	Pandey et al., 1999
399	*Syzygium aromaticum* (L.) Merr. and L. M. Perry (Myrtaceae) L.N.: Clove (Bhiwani), Laung (Banswara)	Clove oil; Dried buds + Jaggery + Sweet oil	Mastitis; Expulsion of placenta	Yadav et al., 2014; Yadav et al., 2015
400	*Syzygium cumini* (L.) Skeels (Myrtaceae) L.N.: Jamun (Jhabua, Jhansi, Tikamgarh, Udaipur, Jaspur and Surguja, Sitapur)	Bark paste; Bark powder + whey + water; Decoction of bark + decoction of bark of *Azadirachta indica*; Bark paste + curd; Seed powder; Bark decoction of *Syzygium cumini + Ficus glomerata + Acacia nilotica* ssp. *indica + Butea monosperma + Acacia catechu*, Fruit juice; New leaves + leaves of *Mangifera indica + Acacia nilotica* ssp. *indica + Euphorbia reticulata + Aegle marmelos + Abutilon indicum* + common salt	Dysentery; Hematuria; Joints pain; Diarrhea; Dysentery; Foot and mouth disease, Abdominal pain; Hematuria	Kadel and Jain, 2006; Tripathi ard Singh, 2010; Nigam and Sharma, 2010; Verma, 2014; Nag et al., 2009; Ekka, 2015; Kumar and Bharati, 2012b; Kumar and Bharati, 2013
401	*Syzygium heyneanum* Wall. ex Gamble (Myrtaceae) L.N.: Chhoti Jamun (Jabalpur)	Stem bark decoction; Stem bark juice	Constipation; Diarrhea and Dysentery	Sikarwar et al., 1994;

TABLE 6.1 (Continued)

S. No.	Name of the Plant	Part(s) used	Diseases	References
402	*Tagetes erecta* L. (Asteraceae) L.N.: Marigold (Gorakhpur), Gaynda (Uttar Pradesh), Gainda (Shivpuri, Jhansi, Tikamgarh), Genda (Narsinghpur)	Grinded leaves, extract of leaves; Flower extract; Leaves; Flower powder + water; Plant paste; Leaf decoction	Blockage of urine, Blood in urine; Otitis; Cuts and wounds; Hydrophobia; Cut and wounds; Hydrophobia	Mangal, 2015; Kumar and Bharati, 2013; Jatav et al., 2014; Nigam and Sharma, 2010; Sanghi, 2014; Verma, 2014
403	*Tamarindus indica* L. (Caesalpiniaceae) L.N.: Imli (Jhabua, Satna, Shivpuri, Tikamgarh, Sitapur), Emli (Rajasthan)	Bark paste; Leaves; Fruits; Boiled leaves; Leaf juice; Extract	Dysentery; Inflamed joints; As carminative, laxative, digestive; Swelling; Diarrhea; Food poisoning	Kadel and Jain, 2006; Gautam and Richhariya, 2015; Jatav et al., 2014; Verma, 2014; Kumar and Bharati, 2012b; Kumar and Bharati, 2013; Gupta et al., 2015
404	*Tamarix aphylla* (L.) H. Karst. (Tamaricaceae) L.N.: Firansh (Bhiwani)	Bark ash + Vaseline	Cracked nipples and burnt injury	Yadav et al., 2014
405	*Taraxacum officinale* (L.) Weber (Asteraceae) L.N.: Rasaad (Jalaun), Rasad (Gorakhpur)	*Taraxacum officinale* + root extract of *Withnia somnifera* + seeds of *Trachyspermum ammi* + Milk fat; Plant + *Withania somnifera* + *Trachyspermum ammi* + Ghee, plant paste + Jaggery + Ghee	Diarrhea; Placenta did not fall	Kumar and Bharti, 2012a; Kumar and Bharati, 2013; Mangal, 2015

S. No.	Name of the Plant	Part(s) used	Diseases	References
406	*Tecomela undulata* (Sm.) Seem (Bignoniaceae) L.N.: Rohida (Rajasthan), Rohira (Southern Rajasthan), Roheda (Bhiwani), Rohira (Arid zone, Rajasthan) (Figure 6.3S)	Bark oil; Stem pieces decoction + Sulphur; Root decoction + root decoction of *Ziziphus nummularia* + Jaggery; Bark paste + Whey	Rashes; Skin irritation; Refreshment of whole body; External parasites	Galav et al., 2013; Takhar, 2004; Kumar et al., 2004; Yadav et al., 2014; Gupta et al., 2015
407	*Tectona grandis* L.*f.* (Verbenaceae) L.N.: Sagon (Jhabua), Sagoan (MIdnapur)	Leaf paste; Wood paste + mustard oil	Wounds; Maggots infested wounds	Kadel and Jain, 2006; Pal, 1980
408	*Tephrosia purpurea* (L.) Pers.(Fabaceae) L.N.: Dhamasa (Morena), Sarponkha (West Nimar), Biyani (Pratapgarh, Southern Rajasthan, Arid zone)	Leaf paste; Crushed leaves; Whole plant decoction; Whole plant extract	Wounds; Skin diseases	Sikarwar, 1996; Satya and Solanki, 2009; Meena, 2014; Takhar, 2004; Kumar et al., 2004
409	*Tephrosia uniflora* Pers. ssp. *petrosa* (Blatt. et Halb.) Gillet et Ali (Fabaceae) L.N.: O (Udaipur, Sirohi)	Leaf paste	External parasites	Sebastian, 1984
410	*Terminalia bellirica* (Gaertn.) Roxb. (Combretaceae) L.N.: Bahera (Satna, Uttar Pradesh), Baheda (Rajasthan)	Fruit powder; Pulp of fresh fruits or dried fruit powder; Oil + Vilayati ghee	Indigestion, flatulence and stomach disorders; Diarrhea; Difficulty breathing, chewing and swelling, loss of appetite, fever	Gautam and Richhariya, 2015; Galav et al. 2013; Kumar and Bharati, 2013

TABLE 6.1 (Continued)

S. No.	Name of the Plant	Part(s) used	Diseases	References
411	*Terminalia chebula* Retz. (Combretaceae) L.N.: Harra (Satna, Shivpuri, Vindhyan region, Jaspur and Surguja), Bari Haar (Sitapur), Harad (Rajasthan, Bhiwani), Bariar (Gorakhpur) (Figure 6.3T)	Fruit powder; Stem bark paste + pepper + garlic; Seed powder; Powder of dried fruits + *Boerhavia diffusa* + seed oil of *Ricinus communis*; Mixture of Fruits + Rock salt + Seeds of *Trachyspermum ammi* + Neero + *Elytaria* + Sodium bicarbonate + Jaggery; Leaf paste + water; Decoction of fruits + *Trigonella foenum-graecum* seeds, powder of Fruits + Ajwain + black salt + human urine	Anthrax; Fever; Cut and injury; Show; Gastro intestinal disorder; Prolapse; Better digestion, stomachache	Gautam and Richhariya, 2015; Jatav et al., 2014; Dwivedi et al., 2009; Ekka, 2015; Kumar and Bharati, 2012b; Galav et al., 2013; Mangal, 2015; Yadav et al., 2014
412	*Terminalia cuneata* Roth (= *T. arjuna* (Roxb. ex DC.) Wt. and Arn. (Combretaceae) L.N.: Arjun (Gonda, Betul, Bhiwani), Sadada (Betul)	Bark decoction; Stem bark paste; Bark paste	Loss of appetite, weakness; Bone fracture; Removal of retained placenta	Pandey et al., 1999; Patil and Deshmukh, 2015; Yadav et al., 2014
413	*Thea sinensis* L. (Theaceae) L.N.: Chai (Arid zone)	Leaf decoction	Gastric problem	Kumar et al., 2004
414	*Thevetia neriifolia* Juss (Apocynaceae) L.N.: kanel (Darbhanga)	Seeds of *Lens culinaris* + *Lallemantia royleana* + pulp of *Aloe vera* + *Acacia catechu* (Kattha) + flower of *Thevetia neriifolia* + turpentine oil	Swelling of foot	Mishra et al., 1996

S. No.	Name of the Plant	Part(s) used	Diseases	References
415	*Tinospora cordifolia* (Willd.) Hook. and Thoms. (Menispermaceae) L.N.: Gorbel (Tikamgarh), Limbbel (Ratlam), Gurich (North Bihar), Giloy (Udaipur, Gorakhpur), Neem Giloe (Mount Abu) (Figure 6.3U)	Stem pieces; Stem extract; Stem, bark decoction; Root paste; Whole plant extract; Plant paste	Syphilis; Dog bite; Anthelmintic, vermifuge, diuretic, blood purifier, cardiac stimulant; Galactagogue; General debility; Fever	Sikarwar, and Kumar, 2005; Jadhav, 2009; Jha et al., 1991; Nag et al., 2009; Galav et al., 2007; Mangal, 2015
416	*Trachyspermum ammi* (L.) Sprague (=*T. copticum* Link) (Apiaceae) L.N.: Ajwain (Jhabua, Bundelkhand, Gwalior, Mount Abu, Gonda, Gorakhpur, Bhiwani, Marwar, Arid zone), Jamain (Darbhanga, North Bihar), Ajvian (Pratapgarh)	Fruit powder; Fruit powder + seeds powder of *Cassia tora* + seed powder of *Trigonella foenum-graecum*; fruit powder of *Piper nigrum* + fruit powder of *Cuminum cyminum* + *Ferula asafoetida* Fruits + bulbs of *Allium sativum* + Jaggery; Seed decoction; Grains; Seeds + *Calotropis gigantea* leaves + salt + water; Seeds + *Zinziber officinale* + *Ferula asafoetida* Decoction of seeds + *Anethum gravelens*; Seed powder + Jaggery; Seed powder + Jaggery; Seeds; Seed powder + black salt; Seed powder + Sugar candy + *Ferula asafoetida* Seed powder + Edible oil; Seed decoction + Jaggery, seed decoction + Rock salt; Seeds + Old jute bags + Hairs and burnt and give smoke, Grinded seeds + jaggery; Seeds + Black salt; Seed powder + Jaggery	Indigestion; Cold and fever; Constipation, Pharyngitis; Antibiotic, carminative, expectorant; Dysentery; Stomachache; Animal does not eat; Crack in palate Ouster induction; Lack of estrus; Indigestion; Constipation, dysentery; Indigestion; Gas problems, Bovine Viral Mammaitis; pneumonia, fever, typhoid, indigestion, diarrhea, measles; Cure canker sores, removal of retained placenta; Bloat (affara); Gastric problem, Bovine Viral Mammaitis; Infertility	Kadel and Jain, 2006; Mishra et al., 2010; Shrivastava et al., 2012; Jha et al., 1991; Mishra et al., 1996; Galav et al., 2007; Pandey et al., 1999; Meena, 2014; Mangal, 2015; Takhar, 2004; Kumar et al., 2004 Yadav et al., 2014; Dudi and Meena, 2015; Gupta et al., 2015

TABLE 6.1 (Continued)

S. No.	Name of the Plant	Part(s) used	Diseases	References
417	*Trianthema portulacastrum* L. (Aizoaceae) L.N.: Pattahr-Choor (Jalaun), Hato (Rajasthan)	Leaves paste + salt; Leaf paste + seeds of *Piper nigrum*	Mastitis; Diarrhea	Kumar and Bharti, 2012a; Kumar and Bharati, 2013; Galav et al., 2013
418	*Tribulus terrestris* L. (Zygophyllaceae) L.N.: Kanti (Udaipur), Gokharu (Pratapgarh, Rajasthan), Bhankhri (Bhiwani)	Leaf juice; Shoot paste, root paste + leaves of *Azadirachta indica*; Fruits; Whole plant extract; Fruits fed	Colic, cough; Internal parasites; external parasites; Increase lactation, weakness; Diarrhea	Verma, 2014; Nag et al., 2009; Meena, 2014; Yadav et al., 2014; Gupta et al., 2015
419	*Trichodesma indicum* (L.) Lehm. (Boraginaceae) L.N.: Tarmudia (raigarh)	Root paste	Wounds to kill worms	Sikarwar et al., 1994.
420	*Tricholepis glaberrima* DC. (Asteraceae) L.N.: Tikatta (West Nimar)	Whole plant paste + whey	Abdominal destination, lack of appetite, Diarrhea	Satya and Solanki, 2009
421	*Trichosanthes cucumerina* L. (Cucurbitaceae) L.N.: Nagphani beldi (Udaipur)	Root decoction	Vulvo-vaginal uterine prolapse	Nag et al., 2009

S. No.	Name of the Plant	Part(s) used	Diseases	References
422	*Tridax procumbens* L. (Asteraceae) L.N.: Baramansi (Chhindwara), Kuradiya (Alirajpur), Ghamra (Shivpuri), Latti (Jaspur and Surguja), Kaliya (Udaipur, Sirohi), Kalali (Udaipur), Larde olapsi (Rajasthan), Kagla Ri Mehndi (Rajasthan)	Whole plant paste + whey; Leaf juice; Leaf paste; Leaf juice; Leaf infusion; Leaf paste	Cut and wounds; Wounds, maggots in wounds; Diarrhea; Wounds	Sikarwar et al., 1994; Chouhan and Ray, 2014; Jatav et al., 2014; Ekka, 2015; Sebastian, 1984; Deora et al., 2004; Galav et al., 2013; Gupta et al., 2015
423	*Trigonella foenum-graecum* L. (Fabaceae) L.N.: Methi (Chitrakoot, Bundelkhand; Gwalior, Jhansi, Tikamgarh, Jalaun, Sitapur, Pratapgarh, Gorakhpur, Bhiwani, Rajasthan), Meethi (Southern Rajasthan)	Seed paste; Seed powder; Seed powder + Fruit powder of *Trachyspermum ammi* + seeds powder of *Cassia tora* + fruit powder of *Piper nigrum* + fruit powder of *Cuminum cyminum* + *Ferula asafoetida* Seeds + seeds of *Trachyspermum ammi* + jaggery; Decoction of seed powder; Sprout seeds, seed powder; 250 g seeds + 250 g seeds of *Lens esculenta*; Seeds paste of *Trigonella foenum-graecum* + Black salt + *Cuminum cyminum* + *Ferula asafoetida* + leaves of *Bombax ceiba*; Fresh mature plant or boiled seeds; Paste of seeds + Jaggery + Mustard oil; Leaf paste + Hair; Boiled seeds + Jaggery; Decoction of seeds + Black salt; Methi + Mustard oil	Respiratory problem; Cold and fever; Indigestion; Constipation, foot and mouth disease; Twitching; Easy delivery, Twitching; Diphtheria; Flatulence, dysentery; Increase lactation; Placenta did not fall, Broken horn; Increase milk, joint pain; Pneumonia	Tripathi and Singh, 2010; Mishra et al., 2010; Shrivastava et al., 2012; Nigam and Sharma, 2010; Verma, 2014; Kumar and Bharti, 2012a; Kumar and Bharati, 2012b; Meena, 2014; Mangal, 2015; Takhar, 2004; Kumar et al., 2004; Yadav et al., 2014; Gupta et al., 2015

TABLE 6.1 (Continued)

S. No.	Name of the Plant	Part(s) used	Diseases	References
424	*Triticum aestivum* L. (Poaceae) L.N.: Genhu (Gwalior, Southern Rajasthan), Gayhu (Jalaun, Gorakhpur), Gehu (Bhiwani)	Poultice; Soaked seeds; Dough of wheat flour; Paste of roasted seeds; Crushed seeds + Tea leaves + Ashwagandha; 1 kg germinated wheat for one week; Bran; Paste of roasted seed + Fodder; Sprouted wheat grains + Jaggery	Worms; Regulation of estrus cycle; Prepare for mating; Loss of teeth; Gas problem; Cold; Anestrous; Internal parasite; Gastric problem; Infertility	Shrivastava et al., 2012; Kumar and Bharti, 2012a; Kumar and Bharti, 2013; Mangal, 2015; Takhar, 2004; Yadav et al., 2014; Yadav et al., 20015; Dudi and Meena, 2015; Kumar et al., 2004; Gupta et al., 2015
425	*Triumfetta rotundifolia* Lam. (Tiliaceae) L.N.: Lapta (Pratapgarh)	Root infusion	Neck sore	Meena, 2014
426	*Urena lobata* L. (Malvaceae) L.N.: Bhayriya (Jalaun), Bhediya gass (Gorakhpur)	Milk fat + Camphor + *Urena lobata*; Plant, plant paste, Plant paste + Kapoor + ghee	Cough, tonsillitis and fever; Vomiting, etching, fever	Kumar and Bharti, 2012a; Kumar and Bharti, 2013; Mangal, 2015
427	*Urginea indica* (Roxb.) Kunth. (Liliaceae) L.N.: Koli kanda (Mount Abu), Kolikando (Udaipur), Jangli piyaz (Marwar)	Bulb fed; Crushed bulb juice, Roasted bulb; Paste	Mastititis (Thanella); Maggots in wounds, external abscess; Skin disease	Galav et al., 2007; Deora et al., 2004; Dudi and Meena, 2015
428	*Vanda parviflora* Lindl. (Orchidaceae) L.N.: Tarwari (Gonda)	Warmed leaves	Swelling and pain	Pandey et al., 1999

S. No.	Name of the Plant	Part(s) used	Diseases	References
429	*Vanda tessellata* (Roxb.) Hook.*f.* ex G.Don (Orchidaceae) L.N.: Banda (Shahdol)	Plant paste	Body swellings	Sikarwar et al., 1994
430	*Vernonia anthelmintica* (L.) Willd. (Asteraceae) L.N.: Garad (Ratlam), Kaljira (Vindhyan region), Kali jiri (Mount Abu, Southern Rajasthan, Arid zone)	Seed extract; Seed decoction; 250 g seeds of *Trachyspermum ammi* + 250 g dried rhizome of *Curcuma longa* + 200 g seeds of Kalijiri + 2 kg jaggery prepared laddu; Seed decoction + Jaggery	Gastric trouble; Fever; Udder disorders	Jadhav, 2009; Dwivedi et al., 2009; Galav et al., 2007; Takhar, 2004; Kumar et al., 2004
431	*Vernonia cinerea* (L.) Less (Asteraceae) L.N.: Kali jiri	Plant + Garlic + Onion + Ginger + Jaggery	Increase appetite	Yadav et al., 2014
432	*Vigna aconitifolia* (Jacq.) Marechal (Fabaceae) L.N.: Moth (Southern Rajasthan, Arid zone)	Chapattis made from seed flour + edible oil	Mouth sores	Takhar, 2004; Kumar et al., 2004
433	*Vigna mungo* (L.) Hepper (Fabaceae) L.N.: Urad (Sitapur)	Seed powder + water	Diarrhea in summer season	Kumar and Bharati, 2012b; Kumar and Bharati, 2013
434	*Vigna radiata* (L.) Wilc. (Fabaceae) L.N.: Mug (Morena), Moong (Jhansi, Tikamgarh)	Leaf powder; Seed powder + seed oil of *Arachis hypogaea*	Wounds; Cough; Cough and cold	Sikarwar, 1996; Nigam and Sharma, 2010; Verma, 2014
435	*Viola biflora* L. (Violaceae) L.N.: Banbasa (Marwar)	Solution of root, leaves and bark + Honey	Skin disease	Dudi and Meena, 2015

TABLE 6.1 (Continued)

S. No.	Name of the Plant	Part(s) used	Diseases	References
436	**Viscum articulatum** Burm.f. (Loranthaceae) L.N.: Vanda (West Nimar)	Leaves and flowers	Fracture	Satya and Solanki, 2009
437	**Vitex negudo** L. (Verbenaceae) L.N.: Nirguri (Ratlam), Nirgur (Betul), Nirgundi (Shivpuri, West Nimar, Udaipur), Negad (Udaipur), Khonkhod (Jaspur and Surguja), Negadia (Pratapgarh)	Leaf paste; Leaf paste + pepper + garlic; Leaf paste; Dried leaves + fodder; Leaf paste; Stem bark paste + Curd; Root juice, leaf juice; Fresh or dried leaves	Wounds; Infectious diseases; Skin diseases; Diarrhea; Skin disease, conjunctivitis; Dysentery; Mastitis, Wounds; Stomachache	Sikarwar, and Kumar, 2005; Jatav et al., 2014; Satya and Solanki, 2009; Verma, 2014; Nag et al., 2009; Ekka, 2015; Deora et al., 2004; Meena, 2014
438	**Wattakaka volubilis** (L. f.) Stapf. (Asclepiadaceae) L.N.: Morash (West Nimar), Kadabadi (Udaipur, Sirohi)	Plant paste; Leaves, roots	Inflamed shoulders; Gastric trouble	Satya and Solanki, 2009; Sebastian, 1984
439	**Wendlandia exserta** DC. (Rubiaceae) L.N.: Tilbad (Shahdol), Nirgundi (Vindhyan region)	Stem bark decoction; Leaf poultice	Skin diseases; Cut and injury	Sikarwar et al., 1994; Dwivedi et al., 2009

S. No.	Name of the Plant	Part(s) used	Diseases	References
440	*Withania somnifera* (L.) Dunal (Solanaceae) L.N.: Ashwagandha (Jhansi, Jalaun, Sitapur, Gorakhpur, Bhiwani), Akshan (Rajasthan)	Decoction of root powder + seed powder of *Hyoscymus niger* + leaf powder of *Bambusa arundinacea* + Jaggery + Rhizome powder of *Zingiber officinale* + milk; 250 g leaves of *Withania somnifera* + 250 g milk fat: Roots + seed of *Myristica fragrans* + seeds of *Piper nigrum* crushed together; Decoction of Pulse of *Cajanus cajan* + Root of *Withania somnifera*; Tuber decoction + seeds oil of *Sesamum indicum*; Leaf paste + Ghee; Root decoction	Retard placenta; Constipation; Diarrhea in winter season; Lumbago; Constipation; Cold and cough	Nigam and Sharma, 2010; Kumar and Bharti, 2012a; Kumar and Bharati, 2012b; Kumar and Bharati, 2013; Galav et al., 2013; Mangal, 2015; Yadav et al., 2014
441	*Woodfordia floribunda* Salisb. (Lythraceae) L.N.: Dhawai (Jaspur and Surguja), Brabmi (Marwar)	Root paste + Egg shell; Leaf paste	Bone fracture; Fever	Ekka, 2015; Dudi and Meena, 2015
442	*Wrightia tinctoria* (Roxb.) R.Br. (Apocynaceae) L.N.: Kalakuda (West Nimar), Khanni (Udaipur, Sirohi) (Figure 6.3V)	Bark paste	Anti-inflammatory; Antidote to snakebite	Satya and Solanki, 2009; Sebastian, 1984
443	*Xanthium strumarium* L. (Asteraceae) L.N.: Chirchita (Morena, Vindhyan region), Andhayo (Udaipur)	Leaf paste; Leaf paste; Leaf juice	Wounds; Shoulder wounds	Sikarwar, 1996; Dwivedi et al., 2009; Deora et al., 2004
444	*Zea mays* L. (Poaceae) L.N.: Makka (Jhabua, Udaipur, Sirohi), Makai (Darbhanga)	Grains decoction; Boiled grains of *Zea mays* + *Cicer arietinum* + *Triticum aestivum* + *Oryza sativa* with salt; Grains + Other grains; 1–2 kg flour + Roughage for 1 week	Antifertility; Increase lactation; Galactagogue; Anestrous	Kadel and Jain, 2006; Mishra et al., 1996; Sebastian, 1984; Yadav et al., 20015

TABLE 6.1 (Continued)

S. No.	Name of the Plant	Part(s) used	Diseases	References
445	*Zingiber cernuum* Dalz. (Zingiberaceae) L.N.: Gaura santh (West Nimar)	Rhizome	Lactation	Satya and Solanki, 2009
446	*Zingiber officinale* Rosc. (Zingiberaceae) L.N.: Adrak (Chitrakoot, Shivpuri, Tikamgarh, Darbhanga, Southern Rajasthan, Bhiwani, Arid zone), Adi (North Bihar), Adrakh (Darbhanga, Uttar Pradesh), Sooth (Jalaun), Sonth (Sitapur, Rajasthan), Saunth (Rajasthan)	Rhizome juice; Rhizome; Rhizome boiled with milk; Rhizome decoction; Rhizome juice; Rhizome powder + 20 g alum + roasted *Solanum melongena*; Decoction of rhizome + plants of *Trachyspermum copticum* + *Withania somnifera* + seeds *Daucus carota*, Rhizome + Seeds of *Piper nigrum* + *Ferula asafoetida* + Black salt; Rhizome powder + Rock salt; Decoction of mixture of dried rhizome powder + milk fat + Black pepper; Mixture of dried rhizome powder + *Allium sativum* + Hen egg; Mixture of dried rhizome + Fruits of *Terminalia chebula* + Rock salt + Jaggery + Sodium bicarbonate; Rhizome + Jaggery; Crushed rhizome + Seeds of *Brassica juncea* + *Bauhinia parviflora* + *Piper nigrum*; rhizome + rock salt; Dried rhizome powder + Jaggery + Luke warm water; Ginger + *Swertia chirayita* + *Trigonella foenum –graegum* seeds + Jaggery	Fever; Cold and fever; Dyspepsia, flatulence, colic, diarrhea: Physically disability; Blood purifier, expectorant, carminative; Cough, cold and fever; Shoulder dislocation; Fever, Animal does not eat; Ansarca; Stomachache, Pneumonia, Paralysis; Indigestion; Cough, throat problem; Diarrhea; Swelling of body; Pneumonia; Anemia	Tripathi and Singh, 2010; Mishra et al., 2010; Jatav et al., 2014; Verma, 2014; Jha et al., 1991; Mishra et al., 1996; Kumar and Bharti, 2012a; Mishra et al., 1996; Kumar and Bharati, 2012b; Galav et al., 2013; Takhar, 2004; Kumar et al., 2004; Kumar and Bharati, 2013; Gupta et al., 2015
447	*Zingiber roseum* Rosc. (Zingiberaceae) L.N.: Jangli ada (Jhabua, Baster)	Rhizome paste	Bone fracture	Sikarwar and Kumar, 2005

S. No.	Name of the Plant	Part(s) used	Diseases	References
448	*Ziziphus mauritiana* Lam. (=*Z. jujuba* (L.) Lam.) (Rhamnaceae) L.N.: Ber (Jhansi, Rajasthan), Pala jhadi (Rajasthan), Deshi ber (Marwar)	Leaf paste + seed oil of *Linum usitatissimum*: Seed powder; Roots of this plant + Sesamum oil + Ammonium chloride + Sugar + Clay + Wheat husk + water (kept in earthen pot for 3 days), Mixture of lac of this plant + seeds of *Foeniculum vulgare* + seeds of *Ocimum americanum*; Plant decoction	Burn, Sun burn; Vulvo-vaginal-uterine-prolapse; Galactagogue, induce lactation; Leucorrhoea, heat stroke; Skin disease	Nigam and Sharma, 2010; Verma, 2014; Galav et al., 2013; Dudi and Meena, 2015
449	*Ziziphus nummularia* (Burm.f.) Wt. and Arn. (Rhamnaceae) L.N.: Jhar (Morena), Jhari-bor (Udaipur), Bordi (Southern Rajasthan), Jhari ber (Bhiwani)	Root decoction, Root paste; Root decoction; Root decoction, whole plant; Ash of bark + Ghee; Tender twigs, Grinded fruits + tea, root decoction + jaggery, root decoction; Leaf paste, Roots + *Crotalaria burhia* + Jaggery + Alum + water kept in a utensil with air tight lid and kept under ground for a few days. The resultant liquid served	Wounds, Foot and Mouth disease; Shoulder wounds; Easy delivery, increasing milk after delivery; Inflammation; Remove intestinal worms, Diarrhea, cold and cough, foot and mouth disease; Wound, as tonic	Sikarwar, 1996; Kadel and Jain, 2006; Dwivedi et al., 2005; Nag et al., 2009; Takhar, 2004; Yadav et al., 2014; Kumar et al., 2004

Abbreviation: L. N.: Local name.

and Pedaliaceae (2 each) and Adiantaceae, Simaurobaceae, Alangiaceae, Balanitaceae, Basellaceae, Begoniaceae, Nyctaginaceae, Cannabinaceae, Lecythidaceae, Chenopodiaceae, Lauraceae, Costaceae, Hypoxidaceae, Cyperaceae, Dioscoreaceae, Ebenaceae, Berberidaceae, Pontederiaceae, Ephedraceae, Equisetaceae, Geraniaceae, Malpighiaceae, Ulmaceae, Hydrocharitaceae, Urticaceae, Scrophulariaceae, Linaceae, Lobeliaceae, Celastraceae, Calophyllaceae, Moringaceae, Oleaceae, Cactaceae, Pandanaceae, Agaricaceae, Piperaceae, Plantaginaceae, Plumbaginaceae, Polygonaceae, Portulacaceae, Dipterocarpaceae, Styraceae, Tamaricaceae, Aizoaceae and Violaceae has single species.

The *Euphorbia* is the largest genus having 12 species which is used in ethnoveterinary medicines. This is followed by *Brassica, Crotalaria, Curcuma Ficus, Sida* and *Solanum* (4 each), *Acacia, Aerva, Amaranthus, Aristolochia, Artemisia, Bambusa, Caesalpinia, Cassia, Citrus, Cleome, Clerodendrum, Datura, Grewia, Indigofera, Ipomoea, Leucas, Ocimum, Phoenix, Phyllanthus, Piper, Saccharum, Sapindus, Senna, Sorghum, Syzygium, Terminalia, Vigna* and *Zingiber* (3 each), *Allium, Alocasia, Annona, Asparagus, Bauhinia, Cajanus, Calligonum, Calotropis, Capparis, Capsicum, Carissa, Coccinia, Cocculus, Coleus, Corchorus, Crinum, Cucumis, Cymbopogon, Dalbergia, Delonix, Equisetum, Erythrina, Gardenia, Gossypium, Hibiscus, Jatropha, Leea, Leptadenia, Luffa, Momordica, Nicotiana, Pennisetum, Prosopis, Rosa, Salvadora, Tephrosia, Vanda, Vernonia* and *Ziziphus* (2 each) and rest genera representing single species.

Acacia nilotica ssp. *indica, Allium cepa, A. sativum, Azadirachta indica, Bombax ceiba, Butea monosperma, Calotropis procera, Cassia fistula, Citrullus colocynthis, Curcuma longa, Ficus racemosa, Justicia adhatoda, Ricinus communis, Trachyspermum ammi, Trigonella foenum-graecum,* and *Zingiber officinale* are used in maximum number of ailments and diseases of almost all states of Indo-Gangetic region.

6.5 CONCLUSION

Worldwide interest in documenting and validating ethnoveterinary practices arose in the early 1980s, as people started to realize that ethnoveterinary knowledge was disappearing. Elderly community members with this knowledge are not able to impart this knowledge to the younger generation and with the introduction of modern practices made it difficult for the younger generations to appreciate and use the beliefs and practices of their ancestors. Animal husbandry sub system is well developed in tribal and rural communities moreover modern medical facilities still eludes them due to various reasons. The indigenous knowledge and practice based on locally available herbs are effective to cure diseases and are easily administrable as seems from the practices that have been carried out. But due to various social, economic and political factors this tradition is facing the threat of rapid erosion. In recent years, interest in ethnoveterinary investigations has been increased enormously on national and international level. Ancient ethnobotanical literature suggests that the tribal, non-tribal and rural populace has been using wild plants since long ago for curing various diseases and disorders in the pet/domesticated animals. All these plants should be screened scientifically in order to investigate newer sources of ethnoveterinary drugs and medicines and need further intensive investigation for their pharmacological activity on the basis of which ethnotherepeutics being practiced by the tribal and rural peoples. This will lead to development of new drugs of herbal origin. Fortunately, since last three to four decades considerable progress has been made in the ethnoveterinary sciences due to recent ethnobotanical and ethnomedicinal explorations.

ACKNOWLEDGEMENTS

The author is thankful to the Organizing Secretary, Deendayal Research Institute, Chitrakoot for providing facilities and to Prof. T. Pullaiah for his fruitful suggestions and encouragement.

KEYWORDS

- **ethnoveterinary medicine**
- **Indo-Gangetic region**
- **prescriptions**
- **tribals and rural communities**

REFERENCES

Ahir, P. C., Hussain, S., & Dhaka, V. (2012). Some traditional ethnoveterinary medicinal plants of Kumbhalgarh wildlife sanctuary, Rajasthan, India. In: Meena, K. L. (Ed.) *Proc. Nat. Conf. Biod. Cons: Caus. Cons. and Sol. MLV GC* Bhlwara, pp. 207–211.

Bandhyopadhyay, S., & Mukherjee, S. K. (2005). Ethnoveterinary medicine from Koch Bihar district, West Bengal. *Indian J. Trad. Knowl. 4*(4), 456–461.

Chouhan, S. S., & Ray, S. (2014). Ethnoveterinary plants used for wounds healing by Bhil, Bhilala and other tribes of Alirajpur district , Madhya Pradesh. *Int. J. Sci. and Res. 3* (12), 2752–2755.

Dar, B. A., Verma, R. K., & Anaiat-ul-Haq (2011). Ethnoveterinary value of some plant species utilized y rural people of Jhansi district, Bundelkhand region. *J. Agric. Sci. 2,* 321–324.

Das, S. K., & Tripathi, H. (2009). Ethnoveterinary practices and socio-cultural values associated with animal husbandry in rural Sundarbans, West Bengal. *Indian J. Trad. Knowl. 8*(2), 201–205.

Deora, G. S., Jhala, G. P. Singh & Rathore, M. S. (2004). Ethnoveterinary medicinal plants in north-west part of Udaipur district (Rajasthan). In: Trivedi, P. C., & Sharma, N. K. (eds.) Ethnomedicinal Plants, Pointer Publishers, Jaipur, India, pp. 75–81.

Dey, A., & De, J. N. (2010). Ethnoveterinary uses of medicinal plants by the aboriginals of Purulia district, West Bengal, India. *Intern. J. Bot. 6*(4), 433–440.

Dwivedi, S., Dwivedi, A. V., & Gupta, P. (2009). Role of plants as veterinary medicine from Madhya Pradesh, India: A status survey. *J. Pharm. Res. 2*(4), 688–690.

Dudi, A., & Meena, M. L. (2015). Ethnoveterinary medicines by goat keepers in Marwar region of Rajasthan, India. . *Indian J. Trad. Knowl. 14* (3), 454–460.

Ekka, A. (2015). Plants used in Ethno-veterinary medicine by Oraon tribals of north-east Chhattisgath, India. *World J. Pharmaceut. Res. 4* (9), 1038–1044.

Galav, P. K., Nag, A., & Katewa, S. S. (2007). Traditional herbal veterinary medicines from Mount Abu, Rajasthan. *Ethnobotany 19,* 120–123.

Galav, P., Jain, A., & Katewa, S. S. (2013). Traditional veterinary medicines used by livestock owners of Rajasthan, India. *Indian J. Trad. Knowl. 12*(1), 47–55.

Gautam, P., & Richhariya, G. P. (2015). Ethnoveterinary medicinal plants used by tribals and rural communities of Chitrakoot, Distt.-Satna (M. P.). *Int. J. Pharm. Life Sci. 6* (4), 4427–4430.

Ghosh, A. K. (2003). Herbal veterinary medicine from the tribal areas of Midnapore and Bankura district, West Bengal. *J. Econ. Taxon. Bot. 27*(3), 573–575.

Gupta, L., Tiwari, G., & Garg, R. (2105). Documentation of Ethnoveterinary remedies of camel diseases in Rajasthan, India. *Indian J. Trad. Knowl. 14* (3), 447–453.

Jain, S. K. (1999). Dictionary of Ethnoveterinary Plants of India, Deep Publications, New Delhi.

Jadhav, D. (2009). Ethnoveterinary plants from tribes inhabited localities of Ratlam district, Madhya Pradesh, India. *J. Econ. Taxon. Bot. 33* (suppl), 64–67.

Jatav, R., Krishna, V. K., & Mehta, R. (2014). Ethnoveterinary use of some medicinal plants of Shivpuri district (M. P.) India. *Golden Research Thoughts 4(6),* 1–5.

Jha, V., Choudhary, U. N., & Saraswati, K. C. (1991). Botanical aspects of an Ethno-veterinary prescription in Mithila, North Bihar (India). *Ethnobotany 3,* 101–104.

Kadel, C., & Jain, A. K. (2006). Plants used in ethnoveterinary practices in Jhabua district, Madhya Pradesh. *Ethnobotany 18*,149–152.

Katewa, S. S., & Chaudhary, B. L. (2000). Ethnoveterinary survey of plants of Rajasamand district, Rajasthan. *Vasundhara 5,* 95.

Kumar, A., Pandey, V. C., & Tewari, D. D. (2011). Documentation and determination of consensus about phytotherapeutic veterinary practices among the Tharu tribal community of Uttar Pradesh, India. *Tropical Animal Health Prod. 44*(4), 863–872.

Kumar, R., & Bharati, A. K. (2012a). Folk veterinary medicines in Jalaun district of Uttar Pradesh, India. *Indian J. Trad. Knowl. 11* (2), 288–295.

Kumar, R., & Bharati, A. K. (2012b). Folk veterinary medicine in Sitapur district of Uttar Pradesh, India. *Indian J. Nat. Prod. Resour. 3* (2), 267–277.

Kumar, R., & Kumar, A. B. (2012). Folk veterinary medicines in Jalaun district of Uttar Pradesh, India. *Indian J. Trad. Knowl. 11*(2), 288–295.

Kumar, R., & Kumar, A. B. (2013). New claims in folk veterinary medicines from Uttar Pradesh, India. *J. Ethnopharmacol. 146,* 581–593.

Kumar, S., Goyal, S., & Parveen, F. (2004). Ethnoveterinary plants in Indian Arid zone. *Ethnobotany 16*, 91–95.

Mandal, M. K., & Chauhan, J. P. S. (2000). A survey of ethnoveterinary medicine practices in West Bengal. *Indian J. Veterinary Medicine 20*(2), 90–91.

Mandal, S. K., & Rahaman, C. H. (2014). Determination of informants' consensus and documentation of ethnoveterinary practices from Birbhum district of West Bengal. *Indian J. Trad. Knowl. 13(4),* 742–751.

Mandal, S. K., & Rahaman, C. H. (2016). Documentation and consensus analysis of traditional knowledge about ethnoveterinary medicinal plants of Birbhum district, West Bengal (India). *Int. J. Livestock Res. 6*(1), 43–56.

Mangal, A. K. (2015). Traditional knowledge in veterinary medicine: Acase study of Gorakhpur district, Uttar Pradesh (India). *Intern. J. Adv. Res. 3*(1), 127–143.

Masika, P. J., Van Averbeeke, W., & Sonandi, A. (2000). Use of herbal remedies by small scale farmers to treat livestock diseases in Central Eastern Cape Province, South Africa. *J. South. Afr. Vet. Assoc. 71,* 87–91.

Mc Corkle, C. M. (1986). An introduction to ethnoveterinary research and development. *J. Ethnobiol. 6* (1), 129–149.

Meena, K. L. (2014). Some traditional Ethno-veterinary Plants of Districts Pratapgarh, Rajasthan, India. *Amer. J. Ethnomedicine 1* (6), 393–401.

Mishra, S., Jha, V., & Jha, S. (1996). Plants in ethnoveterinary practices in Darbhanga (North Bihar). In: Jain, S. K. (Ed.) Ethnobiology in Human Welfare. Deep Publications, New Delhi, India, pp. 189–193.

Mishra, S., Sharma, S., Vsudevan, P., Bhatt, R. K., Pandey, S., Singh, M., Meena, B. S., & Pandey, S. N. (2010). Livestock feeding and traditional healthcare practices in Bundelkhand region of Central India. *Indian J. Trad. Knowl 9*(2), 333–337.

Mitra, S., & Mukherjee, S. K. (2007). Plants used as ethnoveterinary medicine in Uttar and Dakshin Dinajpur district of West Bengal, India. In: A. P. Das & A. K. Pandey (eds.). Advances in Ethnobotany. Bishen Singh Mahendra Pal Singh, Dehra Dun, pp. 117–122.

Mukherjee, A., & Namhata, D. (1988). Herbal veterinary medicine as prescribed by the tribals of Bankura district. *J. Beng. Nat. Hist. Soc. 7,* 69–71.

Nag, A., Galav, P., Katewa, S. S., & Swarnkar, S. (2009). Traditional Herbal veterinary medicines from Kotda tehsil of Udaipur district (Rajasthan). *Ethnobotany 21*, 103–106.

Nigam, G., & Sharma, N. K. (2010). Ethnoveterinary plants of Jhansi district, Uttar Pradesh. *Indian J. Trad. Knowl. 9*(4), 664–667.

Pal, D. C. (1980). Observations on folklore about plants used in veterinary medicine in Bengal, Orissa and Bihar. *Bull. Bot. Surv. India 22,* 96–99.

Pandey, H. P., Verma, V. K., & Narain, S. (1999). Ethnoveterinary plants of Gonda region, U.P., India. *J. Econ. Taxon. Bot. 23*(1), 199–205.

Patil, U. S., & Deshmukh, O. S. (2015). Plants used in ethno-veterinary medicines by Tribal peoples in Betul district, Madhya Pradesh, India. *J. Global Biosci. 4 (8),* 3049–3054.

Prajapati, V. K., & Verma, B. K. (2004). Ethnoveterinary plants of district Mahoba, U. P. *J. Econ. Taxon. Bot., 28* (3), 623–626.

Rahaman, C. H., Ghosh, A., & Mandal, S. (2009). Studies on ethnoveterinary medicinal plants used by the tribals of Birbhum district, West Bengal. *J. Econ. Taxon. Bot. 33*(Suppl.), 333–338.

Raikwar, A., & Maurya, P. (2015). Ethnoveterinary medicine: in present perspective. *Int. J. Agric. Sc. & Vet. Res.3* (1), 44–49.

Saha, M. R., De Sarker, D., & Sen, A. (2014). Ethnoveterinary practices among the tribal community of Malda district of West Bengal, India. *Indian J. Trad. Knowl. 13*(2), 359–367.

Sanghi, S. B. (2014). Study of some ethnoveterinary medicinal plants of Tendukheda, district Narshinghpur, Madhya Pradesh. *Life Science Leaflets 55,* 1–5.

Satya, V., & Solanki, C. M. (2009). Indigenous knowledge of veterinary medicines among the tribes of West Nimar, Madhya Pradesh. *J. Econ. Taxon. Bot. 33,* 896–902.

Sebastian, M. K. (1984). Plants used as veterinary medicines, galactagogues and fodder in the forest areas of Rajasthan. *J. Econ. Taxon. Bot. 5* (4), 785–788.

Sebastian, M. K., & Bhandari, M. M. (1984). Some plants used as veterinary medicines by Bhils. *Int. J. Trop. Afric. 2,* 307–310.

Sharma, S. C. (1996). A medicobotanical study in relation to veterinary medicine of Shahjahanpur district, U. P. *J. Econ. Taxon. Bot. Addl. Ser. 12,* 123–127.

Shrivastava, S., Jain, A. K., & Mathur, R. (2012). Documentation of herbal medicines used in treatment of diseases of goats (*Cypris communis*) in and around Gwalior (M. P.). *Indian J. Nat. Prod. Resour. 3*(2), 278–280.

Shukla, A. N., Singh, K. P., & Kumar, A. (2007). Ethnoveterinary uses of plants from Achanakmar-Amarkantak Biosphere Reserve of Madhya Pradesh and Chhattisgarh. *J. Non-timber For. Prod. 14,* 53–55.

Sikarwar, R. L. S.(1996). Ethno-veterinary herbal medicines in Morena district of Madhya Pradesh, India. In Jain, S. K. (Ed.) Ethnobiology in Human Welfare. Deep Publications, New Delhi, India, pp. 194–196.

Sikarwar, R. L. S. (1999). Less known ethno-veterinary uses of plants in India. In: Mathias, E., Rangnekar, D. V., & McCorkle, C. M. (Eds.) Ethno-veterinary Medicine, Alternatives for Livestock Development (Vol. I), BAIF Development Research Foundation, Pune, India, pp. 103–107.

Sikarwar, R. L. S., Bajpai, A. K., & Painuli, R. M. (1994). Plants used in veterinary medicines by Aboriginals of Madhya Pradesh, India. *Int. J. Pharmacog. 32*(3), 251–255.

Sikarwar, R. L. S., & Kumar, V. (2005). Ethno veterinary Knowledge and practices prevalent among the tribals of Central India. *J. Natural Remedies 5* (2), 147–152.

Singh, D., Kachhawaha, S., Choudhary, M. K., Meena, M. L., & Tomar, P. K. (2014). Ethnoveterinary knowledge of Raikas of Marwar for nomadic pastoralism. *Indian J. Trad. Knowl. 13(1),* 123–131.

Singh, M. P. (1987). Tribal medicinal plants used in animal diseases of Chhotanagpur. *Indian Forester 113* (11), 758–759.

Singh, P. K., Singh, S., Kumar, V., & Krishna, A. (2011). Ethnoveterinary healthcare practices in Marihan sub-division of district Mirzapur, Uttar Pradesh, India. *Life Science Leaflets 16,* 561–569.

Takhar, H. K. (2004). Folk herbal veterinary medicines of southern Rajasthan. *Indian J. Trad. Knowl. 3*(4), 407–418.

Tripathi, M., & Singh, R. (2010). Ethno-veterinary medicines used by Tribal's in Chitrakoot, Satna (M. P.). *Natl. J. Life Sci. 7*(3), 94–96.

Verma, R. K. (2014). An ethnobotanical study of plants used for the treatment of livestock diseases in Tikamgargh district of Bundelkhand, Central India. *Asian Pac. J. Trop. Biomed. 4* (Suppl.), S460–S467.

Yadav, M., Yadav, A., & Gupta, E. (2012). Ethnoveterinary practices in Rajasthan, India-A Review. *Int. Res. J. Bio. Sci. 1*(6), 80–82

Yadav, M. L., Rajput, D. S., & Mishra, P. (2015). Ethnoveterinary practices among the tribes of Banswara District of Rajasthan. *Indian Res. J. Ext. Edu. 15* (2), 87–90.

Yadav, S. S., Bhukal, R. K., Bhandoria, M. S., Ganie, S. A., Gulia, S. K., & Raghav, T. B. S. (2014). Ethnoveterinary medicinal plants of Tosham block of district Bhiwani (Haryana) India. *J. Appl. Pharmaceut. Sci. 4*(6), 40–48.

CHAPTER 7

TRADE IN INDIAN MEDICINAL PLANTS

D. K. VED, S. NOORUNNISA BEGUM, and K. RAVI KUMAR

Centre of Repository of Medicinal Resources, School of Conservation of Natural Resources, Foundation for Revitalization of Local Healthand Traditions, 74/2, Jarakabande Kaval, Attur P.O., Via Yelahanka, Bangalore – 560064, India, E-mail: dk.ved@tdu.edu.in, noorunnisa. begum@tdu.edu.in, k.ravikumar@tdu.edu.in

CONTENTS

ABSTRACT

India is home to a variety of traditional medicine systems that support many health care needs of this country's nearly one billion people. The recent years have witnessed a great deal of interest in medicinal plants for their potential to yield useful drugs. The increasing popularity of herbal medicinal products in developed countries has given new dimension to the demand of medicinal plants at the international market. The health giving properties of medicinal plants have been scientifically validated in a number of diseases. Their effectiveness in such diseases, where modern drugs are hard to find has fueled special interest in these botanicals.

The Global wellness market is currently estimated to be more than 3 trillion USD per year (Bodekar et al. 1997/2003). It includes functional foods, dietary supplements suggesting new dimensions to the usage been discussed. With market balance clearly tilted in favor of products with a tag of 'natural,' the developed countries have generated special demand for herbal products. The present trend is expected to continue in the years to come. Developing countries, which possess a treasure of medicinal plant resources, are to benefit from the arising situation. Despite the rising demand and soaring imports of medicinal plants by developed countries, the source countries possessing rich biodiversity of medicinal plants have not been able to benefit from the situation. There are many hurdles in developing exports of medicinal plants and their products. The international trade in medicinal plants is not well regulated; data is scarce, scattered and inaccessible to those who need it the most; species specific records are hardly available; and true demand, supply, and the price situation is rarely available to the producers.

Due to lack of correct identification, similar looking plants are collected from the field sites along with the genuine medicinal plant by mistake. But many a times similar looking (inferior) cheap alternatives are intentionally mixed along with some quantity of genuine plant material. This may be due to inadequate availability of the genuine medicinal plant material in the large quantity. Medicinal plants collected from the wild may be contaminated by other similar looking species or plant parts through misidentification, accidental contamination or intentional adulteration, all of which may have undesirable consequences. It has also been observed that sometimes the adulteration involves mixing not only similar species but also cheap, inferior and even degraded material.

We at the Foundation for Revitalization of Local Health Traditions (FRLHT) took up a National Medicinal Plants Board (NMPB) supported short-term study to estimate the quantum of demand and supply of medicinal plants in the country in 2006–2007, which was the first of its kind. It for the first time, brought out the diversity of medicinal plants in trade along with the estimates of their domestic consumption as well as exports.

This chapter provides a gist of the trade in medicinal plants in India, including the foreign trade, as revealed through that study.

7.1 TRADE IN MEDICINAL PLANTS: GLOBAL

7.1.1 BACKGROUND

Herbal medicines have become a mainstay of the home health practices of the majority of the world's population over the past two decades. In the industrialized world, studies have found that herbal and other forms of complementary medicine use increases as people become more affluent. At the same time, in the developing world, traditional medicine, with its strong herbal base, has continued to be the first resort to healthcare of the majority of poor and rural populations (Bodekar, 1997, 2003).

Medicinal and Aromatic Plants (MAPs) are produced and offered in a wide variety of products, from crude materials to processed and packaged products like pharmaceuticals, herbal remedies, teas, spirits, cosmetics, sweets, dietary supplements, varnishes and insecticides. The use of botanical raw material is, in many cases, much cheaper than using alternative chemical substances. An estimated number of 70,000 plant species are used in folk medicine worldwide. As a consequence, there is an enormous demand in botanicals – for domestic use and for international trade – resulting in a huge trade on local, regional, national and international level. As the production of botanicals still relies, to a large degree, on wild-collection, profound knowledge of trade, size, structure and streams as well as of commodities, traded quantities and their origin is essential for assessing its impact on the plant populations concerned (Lange, 2006).

Problems identified by industry concerning lead compounds from plants include that these can be overly complex for ease of chemical manufacturing, and that there are challenges of supply of the medicinal plant raw material required for reliable scaling up of production. In addition, legal challenges over intellectual property rights from traditional knowledge holders created a potential minefield for companies and this acted as a deterrent to many who had seen traditional knowledge as a form of global commons that could be harvested freely. High Throughput Screening (HTS) is still being used by a number of smaller R & D companies in industrialized countries (e.g., www.alkion-biopharma.com; www.a-r.com), and increasingly in developing countries with a history of traditional medicine use (Zhu et al., 2010).

7.1.2   GLOBAL TRADE FIGURES: AN OVERVIEW

The global wellness economy in the year 2013 was valued at US$3.4 trillion (www.globalwellness-summit.com/industry-resource). And in almost all of the categories in which growth was documented, medicinal plants play a role: Healthy Eating/Nutrition/Weight Loss (US$574.2 billion); Fitness & Mind/Body ($US446.4 billion); Beauty & Anti-Aging ($US1.025 trillion); Preventative/Personalized Health ($US432.7 billion); Complementary/ Alternative Medicine (US$186.7 billion); Workplace wellness (US$40.7 billion); Wellness Lifestyle Real Estate (US$100 billion) (www.globalwell-nesssummit.com/industry-resource).

7.1.3   FROM ALTERNATIVE MEDICINES TO HERBS FOR WELLNESS

This new evidence about the size and value of the global wellness market puts medicinal plants in an entirely fresh light. From filling a niche as a source of "alternative medicines" in the past, medicinal plants have now become the source of a whole new range of products and approaches designed to promote wellbeing, reduce the effects of aging, increase energy levels and general vitality, promote skin health, add power to nutrition, and combat stress. In short, they are a core foundation of the multi-trillion dollar global wellness market.

This shift from herbal medicine to herbs for wellness has come about in part in response to regulatory environments. The EU, the USA, Canada, Australia and other industrialized economies with high herbal use have seen increased barriers raised against the sale of herbal medicines and claims of their traditional uses. So, a shift into the functional food market or the natural beauty and anti-aging markets has allowed companies to grow their presence with less regulatory demand for expensive and lengthy gathering of research data (www.globalwellness-summit.com/industry-resource).

7.1.4   FUNCTIONAL FOODS

In Asia, what are now termed "functional foods" have been a part of Eastern cultures for millennia. Food has always been used medicinally in Ayurveda, Chinese medicine and other traditional medicine systems in Asia. Asian health systems have used foods for both preventive and therapeutic health effects, a view that is now being increasingly recognized around the world. Accordingly, as the wellness market has surged in Asia, the concept of wellness products and foods has been an easy one for companies to convey to consumers. In the West, by contrast, functional foods are considered revolutionary and represent a rapidly growing segment of the food industry. Food and pharmaceutical companies alike are competing to bring functional foods into the mass market.

7.1.5 DIETARY SUPPLEMENTS

Total retail sales of herbal and botanical Dietary Supplements (DS) in the United States increased by an estimated 7.9% in 2013 – the highest observed growth percentage since the late 1990s – according to aggregated market statistics calculated by the Nutrition Business Journal (NBJ). These sales did not include sales of herbal teas, herbs sold in natural cosmetic products, or herbs sold as government – approved ingredients in nonprescription medications (aka over – the-counter [OTC] drugs), for example, senna leaf or fruit extract, or slippery elm bark. The 40 top-selling herbal dietary supplements in the mainstream multi-outlet channel in the United States in 2013 are *Marrubium vulgare* (Horehound), *Pausinystalia johimba* (Yohimbe), *Vaccinium macrocarpon* (Cranberry), *Actaea racemosa* (Black Cohosh), *Senna alexandrina* (Senna), *Cinnamomum* sp. (Cinnamon), *Linum usitatissimum* (Flax seeds and/or oil), *Echinacea* spp. (Echinacea), *Valeriana officinalis* (Valerian), *Silybum marianum* (Milk thistle), *Ginkgo biloba* (Ginkgo), *Hypericum perforatum* (St. John's Wort), *Tribulus terrestris* (Tribulus), *Gymnema sylvestre* (Gymnema), *Ulmus rubra* (Slippery Elm Bark), *Euterpe oleracea* (Acal), etc. (Lindstrom et al., 2014).

7.1.6 FROM TRADITIONAL USE TO GLOBAL MARKETS: EXAMPLES

- The Vietnamese fruit "gac" (*Momordica cochinchinensis* Spreng.) used during the Vietnamese New Year has reached out in a new format to a global wellness market as powerful anti-oxidant, high in Vitamin C, lycopene and beta-carotene.
- A well-known "wonder herb", turmeric, is one of nature's most powerful anti-inflammatory agents and has been shown in over 2,000

studies to reduce and prevent conditions associated with inflammation, such as heart disease, rheumatoid conditions and even potentially Alzheimer's disease.

7.1.7 MAJOR TRADED BOTANICALS

There are no or only few reliable trade data available for single botanicals (Lange, 1998). The commodity group of Pharmaceutical plants includes those used only in small quantities as well as bulk material with great industrial importance. In Germany, the most used medicinal plant is Gingko (*Gingko biloba*), followed by Horse-chestnut (*Aesculus hippocastanum*), Hawthorne (*Crataegus* spp.), St John's-Wort (*Hypericum perforatum*), Nettle (*Urtica dioca*), Echinacea (*Echinacea* spp.), Saw Palmetto (*Serenoa repens*), and Milk Thistle (*Silybum marianum*) (Grünwald and Büttle, 1996). Some of these plants are also highly used in the USA, like *Echinacea*, St. John's-wort, and Saw Palmetto, but the preferences are somewhat different: Siberian Ginseng (*Eleutherococcus senticosus*), Goldenseal (*Hydrastis canadensis*), Cat's Claw (*Uncaria* species), *Astragalus membranaceus*, Dong Quai (*Angelica sinensis*) and Cascara Sagrada (*Rhamnus purshiana*) are listed among the top-selling botanicals (Laird, 1999).

7.2 TRADE IN MEDICINAL PLANTS: INDIA

7.2.1 BACKGROUND

India has one of the oldest, richest and most diverse cultural traditions associated with the use of medicinal and aromatic plants. The country has a great heritage of medicinal plants use dating back to the early Vedic period. Like in many other indigenous cultures of civilizations across the world, Indian indigenous communities have possessed/accumulated vast knowledge on multifarious uses of plants and other natural resources found around them. By empirical reasoning and experimentation, the indigenous societies have developed their own unique wealth of knowledge pertaining to conservation and sustainable use of plants, animals and other natural resources. The ancient Indians had an incredible knowledge about several plants, which they utilized in so many ways for meeting their day-to-day requirements of life. Medicinal plants constituted one of the major groups of plant resources used by Indians since four millennia. Biodiversity provides number of essential natural services such as food production, soil fertility, climate regulation, carbon storage that are foundation of human well-being. It also provides building blocks for sustainable food, health and livelihood security systems. It is the feedstock for the biotechnology industry and climate resilient farming system.

Earlier, All India Co-ordinated Research Project on Ethnobiology (AICRPE) 1982–1998 estimated that, around 8000 species of plants are used by several ethnic communities (Sreedevi et al., 2013). *Foundation for Revitalization of Local Health Traditions* (FRLHT) has been developing over the past 20 years, a comprehensive database on Indian Medicinal Plants and currently the database incorporates 8361 botanical names referring to 6560 taxa.

7.2.2 BOTANICALS TRADED IN INDIA: STUDY BY FRLHT

The study commissioned by National Medicinal Plant Board to undertake *Study of Demand and Supply of Medicinal Plants in India*, in 2006, resulted in generation of a comprehensive inventory of 960 medicinal plant species that form source of 1289 botanical raw drugs recorded in trade in the Indian raw drug markets (Ved and Goraya, 2008). This list provided correlation of the popular trade names with the updated botanical nomenclature and formed a reliable base for any study on traded medicinal plants of India. Out of these 960 species, 688 were identified to be part of the classical *Materia Medica* of 'Ayurveda,' 501 species of 'Siddha' and 328 of the 'Unani' system, with many species overlapping across these systems. 41% of these 960 species are herbs, 26% are trees, 18% are shrubs and the remaining 15% are climbers. Whole plants, roots, wood and bark constitute more than 50% of the 1289 botanical raw drugs in trade and as such their collection involves damage/destruction of the specific plant entities. 81% (780) of these species in trade are sourced entirely or largely from the wild, the remaining traded species being obtained from either cultivation or imports.

The study also resulted in assessment of total annual demand of botanical raw drugs and its corresponding monetary valuation. Inferences drawn from the analyzed data provided guidance for shifts in the strategies and focus for management of medicinal botanicals both in the agricultural and forestry sector. It has been a definite step forward towards clearer understanding of the priorities for management of the medicinal plant resources of the country.

7.2.3 IDENTIFICATION OF 178 SPECIES TRADED IN HIGH VOLUMES (>100 MT/YR.)

Out of the 960 medicinal plant species recorded in trade in the country, 178 species were assessed to be in high consumption, i.e., volumes exceeding 100 Metric Ton (dry weight) per year. It is interesting to note that the consolidated annual consumption of these high trade species, by the domestic herbal industry, accounted for nearly 80% of the total industrial demand of all botanical raw drugs. These species were, thus, identified as the mainstay of domestic herbal industry needing continued availability for sustenance and growth of this industry.

A key question is whether all these 178 species in high volume trade species need management interventions for their sustained availability? To find an answer to this question we subjected this list of species to an analysis of their major sources of supply and the results are quite revealing. Out of these, 36 species (20%) were sourced largely from cultivation. The wastelands, fallow lands, roadsides and farm bunds formed the source of 46 species (25%) and 5 species (3%) were being imported. The remaining 91 species (52%) were being sourced from the forests, both tropical and temperate (Figure 7.1).

Critical evaluation of these 178 species in view of their demand and the major source of supply brings out that not all these species need immediate management focus. An assessment of the criticality of the species needing management interventions in view of their supply source and demand is presented in the following subsections.

7.2.3.1 Ninety-One Wild (Forest) Species for Prioritized Management Interventions

Analysis of the 91 species that are sourced primarily from forests shows that 21 species are from Himalayan (temperate) forests and 70 from tropical forests. Additionally, the raw drugs obtained from two very important tropical forest species namely *Commiphora mukul* (Hook. *ex* Stocks) Engl. [= *C. wightii* (Arn.) Bhandari] (guggul) and *Aquilaria malaccensis* Lam. (agar)

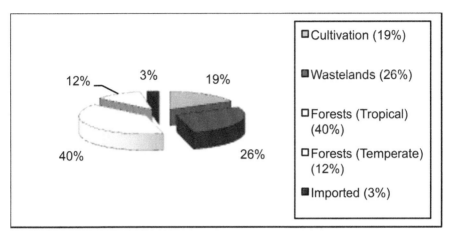

FIGURE 7.1 Supply source wise analysis of 178 species in high demand.

are being imported in sizeable quantities to meet the demand of domestic herbal industry. These two species need a special management focus in their natural range of distribution in the country.

A closer look at the 21 Himalayan species (Table 7.1) reveals that barring six tree species viz. *Cedrus deodara* (Roxb.) G. Don, *Cinnamomum tamala* (Buch.-Ham.) T. Nees & Eberm., *Juniperus communis* L., *Pistacia integerrima* Stew. *ex* Brand, *Taxus wallichiana* Zucc. and *Abies spectabilis* (G. Don) Mirb., and two shrubs, that is, *Berberis aristata* DC, and *Rhododendron anthopogon* D.Don. the remaining species are temperate and alpine herbs. The wild populations of these temperate and alpine herbs are going down drastically due to indiscriminate removals to meet the high industrial demand. It is these species that need the most urgent and immediate management focus if their sustained supplies over a long-term period are to be ensured. An obvious strategy for increasing their supplies should be to build up their wild populations by putting in place effective management interventions to regulate their wild collections. The domestication and commercial cultivation of these species also needs to complement the conservation initiatives. However, domestication and cultivation would need intensive research inputs as most of these species produce harvestable parts after more than one year and, therefore, would need meticulous working out of their economic viability vis-a-vis conventional crops. Each of the eight woody perennials of temperate Himalayan forests will also need careful and immediate appraisal of the status of their wild populations in order to have an objective basis to plan appropriate interventions.

TABLE 7.1   List of 21 Medicinal Plant Species in High Trade Sourced from Temperate Forests

S. No.	Plant Raw Drug Consumption (in tonnes)	Name of Plant Species
1	200–500 (11 species)	*Aconitum ferox* Wall. *ex* Ser., *Aconitum heterophyllum* Wall. *ex* Royle, *Bergenia ciliata* (Haw.) Sternb., *Ephedra gerardiana* Wall. *ex* C. A. Mey., *Nardostachys jatamansi* (D. Don) DC., *Picrorhiza kurroa* Royle *ex* Benth., *Pistacia integerrima* Stew. *ex* Brand., *Rhododendron anthopogon* D.Don, *Taxus wallichiana* Zucc., *Valeriana hardwickii* Wall., *Viola pilosa* Blume
2	500–1000 (8 species)	*Abies spectabilis* (G. Don) Mirb., *Berberis aristata* DC., *Cedrus deodara* (Roxb.) G. Don, *Cinnamomum tamala* (Buch.-Ham.) T. Nees & Eberm., *Juniperus communis* L., *Onosma hispida* Wall. *ex* G. Don, *Rheum australe* D. Don, *Swertia chirayta* (Roxb. *ex* Fleming) H.Karst.
3	Above 1000 (2 species)	*Jurinea macrocephala* DC., *Parmelia perlata* (Huds.) Ach.

As far as the 70 species mainly sourced from the tropical forests (Table 7.2) are concerned, it is alarming to note that most of these are trees or woody perennials with many of them occurring in specialized niches. Of special concern are the tree species like *Oroxylum indicum* (L.) Benth.*ex* Kurz, *Stereospermum chelonoides* (L.f.) DC., & *Premna serratifolia* L. [= *P. integrifolia* L.] the roots of which form part of the popular Ayurvedic raw drug group 'Dashamoola'. With no current management focus for promoting populations of such species, their populations are on the decline and would need to be urgently built up if herbal industry is to be fed with regular supplies of these species, as the current level of estimated consumption exceeds 1000 MT in respect of each of these species. All the 70 species needs focused and urgent attention by NMPB.

7.2.3.2   Wild Species in High Trade with Adequate Production from Self-Grown Populations

Amongst the high volume trade category we have identified 46 such species, which grow in abundance along roadsides, farm bunds, fallow lands and wastelands (Table 7.3) across the tropical regions of the country. These

TABLE 7.2   List of 70 Medicinal Plant Species in High Trade Sourced from Tropical Forests

S. No.	Plant Raw Drug Consumption (In tonnes)	Name of Plant Species
1	200–500 (47 species)	*Acacia catechu* (L.f.) Willd., *Acacia concinna* (Willd.) DC, *Acacia nilotica* (L.) Willd. *ex* Delile subsp. *indica* (Benth.) Brenan, *Albizia amara* (Roxb.) Boivin, *Alstonia scholaris* (L.) R. Br., *Anogeissus latifolia* (Roxb. *ex* DC.) Wall. *ex* Guill. & Perr., *Baliospermum montanum* (Willd.) Muell.-Arg., *Bombax ceiba* L., *Boswellia serrata* Roxb. ex Colebr., *Buchanania lanzan* Spreng., *Careya arborea* Roxb., *Celastrus paniculatus* Willd., *Chlorophytum tuberosum* (Roxb.) Baker, *Cinnamomum sulfuratum* Nees, *Clerodendrum phlomidis* L.f., *Coscinium fenestratum* (Gaertn.) Colebr., *Cyclea peltata* (Lam.) Hook.f. & Thomson, *Decalepis hamiltonii* Wight & Arn., *Gardenia resinifera* Roth, *Hedyotis corymbosa* (L.) Lam., *Holoptelea integrifolia* (Roxb.) Planch., *Holostemma annulare* (Roxb.) K. Schum., *Ixora coccinea* L., *Lannea coromandelica* (Houtt.) Merr., *Litsea glutinosa* (Lour.) C. B. Rob., *Lobelia nicotianifolia* Roth *ex* Roem. & Schult., *Madhuca indica* J. F. Gmel., *Mesua ferrea* L., *Mimusops elengi* L., *Morinda coreia* Buch.-Ham., *Mucuna pruriens* (L.) DC., *Operculina turpethum* (L.) Silva Manso, *Pterocarpus santalinus* L.f., *Rauvolfia serpentina* (L.) Benth. *ex* Kurz, *Santalum album* L., *Sapindus mukorossi* Gaertn., *Schrebera swietenioides* Roxb., *Semecarpus anacardium* L.f., *Shorea robusta* Gaertn., *Smilax glabra* Roxb., *Soymida febrifuga* (Roxb.) A. Juss., *Sterculia urens* Roxb., *Strobilanthes ciliata* Nees, *Strychnos potatorum* L.f., *Vateria indica* L., *Wrightia tinctoria* (Roxb.) R. Br. and *Ziziphus xylopyrus* (Retz.) Willd.
2	500–1000 (10 species)	*Butea monosperma* (Lam.) Taub., *Cassia fistula* L., *Desmodium gangeticum* (L.) DC, *Embelia tsjeriam-cottam* (Roem. & Schult.) A. DC, *Garcinia indica* (Thouars) Choisy, *Gymnema sylvestre* R. Br. *ex* Schult., *Ipomoea mauritiana* Jacq., *Pterocarpus marsupium* Roxb., *Rubia cordifolia* L. and *Symplocos racemosa* Roxb.
3	Above 1000 (13 species)	*Aegle marmelos* (L.) Correa, *Asparagus racemosus* Willd., *Gmelina arborea* Roxb., *Holarrhena pubescens* (Buch.-Ham.) Wall. *ex* G. Don, *Oroxylum indicum* (L.) Benth. *ex* Kurz, *Phyllanthus emblica* L., *Premna serratifolia* L., *Saraca asoca* (Roxb.) W. J. de Wilde, *Stereospermum chelonoides* (L.f.) DC, *Strychnos nux-vomica* L., *Terminalia arjuna* (Roxb. *ex* DC.) Wight & Arn., *Terminalia bellirica* (Gaertn.) Roxb. and *Terminalia chebula* Retz.

* *Commiphora mukul* (Hook. *ex* Stocks) Engl. [= *C.wightii* (Arn.) Bhandari] (Guggul) and *Aquilaria malaccensis* Lam. [= *A. agallocha* Roxb.] (agar), largely sourced through imports at present, are also native tropical species and need special management focus.

TABLE 7.3   List of 46 Plant Species, in High Trade, Largely/Entirely obtained from Wastelands/Roadsides

S. No.	Plant Raw Drug Consumption (in tonnes)	Name of Plant Species
1	200–500 (22 species)	*Abrus precatorius* L., *Achyranthes aspera* L., *Aerva lanata* (L.) Juss., *Cardiospermum halicacabum* L., *Chamaecrista absus* (L.) H. S. Irwin & Barneby, *Chrysopogon zizanioides* (L.) Roberty, *Cullen corylifolium* (L.) Medik., *Curculigo orchioides* Gaertn., *Cynodon dactylon* (L.) Pers., *Datura metel* L., *Fumaria indica* (Hausskn.) Pugsley, *Hedyotis corymbosa* (L.) Lam., *Hygrophila schulli* (Buch.-Ham.) M. R. Almeida & S. M. Almeida, *Ipomoea nil* (L.) Roth, *Merremia tridentata* (L.) Hallier f., *Peganum harmala* L., *Pluchea lanceolata* (DC.) Oliv. & Hiern., *Pseudarthria viscida* (L.) Wight & Arn., *Sisymbrium irio* L., *Sphaeranthus indicus* L., *Tragia involucrata* L. and *Withania coagulans* (Stocks) Dunal.
2	500–1000 (5 species)	*Centella asiatica* (L.) Urb., *Hemidesmus indicus* (L.) R. Br. ex Schult., *Ocimum americanum* L., *Tephrosia purpurea* (L.) Pers. and *Trichosanthes cucumerina* L.
3	Above 1000 (19 species)	*Andrographis paniculata* (Burm.f.) Wall. *ex* Nees, *Baccharoides anthelmintica* (L.) Moench, *Bacopa monnieri* (L.) Wettst., *Boerhavia diffusa* L., *Citrullus colocynthis* (L.) Schrad., *Convolvulus prostratus* Forssk., *Cyperus esculentus* L., *Cyperus rotundus* L., *Eclipta prostrata* (L.) L., *Phyllanthus amarus* Schumach. & Thonn., *Plumbago zeylanica* L., *Senna tora* (L.) Roxb., *Sida rhombifolia* L., *Solanum anguivi* Lam., *Solanum nigrum* L., *Solanum virginianum* L., *Tinospora cordifolia* (Willd.) Miers *ex* Hook.f. & Thomson, *Tribulus terrestris* L. and *Woodfordia fruticosa* (L.) Kurz

species have high regeneration potential, are very hardy with and are known to come up even in high stress conditions.

With large tracts of land under these land-use categories and the innate capacity of these species to regenerate, there does not appear to be the need for concern regarding sustenance of their supplies from such self-grown populations and/or the need to focus attention towards their domestication. Only in case of some specific species, the quality requirements and ease of collection may, in future, require their domestication and cultivation.

7.2.3.3   Species of High Trade and Stable Cultivation That May Need Development of Improved Varieties (Selected Species)

A closer look at the 36 species sourced from cultivation (Table 7.4) reveals that whereas species like *Plantago ovata* Phil., *Senna alexandrina* Gars. *ex* Mill. [=*Cassia angustifolia* Vahl], *Aloe vera* (L.) Burm.f. [=*Aloe barbadensis* Mill.], *Lawsonia inermis* L., *Ocimum tenuiflorum* L., *Ocimum basilicum* L., *Inula racemosa* Hook.f., *Saussurea costus* (Falc.) Lipsch. and *Caesalpinia sappan* L. etc. are sourced almost exclusively from cultivation, major proportion of supplies of some of the indigenous species like *Justicia adhatoda* L. [= *Adhatoda zeylanica* Nees], *Acorus calamus* L., *Gloriosa superba* L., *Withania somnifera* (L.) Dunal are also obtained from cultivated sources with part supplies coming from the wild.

The point that needs to be taken note of in respect of these 36 species is that the cultivation of these species has already stabilized and got incorporated into the local agricultural systems. The farmers have come to understand the market dynamics of these species and are able to decide the extent

TABLE 7.4   Thirty Six Medicinal Plant Species in High Trade sourced largely from Cultivation

S. No.	Plant Raw Drug Consumption (in tonnes)	Name of Plant Species
1	200–500 (20 species)	*Abelmoschus moschatus* Medik., *Alpinia calcarata* (Haw.) Roscoe, *Caesalpinia sappan* L., *Catharanthus roseus* (L.) G. Don, *Croton tiglium* L., *Curcuma angustifolia* Roxb., *Ficus benghalensis* L., *Ficus religiosa* L., *Gloriosa superba* L., *Indigofera tinctoria* L., *Inula racemosa* Hook.f., *Jatropha curcas* L., *Kaempferia galanga* L., *Lawsonia inermis* L., *Lepidium sativum* L., *Prunus cerasoides* Buch.-Ham. *ex* D.Don, *Saussurea costus* (Falc.) Lipsch., *Vitex negundo* L., *Zingiber zerumbet* (L.) Roscoe *ex* Sm. and *Ziziphus jujuba* Mill.
2	500–1000 (4 species)	*Acorus calamus* L., *Cichorium intybus* L., *Silybum marianum* (L.) Gaertn., *Simmondsia chinensis* (Link) C. K. Schneid.
3	Above 1000 (12 species)	*Aloe vera* (L.) Burm.f., *Azadirachta indica* A. Juss., *Justicia adhatoda* L., *Ocimum basilicum* L., *Ocimum tenuiflorum* L., *Piper longum* L., *Plantago ovata* Forssk., *Plectranthus barbatus* Andrews, *Pongamia pinnata* (L.) Pierre, *Senna alexandrina* Mill., *Trachyspermum roxburghianum* (DC.) H. Wolff and *Withania somnifera* (L.) Dunal

of area to be put under cultivation of these species in response to the market demands. Thus, cultivation of these species does not need special promotional incentives.

The focus in respect of these species needs to be on development of new and improved varieties, multiply the germplasm of such improved varieties, and supply the same to the growers for enhancing farm income.

7.2.3.4   Five Currently Imported Species Needing Support for Indigenous Production

The industrial demand of botanical raw drugs derived from 5 species (Table 7.5) is being largely met through imports.

Of these five species, limited wild populations of *Aquilaria malaccensis* Lam. occur in the north-eastern part of the country. Due to rapid decline of wild populations of this species, it is already included in appendix II of CITES. Similarly, much reduced wild populations of *Commiphora wightii* (Arn.) Bhandari occur in the states of Gujarat and Rajasthan. The government of Rajasthan has put a ban on the tapping of guggul in the state to save the remnant guggul plants. Some tapping of guggul plants does take place in Gujarat, but the total yield is too little in relation to the huge trade demand. It is estimated that less than 10% of the domestic industrial demand for guggul gum is currently being met from domestic production. These two species will, therefore, need afforestation efforts to augment their existing populations as well as research on sustainable harvest for utilization. The remaining three species are exotics, which are currently imported. However, given their high level of demand it will be prudent for NMPB to support initiatives to bring them under cultivation in suitable areas for meeting the high level of domestic demand.

TABLE 7.5   List of 5 Species, in High Demand, Largely/Entirely Obtained Through Imports

S. No.	Plant Raw Drug Consumption (in tonnes)	Name of Plant Species
1	200–500 (1 species)	*Aquilaria malaccensis* Lam.
2	500–1000 (3 species)	*Commiphora wightii* (Arn.) Bhandari, *Piper chaba* Hunter and *Quercus infectoria* G. Olivier
3	Above 1000 (1 species)	*Glycyrrhiza glabra* L.

7.2.4 DYNAMIC NATURE OF THE TOP TRADED PLANTS LIST AND NEED FOR UPDATES

Although the list of 178 species in high demand is fairly comprehensive, two issues need to be appreciated. Firstly, some species in high trade might have been missed out due to limitations of the sampling size in the given time frame (e.g., trade of *Mentha* sp. could not be captured in this survey because the ISM based herbal industries which were sampled did not consume this species in sizeable quantities. Yet it is a medicinal plant in high demand on account of exports of its derivatives and it is commercially cultivated on a large scale in U.P.). Secondly, the changes in use pattern brought about by the technology advancement, preference of certain formulations, or spurt in exports may cause a change in the list of top traded medicinal plant species in the country. This list, thus, needs continuous monitoring and updating in order to have accurate picture of the demand of botanicals.

Further, in many cases, several species contribute to a particular raw drug entity in trade, whereas the list of 178 species includes names of only the main species forming the source of such raw drug entity. For example, 'Bala' of trade is traditionally correlated to *Sida rhombifolia* L. but in practice it also includes material obtained from *S. acuta* Burm.f. and *S. cordifolia* L. Similarly, 'Gokshura,' traditionally linked to *Tribulusterrestris* L., is also obtained from other species of this genus viz. *T. alatus* Delile, *T. lanuginosus* L. and *T. subramanyamii* P. Singh, Giri & V. Singh. 'Daruharidra' is obtained from different species of *Berberis* viz. *B. aristata* DC., *B. lycium* Royle, *B. asiatica* Roxb., or *B. chitria* Lindl., etc. Whereas the name of only one species for each such raw drug entity has been included in the list of 178 species in high demand, it is implied in such cases that the one name *ipso facto* includes the allied or equivalent species contributing to the traded raw drug material (Table 7.6).

7.2.5 HS CODES AND ITC (HS) CODES

Millions of trade transactions, across different countries, occurring each year are classified under approximately 8000 item codes. Each such item code known as HS-code (Harmonized System) consists of progressively more specific identifiers and consists of 6 to 10 digits. As per this system most of the medicinal plants related materials fall under the major code 12 which is linked to the description "oil seeds and oleaginous fruits; miscellaneous grains, seeds and fruits; industrial or medicinal plants; straw and fodder." A

TABLE 7.6 List of the Top Traded Species along with their Substitutes and Adulterants

S. No	Name of the Botanical Drug	Name of Accepted Plant Source	Substituents/Known Adulterants
1	Mushkdana/ Latakasturi	*Abelmoschus moschatus* Medik. [=*Hibiscus abelmoschus* L.]	*Abelmoschus ficulneus* (L.) Wight & Arn.
2	Talispatra	*Abies spectabilis* (Royle *ex* D.Don) Mirb.	*Abies pindrow* Royle, *A. densa* Griff., *Rhododendron anthopogon* D. Don and *Taxus wallichiana* Zucc.
3	Katha	*Acacia catechu* (L.f.) Willd.	*Acacia chundra* (Roxb. ex Rottler) Willd.
4	Babul/Keekar	*Acacia nilotica* (L.) Willd. *ex* Delile subsp. *indica* (Benth.) Brenan	*Acacia farnesiana* Willd. and *A. leucophloea* Willd.
5	Apamarga	*Achyranthes aspera* L.	*Achyranthes bidentata* Blume
6	Vachnag	*Aconitum ferox* Wall. *ex* Ser.	*Aconitum balfourii* Stapf., *A. deinorrhizum* Stapf., *A. chasmanthum* Stapf *ex* Holmes and *A. falconeri* Stapf
7	Atis/Ativish	*Aconitum heterophyllum* Wall. *ex* Royle	*Aconitum kashmerinum* Royle, *Delphinium denudatum* Wall., *Chaerophyllum villosum* Wall.
8	Bael/Bilva	*Aegle marmelos* (L.) Corr.	*Limonia acidissima* L.
9	Cheroola	*Aerva lanata* (L.) Juss.	*Bergenia ciliata* (Waw.) Sternb.
10	Chittartha	*Alpinia calcarata* (Haw.) Roscoe	*Alpinia galanga* (L.) Willd. and *A. officinarum* Hance
11	Dhawada/Ghatti Gum	*Anogeissus latifolia* (Roxb. ex DC.) Wall. *ex* Bedd.	*Astragalus tragacantha* L., *Acacia nilotica* (L.) Willd. *ex* Delile subsp. *indica* (Benth.) Brenan
12	Shatavari/ Satavari	*Asparagus racemosus* Willd.	*Asparagus sarmentosus* Grah., *A. adscendens* Roxb.
13	Brahmi/Jal/ Brahmi	*Bacopa monnieri* (L.) Wettst.	*Centella asiatica* (L.) Urb.
14	Dantimool/ Nagadanti	*Baliospermum solanifolium* (Burm.) Suresh	*Ricinus communis* L., *Croton oblongifolius* Roxb.
15	Daruhaldi/ Daruharidra/ Rasanjan/Zarisk	*Berberis aristata* DC.	*B. lyceum* Royle, *B. asiatica* Roxb., *B. chitria* Lindl., *B.tinctoria* Lesch and *B. umbellata* Wall.
16	Pashanabheda	*Bergenia ciliata* (Haw.) Sternb.	*Bergenia stracheyi* (Hook.f. & Thoms.) Engl.

TABLE 7.6 (Continued)

S. No	Name of the Botanical Drug	Name of Accepted Plant Source	Substituents/Known Adulterants
17	Punarnava/ Punarnava raktha	*Boerhavia diffusa* L.	*Boerhavia repanda* Willd. and *Trianthema portulacastrum* L.
18	Salai guggul/ Kundur	*Boswellia serrata* Roxb. *ex*Colebr.	*Bowellia carteri*Brdw., *B. frereana* Birdw. and *Garuga pinnata* Roxb.
19	Sadabahar	*Catharanthus roseus* (L.) G.Don [=*Vinca rosea* L.]	*Solanum melongena* L., & *Lycopersicon esculentum* Mill.
20	Malkangani/ Jyothismathi	*Celastrus paniculatus* Willd.	*Cardiospermum halicacabum* L., & *Duranta plumieri* Jacq.
21	Brahmi/Brahmi booti	*Centella asiatica* (L.) Urb.	*Bacopa monnieri* (L.) Wettst.
22	Safed musli	*Chlorophytum tuberosum* (Roxb.) Baker	*Chlorophytum arundinaceum* Baker and *C.borivilianum* Santapau & Fernandes
23	Dalchini/Tejpatta	*Cinnamomum sulfuratum* Nees	*Cinnamomum cassia* (Nees & T.Nees) J.Presl, *C. zeylanica* Blume, *C. malabathrum* (Lam.) J. Presl., *C. verum* J. Presl and *C. wightii* Meisn.
24	Tejpatta/ Tamalpatra	*Cinnamomum tamala* (Buch.-Ham.) T. Nees & Eberm.	*Cinnamomum zeylanica* Blume and *C. malabathrum* (Lam.) J. Presl.
25	Guggul/Guggulu	*Commiphora mukul* (Hook. *ex* Stocks) Engl. [=*C. wightii* (Arn.) Bhandari]	*Boswellia serrata* Roxb. and *Commiphora myrrha* Engl.
26	Shankhapushpi	*Convolvulus prostratus* Forssk. [=*C. pluricaulis* Chois., *C. microphyllus* Sieber *ex* Spreng.]	*Evolvulus alsinoides* L. *Clitorea ternatea* L., & *Canscora deccusata* (Roxb.) Schult. & Schult.f.
27	Maramanjal	*Coscinium fenestratum* (Goetgh.) Colebr.	*Mahonia leschenaultii* Wall. *ex* Wight & Arn., *Berberis aristata* DC.
28	Jamalghota/ Japala	*Croton tiglium* L.	*Jatropha curcas* L.
29	Thikhur	*Curcuma angustifolia* Roxb.	*Maranta arundinacea* L.
30	Kachur/Kachora	*Curcuma zanthorrhiza* Roxb.	*Curcuma caesia* Roxb.

TABLE 7.6   (Continued)

S. No	Name of the Botanical Drug	Name of Accepted Plant Source	Substituents/Known Adulterants
31	Nagarmotha/ Mustha	*Cyperus rotundus* L.	*Cyperus scariosus* R.Br.
32	Magali	*Decalepis hamiltonii* Wight & Arn.	*Hemidesmus indicus* (L.) R.Br.
33	Salparni/ Prshniparni	*Desmodium gangeticum* (L.) DC.	*Desmodium pulchellum* Benth. *ex* Baker, *Uraria hamosa* Wall. ex Wight and *Pseudarthria viscida* (L.) Wight & Arn.
34	Vaividang	*Embelia tsjeriam-cottam* A.DC. [=*E. basaal* sensu (Roem. & Schult.) A.DC, non Mez	*Maesa indica* (Roxb.) A.DC,*Myrsine africana* L., & *Myrsine capitellata* Wall. Includes fruits of *Embelia ribes*, the most accepted candidate for 'Vaividang'
35	Somlata	*Ephedra gerardiana* Wall. *ex* Stapf	*Equisetum arvense* L., *Ephedra sinica* Stapf. and *Ephedra equisetina* Bunge
36	Shahtara	*Fumaria indica* (Hausskn.) Pugsley	*Fumaria officinalis* L., *F. parviflora* Lam., *F. vaillantii* Loisel
37	Kokam/ Cambogia	*Garcinia indica* (Thouars) Choisy	*Garcinia gummi-gutta* (L.) Robs.
38	Dikamali	*Gardenia gummifera* L.f.	*Gardenia resinifera* Roth
39	Mulethi/ Jeshtimadhu/ Gulegafis	*Glycyrrhiza glabra* L.	*Abrus precatorius* L. and *Glycyrrhiza uralensis* Fisch.
40	Gambar chal/ Shivan Chal	*Gmelina arborea* Roxb.	*Gmelina asiatica* L., *Premna flavescens* Ham.
41	Pitpapra	*Hedyotis corymbosa* (L.) Lam. [=*Oldenlandia corymbosa* L.]	*Mollugo cerviana* Ser.
42	Anantmool/ Sariwa/ Sarasaparilla	*Hemidesmus indicus* (L.) R.Br. Schult.	*Cryptolepis buchananii* Roem. & Schult., *Ichnocarpus frutescens* R. Br. and *Decalepis hamiltonii* Wight & Arn.
43	Inderjao Kadwa/ Kutja	*Holarrhena pubescens* (Buch.-Ham.) Wall.*ex*G. Don [=*H. antidysenterica* Wall. *ex* DC.]	*Wrightia tomentosa* Roem. & Schult. and *W. tinctoria* (Roxb.) R.Br.

TABLE 7.6 (Continued)

S. No	Name of the Botanical Drug	Name of Accepted Plant Source	Substituents/Known Adulterants
44	Pushmarmool/ Pokharmool	*Inula racemosa* Hook.f.	*Saussurea lappa* (Decne.) Sch. Bip., *Inula royleana* DC., & *Carduus nutans* L.
45	Kachora/ Narkachur/ Kapoor kachri No.1	*Kaempferia galanga* L.	*Hedychium spicatum* Ham. *ex* Smith
46	Lobelia leaves	*Lobelia nicotianifolia* Roth *ex* Roem. & Schult.	*Lobelia pyramidalis* Wall., *L. leschenaultiana* (Presl) Skottsb. and *Verbascum thapsus* L.
47	Madhuka	*Madhuca indica* J. F.Gmel.	*Madhuca longifolia* (Koen.) Macler.
48	Nagekesar	*Mesua ferrea* L. [=*M. nagassarium* (Burm.f.) Kosterm.]	*Mammea suriga* Buch.-Ham. *ex* Roxb., *Cinnamomum* spp. and *Dillenia pentagyna* Roxb.
49	Balchad/ Jatamansi	*Nardostachys grandiflora* DC.	*Selinum vaginatum* C. B. Clarke and *S. candolei* DC.
50	Kali tulsi/ Bantulsi	*Ocimum americanum* L.	*Ocimum tenuiflorum* L.
51	Tetu chaal/Arlu	*Oroxylum indicum* (L.) Benth. ex Kurz	*Ailanthes excelsa* Roxb.
52	Bhumiamla	*Phyllanthus amarus* Schumach. & Thonn.	*P. urinaria* L., *P.reticulatus* Poir., *P.virgatus* G. Forst., *P.debilis* Klein ex Willd., *P. fraternus* Webster and *P. maderaspatensis* L.
53	Kutaki	*Picrorhiza kurroa* Royle *ex* Benth.	*Picrorhiza scrophulariiflora* Pennel D. Y. Hong, *Lagotis glauca* Gaertn.
54	Pipal/Thippili/ Pippali	*Piper longum* L.	*Piper retrofractum* Vahl, *P. betle* L., & *P. peepuloides* Roxb.
55	Isabgol	*Plantago ovata* Phil.	*Plantago major* L., & *P. lanceolata* L.
56	Arnimool/ Agnimantha	*Premna serratifolia* L.	*Premna corymbosa* Rottl., *P. flavescens* Ham. and *P. latifolia* Roxb.
57	Damul-akhwain/ Vijaysar	*Pterocarpus marsupium* Roxb.	*Terminalia tomentosa* Wight & Arn. and *Bridelia montana* Willd.

TABLE 7.6   (Continued)

S. No	Name of the Botanical Drug	Name of Accepted Plant Source	Substituents/Known Adulterants
58	Raktachandana/ Lal chandan	*Pterocarpus santalinus* L.f.	*Adenanthera pavonina* Willd.
59	Sarpagandha/ Pagal buti	*Rauvolfia serpentina* (L.) Benth. *ex* Kurz	*Rauvolfia densiflora* (Wall.) Benth. *ex* Hook.f., *R. micrantha* Hook.f., *R. tetraphylla* L., *Tabernaemontana divaricata* (L.) R.Br.
60	Revan chini	*Rheum australe* D.Don [=*R. emodi* Wall. ex Meissn.]	*R. moorcroftianum* Royle, *R. palmatum* L., *R. spiciforme* Royle and *R. webbianum* Royle
61	Majith/Manjistha	*Rubia cordifolia* L.	*Rubia sikkimensis* Kurz.
62	Chandan	*Santalum album* L.	*Erythroxylum monogynum* Roxb.
63	Reetha	*Sapindus mukorossi* Gaertn.	*Sapindus emarginatus* Vahl. and *S. laurifolius* Vahl
64	Ashoka chhal	*Saraca asoca* (Roxb.) W. J. de Wilde	*Humboldtia vahliana* Wight, *Mallotus nudiflorus* (L.) Kulju & Welzen and *Shorea robusta* Gaertn.
65	Kuth/Uplet	*Saussurea costus* (Falc.) Lipsch. [=*S. lappa* C. B. Clarke]	*Saussurea hypoleuca* Sprengl, *Inula royleana* DC., *I. racemosa* Hook. f. and *Carduus nutans* L.
66	Raal	*Shorea robusta* Gaertn.f.	*Vateria indica* L.
67	Bala	*Sida rhombifolia* L.	*Sida acuta* Burm.f., *S.cordifolia* L. and *S. cordata* L.
68	Chopchini	*Smilax glabra* Roxb.	*Smilax china* L., *Smilax aspera* L., *S.china* L. & *S.zeylanica* L.
69	Gorakh mundi	*Sphaeranthus indicus* L.	*Sphaeranthus africanus* L. & *S. amaranthoides* Burm.f.
70	Patala	*Stereospermum chelonoides* (L.f.) DC. [=*S. suaveolens* (Roxb.) DC.]	*Stereospermum colais* (Buch.-Ham. *ex* Dillw.) Mabberley and *Radermachera xylocarpa* (Roxb.) K.Schum.
71	Kuchla/Itti beeja	*Strychnos nux-vomica* L.	*Strychnos nux-blanda* A. W. Hill
72	Chiraiyata	*Swertia chirayita* (Roxb. *ex* Fleming) H.Karst.	*Andrographis paniculata* (Burm.f.) Nees, *Swertia angustifolia* Buch.-Ham. *ex* D.Don. and *S. alata* Royle *ex* D.Don, *Rubia cordifolia* L.
73	Pathani Lodh	*Symplocos racemosa* Roxb.	*Symplocos cochinchinensis* S. Moore and *Symplocos paniculata* (Thunb.) Miq.

TABLE 7.6 (Continued)

S. No	Name of the Botanical Drug	Name of Accepted Plant Source	Substituents/Known Adulterants
74	Talispatra	*Taxus wallichiana* Zucc.	*Abies spectabilis* (Royle ex D.Don) Mirb., *A. pindrow* Royle, *A. densa* Griff. and *Rhododendron anthopogon* D. Don.
75	Giloy/Galo/ Amrithballi	*Tinospora cordifolia* (Willd.) *ex* Hook.f. & Thomson	*Tinospora crispa* (L.) Hook.f. & Thomsonand *Tinospora sinensis* (Lour.) Merr.
76	Ajwain	*Trachyspermum ammi* (L.) Sprague	*Trachyspermum roxburghianum* (DC.) Craib.
77	Barhanta	*Tragia involucrata* L.	Vrscikalli
78	Gokhru/ Gokshura	*Tribulus terrestris* L.	*T. lanuginosus* L., *T. subramanyamii* P.Singh, Giri & V. Singh, *T. alatus* Delile, *Pedalium murex* L.
79	Patol panchang	*Trichosanthes cucumerina* L.	*Trichosanthes dioica* Roxb. and *T.lobata* Roxb.
80	Musakbala/ Tagara	*Valeriana jatamansi* Jones	*Valeriana hardwickii* Wall., *V. officinalis* L., *Cryptocoryne spiralis* Fisch.
81	Manda dhupa/ Dupa	*Vateria indica* L.	*Shorea robusta* Gaertn.f.
82	Banafsha	*Viola pilosa* Blume	*V. odorata* L., *V. canescens* Wall., *V. biflora* L. and *V. betonicifolia* Sm.

subset of this code is the more specifically defined 4 digit code 1211 pertaining to the description "Plants and plant parts for pharmacy, perfumery, insecticides, fungicidal; fresh, dried, cut or not, crushed, powdered herbs, licorice, mint, ginseng." This four digit code gets further refined to 6 or more digits for referring to more sharply defined entities/items. This system for classification of internationally traded goods has been developed, and is maintained, by the World Customs Organization (WCO) since 1st January, 1988 and it has been adopted by most trading nations, including India.

The international HS codes are subject to an on-going periodic review by WCO. More detailed classification at the country level, to identify goods for tariff or statistical purposes, uses additional digits added to these codes (upto a total of 10 digits).

The data relating to India's exports by commodities is compiled and published by the Directorate General of Commercial Intelligence and Statistics (DGCIS) of Government of India. In these compilations the items are listed with their HS (ITC) codes along with annual quantities in Kilograms and trade value in Rupees (Table 7.7). The ITC (HS) codes, referring to the Indian Trade Classification (HS) codes, are derived from international HS codes and can have up to 10 digits. ITC (HS) is primarily used as a tool for (a) levying of excise, (b) import tariffs and (c) generation of EXIM data. The medicinal plant materials do not get enlisted fully under any specific major category of HS (Harmonized System) codes followed internationally.

7.2.6 SPECIES OF CONSERVATION CONCERN IN TRADE

A global analysis of a representative sample of the world's plants, conducted in 2010 by the Royal Botanic Garden, Kew together with the National History Museum, London and IUCN had revealed that *one* in *five* of the worlds plant species are threatened with extinction.

Out of around 5000 wild Indian medicinal plants in India (as per FRLHT database) 20% of these, that is, around 1000 species may be threatened with

TABLE 7.7   A Few Examples of ITC (HS) Codes (Recorded in Reports Produced by DGCIS)

S. No.	ITC (HS) Code	Description
1	12119012	Nux vomica dried ripe seeds
2	12119013	Psyllium seed
3	12119019	Other seeds fresh/dried W/M cut, crushed/powdered, perfumery, pharmaceuticals, etc.
4	12119024	Gymnema powder
5	12119029	Other leaves, powder, flowers and pods fresh/dried W/N, cut crushed/powdered
6	12119032	Psyllium husk (Isabgol husk)
7	12119021	Belladona leaves
8	12119041	Belladona roots
9	12119039	Other bark, husk and rind fresh/dried, W/N cut crushed/powdered
10	12119048	Sweet flag rhizome
11	12119049	Other roots & rhizomes fresh/dried W/M crushed/powdered

extinction. FRLHT has coordinated rapid threat assessment processes and organized Conservation Assessment Management Prioritization (CAMP) workshop based on IUCN Red List Categories and Criteria, resulting in assessment of around 350 species across 18 states. This assessment of Threat status of Indian medicinal plants, undertaken so far, is far from completion and there is a need for undertaking rapid assessments to identify and assess the species needing urgent conservation action.

Analysis of the list of 178 species in high trade reveals that a number of these are threatened in the wild, for example, *Aconitum heterophyllum* Wall. *ex* Royle, *Coscinium fenestratum* (Goetgh.) Colebr., *Decalepis hamiltonii* Wight & Arn., *Nardostachys jatamansi* (D.Don) DC., *Oroxylum indicum* (L.) Benth. *ex* Kurz, *Picrorhiza kurroa* Royle *ex* Benth., *Saraca asoca* (Roxb.) W. J. de Wilde, *Swertia chirayta* (Roxb. *ex* Fleming) H. Karst. and *Vateria indica* L. etc. Concerns about global depletion of populations of some of these species have prompted their inclusion in the CITES lists. Government of India, in 1994, had also drawn a 'negative list of exports' wherein some of these species were included. However, wild populations of many of these threatened species continue to deplete in the face of indiscriminate harvesting. Exports of some of the species included in the 'negative list' still continue due to nomenclature issues. This is illustrated with example of temperate Himalayas herb traded in the name of Kutki, found growing at altitudes ranging from 3000 to 3500 m was hitherto believed to have wide distribution extending from North-west Himalayas to Nepal and Bhutan in the east. Its increasing domestic and global trade coupled with concerns about its fast shrinking wild populations prompted the inclusion of *Picrorhiza kurroa* Royle *ex* Benth. in the CITES Appendix II to regulate its foreign trade. However, typification of a closely resembling taxon growing in Uttarakhand, Nepal and Bhutan as *Picrorhiza scrophulariiflora* Pennel (=*Neopicrorhiza scrophulariiflora* Pennel D. Y. Hong) has added a new dimension to the trade of 'Kutki,' as this taxon is not included in the CITES list even though it is also used as Kutki without any discrimination.

The estimated consumption of 'Kutki' by the domestic herbal industries was 416 MT during 2005–06. Since rhizomes of both these species are freely traded as 'Kutki,' it is difficult to link a specific quantitative estimate to any one of these. The foreign trade of 'Kutki,' derived from both the species, can also continue under the name of *P. scrophulariiflora* Pennel. The net result is further decline in wild populations of both *P. kurroa* Royle *ex* Benth. and *P. scrophulariiflora* Pennel.

In addition to the medicinal plant species in high trade, there are other threatened medicinal plant species in trade, presently traded in lower volumes but with high trade potential, that needs focused management interventions. Some such species like *Podophyllum hexandrum* Royle are subjected to indiscriminate harvesting because of their high demand in international market for their chemotherapeutic use.

Podophyllum hexandrum Royle (=*P. emodi* Wall. *ex* Honig.) is a temperate Himalayan herb found at altitudes ranging from 2800 to 3500 m. It is commercially collected for its rhizomes that are processed to extract 'podophyllin,' derivatives of which are used in treatment of tumors. The species, once known to form extensive dense populations in its natural zone, has borne the brunt of heavy exploitation over the past more about 50 years and has become endangered now. Since the rhizomes take 5–6 years to mature, efforts to domesticate the species and cultivate it on commercial scale have proved to be non-viable and abandoned.

In view of threat to its wild populations, export of this species was banned in 1994. However, since it is preferred over its American allied species (*P. peltatum* L.) for its higher alkaloid content, it continues to be indiscriminately harvested from the wild, further endangering even its residual populations. A few others like *Gentiana kurroo* Royle and *Dactylorhiza hatageria* (D.Don) Soo are already on the verge of extinction and their populations have drastically declined.

There is an urgent need to initiate urgent measures for management of high-risk wild medicinal plant species and to take up re-assessment of wild populations of these species at the regional, national and global level for guiding comprehensive *in situ* and *ex situ* conservation including species recovery and commercial cultivation.

KEYWORDS

- conservation
- demand and supply
- HS codes and ITC (HS) codes
- medicinal plants
- plant raw drugs
- substitutes and adulterants
- trade

REFERENCES

Bodeker, G., Bhat, K. K. S., Burley, J. & Vantomme P (Eds.). (1997/2003). Medicinal plants for forest conservation and health care. UN Food and Agriculture Organization (FAO). Non-Wood Forest Products Series, No. 11. Rome.

Grünwald, J., & Büttle, K. (1996). The European phytotherapeutics market. Drugs made in Germany 39, 6–11.

Laird, S. A. (1999). The botanical medicine industry. p. 78–116. In: ten Kate, K., & Laird, S. A. (eds.). The commercial use of biodiversity. Earthscan, London.

Lange, D. (1998). Europe's medicinal and aromatic plants: their use, trade and conservation. TRAFFIC International, Cambridge.

Lange, D. (2006). International Trade in Medicinal and Aromatic Plants – Actors, volumes and commodities. pp. 155–170. In: Bogers, R. J., Craker, L. E., & Lange, D. (eds.). Medicinal and Aromatic Plants, 155–170. Springer. Printed in the Netherlands.

Lindstrom, C. O., Lynch, M. E., Blumenthal, M., & Kawa, K. (2004). Market Report. Issue 103. www.herbalgram.org.

Sreedevi, P., Ijinu, T. P., Anzar, S., Bincy, A. J., George, V., Rajasekharan, S. & Pushpangadan, P. (2013). Ethnobiology, ethnobotany, ethnomedicine and traditional knowledge with special reference to India. *Annals of Phytomedicine 2(2),* 4–12.

Ved, D. K. & Goraya, G. S. (2007). Demand and Supply of Medicinal Plants in India. NMPB, New Delhi & FRLHT, Bangalore, India.

Zhu, Z. Y., Zhang, Z., Zhang, M., Mais, D. E., & Wang, M. W. (2010). High Throughput Screening for Bioactive Components from Traditional Chinese Medicine. *Combinatorial Chemistry and High Throughput Screening. 13,* 837–848.

CHAPTER 8

ETHNOBOTANY OF USEFUL PLANTS IN INDO-GANGETIC PLAIN AND CENTRAL INDIA

VIJAY V. WAGH

Plant Diversity, Systematics and Herbarium Division, CSIR – National Botanical Research Institute, Rana Pratap Marg, Lucknow – 226001, Uttar Pradesh, India, E-mail: vijaywagh65@gmail.com

CONTENTS

ABSTRACT

The Indo-Gangetic plain is the fertile land of India, the plain supports the highest population densities depending upon purely agro-based economy on the other hand Central India represented forest rich area. The local tribals of both these areas are dependent on the forest resources for food, medicine, timber, gum, fiber, ornamental, sacred plants, fodder, oil, etc. These plant resources also support socio-economic development of the local tribal communities. The aim of the present review is to present the list of useful plants other than medicinal and wild edible plants of Indo-Gangetic plain and Central India and also spread awareness about the sustainable use of the biodiversity of the region.

8.1   INTRODUCTION

The Indo-Gangetic plain is the largest unit of the Great Plain of India stretching from Delhi to Kolkata in the states of Uttar Pradesh, Bihar and West Bengal covering an area of about 3.75 lakh sq. km. The Ganga is the master river after whose name this plain is named. The Ganga along with its large number of tributaries originating in the Himalayan ranges viz., the Yamuna, the Gomati, the Ghaghara, the Gandak, the Kosi, etc. have brought large quantities of alluvium from the mountains and deposited it here to build this extensive plain. Depending upon its geographical variations, this plain can be further subdivided into the following three divisions: The Upper Ganga Plain: Comprising the upper part of the Ganga Plain, this plain is delimited by the 300 m contour in Shiwaliks in the north, the Peninsular boundary in the south and the course of the Yamuna river in the west. This plain is about 550 km long in the east-west direction and nearly 380 km wide in north-south direction, covering an approximate area of 1.49 lakh sq. km. Its elevation varies from 100 to 300 m above mean sea level. The Middle Ganga Plain: To the east of the Upper Ganga plain is Middle Ganga plain occupying eastern part of Uttar Pradesh and Bihar. It measures about 600 km in east-west and nearly 330 km in north-south direction accounting for a total area of about 1.44 lakh sq. km. The Lower Ganga Plain: This plain includes the Kishanganj tehsil of Purnea district in Bihar, the whole of West Bengal. It measures about 580 km from the foot of the Darjeeling Himalaya in the north to Bay of Bengal in the south and nearly 200 km from the Chotanagpur

Highlands in the West to the Bangladesh border in the east. Central India is an area of great physical complexity and immense geopolitical significance. It is an area transitional the passageway between the north and south India. It includes the states Madhya Pradesh, Chhattisgarh, some part of Maharashtra, Rajasthan. The dominant tribal communities of this region are primitive tribes like Gonds, Bhils and Santhals.

Plants and human beings have intrinsic relationship since ancient times and were evolved along parallel lines for their existence, cooperating and depending upon each other. This intimate relationship had progressed over generations of experiences and practices. Apart from their nutritional, ritual and magical value, plants have important contributions in the health care system of human being (Shil et al., 2014). Due to the close association with nature and its various components, the tribal and local communities have effectively developed their traditional knowledge system which incorporates the use of locally available plants and its products for treatment of various ailments (Kala, 2005).

The tribals of Indo-Gangetic plain and Central India depend upon natural resources because of the remoteness and availability of the resources in their vicinity. The ethnomedicinal and wild edible plants found in Indo-Gangetic plain and Central India are reported well (Ahirwar, 2013; Nargas and Trivedi, 2003; Pandey and Tiwari, 2014; Oommachan et al., 1989; Rai et al., 2004; Jha and Gurudatta, 2015; Mahto et al., 2007; Mairh et al., 2010; Singh, 2014; Jain et al., 2004; Panghal et al., 2010; Wagh and Jain, 2010a,b; Singh et al., 1999; Sikarwar, 1994; Saxena, 1986; Siddiqui and Dixit, 1975; Kala, 2009; Kumar and Jain, 1998). It is equally important to document the other uses of plants (fiber, wood, dye, gums, resins, latex) etc. An account of these different uses of plants used by the tribals in Indo-Gangetic plains and Central India is given in Table 8.1.

8.2 DYE YIELDING PLANTS

The making of natural dye is one of the oldest known to man and dates back to the dawn of human civilization. Color on clothing has been extensively used since 5000 years back (Kar and Borthakur, 2008). It was practice during the Indus river valley civilization at Mohenjodaro and Harappa (3500 BC), former Egyptian and China period (Siva, 2007). Moldenke and Moldenke (1983) reports that an orange or yellow impermanent dye is made from corolla tubes of *Nyctanthes arbortristis* Linn. for Buddhist robes in Sri Lanka (Panigrahi and Murti, 1989–1999). In the making of natural dyes the uses of mordant to hold fast the dye and to prevent them from touching the cloth were printed bales of soft textile. In India there are more than 450 plants out of 17,000 plants have been recorded that can produce dye. In 19th century the discovery of synthetic dyes has been dealt a massive blow to Indian textile industry. Research has been shown that the vast uses of synthetic dyes associated with hazards effecting human body system; it causes skin cancer, temporary or permanent blindness and also the respiratory system, etc. (Dubey, 2007; Singh, 2001).

Earliest evidence for the use of natural dyes dates back to more than 5000 years, with Madder (*Rubia cordifolia*) dyed cloth found in the Indus river valley at Mohenjodaro. India is endowed with a wealth of natural flora and fauna, which provide the basic resources for a rainbow of natural dyes. Natural dyes are environment friendly; for example, turmeric, the brightest of naturally occurring yellow dyes is a powerful antiseptic and revitalizes the skin, while indigo yields a cooling sensation. Research has shown that synthetic dyes are suspected to release harmful chemicals that are allergic, carcinogenic and detrimental to human health. Ironically, Germany that discovered azo dyes, became the first country to ban certain azo dyes in 1996 (Singh and Singh, 2002).

8.3   FIBER YIELDING PLANTS

Fiber-yielding plants have been of great importance to man and they rank second only to food plants in their usefulness. In ancient times, plants were of considerable help in satisfying man's necessities in respect of food, clothing and shelter. Although other materials like animal skin and hides were also used to meet the demands with regard to clothing, they were quite insufficient for the purpose. Further, the need for some lighter and cooler substance was keenly felt. In those days, man also required some form of cordage for his snares, bow-strings, nets, etc., and also for better types of covering for his shelter. Tough, flexible fibers obtained from stems, leaves, roots, etc., of various plants served the above purposes very well. With the advancement of civilization, the use of plants fibers has gradually increased and their importance today is very great. Although many different species of plants, roughly about two thousand or more, are now known to yield fibers, commercially important ones are quite small in number. The use of plants fibers preferred from time immemorial due to its easy availability, durability and flexibility. The use of cotton fiber and silk is known to occur since 5000 B.C. Many fiber yielding plants, including *Boehmeria nivea* Gaud. (Ramie), *Crotolaria juncea* L. (Sunhemp), *Corchorus capsularis* L. (Jute), *Gossypium arboretum* L. (cotton), *Hibiscus cannabinus* L. (Kenaf), *Linum usitatissium* L. (Flax) are the best known commercial plants which provide durable and flexible fiber. The utility of plant fibers is manifested in a diverse range of products which includes, making ropes, paper and various household materials. The fiber production also contributes signifi-cantly to the economy of a region in various ways, including agricultural, clothing, small scale industry and products for other household operations. It has been estimated that over a thousand species of plants are yielding fibers in America alone, over 800 in Philippines and about 790 species in India (Pandey and Gupta, 2003). However, use of plant fibers in commer-cial sector is relatively little. Information on various fiber yielding plants is needed for maximum utilization and this would help in improving the socio-economic status by supporting the livelihood and income generation opportunity.

8.4   GUM AND RESIN YIELDING PLANTS

Resins and gums are metabolic by-products of plant tissues either in normal course or often as a result of disease or injury to the bark or wood of certain tree species. There are a large number of trees in India which exude gums and resins. Some of these are of local or limited interest, while a few are used extensively all over India and also entered the export trade of the country. Annual average export of gum and resin during 2001–2002 to 2005–2006 was Rs. 7848 million. This included Rs. 1,371 million of resins and Rs. 6,363 million of gums. The gums and gum-resins of commercial importance collected from the forest are gum karaya, gum ghatti, salai gum, guggul, and gums from various species of *Acacia*, including Indian gum Arabic from *Acacia nilotica* and gum arabic from *A. senegal*. The important commercial resins are obtained from Pinaceae (Resin, Amber), Leguminosae (Copal) and Dipterocarpaceae (Dammar) families. Gums and resins are perhaps the most widely used and traded non-wood forest products other than items con-sumed directly as food, fodder and medicine. Human beings have been using gums and resins in various forms for ages. The history of Gum Arabic, long recognized as an ideal adhesive, dates back 2000 years. In modern times, gums and resins have been used the world over as embalming chemicals, incense, medicines (mainly anti-septic properties and balms), cosmetics in paints and for waterproofing and caulking ships. Certain natural gums and resins are approved by the U.S. Food and Drug Administration for use in food and pharmaceuticals. The present day uses of natural gums and resins are numerous and they are employed by a large number of manufacturing industries including food and pharmaceutical industries.

Use of gums and resins for domestic consumption and for sale to earn some cash is very common among the forest dwelling communities, particularly tribals, in India. Thousands of forest dwellers in the central and western Indian states depend on gums and resins as a viable income source. However, if we go through the statistics, the developments are not encouraging, as the market for local gums and resins at the national level has largely remained stagnant or has decreased over years. This development has made an adverse impact on the gum dependent communities.

8.5 SACRED/SOCIO-RELIGIOUS PLANTS

India has a vast variety of flora that feature in our myths, our epics, our rituals, our worship and our daily life. There is the pipal, under which the Buddha meditated on the path to enlightenment; the banyan (*Ficus religiosa*), in whose branches hide spirits; the Ashoka (*Saraca indica*), in a grove of which Sita sheltered when she was Ravana's prisoner; the tulsi (*Ocimum tenuiflorum*), without which no Hindu house is considered complete; the bilva (*Aegle marmeols*), with whose leaves it is possible to inadvertently worship Shiva. Before temples were constructed, trees were open-air shrines sheltering the deity, and many were symbolic of the Buddha himself. Sacred Plants of India systematically lays out the socio-cultural roots of the various plants found in the Indian subcontinent, while also asserting their ecological importance to our survival.

Indo-Gangetic plain and Central India encompasses many plant species which are being used as food, shelter, clothing and medicines by the people of village communities. Besides these, some plants are used by the people in different social and religious customs, are known as Socio-religious plants (Ahirwar, 2010). The relationship between man and plant communities is as old as his hunger, and long before science was born, our ancestors studied the plants around them to meet their basic requirements, which laid the foundation of civilization (Pandey and Verma, 2005). Many festivals are associated with the significance of plants in India (Dashora ct al., 2010). The tribal devotion to these plants is so high that they never think to cut these plants. If it happens so they try to expiation.

8.6   TIMBER YIELDING PLANTS

India is blessed with a vast variety of timber yielding tree species and as many as 1500 species are commercially utilized for diverse purposes. Some of the important plantation tree species grown in India are *Tectona grandis, Eucalyptus* spp., *Acacia* spp., *Dalbergia sissoo, Swietenia* spp., *Santalum album, Melia dubia, Ailanthus excelsa, Leucaena leucocephala,* etc. Since time immemorial human beings are depending on the plants for food, shelter and medicine. Besides this wood had considerable importance in the livelihood of ancient people, use of wood in making several things such as agricultural implements, boat building, handicrafts, packing cages, toys, construction, furniture, musical instruments, turnery, carving, etc. (Ambasta, 1992; Anonymous. 1948–1976; Asolkar et al., 1992). The wood is considered as most important forest product till date and has contributed a lot to advancement of civilization. Though the forests are vanishing at alarming rates the requirement of the wood has not declined and even today wood is the most widely used commodity other than food and clothing (Cooke, 1958; Jain, 1991, 1996). The most commonly used wood in India is from following plants viz., *Acacia nilotica* (L.) Del., *Bombax ceiba* L., *Albizia lebbeck* (L.) Bth., *Toona ciliata* Roem., *Juglans regia* L., *Salix alba* L., *Morus alba* L., *Cedrus deodara* (Roxb. ex Lamb) G. Don, *Picea smithiana* (Wall.) Boiss., *Pinus roxburghii* Sarg., *Dalbergia latifolia* Roxb., *Dalbergia sissoo* Roxb., *Pterocarpus marsupium* Roxb., *Pterocarpus santalinus* L.f., *Diospyros ebenum* Koenig, *Haldina cordifolia* (Roxb.) Ridsdale, *Tectona grandis* L.f., *Shorea robusta* Roxb. ex Gaertn. f. etc. Productivity of forests in general and particularly that of commercial forest plantations is very much affected by frequent outbreak of pests and diseases, besides human interventions and various natural calamities. The total production of timber in India from forests is reported at an average 2.3 million cu.m in 2010. The wood and wood products imports to India have gradually increased since 1998 and have reached 6.3 million m³ in 2011 with a total import value of Rs. 9800 crores. Though wood is imported from about 100 countries, six countries namely Malaysia, Myanmar, New Zealand, Ghana, Ivory Coast, and Gabon constitute bulk of the timber imports to India (about 80%). Teak constitutes about 15% of total timber imports to India and the major teak exporting countries to India include Myanmar, Ivory Coast, Ghana, Ecuador, Costa Rica and Benin.

The total fuel-wood consumption estimated in household sector is 248 million m³ and about 13 million m³ additional fuel-wood is consumed in hotels and restaurants, cottage industries and cremation of dead human bodies. This makes the total annual consumption of fuel-wood to be 261 million m³ which comes from different sources. The production of fuel-wood from forests has been estimated to be 52 million m³ (FSI, 2009) and remaining 209 million m³ from farmland, community land, homestead, roadside, canal side and other wastelands (ICFRE, Forest Sector Report India, 2010). Meena et al. (2013) gave an account of traditional uses of ethnobotanical plants for construction of the hut and hamlets in the Sitamata Wildlife Sanctuary of Rajasthan.

8.7 ORNAMENTAL PLANTS

Nature has been generous and has given abundant wealth of wild ornamental flowers and they vary in composition and density in contrast with domesticated plants. Ornamental plants are grown usually for the purpose of beauty for their fascinating foliage, flowers, fruits and pleasant smell (Swaroop, 1998). They are very important in view of esthetic and recreational value for human beings. Most of the present day ornamental plants are coming from wild resources, few of which still exist in natural habitat (Thomas et al., 2011). Wild plants are a striking feature of the land surface; they vary greatly in composition and density in marked contrast with domesticated plants (Raju, 1998). The more attractive wild flowers have long been prized for the beauty and planted in the garden around mankinds dwelling places Wild ornamental plants are those which occur naturally in the field and have highly ornamental features such as ornamental flowers, foliage and fruits (Li and Zhou, 2005). They play an important role in environmental planning of urban and rural areas for abatement of pollution, social and rural forestry, wasteland development, afforestation and landscaping of outdoor and indoor spaces (Kapoor and Sharga, 1993). Though nature has given a wealth of wild flowers and ornamental plants many of them have been destroyed and several have become extinct and survival of many endangered by our exploitation by human beings (Arora, 1993).

India is rich in wild ornamental plants and has made great contribution to the collection, introduction and cultivation of ornaments from the mid-19th century. Large numbers of exotic trees, shrubs and herbaceous plants with high ornamental value have been distributed and grown in botanical garden, arboreta and many collections from foreign countries. However, due to large area and the complex topography and climate, a large number of wild ornamental plants still remain hidden in the depths of the forests and on the highland plateaus, places that are difficult for human reach and exploration. The trade of ornamental horticulture runs into many millions of rupees annually and there is considerable potential for further development and introduction of many new species from the wild into trade.

In this chapter the details regarding some of the important traditional useful plants that are available in forests of Indo-Gangetic plain and Central India are listed with their botanical names, family, parts used, purpose of use in Table 8.1.

TABLE 8.1   Useful Plants of Indo-Gangetic Region and Central India

S. No.	Botanical Name and family	Part	Uses	References
1.	*Abelmoschus manihot* (L.) Medik. (Malvaceae)	Stem	Fiber	Srivastava and Varma, 1981
2.	*Abrus precatorius* L. (Fabaceae)	Root	Symb in talisman to ward off evil spirits	Kamble and Pradhan, 1980
		Seed	Sacred garland	Joshi, 1982
		Fruit	Socio-religious	Ahirwar, 2013
3.	*Abutilon indicum* (L.) Sweet (Malvaceae)	Fruit	Ear pendant	Jain, 1963
		Flower	Brown dye	Upadhay and Choudhary, 2012
4.	*Acacia caesia* (L.) Willd. (Mimosaceae)	Root	Fish poison	Kala, 2009

TABLE 8.1   (Continued)

S. No.	Botanical Name and family	Part	Uses	References
5.	*Acacia catechu* (L.f.) Willd. (Mimosaceae)	Heart wood	Dye and tannin	Maheshwari et al., 1990; Ghosh, 2004; Rai and Nath, 2006; Kapoor et al., 2008; Alawa et al., 2013; Jha and Gurudatta, 2015; Singh and Bharti, 2015
		Bark	Dye brown, black and pink	Tiwari and Bharat, 2008; Upadhay and Choudhary, 2012
		Bark	Gum	
6.	*Acacia farnesiana* (L.). Willd. (Mimosaceae)	Whole plant	Repels rats, snakes	Pal and Jain, 1989
7.	*Acacia leucophloea* (Roxb.) Willd. (Mimosaceae)	Wood	Symbol of charms	Joshi, 1982
		Bark and leaves	Dye yielding	Tiwari and Bharat, 2008; Upadhyay and Chaudhary, 2014
8.	*Acacia nilotica* (L.) Delile (Mimosaceae)	Seed	Dye brown black	Tiwari and Bharat, 2008
		Bark	Pink dye	Kala, 2009; Upadhyay and Chaudhary, 2014; Bodane, 2015
		Flower	Yellow	Alawa et al., 2013
		Wood	Fuel wood	Singh et al., 2013
		Branches	Fencing	Jha and Gurudatta, 2015
		Wood	Preparing handles of tools	Yadav and Bhandoria, 2013; Jha and Gurudatta, 2015; Singh and Bharti, 2015
		Pods	Fodder	Jha and Gurudatta, 2015
		Root	Sacred thread	Jadhav, 2008
9.	*Acacia pennata* (L.) Willd. (Mimosaceae)	Bark	Detergent	Jain, 1963
		Bark	Fish poison	Joshi, 1986.
10.	*Achillea millefolium* L. (Asteraceae)	Flower	Eardrop	Koelz, 1979

TABLE 8.1 (Continued)

S. No.	Botanical Name and family	Part	Uses	References
11.	*Acorus calamus* L. (Araceae)	Leaves	Insecticide	Mishra et al., 2012
		Rhizome	Water purifier	Mishra et al., 2012
12.	*Achyranthes aspera* L. (Amaranthaceae)	Root	Sacred thread	Jadhav, 2008
13.	*Aegle marmelos* (L.) Corr. (Rutaceae)	Twig and Fruit	Sacred (Keep evil spirit away)	Kumar et al., 2006;
		Leaves	Used worshipped lord Shiva	Singh et al., 2011; Shukla and Chakaravarty, 2012; Sahu et al., 2013; Pandey and Tiwari, 2014; Agarwal, 2014; Sandhya and Ahirwar, 2015; Ahirwar, 2013
		Fruit pulp	Roof plastering	Jain et al., 2010
		Fruit	Mucilaginous	Kumar, 2014
		Fruit shell	Yellow dye	Das and Mondal, 2012; Bodane, 2015
14.	*Aesandra butyracea* Bachni. (Sapotaceae)	Fruit (Beverage)	Used in rituals	Mairh et al., 2010
15.	*Aeschynomene indica* L. (Fabaceae)	Pith	Toys, flowers, models	Mishra et al., 2012
		Whole plant	Fodder and fuel	Mishra et al., 2012
		Stem	Floater for fish nets; green manure	Singh et al., 2013
16.	*Agave vera-cruz* Mill. (Agavaceae)	Leaves	Fibers for making rope	Rai and Nath, 2006
17.	*Ageratum conyzoides* (L.) L. (Asteraceae)	Flower	Tilak	Jain, 1963
		Leaves	Green dye	Kumar, 2014
18.	*Ailanthus excelsa* Roxb. (Simaroubaceae)	Seed oil	Insecticide	Yadav and Bhandoria, 2013
19.	*Alangium salvifolium* (L.f.) Wang. (Cornaceae)	Stem	Basketry	Singh and Maheshwari, 1985

TABLE 8.1   (Continued)

S. No.	Botanical Name and family	Part	Uses	References
20.	*Albizia lebbeck* (L.) Benth. (Mimosaceae)	Leaf	Fodder	Mishra et al., 2010
		Fruit	Dye	Ghosh, 2004
		Branch	Fuel wood	Shukla and Chakaravarty, 2012
		Wood	Timber for carpentry	Singh and Bharti, 2015
21.	*Albizia odoratissima* (L.f.) Benth. (Mimosaceae)	Bark	Fish poison	Bedi, 1978
22.	*Albizia procera* (Roxb.) Benth. (Mimosaceae)	Branch	Fuel wood	Shukla and Chakaravarty, 2012
23.	*Allium sativum* L. (Liliaceae)	Clove	Sacred thread	Jadhav, 2008
24.	*Aloe vera* L. (Liliaceae)	Leaves	Hedge, ornamental and soil binder	Singh et al., 2013
25.	*Alternanthera philoxeroides* (Mart.) Griseb. (Amaranthaceae)	Stem and leaves	Green manure	Mishra et al., 2012
26.	*Alternanthera sessilis* (L.) R.Br. ex DC. (Amaranthaceae)	Whole plant	Fodder	Mishra et al., 2012
27.	*Alysicarpus longifolius* (Spreng.) Wight & Arn. (Fabaceae)	Root	Soil improver	Singh et al., 2013
28.	*Alysicarpus rugosus* (Willd.)DC. (Fabaceae)	Leaves and root	Beverages	Singh et al., 2013
29.	*Amaranthus viridis* L. (Amaranthaceae)	Whole plant	Fodder	Singh and Bharti, 2015
30.	*Ampelocissus latifolia* (Roxb.) Planch. (Vitaceae)	Leaves	Green dye	Rai and Nath, 2006
		Stem	Sacred thread	Jadhav, 2008
31.	*Amorphophallus bulbifer* (Roxb.) Blume (Araceae)	Flower	Decorative article preparation	Shukla and Chakaravarty, 2012

TABLE 8.1   (Continued)

S. No.	Botanical Name and family	Part	Uses	References
		Tuber	Socio-religious	Hemrom and Yadav, 2015
32.	*Anacardium occidentale* L. (Anacardiacaeae)	Leaves	Keep away evil spirit	Mairh et al., 2010.
33.	*Anthocephalus chinenesis* Walp. (Rubiaceae)	Whole tree	Socio-religious	Ahirwar, 2013; Sandhya and Ahirwar, 2015
34.	*Anogeissus latifolia* (Roxb. ex DC.) Wall. ex Guillem. & Perr. (Combretaceae)	Bark	Gum ghatti	Rai and Nath, 2006
		Leaves and shoot	Tanin	Rai and Nath, 2006
		Wood	Socio-religious	Ramsankar, 2001
35.	*Antigonon leptopus* Hook. & Arn. (Polygonaceae)	Whole plant	Ornamental	Agarwal, 2013
36.	*Apluda mutica* L. (Poaceae)	Whole plant	Thatching	Mitra and Mukherjee, 2005
37.	*Areca catechu* L. (Areaceae)	Nut	Copper red dye	Kumar, 2014
		Nut	Sacred	Sahu et al., 2013
38.	*Artocarpus heterophyllus* Lam. (Moraceae)	Leaves	Sweetness in local drink	Ghosh and Das, 2004
		Wood	Yellow dye	Tiwari and Bharat, 2008
		Leaf	Goat fodder	Shukla and Chakravarty, 2012
		Wood	Drum preparation	Ramsankar, 2001
39.	*Artocarpus lakoocha* Roxb. (Moraceae)	Wood	Yellow dye	Tiwari and Bharat, 2008
40.	*Arundinella setosa* Trin. (Poaceae)	Inflorescence	Broom	Mairh et al., 2010.
41.	*Arundo donax* L. (Poaceae)	Culm	Walking stick, cleaning roof, flutes, etc.	Mishra et al., 2012
42.	*Asparagus racemosus* Willd. (Liliaceae)	Whole plant	Socio-religious	Jain, 1963; Hemrom and Yadav, 2015

TABLE 8.1 (Continued)

S. No.	Botanical Name and family	Part	Uses	References
		Roots	Fodder; sacred thread	Jain et al., 2010; Jadhav, 2008
43.	*Atylosia scarabaeoides* (L.) Benth. (Fabaceae)	Root	Soil binder	Singh et al., 2013
44.	*Avena sativa* L. (Poaceae)	Grain	Fodder	Mishra et al., 2010.
45.	*Azadirachta indica* A. Juss. (Meliaceae)	Bark	Bhill clan name	Joshi, 1982
		Whole plant	Religious	Heda, 2012; Sandhya and Ahirwar, 2015
		Stem	Gum yielding	Kumar, 2014
		Bark	Gum is used as adhesive	Rai and Nath, 2006
		Leaves	Keep away bad effect of soul	Ahirwar, 2013; Pandey and Tiwari, 2014; Agarwal, 2014; Hemrom and Yadav, 2015;
		Wood	Furniture making	Singh and Bharti, 2015
		Bark	Brown dye	Bodane, 2015
46.	*Balanites aegyptiaca* (L.) Del. (Zygophyllaceae)	Root	Fish poison	Joshi, 1982
47.	*Bambusa bambos* (L.) Voss (Poaceae)	Whole plant	Religious	Heda, 2012; Ahirwar, 2013
		Whole plant	Basketry, artifacts, musical instrument, etc.	Kumar and Yadav, 2010; Prasad et al., 2013; Singh and Bharti, 2015
48.	*Bambusa nutans* Wall. ex Munro (Poaceae)	Whole plant	Basketry, artifacts, Musical instrument, etc.	Kumar and Yadav, 2010
49.	*Basella alba* L. (Basellaceae)	Flowers and fruit	Scarlet red dye	Das and Mondal, 2012; Kumar, 2014
50.	*Bauhinia purpurea* L. (Fabacaeae)	Stem fiber	Cordage and thatching	Goel et al., 1984,
		Whole plant	Furl wood and Ornamental	Singh et al., 2013

TABLE 8.1   (Continued)

S. No.	Botanical Name and family	Part	Uses	References
		Bark	Red dye	Upadhyay and Chaudhary, 2014
		Leaves	Thatching	Rai and Nath, 2006
		Flower	Violet dye	Alawa et al., 2013
51.	*Bauhinia racemosa* Lamk. (Fabaceae)	Stem fiber	Basketry	Joshi, 1982
52.	*Bauhinia vahlii* Wight & Arn. (Fabaceae)	Leaf	Local cigarettes,	Jain, 1963
		Leaf	Basketry and umbrella	Mairh et al., 2010.
		Stem bark	Fiber	Jain and De, 1966; Ramsankar, 2001; Mairh et al., 2010
		Fiber	Magico-religious	Jain, 1963
		Stem	Fibers for mats and rope	Rai and Nath, 2006
53.	*Bauhinia variegata* L. (Fabaceae)	Whole plant	Ornamental and fuel wood	Singh et al., 2013
		Bark	Dye	Kala, 2009
54.	*Berberis aristata* DC. (Berberidaceae)	Roots and tubers	Yellow dye	Tiwari and Bharat, 2008
55.	*Beta vulgaris* L. (Amaranthaceae)	Root	Dye	Kapoor et al., 2008
56.	*Blepharispermum subsessile* DC. (Asteraceae)	Whole plant	Magico-religious	Jain, 1963
57.	*Campsis radicans* (L.) Seem. (Bignoniaceae)	Whole plant	Ornamental	Agarwal, 2013
58.	*Bixa orellana* L. (Bixaceae)	Seed	Red and pink dye	Rai and Nath, 2006; Tiwari and Bharat, 2008; Kapoor et al., 2008; Das and Mondal, 2012
59.	*Boerhavia diffusa* L. (Nyctaginaceae)	Root	Sacred thread	Jadhav, 2008
60.	*Bombax ceiba* L. (Bambaceae)	Seed and fiber	For pillow and blanket	Shukla and Chakaravarty, 2012
		Flower	Red dye	Alawa et al., 2013
		Wood	Agricultural implements	Rai and Nath, 2006

TABLE 8.1   (Continued)

S. No.	Botanical Name and family	Part	Uses	References
61.	*Borassus flabellifer* L. (Arecaceae)	Sap	Local drink	Goel et al., 1984,
		Fiber	Cordage	Jain and De, 1966
		Leaf	Miscellaneous article and thatching	Goel et al., 1984,
62.	*Boswellia serrata* Roxb. ex Colebr. (Burseraceae)	Resin	Incense	Joshi, 1982
		Wood pole	Socio-religious in marriage	Joshi, 1982; Sandhya and Ahirwar, 2015; Ahirwar, 2013
63.	*Bougainvillea glabra* Choisy (Nyctaginaceae)	Whole plant	Ornamental	Agarwal, 2013
64.	*Buchanania lanzan* Spreng. (Anacardiaceae)	Wood	Musical instrument	Jain, 1965
		Fruit	Taboo	Jain, 1963
		Bark	Red dye	Upadhyay and Chaudhary, 2014
65.	*Butea monosperma* (Lamk.) Taub. (Fabaceae)	Flower	Dye	Goel et al., 1984; Ghosh, 2004; Rai and Nath, 2006; Tiwari and Bharat, 2008; Das and Mondal, 2012; Upadhay and Choudhary, 2012; Alawa et al., 2013; Singh and Bharti, 2015
		Bark	Fiber	Joshi, 1982; Goel et al., 1984
		Leaf	Plates and Bowls	Jain and De, 1966; Rai and Nath, 2006; Mairh et al., 2010
		Leaf	Fodder	Mishra et al., 2010.
		Flower	Orange dye and religious	Kumar, 2014; Sandhya and Ahirwar, 2015
		Leaf	Thatching	Sandhya and Ahirwar, 2015
		Whole tree	Sacred	Ahirwar, 2013; Sahu et al., 2013; Singh and Bharti, 2015

TABLE 8.1 (Continued)

S. No.	Botanical Name and family	Part	Uses	References
66.	*Butea superba* Roxb. (Fabaceaae)	Root tubers	Red dye	Tiwari and Bharat, 2008
		Flower	Yellow dye	Upadhay and Choudhary, 2012
67.	*Brassica campestris* L. (Brassicaceae)	Seed	Socio-religious	Ahirwar, 2013
68.	*Cajanus cajan* (Fabaceae)	Twig	Fish trap	Prasad et al., 2013.
		Stem	Fuel wood and basketry	Singh et al., 2013
69.	*Calotropis gigantea* (L.) Dryand. (Asclepiadaceae)	Flower	Worshipped Lord Shiva	Agarwal, 2014
70.	*Calotropis procera* (Aiton) Dryand. (Asclepiadaceae)	Twig	Magico-religious	Joshi, 1982
		Flowers and fruits	Worshipped lord Shiva	Ahirwar, 2013; Sahu et al., 2013; Sandhya and Ahirwar, 2015
71.	*Camellia sinensis* (L.) Kuntze (Theaceae)	Leaves	Dye	Ghosh, 2004
72.	*Cannabis sativa* L. (Cannabaceae)	Leaves	Socio-religious	Ahirwar, 2013; Sahu et al., 2013; Pandey and Tiwari, 2014
73.	*Canscora diffusa* (Vahl) R.Br. ex Roem. & Schult. (Gentianaceae)	Whole plant	Socio-religious	Goel et al., 1984,
74.	*Capsicum annuum* L. (Solanaceae)	Fruit	Red and brown dye	Kapoor et al., 2008
		Fruit	Socio-religious	Ahirwar, 2013
75.	*Careya arborea* Roxb. (Lecythidaceae)	Root	Fish poison	Jain and De, 1966
		Root bark	Fish Poison	Heda and Kulkarni, 2009
76.	*Carica papaya* L. (Caricaceae)	Leaf and fruit	Dye	Bodane, 2015
77.	*Caryota urens* L. (Arecaceae)	Wood	Musical instrument	Jain, 1965

TABLE 8.1   (Continued)

S. No.	Botanical Name and family	Part	Uses	References
78.	*Casearia elliptica* Willd. (Salicaceae)	Fruit	Fish poison	Goel et al., 1984,
79.	*Casearia graveolens* Dalz. (Salicaceae)	Part not specified	Fish poison	Jain, 1963; Singh et al., 2011
		Fruit	Bead preparation	Singh et al., 2011
80.	*Cassia auriculata* L. (Caesalpiniaceae)	Bark	Red-orange dye	Alawa et al., 2013
81.	*Cassia fistula* L. (Caesalpiniaceae)	Bark	Tannin	Rai and Nath, 2006
		Wood	Agricultural implements	Rai and Nath, 2006
		Fruit	Sacred	Jadhav, 2008
82.	*Casuarina equisetifolia* L. (Casuarinaceae)	Bark	Light red	Upadhyay and Chaudhary, 2014
83.	*Ceriops decandra* (Griff.) W.Theob. (Rhizophoraceae)	Bark	Dye	Ghosh, 2004
84.	*Cicer arietinum* L. (Fabaceae)	Seed	Socio-religious	Pandey and Tiwari, 2014; Ahirwar, 2013
85.	*Cissampelos pareira* L. (Menispermaceae)	Stem	Fiber	Das et al., 1983
86.	*Citrus limon* (L.) Osbeck (Rutaceae)	Fruit	Offered goddess Durga	Ahirwar, 2013; Pandey and Tiwari, 2014
87.	*Chloroxylon swietenia* DC (Rutaceae)	Bark	Yellow dye	Upadhayay and Choudhary, 2014
		Leaf	Mosquito repellent and for fishing	Kala, 2009
88.	*Chukrasia tabularis* A. Juss. (Meliaceae)	Wood	Hut construction, furniture, fuel wood and agricultural equipments	Shukla and Chakaravarty, 2012
89.	*Cleianthus coccineus* Lour. ex B. A.Gomes (Verbenaceae)	Root	Arrow poison	Singh and Maheshwari, 1985

TABLE 8.1 (Continued)

S. No.	Botanical Name and family	Part	Uses	References
		Tender leaves	Fish poison	Singh et al., 2011; Heda and Kulkarni, 2009
		Fruit	Fish poison	Jain and De, 1966
		Fruit	Tanin	Jain and De, 1966
		Leaf	Insecticide	Lakra et al., 2010; Singh et al., 2011
90.	*Clerodendrum viscosum* Vent. (Verbenaceae)	Flower	Socio-religious	Jain, 1963
		Tender leaves	Bitter taste in local drink	Ghosh and Das, 2004
91.	*Clitoria ternatea* L. (Fabaceae)	Flowers	Blue dye	Das and Mondal, 2012; Kumar, 2014; Singh et al., 2013
		Whole plant	Green manure, ornamental	Singh et al., 2013
		Whole plant	Socio-religious	Ahirwar, 2013
92.	*Cochlospermum religiosum* (L.) Alston (Cochlospermaceae)	Flower	Religious	Jain, 1963
93.	*Coccinia grandis* (L.) Voigt (Cucurbitaceae)	Tuberous root	Sweeten the local drink	Ghosh and Das, 2004
94.	*Cocculus hirsutus* (L.) W.Theob. (Menispermaceae)	Fruit	Red dye	Alawa et al., 2013
95.	*Cocos nucifera* L. (Arecaceae)	Hard shell	For hookah hubble-bubble	Joshi, 1982
		Fruit	Socio-religious	Ahirwar, 2013; Agarwal, 2014
96.	*Coix aquatica* Roxb. (Poaceae)	Seeds	Chain preparation	Mishra et al., 2012
97.	*Cordia dichotoma* G. Forst. (Boraginaceae)	Bark	Red dye	Upadhay and Choudhary, 2012, 2014
98.	*Costus speciosus* (J.Koenig.) Sm. (Costaceae)	Rhizome	Fish poison	Heda and Kulkarni, 2009; Kala, 2009
		Rhizome	Socio-religious	Hemrom and Yadav, 2015

TABLE 8.1 (Continued)

S. No.	Botanical Name and family	Part	Uses	References
99.	*Crotalaria bilata* Schr. (Fabaceae)	Whole plant	Fodder	Goel et al., 1984,
100.	*Crotalaria pallida* Aiton (Fabaceae)	Stem	Fiber	Das et al., 1983
101.	*Crotalaria juncea* L. (Fabaceae)	Stem	Fiber	Chaudhury et al., 2000; Ramsankar, 2001; Prasad et al., 2013, Bhat and Saha, 2014
102.	*Cryptolepis buchanani* Roem & Schutt. (Asclepiadaceae)	Stem	For tying as a rope	Mairh et al., 2010.
103.	*Chrysopogon zizanioides* (L.) Roberty (Poaceae)	Whole plant	Basketry, artifacts	Tripathi et al., 2014
104.	*Cucurbita maxima* Duch. (Cucurbitaceae)	Fruit	Utensil	Goel et al., 1984
105.	*Cucurbita pepo* L. (Cucurbitaceae)	Fruit	Utensil	Goel et al., 1984; Agarwal, 2013
106.	*Cucurbita maxima* Duch. (Cucurbitaceae)	Fruit	Utensil, musical instruments	Agarwal, 2013
107.	*Curcuma angustifolia* Roxb. (Zingiberaceae)	Tuber	Yellow dye	Tiwari and Bharat, 2008
		Tuber	Starch extraction	Patel et al., 2015
		Tuber	Used in confectionary	Singh et al., 2011
108.	*Curcuma aromatica* Salisb. (Zingiberaceae)	Tuber	Yellow dye	Tiwari and Bharat, 2008
109.	*Curcuma longa* L. (Zingiberaceae)	Rhizome	Yellow Dye	Sachan and Kapoor, 2007; Tiwari and Bharat, 2008; Das and Mondal, 2012; Alawa et al., 2013; Kumar, 2014
		Rhizome	Socio-religious	Ramsankar, 2001; Ahirwar, 2013; Pandey and Tiwari, 2014; Agarwal, 2014
		Rhizome	Sacred	Sahu et al., 2013
110.	*Cuscuta reflexa* Roxb. (Cuscutaceae)	Whole plant	In Sorcery	Joshi, 1982

TABLE 8.1 (Continued)

S. No.	Botanical Name and family	Part	Uses	References
111.	*Cymbopogon martini* (Roxb.) Wats. (Poaceae)	Whole plant	Thatching roof	Mitra and Mukherjee, 2005
112.	*Cyperus alopecuroides* Rottb. (Cyperaceae)	Inflorescence stalk	Mat preparation	Mishra et al., 2012
		Leaves and inflorescence stalk	Thatching	Mishra et al., 2012
113.	*Cynodon dactylon* (L.) Pers. (Poaceae)	Whole plant	Religious	Ramsankar, 2001; Ahirwar, 2013; Sahu et al., 2013; Pandey and Tiwari, 2014; Agarwal, 2014
		Whole plant	Fodder	Singh and Bharti, 2015
		Whole plant	Paper making	Singh and Bharti, 2015
114.	*Cyperus corymbosus* Rottb. (Cyperaceae)	Flowering stalk	Mats	Mishra et al., 2012
115.	*Cyperus pangorei* Rottb. (Cyperaceae)	Flowering stalk	Mat, bag, baskets and household decorative materials	Mishra et al., 2012
116.	*Dalbergia latifolia* Roxb. (Fabaceae)	Wood	Wood for making idol	Jain, 1963
		Wood	Timber, fuel wood	Singh et al., 2013
		Wood	Agricultural implements	Kala, 2009
117.	*Dalbergia paniculata* Roxb. (Fabaceae)	Wood	Timber for making door	Kala, 2009
118.	*Dalbergia sissoo* DC. (Fabaceae)	Wood	Timber, agriculture implements and fuel wood	Singh et al., 2013; Singh and Bharti, 2015
119.	*Datura metel* L. (Solanaceae)	Fruit	Offer to Lord Shiva	Ahirwar, 2013; Pandey and Tiwari, 2014;
		Root	Sacred thread	Jadhav, 2008
120.	*Datura stramonium* L. (Solanaceae)	Flower and fruit	Offered to Lord Shiva	Agarwal, 2014; Singh and Bharti, 2015
121.	*Daucus carota* L. (Apiaceae)	Roots	Dye	Ghosh, 2004

TABLE 8.1   (Continued)

S. No.	Botanical Name and family	Part	Uses	References
122.	*Delonix regia* Bij. (Caesalpiniaceae)	Flower	Dye yielding	Singh and Bharti, 2015; Bodane, 2015
123.	*Dendrocalamus strictus* Nees (Poaceae)	Whole plant	Basketry, cordage, mats brooms	Goel et al., 1984; Rai and Nath, 2006; Kala, 2009; Kumar and Yadav, 2010
		Stem	Building material	Rai and Nath, 2006
		Culm	Socio-religious	Ramsankar, 2001
		Root	Sacred thread	Jadhav, 2008
124.	*Dendrocnide sinuata* (Blume) Chew. (Urticaceae)	Bark	Fiber	Das et al., 1983
125.	*Desmodium gangeticum* (L.) DC. (Fabaceae)	Whole plant	Green manure	Singh et al., 2013
126.	*Desmodium pulchellum* (L.) Benth. (Fabaceae)	Twig	Repel bed bug	Jain, 1963
127.	*Desmodium velutinum* (Willd.) DC. (Fabaceae)	Dry shoot	Broom	Jain, 1963
128.	*Dillenia indica* L. (Dilleniaceae)	Leaf	Plates making	Shukla and Chakaravarty, 2012
129.	*Dillenia pentagyna* Roxb. (Dillenaceae)	Fruit	Religious	Jain, 1963; Dubey et al., 2009
		Wood	Cot and bed preparation	Dubey et al., 2009
130.	*Dioscorea esculenta* (Lour.) Burl. (Dioscoreaceae)	Tuber	Religious ceremony	Goel et al., 1984,
131.	*Diospyros melanoxylon* Roxb. (Ebeanaceae)	Wood	Symbol for idol	Jain, 1963
		Bark	Pink dye & black	Upadhayay and Choudhary, 2014; Alawa et al., 2013
132.	*Diplocyclos palmatus* (L.) C. Jeffery (Cucurbitaceae)	Whole plant	Fodder	Goel et al., 1984,

TABLE 8.1 (Continued)

S. No.	Botanical Name and family	Part	Uses	References
133.	*Echinops echinatus* Roxb. (Asteraceae)	Root	Sacred thread	Jadhav, 2008
134.	*Echinochloa crusgalli* (L.) P. Beauv. (Poaceae)	Whole plant	Fodder	Mishra et al., 2012
135.	*Echinochloa stagnina* (Retz.) P. Beauv. (Poaceae)	Whole plant	Fodder	Mishra et al., 2012
136.	*Eclipta prostrata* (L.) L. (Asteraceae)	Leaves	Black dye	Alawa et al., 2013
137.	*Eichhornia crassipes* (Mart.) Solms. (Pontederiaceae)	Whole plant	Fodder; Green manure	Mishra et al., 2012,
138.	*Ehretia laevis* Roxb. (Ehretiaceae)	Stem	Thatching	Goel et al., 1984,
139.	*Elephantopus scaber* L. (Asteraceae)	Not specified	Fermentation	Maheshwari et al., 1990.
140.	*Eleocharis caduca* (Delile) Schult. (Cyperaceae)	Tuber	Fodder	Mishra et al., 2012
141.	*Elaeocarpus ganitrus* Roxb. ex G. Don (Elaeocarpaceae)	Fruit	Socio-religious	Agarwal, 2014
142.	*Enydra fluctuans* DC. (Asteraceae)	Leaves	Light green dye	Das and Mondal, 2012
143.	*Erythrina suberosa* Roxb. (Fabaceae)	Bark	Fiber	Singh et al., 2013
		Whole plant	Live fence, liquor, green manure, timber, ornamental	Singh et al., 2013
		Wood	Auspicious in weddings	Sandhya and Ahirwar, 2015
		Stem bark	Brown dye	Das and Mondal, 2012
144.	*Erythrina stricta* Roxb. (Fabaceae)	Whole plant	Live fence, green manure, ornamental	Singh et al., 2013
145.	*Erythrina indica* Lam. (Fabaceae)	Whole plant	Live fence, green manure, timber, ornamental	Singh et al., 2013

TABLE 8.1   (Continued)

S. No.	Botanical Name and family	Part	Uses	References
146.	*Eriophorum comosum* (Wall.) Nees (Cyperaceae)	Whole plant	Cordage	Jain, 1984
147.	*Eulaliopsis binata* (Retz.) C. E. Hubb. (Poaceae)	Whole plant	Cordage	Goel et al., 1984,
		Whole plant	Keeps evil away	Srivastava and Varma, 1981
148.	*Euphorbia elegans* Spreng. (Euphorbiaceae)	Whole plant steam bath	Keeps evil away	Jain, 1963
149.	*Euphorbia nivulia* Buch.-Ham. (Euphorbiaceae)	Latex	Fish poison	Srivastava and Varma, 1981
150.	*Evolvulus alsinoides* (L.) L. (Convolvulaceae)	Whole Plant	Religious ceremony	Goel et al., 1984,
151.	*Ficus benghalensis* L. (Moraceae)	Aerial root	Fragrance	Singh et al., 2013
		Bark	Dye; religious	Singh et al., 2013; Pandey and Tiwari, 2014; Agarwal, 2014
		Whole plant	Sacred and worshipped on Bara Barsat festival	Sandhya and Ahirwar, 2015; Sahu et al., 2013; Ahirwar, 2013
152.	*Ficus carica* L. (Moraceae)	Bark	Light pink dye	Choudhary et al., 2012
153	*Ficus cupulata* Haines (Moraceae)	Bark	Light pink dye	Choudhary et al., 2012
154.	*Ficus hispida* L.f. (Moraceae)	Wood	Fuel wood	Singh et al., 2013
155.	*Ficus neriifolia* Smith (Moraceae)	Leaf and branches	Fodder & fuel wood	Shukla and Chakaravarty, 2012
156.	*Ficus racemosa* L. (Moraceae)	Bark	Red dye	Upadhayay and Choudhary, 2014; Choudhary et al., 2012
		Whole plant	Religious	Singh et al., 2013
157.	*Ficus religiosa* L. (Moraceae)	Bark	Fiber	Jain and De, 1966

TABLE 8.1 (Continued)

S. No.	Botanical Name and family	Part	Uses	References
		Leaf	Fodder	Mishra et al., 2010; Shukla and Chakaravarty, 2012; Singh and Bharti, 2015
		Whole plant	Sacred	Shukla and Chakaravarty, 2012; Singh et al., 2013; Ahirwar, 2013; Sandhya and Ahirwar, 2015; Pandey and Tiwari, 2014; Agarwal, 2014; Sahu et al., 2013; Singh and Bharti, 2015
158.	*Ficus retusa* L. (Moraceae)	Bark	Light pink	Upadhayay and Choudhary, 2014; Choudhary et al., 2012
159.	*Ficus virens* Aiton (Moraceae)	Whole plant	Socio-religious	Ahirwar, 2013
160.	*Gardenia latifolia* Aiton (Rubiaceae)	Fruit	Perfume	Kala, 2009
161.	*Gardenia turgida* Roxb. (Rubiaceae)	Fruit	Detergent	Srivastava and Varma, 1981
162.	*Gliricidia sepium* (Jacq.) Walp. (Fabaceae)	Wood	Timber, fuel wood	Singh et al., 2013
163.	*Gmelina arborea* Roxb. (Verbenaceae)	Wood	Comb and House building	Goel et al., 1984, Singh and Bharti, 2015
		Wood	Drum	Jain, 1965; Ramsankar, 2001
164.	*Gloriosa superba* L. (Liliaceae)	Root	Socio-religious	Ahirwar, 2013
165.	*Habenaria marginata* Colebr. (Orchidaceae)	Tuber	Socio-religious	Hemrom and Yadav, 2015
166.	*Haldina cordifolia* (Roxb.) Rids. (Rubiaceae)	Wood	Combs, tools, auspicious	Jain and De, 1966; Sandhya and Ahirwar, 2015
167.	*Hardwickia binata* Roxb. (Caesalpiniaceae)	Wood, bark	Fuel wood, timber, farm implements, fiber	Singh et al., 2013
168.	*Helicteres isora* L. (Sterculiaceae)	Bark	Fiber	Jain and De, 1966, Jain, 1964; Kala, 2009

TABLE 8.1 (Continued)

S. No.	Botanical Name and family	Part	Uses	References
		Bark and leaf	Fans and Miscellaneous article	Singh et al., 1985.
169.	*Hemidesmus indicus* (L.) Br. ex Schult. (Apocynaceae)	Root	Keep away evil spirit	Jain, 1963
170.	*Hibiscus sabdariffa* L. (Malvaceae)	Stem	Fiber	Maheshwari et al., 1990.
171.	*Hibiscus rosa-sinensis* L. (Malavaceae)	Flower	Worship	Singh and Bharti, 2015
172.	*Holarrhena antidysenterica* (Roth) Wall. ex A. DC. (Apocynaceae)	Wood	Combs and miscellaneous article	Jain, 1963
		Flower	Ornaments	Srivastava and Varma, 1981
		Sticks	Drum beating (Religious)	Ramsankar, 2001
173.	*Holoptelea intergrifolia* Planch. (Ulmaceae)	Leaf and bark	Fish poison	Sandhya and Ahirwar, 2015
174.	*Hordeum vulgare* L. (Poaceae)	Whole plant	Used in Hawan	Pandey and Tiwari, 2014; Ahirwar, 2013
		Whole plant	Pisciculture	Mishra et al., 2012
		Whole plant	Storing, packing and manure	Mishra et al., 2012
175.	*Hygroryza aristata* (Retz.) Nees ex Wight & Arn. (Poaceae)	Whole plant	Fodder	Mitra and Mukherjee, 2005
176.	*Hyptis suaveolens* (L.) Poit. (Lamiaceae)	Twig	Repel bed bug	Chaudhuri et al., 1975
177.	*Ichnocarpus frutescens* (L.) W. T. Aiton (Apocynaceae)	Stem fiber	Cordage, Basketry	Singh and Maheshwari, 1985; Goel et al., 1984,
178.	*Indigofera cassioides* DC. (Fabaceae)	Leaves and flower	Blue dye	Tiwari and Bharat, 2008; Alawa et al., 2013
179.	*Indigofera tinctoria* L. (fabaceae)	Leaves and flower	Blue dye	Tiwari and Bharat, 2008

TABLE 8.1   (Continued)

S. No.	Botanical Name and family	Part	Uses	References
180.	*Ipomoea batatas* (L.) Lam. (Convolvulaceae)	Roots	Sacred	Ahirwar, 2013
181.	*Ipomoea carnea* Jacq. (Convolvulaceae)	Stem	Fencing	Mishra et al., 2012
182.	*Ipomoea palmata* Forssk. (Convolvulaceae)	Whole plant	Ornamental	Agarwal, 2013
183.	*Isachne globosa* (Thunb.) Kuntze (Poaceae)	Whole plant	Green manure	Mishra et al., 2012
184.	*Ixora pavetta* Andr. (Rubiaceae)	Branches	Basketry	Singh and Maheshwari, 1985
185.	*Jasminum auriculatum* Vahl. (Oleaceae)	Whole plant	Ornamental	Agarwal, 2013
186.	*Jasminum grandiflorum* L. (Oleaceae)	Flower	Perfumery	Singh and Bharti, 2015
187.	*Jasminum multiflorum* (Burm.f) Andr. (Oleaceae)	Whole plant	Ornamental	Agarwal, 2013
188.	*Jasminum officinale* L. (Oleaceae)	Whole plant	Ornamental	Agarwal, 2013
189.	*Jatropha curcas* L. (Euphorbiaceae)	Leaves	Dye yielding	Srivastava et al., 2008
190.	*Justicia betonica* L. (Acanthaceae)	Whole plant	Keep evil spirit away	Srivastava and Varma, 1981
191.	*Lablab purpureus* (L.) Sw. (Fabaceae)	Leaf	Green dye for tattoo	Jain, 1963; Kumar, 2014
		Whole plant	Live fence, ornamental	Singh et al., 2013
192.	*Lagenaria siceraria* (Molina) Standl. (Cucurbitaceae)	Fruit	Bottles, bowls, utensils, container s	Jain, 1964
193.	*Lagerstroemia parviflora* Roxb. (Lythraceae)	Wood	Musical instruments	Jain, 1965
		Bark	Tanning	Singh et al., 2011
		Bark	Dye	Goel et al., 1984,
		Stem	Thatching	Goel et al., 1984,

TABLE 8.1 (Continued)

S. No.	Botanical Name and family	Part	Uses	References
		Wood	Hut construction, furniture, fuel wood, agricultural equipments	Shukla and Chakaravarty, 2012
194.	*Lagerstroemia speciosa* (L.) Pers. (Lythraceae)	Wood	Hut construction, furniture, fuel wood, agricultural equipments	Shukla and Chakaravarty, 2012
195.	*Lantana camara* L. (Verbenaceae)	Whole plant	Live fencing	Shukla and Chakaravarty, 2012
196.	*Lannea coromandelica* (Houtt.) Merr. (Anacardiaceae)	Wood	Combs, tobacco pouch	Jain, 1964
		Fruit	Fish poison	Heda and Kulkarni, 2009
		Gum	Textile industries	Singh et al., 2011
		Bark	Light red	Upadhayay and Choudhary, 2014.
		Stem	Socio-religious	Ahirwar, 2013
197.	*Lathyrus odoratus* L. (Fabaceae)	Whole plant	Ornamental	Agarwal, 2013
198.	*Lathyrus sativus* L. (Fabaceae)	Whole plant	Ornamental	Agarwal, 2013
199.	*Launaea procumbens* (Roxb.) Ram. & Raj. (Asteraceae)	Root	Keep away evil spirit	Jain, 1971
200.	*Perilla ocymodies* L. (Lamiaceae)	Twig	Keep evil spirit away	Mairh et al., 2010.
		Whole plant	Fencing	Mairh et al., 2010.
201.	*Lawsonia inermis* L. (Lythraceae)	Leaf	Green/red dye	Tiwari and Bharat, 2008; Das and Mondal, 2012; Alawa et al., 2013; Kumar, 2014; Parul and Vashishta, 2015; Sandhya and Ahirwar, 2015; Singh and Bharti, 2015; Bodane, 2015

TABLE 8.1 (Continued)

S. No.	Botanical Name and family	Part	Uses	References
		Leaf	Socio-religious	Ahirwar, 2013
202.	*Leucas zeylanica* (L.) W. T. Aiton (Lamiaceae)	Seed	Oil yielding	Saxena, 1986.
203.	*Leucas plukenetii* (Roth.) Spr. (Lamiaceae)	Flower	Socio-religious	Jain, 1971
204.	*Machilus villosa* (Roxb.) Hook.f. (Lauraceae)	Leaves	Fragrance	Shukla and Chakravarty, 2012
205.	*Madhuca longifolia* (J. Koen. ex L.) J. F.Macbr. (Sapotaceae)	Seed	Oil yielding	Jain, 1964; Goel et al., 1984; Sikarwar, 2002; Singh et al., 2011
		Wood	Timber yielding	Jain, 1964; Sikarwar, 2002; Singh and Bharti, 2015
		Leaf	Fodder	Sikarwar, 2002
		Flower	Country liquor	Sikarwar, 2002; Rai and Nath, 2006; Kumar and Rao, 2007; Singh et al., 2011, 2013; Kala, 2009; Singh and Bharti, 2015
		Seed cake	Fish poison	Heda and Kulkarni, 2009
		Whole plant	Religious	Heda, 2012; Ahirwar, 2013; Pandey and Tiwari, 2014; Hemrom and Yadav, 2015; Singh and Bharti, 2015; Sandhya and Ahirwar, 2015
		Flower	Red dye	Alawa et al., 2013
		Seed oil	Socio-religious	Ramsankar, 2001
206.	*Mallotus philippensis* (Lam.) Mull. Arg. (Euphorbiaceae)	Fruit	Dye yielding	Jain, 1964; Tiwari and Bharat, 2008; Alawa et al., 2013
		Seed	Varnish (Oil)	Jain, 1964
207.	*Mallotus roxburghianus* Mull. Arg. (Euphorbiaceae)	Flowers and fruits	Dye yielding	Mairh et al., 2010.

TABLE 8.1   (Continued)

S. No.	Botanical Name and family	Part	Uses	References
		Leaves	Thatching roof	Mairh et al., 2010.
208.	*Malvastrum coromandelianum* (L.) Garcke (Malvaceae)	Twig	Broom	Joshi, 1982
209.	*Mangifera indica* L. (Anacardiaceae)	Wood	Drum preparation	Jain, 1965; Goel et al., 1984; Ramsankar, 2001
		Wood	Timber	Singh and Bharti, 2015
		Fruit and Leaves	Religious ceremony	Singh et al., 2011; Shukla and Chakaravarty, 2012; Ahirwar, 2013; Sahu et al., 2013; Pandey and Tiwari, 2014; Agarwal, 2014; Sandhya and Ahirwar, 2015;
		Bark	Light yellow dye	Upadhayay and Choudhary, 2014; Bodane, 2015
210.	*Manilkara hexandra* (Roxb.) Dubard (Sapotaceae)	Bark	Pink dye	Upadhayay and Choudhary, 2014.
211.	*Marsdenia tenacissima* (Roxb.) Moon. (Asclepiadaceae)	Stem	Fiber	Goel et al., 1984; Singh and Maheshwari, 1985
212.	*Melastoma malabathricum* L. (Melastomaceae)	Flower and Leaves	Decorative; fodder	Shukla and Chakaravarty, 2012
213.	*Melilotus indica* (L.) All. (Fabaceae)	Aerial part	Flavoring agent	Singh et al., 2013
214.	*Merremia hederacea* (Burm.f.) Hallier f. (Convolvulaceae)	Whole plant	Fodder	Goel et al., 1984
215.	*Michelia champaca* L. (Magnoliaceae)	Wood	Yellow dye	Tiwari and Bharat, 2008
		Branches	Fuel wood	Shukla and Chakaravarty, 2012
216.	*Millettia extensa* (Benth.) Baker. (Fabaceae)	Whole plant	Fodder	Das et al., 1983; Goel et al., 1984

TABLE 8.1 (Continued)

S. No.	Botanical Name and family	Part	Uses	References
217.	*Mimosa pudica* L. (Mimosaceae)	Whole plant	Ornamental	Singh et al., 2013
218.	*Mimusops elengi* L. (Sapotaceae)	Seed	Yellow dye	Tiwari and Bharat, 2008
219.	*Mitragyna parvifolia* (Roxb.) Korth. (Rubiaceae)	Whole plants	Socio-religious	Ahirwar, 2013
220.	*Morus alba* L. (Moraceae)	Wood	Fuel wood, farm implements	Singh et al., 2013
		Branches	Baskets	Yadav and Bhandoria, 2013
221.	*Morinda citrifolia* L. (Rubiaceae)	Bark	Red dye	Upadhayay and Choudhary, 2014.
222.	*Morinda pubescens* Sm. (Rubiaceae)	Root	Yellow dye	Goel et al., 1984
223.	*Moringa oleifera* Lam. (Moringaceae)	Leaves	Fodder	Goel et al., 1984
		Leaves, stem	Paper making	Singh et al., 2013
224.	*Mucuna pruriens* (L.) DC. (Fabaceae)	Whole plant	Ornamental, live fence, green manure	Singh et al., 2013
225.	*Musa paradisiaca* L. (Musaceae)	Leaves, fruit	Religious	Ahirwar, 2013; Pandey and Tiwari, 2014
226.	*Murraya koenigii* (L.) Spreng. (Rutaceae)	Bark	Blue dye	Upadhayay and Choudhary, 2014.
227.	*Mussaenda roxburghii* Hook.f. (Rubiaceae)	Root	Yellow color in local drink	Ghosh and Das, 2004
228.	*Myrica esculenta* Buch.-Ham. x D. Don (Myricaceae)	Bark	Red dye	Tiwari and Bharat, 2008
229.	*Naringi crenulata* (Roxb.) Nicol. (Rutaceae)	Leaf	Keep away evil spirit	Jain, 1971
230.	*Nelumbo nucifera* Gaertn. (Nelumbonaceae)	Leaf	Plates and packing material	Mishra et al., 2012
		Flower	Religious	Mishra et al., 2012; Ahirwar, 2013; Agarwal, 2014

TABLE 8.1   (Continued)

S. No.	Botanical Name and family	Part	Uses	References
		Seed	Rosaries, sacred	Mishra et al., 2012
231.	*Nerium indicum* Mill. (Apocynaceae)	Flower	Pink dye	Bodane, 2015
232.	*Nyctanthes arbor-tritis* L. (Oleaceae)	Twig	Basketry	Singh and Maheshwari, 1985
		Flower	Orange dye	Das and Mondal, 2012
		Flower	Worship	Ahirwar, 2013; Singh and Bharti, 2015
233.	*Nymphaea nouchali* Burm.f. (Nymphaeaceae)	Flower	Sacred	Jain and De, 1966; Mishra et al., 2012
234.	*Ocimum tenuiflorum* L. (Lamiaceae)	Whole plant	Sacred	Ramsankar, 2001; Ahirwar, 2013; Yadav and Bhandoria, 2013; Sahu et al., 2013; Pandey and Tiwari, 2014; Agarwal, 2014; Sandhya and Ahirwar, 2015
235.	*Ocimum americanum* L. (Lamiaceae)	Whole plant	Sacred	Jain, 1971; Ahirwar, 2013
236.	*Ougeinia oogeinensis* (Roxb.) Hochr. (Fabaceae)	Gum	Fish poison	Singh et al., 2011
		Stem	Furniture	Kala, 2009
		Bark	Fish poison	Kala, 2009
237.	*Oroxylum indicum* (L.) Kurz (Bignoniaceae)	Bark	Bitter taste in local drink	Ghosh and Das, 2004
238.	*Oryza sativa* L. (Poaceae)	Whole plant	Socio-religious	Ramsankar, 2001; Ahirwar, 2013; Jain, 1963; Heda, 2012; Agarwal, 2014
		Grain	Local drink	Ghosh and Das, 2004; Kumar and Rao, 2007.
		Residual material	Fodder	Mishra et al., 2012
239.	*Ostodes paniculata* Blume (Euphorbiaceae)	Leaf	Fodder	Das et al., 1983

TABLE 8.1   (Continued)

S. No.	Botanical Name and family	Part	Uses	References
		Bark	Gum	Das et al., 1983
240.	*Pandanus odoratissimus* L.f. (Pandanaceae)	Flower and root	Perfumes, bouquets, etc.	Rahangdale et al., 2014
241.	*Panicum paludosum* Roxb. (Poaceae)	Whole plant	Fodder	Mishra et al., 2012
242.	*Paspalum scrobiculatum* Linn. (Poaceae)	Grain	Local drink	Mitra and Mukherjee, 2005
243.	*Paspalidium flavidum* (Retz.) A. Camus (Poaceae)	Whole plant	Fodder	Mishra et al., 2012
244.	*Pavetta crassicaulis* Bremek. (Rubiaceae)	Flower	Ear ring	Jain and De, 1966
245.	*Pennisetum glaucum* (L.) R.Br. (Poaceae)	Whole plant	Fodder	Mishra et al., 2010.
246.	*Peristrophe tinctoria* (Roxb.) Nees. (Acanthaceae)	Whole plant	Red dye	Das and Mondal, 2012
247.	*Phlogacanthus thyrsiformis* (Roxb. ex Hardw.) Mabb. (Acanthaceae)	Flower	Fodder	Das et al., 1983
248.	*Pithecellobium dulce* (Roxb.) Benth. (Mimosaceae)	Bark	Light pink dye	Upadhayay and Choudhary, 2014.
249.	*Phoenix acaulis* Roxb. (Arecaceae)	Leaf	Basketry and for thatching	Maheshwari et al., 1990.
		Leaf	Broom	Singh and Maheshwari, 1985; Rai and Nath, 2006
		Fruit	Country liquor	Rai and Nath, 2006
250.	*Phoenix sylvestris* (L.) Roxb. (Arecaceae)	Leaf	Basketry	Joshi, 1982; Rai and Nath, 2006
		Leaf	Headgear and broom	Sandhya and Ahirwar, 2015
		Leaf	Socio-religious	Ahirwar, 2013
251.	*Phyllanthus emblica* L. (Euphorbiaceae)	Fruit	Black dye; Brown dye	Tiwari and Bharat, 2008; Alawa et al., 2013; Kumar, 2014

TABLE 8.1 (Continued)

S. No.	Botanical Name and family	Part	Uses	References
		Fruit and twig	Tanin	Rai and Nath, 2006
		Whole plant	Sacred	Ahirwar, 2013; Sahu et al., 2013; Pandey and Tiwari, 2014; Agarwal, 2014; Sandhya and Ahirwar, 2015
252.	*Phragmites karka* (Retz.) Trin. ex Steud. (Poaceae)	Culm	Mats and Baskets preparation	Mishra et al., 2012
253.	*Piper betle* L. (Piperaceae)	Leaf	Socio-religious	Agarwal, 2014; Ahirwar, 2013
254.	*Piliostigma malabaricum* (Roxb.) Benth. (Caesalpiniaceae)	Bark	Fermentation	Jain, 1963
255.	*Plumbago zeylanica* L. (Plumbaginaceae)	Leafy branches	Used as enhancer in local drink	Ghosh and Das, 2004
256.	*Piper longum* L. (Piperaceae)	Root	Local drink	Goel et al., 1984
257.	*Pogostemon benghlensis* (Burm.f.) Kutze (Lamiaceae)	Root	Local drink	Goel et al., 1984
258.	*Polyalthia longifolia* (Sonn.) Thwaites (Annonaceae)	Whole plant	Socio-religious	Ahirwar, 2013
259.	*Polygonum glabrum* Willd. (Polygonaceae)	Leaf	Fish poison	Goel et al., 1984
260.	*Polygonum hydropiper* L. (Polygonaceae)	Leaf	Fish poison	Goel et al., 1984
261.	*Pothos scandens* L. (Araceae)	Whole plant	Ornamental	Agarwal, 2013
262.	*Prosopis cineraria* (L.) Druce (Mimosaceae)	Leaf	Fodder	Yadav and Bhandoria, 2013
		Whole plant	Socio-religious	Agarwal, 2014
263.	*Psoralea corylifolia* L. (Fabaceae)	Whole plant	Green manure	Singh et al., 2013
264.	*Pseudoraphis spinescens* (R.Br.) Vickery (Poaceae)	Whole plant	Fodder	Mishra et al., 2012

TABLE 8.1 (Continued)

S. No.	Botanical Name and family	Part	Uses	References
265.	*Psidium guajava* L. (Myrtaceae)	Bark	Brown dye	Bodane, 2015
266.	*Pterocarpus marsupium* Roxb. (Fabaceae)	Wood	Drum preparation	Jain, 1965
		Stem bark	Fish poison	Heda and Kulkarni, 2009
		Bark	Red dye	Tiwari and Bharat, 2008; Upadhayay and Choudhary, 2014.
		Wood	Door preparation	Kala, 2009
267.	*Punica granatum* L. (Punicaceae)	Fruit rind	Yellow Dye	Ghosh, 2004; Tiwari and Bharat, 2008; Upadhay and Choudhary, 2012
		Flower	Red dye	Alawa et al., 2013
268.	*Putranjiva roxburghii* Wall. (Euphorbiaceae)	Stem	Socio-religious	Ahirwar, 2013
269.	*Quisqualis indica* L. (Combretaceae)	Whole plant	Ornamental	Agarwal, 2013
270.	*Randia dumetorum* (Retz.) Lam. (Rubiaceae)	Bark	Pink dye	Upadhayay and Choudhary, 2014.
		Fruit	Fish poison	Kala, 2009
271.	*Rauvolfia serpentina* (L.) Benth. ex Kurz (Apocynaceae)	Root bark	Develop bitter taste in local drink	Ghosh and Das, 2004
272.	*Rhizophora apiculata* Blume (Rhizophoraceae)	Bark	Dye	Ghosh, 2004
273.	*Rhynchosia minima* (L.) DC. (Fabaceae)	Leaves	Green manure	Singh et al., 2013
274.	*Ricinus communis* L. (Euphorbiaceae)	Seed oil	Lubricant	Jain, 1964
		Twig	Ward off evil	Jain, 1963
275.	*Rubia cordifolia* L. (Rubiaceae)	Whole plant	Red dye	Tiwari and Bharat, 2008; Kapoor et al., 2008

TABLE 8.1 (Continued)

S. No.	Botanical Name and family	Part	Uses	References
276.	*Rumex dentatus* L. (Polygonaceae)	Leaf	Fodder	Parul and Vashishta, 2015
277.	*Saccharum munja* Roxb. (Poaceae)	Stem	Fish trap	Prasad et al., 2013.
278.	*Saccharum officinarum* L. (Poaceae)	Stem	Religious	Ahirwar, 2013; Pandey and Tiwari, 2014
279.	*Saccharum spontaneum* L. (Poaceae)	Young shoot	Fodder	Mishra et al., 2012
		Whole plant	Thatching	Mishra et al., 2012
		Flowering culms	Making ropes	Mishra et al., 2012
		Whole plant	Socio-religious	Ahirwar, 2013
280.	*Sacciolepis interrupta* (Willd.) Stapf (Poaceae)	Whole plant	Fodder	Mishra et al., 2012
281.	*Santalum album* L. (Santalaceae)	Whole plant	Socio-religious	Ahirwar, 2013; Agarwal, 2014
282.	*Saraca indica* L. (Caesalpiniaceae)	Whole plat	Socio-religious	Agarwal, 2014
283.	*Schleichera oleosa* (Lour.) Merr. (Sapindaceae)	Seed	Oil	Jain and De, 1966; Pal and Srivastava, 1976
284.	*Scirpus grossus* L.f. (Cyperaceae)	Inflorescence stalk	Mats	Mishra et al., 2012
285.	*Scoparia dulcis* L. (Plantaginaceae)	Leafy twig	Sweetness in local drink	Ghosh and Das, 2004
		Seed oil	Luminant	Jain, 1964
286.	*Sesbania javanica* Miq. (Fabaceae)	Whole plant	Green manure	Mishra et al., 2012
287.	*Sesbania sesban* (L.) Merr. (Fabaceae)	Whole plant	Green manure	Singh et al., 2013
288.	*Semecarpus anacardium* L.f. (Anacardiaceae)	Seed oil	Tattooing	Jain, 1963
		Fruit	Black dye	Tiwari and Bharat, 2008
		Leaf	Keep evil spirit away	Hemrom and Yadav, 2015

TABLE 8.1 (Continued)

S. No.	Botanical Name and family	Part	Uses	References
289.	*Sesamum indicum* L. (Pedaliaceae)	Seed	Oil	Maji and Sikdar, 1982; Jain and De, 1966
		Seed	Social and religious occasion in Havan samgri	Ahirwar, 2013; Pandey and Tiwari, 2014
290.	*Setaria glauca* (L.) P.Beauv. (Poaceae)	Achenes	Country liquor	Mitra and Mukherjee, 2005
		Grain	Fodder	Mitra and Mukherjee, 2005
291.	*Setaria verticillata* (L.) P. Beauv. (Poaceae)	Inflorescence	Expel rodents	Mitra and Mukherjee, 2005
292.	*Shorea robusta* Gaertn. (Dipterocarpaceae)	Seed	Oil	Maji and Sikdar, 1982
		Leaf and stem	House building	Goel et al., 1984; Rai and Nath, 2006; Kala, 2009
		Wood	Incense	Jain and De, 1966
		Resin	Incense	Gupta, 1981; Singh et al., 2011
		Wood	Musical Instrument	Goel et al., 1984
		Leaf	Plates	Goel et al., 1984
		Twig	Sacred	Jain, 1963; Ramsankar, 2001; Sahu, 2008
		Stem and wood	Hut and furniture making	Shukla and Chakaravarty, 2012; Singh and Bharti, 2015
		Bark	Gum used in ceramics	Rai and Nath, 2006
		Leaves	Thatching	Rai and Nath, 2006
293.	*Sida acuta* Burm.f. (Malvaceae)	Stem	Fiber for mat weaving	Rai and Nath, 2006
294.	*Sida cordata* (Burm.f.) Borss. Waalk. (Malvaceae)	Stem	Fiber	Goel et al., 1984
295.	*Sida rhombifolia* L. (Malvaceae)	Fiber	Cordage	Goel et al., 1984

TABLE 8.1   (Continued)

S. No.	Botanical Name and family	Part	Uses	References
296.	*Smilax ovalifolia* Roxb. ex D. Don (Smilaceae)	Stem	Fiber	Goel et al., 1984
297.	*Smilax wightii* A.DC. (Smilaceae)	Root	Stops bad dreams	Jain, 1971
298.	*Sorghum halepens* (L.) Pers. (Poaceae)	Grain	Local drink	Chaudhuri and Pal, 1978
299.	*Sporobolus diandrus* (Retz.) P. Beauv. (Poaceae)	Stem	Chopped (Fooder)	Mitra and Mukherjee, 2005
300.	*Stereospermum chelonoides* (L.f.) DC. (Bignoniaceae)	Wood	Magico-religious	Jain, 1963
301.	*Stephania japonica* (Thunb.) Miers (Menispermaceae)	Roots	Local drink preservative	Ghosh and Das, 2004
302.	*Stephania glabra* (Roxb.) Miers (Menispermaceae)	Roots	Local drink preservative	Ghosh and Das, 2004
303.	*Sterculia villosa* Roxb. (Steculiaceae)	Leaf	Plates making	Shukla and Chakaravarty, 2012
304.	*Strychnos nux-vomica* L. (Loganiaceae)	Fruit	Fish poison	Goel et al., 1984
305.	*Swietenia mahogani* (L.) Jacq. (Meliaceae)	Fruit	Dye	Ghosh, 2004
306.	*Symplocos racemosa* Roxb. (Symplocaceae)	Bark	Dye	Kala, 2009
307.	*Syzygium cumini* (L.) Skeels (Myrtaceae)	Leaf	Thatching	Joshi, 1982.
		Bark and leaves	Red dye	Tiwari and Bharat, 2008; Upadhayay and Choudhary, 2014.
		Seed	Indigo dye	Alawa et al., 2013
		Wood	Furniture	Singh and Bharti, 2015
308.	*Syzygium heyneanum* (Duthie) Gamble (Myrtaceae)	Bark	Dark blue dye	Upadhayay and Choudhary, 2014.
309.	*Tagetes erecta* L. (Asteraceae)	Flower	Yellow Dye	Ghosh, 2004; Das and Mondal, 2012

TABLE 8.1 (Continued)

S. No.	Botanical Name and family	Part	Uses	References
310.	*Tamarindus indica* L. (Fabaceae)	Wood	Fuel wood	Singh et al., 2013
		Wood	Furniture	Singh and Bharti, 2015
311.	*Tectona grandis* L.f. (Verbenaceae)	Wood	Musical instrument	Goel et al., 1984
		Leaf	Umbrella	Goel et al., 1984
		Leaf Juice	Red dye (Paint furniture)	Jain and De, 1966; Das and Mondal, 2012; Singh and Bharti, 2015
		Wood	Agricultural implements	Shukla and Chakaravarty, 2012
312.	*Tephrosia purpurea* (L.) Pers. (Fabaceae)	Whole plant	Green manure	Singh et al., 2013
313.	*Tephrosia villosa* (L.) Pers. (Fabaceae)	Whole plant	Green manure	Singh et al., 2013
314.	*Terminalia alata* Wall. (Combretaceae)	Leaf	Fodder	Maheshwari et al., 1990; Jain et al., 2010
		Wood	Musical instrument	Jain, 1965
		Wood	Sacred	Jain, 1963
		Bark	Red and brown dye	Tiwari and Bharat, 2008
315.	*Terminalia arjuna* (Roxb. ex DC.) Wight & Arn. (Combretaceae)	Whole plant	Resin	Goel et al., 1984
		Bark	Red and black dye	Tiwari and Bharat, 2008; Kumar, 2014; Upadhayay and Choudhary, 2014
		Leaf	Fodder	Mishra et al., 2010
		Branches	Fuel wood	Shukla and Chakaravarty, 2012
		Whole plant	Sacred	Sandhya and Ahirwar, 2015; Jha and Gurudatta, 2015
		Gum	Book binding	Jha and Gurudatta, 2015
316.	*Terminalia bellirica* (Gaertn.) Roxb. (Combreataceae)	Fruit	Black dye	Tiwari and Bharat, 2008; Kumar, 2014; Jha and Gurudatta, 2015

TABLE 8.1   (Continued)

S. No.	Botanical Name and family	Part	Uses	References
		Bark	Brown dye	Upadhayay and Choudhary, 2014.
		Fruit	Tanin	Rai and Nath, 2006
317.	*Terminalia chebula* Retz. (Combretaceae)	Whole plant	Tannin	Gupta, 1981
		Fruit	Black dye	Tiwari and Bharat, 2008; Kumar, 2014
		Fruits	Tanin	Rai and Nath, 2006
318.	*Thespesia lampas* (Cav.) Dalzell (Malvaceae)	Stem fiber	Cordage	Goel et al., 1984; Kala, 2009
319.	*Thespesia populnea* (L.) Sol. ex.Correa (Malvaceae)	Tem bark	Fiber yielding	Singh and Bharti, 2015
320.	*Thevetia peruviana* (Pers.) K. Schum. (Apocyanaceae)	Flower	Pink dye	Bodane, 2015
321.	*Thysanolaena maxima* (Roxb.) Kuntze (Poaceae)	Culms	Bows	Maheshwari et al., 1990
		Inflorescence	Brooms	Jain and De, 1966; Gupta, 1981
322.	*Tinospora cordifolia* (Willd.) Miers (Menispermaceae)	Leaf	Fodder	Goel et al., 1984
323.	*Toona ciliata* M. Roem. (Meliaceae)	Wood	Hut construction, furniture, fuel wood, agricultural equipments	Shukla and Chakravarty, 2012
324.	*Trapa bispinosa* Roxb. (Lythraceae)	Fruit powder	Socio-religious	Ahirwar, 2013
325.	*Trifolium alexandrium* L. (Poaceae)	Whole plant	Fodder	Mishra et al., 2010.
326.	*Triticum aestivum* L. (Poaceae)	Grain	Socio-religious	Ahirwar, 2013
327.	*Typha angustata* Bory & Chaub. (Typhaceae)	Whole plant	Thatch, brooms and mats	Maheshwari et al., 1990; Mishra et al., 2012

TABLE 8.1 (Continued)

S. No.	Botanical Name and family	Part	Uses	References
		Inflorescence	Decorative	Mishra et al., 2012
328.	Urena lobata L. (Malvaceae)	Stem	Fiber	Tarafder, 1986
329.	Vallaris solanacea (Roth) Kuntze (Apocynaceae)	Fiber	Cordage	Goel et al., 1984
330.	Vanda tessellata (Roxb.) Hook. ex G. Don (Orchidaceae)	Leaf	Anklets	Jain, 1987
331.	Ventilago denticulata Willd. (Rhamnaceae)	Fiber	Cordage	Goel et al., 1984
		Seed	Oil	Jain, 1964
		Bark and root	Violets dye	Tiwari and Bharat, 2008
		Bark	Red dye	Upadhayay and Choudhary, 2014
332.	Ventilago maderaspatana Gaertn. (Rhamnaceae)	Root bark	Chocolate and red dye	Kala, 2009
333.	Vernonia cinerea (L.) Less. (Asteraceae)	Whole plant	Sweetens the local drink	Ghosh and Das, 2004
334.	Vigna aconitifolia (Jacq.) Marechal (Fabaceae)	Root	Soil improver	Singh et al., 2013
335.	Vigna trilobata (L.) Verd. (Fabaceae)	Whole plant	Green manure	Singh et al., 2013
336.	Vicia hirsuta (L.) Gray (Fabaceae)	Leaves	Fodder	Jain et al., 2010
337.	Vitex negundo L. (Verbinaceae)	Twig	Basketry	Singh and Maheshwari, 1985
		Twig	Insect repellant	Jain and De, 1966
		Leaf	Pest control	Lakra et al., 2010
338.	Wattakaka volubilis (L.f.) Stapf (Asclepiadaceae)	Stem bark	Develop bitter taste in local drink	Ghosh and Das, 2004
339.	Wedelia chinensis (Osbeck) Merr. (Asteraceae)	Root	Black dye	Das and Mondal, 2012
340.	Wendlandia tinctoria (Roxb.) DC. (Rubiaceae)	Stem bark	Dye	Goel et al., 1984

TABLE 8.1 (Continued)

S. No.	Botanical Name and family	Part	Uses	References
		Leaf	Fodder	Goel et al., 1984
341.	*Woodfordia fruticosa* (L.) Kurz (Lythraceae)	Flower	Dye, Tanin	Goel et al., 1984; Maheshwari et al., 1990; Tiwari and Bharat, 2008; Kala, 2009; Singh et al., 2013; Alawa et al., 2013
342.	*Wrightia tinctoria* R.Br. (Apocynaceae)	Seed	Blue dye	Tiwari and Bharat, 2008
		Bark	Pumice	Upadhayay and Choudhary, 2014
343.	*Zea mays* L. (Poaceae)	Root	Local drink	Jain and De, 1964
344.	*Ziziphus jujuba* Mill. (Rhamnaceae)	Ash of the twig	Dye	Ghosh, 2004
		Leaves and bark	Pink red dye	Tiwari and Bharat, 2008
		Wood	Fuel wood	Singh et al., 2013
		Root bark	Used to extract liquor	Rai and Nath, 2006
345.	*Ziziphus xylopyrus* (Retz.) Willd. (Rhamnaceae)	Fruit	Dye	Maheshwari et al., 1990; Jain et al., 2010
346.	*Ziziphus nummularia* (Burm.f.) Wight & Arn. (Rhamnaceae)	Leaf	Fodder	Mishra et al., 2010
		Whole plant	Worship	Ahirwar, 2013
347.	*Ziziphus oenopolia* Mill. (Rhamnaceae)	Ash of the twig	Dye	Ghosh, 2004
348.	*Zornia diphylla* (L.) Pers. (Fabaceae)	Whole plant	Soil improver	Singh et al., 2013

8.8　CONCLUSION

From the perusal of literature it was found that the tribals of the Indo-Gangetic plain and Central India still practicing the age old traditional practices for their livelihood like, extraction of dye from various plants, preparation of baskets and other artifacts, broom, etc. The tribals of this region are imparting training to their children to keep this art alive. In this era of economic transformation these arts have to be preserved. The plant resources are depleting rapidly due to various reasons, jeopardizing the livelihood of the tribal dependent on plant resources. Therefore, measures should be taken up on priority by different Government and non-government organizations involving the stakeholders for the benefit of the humanity.

ACKNOWLEDGEMENTS

The author is thankful to Director, CSIR – National Botanical Research Institute Lucknow, for encouragement and providing facilities to carry out the work.

KEYWORDS

- **dye yielding plants**
- **fiber**
- **fodder**
- **gums**
- **ornamental**
- **sacred plants**
- **timber**

REFERENCES

Agarwal, P. (2013). Study of useful climbers of Fatehpur, Uttar Pradesh, India. *Int. J. Pharm. & Life Sci. 4*(9), 2957–2962.

Agarwal, P. (2014). Study of sacred plants used by people in Fatehpur district of Uttar Pradesh (India). *Life Sciences Leaflets 54,* 91–98.

Ahirwar, J. R. (2010). Diversity of socio-religious plants of Bundelkhand region of India. *Proc. of Nat. Seminar on Biological Diversity and its Conservation* at Govt. P. G. College, Morena (M. P.) pp. 21

Ahirwar, J. R. (2013). Socio-religious importance of plants in Bundelkhand region of India. *Res. Jour. Rece. Sci. 2*, 1–4.

Alawa, K. S., Ray, S., & Dubey, A. (2013). Dye yielding plants used by tribals of Dhar District, Madhya Pradesh, India. *Science Research Reporter 3*(1), 30–32.

Ambasta, S. P. (1992). The useful plants of India. Publication & Information Directorate, CSIR, New Delhi.

Anonymous. (1948–1976). The Wealth of India- Raw Materials, Vol. I–XI. Publicatin and Informatin Diectorate, New Delhi.

Arora, J. S. (1993). Introductory Ornamental Horticulture. Kalyani publuishers, Ludhiana.

Asolkar, L. V., Kakkar, K. K., & Chakra, O. J. (1992). Second supplement to glossary of Indian Medicinal plants with Active principles. Part I (A-K), (1965–81). Publications & Information Directorate, CSIR, New Delhi, India.

Bedi, S. J. (1978). Ethnobotany of the Ratan Mahal hills, Gujarat, India. *Econ. Bot. 32,* 278–284.

Bhatt, K. C., & Saha, D. (2014). Indigenous knowledge on fiber extraction of Sunnhemp in Bundelkhand Region, India. *Indian J. Nat. Prod. Reso.* 5(1), 92–96.

Bodane, A. K. (2015). Some ethno-medicinal plants and eco-friendly natural colors yielding flowering plants of B. S. N. Govt. P. G. college campus, Shajapur (M. P.) – A survey report. *Intern. J. Res. Granthaalayah. 3*(4), 1–6

Chaudhuri, H. N. R., & Pal, D. C. (1978). Less known uses of some grasses of India. *Bull. Bot. Soc. Bengal 32,* 48–53.

Chaudhuri, Rai, H. N., Pal, D. C., & Tarafder C. R. (1975). Less known uses of some plants from the tribal areas of Orissa. *Bull. Bot. Surv. India 17,* 132–136.

Chaudhury, J., Singh, D. P., & Hazra, S. K. (2000). Sunhemp *(Crotalaria juncea* L.) Central Research Institute for Jute and Allied Fibers, Barrackpore, West Bengal, India.

Choudhary, M. S., Upadhyay, S. T., & Upadhyay, R. (2012). Observation of natural dyes in *Ficus* species from Hoshangabad District of Madhya Pradesh. *Bull. Environ. Pharmacol. Life Sci.*1 (10), 34–37.

Cooke, T. (1958). The Flora of the Presidency of Bombay, Vols. 1–3 Reprinted Edition, Government of India.

Das, P. K., & Mondal, A. K. (2012). The dye yielding plants used in traditional art of 'patchitra' in pingla and mat crafts in sabang with prospecting proper medicinal value in the Paschim Medinipur District, West Bengal, India. *Int. J. Life Sc. Bt. & Pharm. Res. 1*(2), 158–171.

Das, S. N., Janardhanan, K. P., & Roy, S. C. (1983). Some observation on the ethnobotany of the tribes of Totopara and adjoining areas in Jalpaiguri district, West Bengal. *J. Econ. Taxon. Bot., 4,* 453–474.

Dashora, K., Bharadwaj, M., & Gupta, A. (2010). Conservation ethics of plants in India. *Indian Forester 136*(6), 837–842.

Dubey, A. (2007). Splash the colors of holi, Naturally. *Sci. Rep. 44(*3): 9- 13.

Dubey, P. C., Sikarwar, R. L. S., Khanna, K. K., & Tiwari, A. P. (2009). Ethnobotany of *Dillenia pentagyna* Roxb. in Vindhya region of Madhya Pradesh, India. *Indian J. Nat. Prod. Reso. 8*(5), 546–548.

Ghosh, A. (2004). Plant and clay dyes used by weavers and potters in West Bengal. *Indian J. Nat. Prod. Reso. 3*(2): 91.

Ghosh, C., & Das, A. P. (2004). Preparation of rice beer by the tribal inhabitants of tea gardens in Terai of West Bengal. *Indian J. Trad. Knowl. 3*(4), 373–382.

Goel, A. K., Sahoo, A. K., & Mudgal, V. (1984). A contribution to the ethnobotany of Santal Pargana Bihar. Bot. Surv. India Howrah

Gupta, S. P. (1981). Folklore about plants with reference to Munda culture. In: S. K. Jain (Ed.): Glimpses of Indian Ethnobotany. pp. 199–207.

Heda, K. N., & Kulkarni K. M. (2009). Fish stupefying plants used by the Gond tribal of Mendha village of Central India. *Indian J. Trad. Knowl. 8*(4), 531–534.

Heda, N. (2012). Folk conservation practices of the Gond tribal of Mendha (Lekha) village of Central India. *Indian J. Trad. Knowl. 12*(4), 727–732.

Hemrom, A., & Yadav, K. C. (2015). Festivals, traditions & rituals associated with sacred groves of Chhattisgarh. *Int. J. Multidis.Res. & Deve. 2*(2), 15–21.

ICFRE. (2010). Forest Sector Report India 2010. Indian Council of Forestry Research and Education, Dehradun (Ministry of Environment and Forests). Government of India.

Jadhav, D. (2008). Amulets and other plants wearing believed to be contact therapy among tribals of Ratlam district (MP) India. *Ethnobotany 20,* 144–146.

Jain, A., Katewa, S. S., Chaudhary, B. L., & Galav, P. (2004). Folk herbal medicine used in birth control and sexual diseases by tribals of southern Rajasthan, India. *J. Ethnopharmacol. 90,* 171–177.

Jain, A. K., Vairale, M. G., & Singh, R. (2010). Folklore claims on some medicinal plants used by Bheel tribe of Guna district Madhya Pradesh. *Indian J. Trad. Knowl. 9*(1), 105–107.

Jain S.K, & De, J. N. (1966). Observations on ethnobotany of Purulia district, West Bengal. *Bull. Bot. Surv. India 8*, 237–251.

Jain, S. K. (1963). Studies in Indian ethnobotany: Less known uses of 50 common plants from tribal areas of Madhya Pradesh. *Bull. Bot. Surv. India 5*, 223–226.

Jain, S. K. (1963). Magico-religious beliefs about plants among the Adivasis of Bastar. *Quart. J. Myth. Soc. 54*, 73–94.

Jain S. K. (1964). Plant resources in tribal areas of Bastar. *Vanyajati 12,* 147–173.

Jain, S. K. (1964). An indigenous water bottle. *Indian Forester 90,* 109.

Jain, S. K. (1965). Wooden musical instruments of the Gonds of Central India. *Ethnomusicol. 9*, 39–42.

Jain, S. K. (1971). Some magico-religious beliefs about plants among Adibasis or Orissa. *Indian J. Orthopaed. 1*, 95–104.

Jain, S. K. (1984). Wild plants food of the tribals of Bastar (Madhya Pradesh). *Proc. Nat. Inst. Sci. India 30*B, 56–80.

Jain, S. K. (1987). Plants in Indian medicine and folklore associated with healing of bones. *Indian J. Orthropaed. 1*, 95–104.

Jain, S. K. (1991). Dictionary of Indian folk medicine and Ethnobotany, Deep publications, New Delhi.

Jain, S. K. (1996*).* Ethnobiology in Human Welfare. Deep publications, New Delhi.

Jain, S. K., & De, J. N. (1964). Some less known plant foods among the tribals of Purulia (West Bengal*). Sci. & Cult.* 30, 285–286.

Jain S. P. (1984). Ethnobotany of Morni and Kalesar (Ambala-Haryana). *J. Econ. Taxon. Bot. 5,* 809–813.

Jha, A. K., & Gurudatta, Y. (2015). Some Wild Trees of Bihar and Their Ethnobotanical Study. *J. Research & Method in Education 5*(6), 74–76.

Joshi, P. (1982). Ethnobotanical study of Bhils-A Preliminary survey. *J. Econ. Taxon. Bot. 3,* 257–266.

Joshi, P. (1986). Fish stupefying plants employed by tribals of southern Rajasthan: A probe. *Curr Sci. 55,* 647–650.

Kala, C. P. (2005). Indigenous uses, population density, and conservation of threatened medicinal plants in protected areas of the Indian Himalayas. *Conservation Biol. 19* (2), 368–378.

Kala, C. P. (2009). Aboriginal uses and management of ethnobotanical species in deciduous forests of Chhattisgarh state in India. *J. Ethnobiol. Ethnomedicine 5*, 20 doi: 10.1186/1746–4269–5-20.

Kamble, S. Y., & Pradhan, S. G. (1980). Ethnobotany of korkus in Maharashtra. *Bull. Bot. Surv. India 22,* 201–202.

Kapoor, S. L., & Sharga, A. N. (1993). House plant. Vatika Prakashnan, India.

Kapoor, V. P., Katiyar, K., Pushpangadan, P., & Singh, N. (2008). Development of natural dye based sindoor. *Indian J. Nat. Prod. Reso. 7*(1), 22–29.

Kar, A., & Borthakur, S. K. (2008), Dye yielding plants of Assam for dyeing handloom textile products. *Indian J. Trad. Knowl. 7*(1), 166–171.

Koelz, W. H. (1979). Notes on the ethnobotany of Lahul province of Punjab. *Quart. J. Crude drug Res. 17*, 1–56.

Kumar, A., Tewari, D. D., & Tewari, J P. (2006). Ethnomedicinal knowledge among Tharu tribe of Devipatan division. *Indian J. Trad. Knowl. 5*(3), 310–313.

Kumar, A., & Yadav, D. K. (2010). Ethnomedicinal, mythological & Socio-ecological aspects of bamboos in Hosagngabad district (Madhya Pradesh). *Ethnobotany 22*, 97–101.

Kumar, S. (2014). Use of plants as color in Pytkar and Jadopatia folk arts of Jharkhand. *Indian J. Trad. Knowl. 13*(1), 202–207.

Kumar, V., & Jain, S. K. (1998). A contribution to ethnobotany of Surguja district in Madhya Pradesh, India. *Ethnobotany 10,* 89–96.

Kumar, V., & Rao, R. R. (2007). Some interesting indigenous beverages among the tribals of Central India. *Indian J. Trad. Knowl. 6*(1), 141–143.

Lakra, V., Singh, M. K., Sinha, R., & Kudada, N. (2010). Indigenous technology of tribal farmers in Jharkhand. *Indian J. Trad. Knowl. 9(*2), 261–263.

Li, X. X., & Zhou Z. K. (2005). Endemic wild ornamental plants from North Western Yunnan, China. *Hort. Sci. 40,* 1612–1619.

Maheshwari, J. K., Painuli, R. M., & Dwivedi, R. P. (1990). Notes on ethnobotany of Oraon and Korwa tribes ok Madhya Pradesh. In: S. K. Jain (Ed.): Contribution to Indian Ethnobotany. Scientific Publisher, Jodhpur, India. pp. 75–90.

Mahto, M., Singh, C. T. N., & Kumar, J. (2007). Some religious plants of Jharkhand and their medicinal uses. *Int. J. Mendel 24*, 47–48.

Mairh, A. K., Mishra, P. K., Kumar, J., & Mairh, A. (2010). Traditional botanical wisdom of Birhore tribes of Jharkhand. *Indian J. Trad. Knowl. 9*(3), 467–470.

Maji, S., & Sikdar, J. K. (1982). A taxonomic survey and systematic census on the edible wild plants of Midnapore district, West Bengal. *J. Econ. Taxon. Bot. 3*, 717–737.

Meena, K. L., Dhaka, V., & Ahr, P. C. (2013). Traditional uses of ethnobotanical plants for construction of the hut and hamlets in the Sitamata wildlife sanctuary of Rajasthan, India. *J. Energy Natural Resources 2*(5), 33–40.

Mishra, S., Sharma, S., Vasudevan, P., Bhatt, R. K., Pandey, S., Singh, M., Meena, B. S., & Pandey, S. N. (2010). Livestock feeding and traditional healthcare practices in Bundelkhand region of Central India. *Indian J. Trad. Knowl. 9*(2), 333–337.

Mishra, M. K., Panda, A., & Sahu, D. (2012). Survey of useful wetland plants of South Odisha, India. *Indian J. Trad. Knowl. 11*(4), 658–666.

Misra, M. K., Panda, A., & Sahu, D. (2014). Survey of useful wetland plants of South Odisha, India. *Indian J. Trad. Knowl. 11*(4), 658–666.

Mitra, S., & Mukherjee, S. K. (2005). Ethnobotanical usages of grasses by the tribals of West Dinajpur district, West Bengal. *Indian J. Trad. Knowl. 4*(4), 396–402.

Mohanta, D., & Tiwari, S. C. (2005). Natural dye yielding plants indigenous knowledge on dye preparation in Arunachal Pradesh, Northeast India. *Curr. Sci. 88*(9), 1474–1480.

Moldenke, H. N., & Moldenke, A. L. (1983). Nyctanthaceae. In: Dassanayake, M. D., & Fosberg, F. R. (Editors): A revised handbook to the Flora of Ceylon. Vol. 4. Smithsonian Institution, Washington D. C. pp. 178–181.

Nargas, J., & Trivedi, P. C. (2003). Traditional and medicinal importance of *Azadirachta indica* Juss. in India. In: Maheshwari, J. K. (ed.). Ethnobotany and medicinal plants of the Indian Subcontinent. Scientific Publishers (India) Jodhpur. pp. 33–37.

Oommachan, M., Masih, S. K., & Shrivastatva, J. L. (1989). Ethnobotanical studies in certain forest areas of Madhya Pradesh. *J. Trop For 5*, 192–196.

Pal, D. C., & Jain, S. K. (1989). Notes on Lodha medicine in Midnapur district, West Bengal, India. *Econ. Bot. 43*, 464–470.

Pal, D,C., & Srivastava, J. N. (1976). Preliminary notes of ethnobotany of Singhbhum district, Buhar. *Bull. Bot. Surv. India 18*, 247–250.

Pandey, A., & Gupta, R. (2003). Fiber yielding plants of India-Genetic resources, perspective for collection and utilization. *Nat. Prod. Rad. 2*(4), 194–204.

Pandey, H. P., & Verma, B. K. (2005). Phytoremedial wreath: A traditional excellence of healing. *Indian Forester 131*(3), 437–441.

Pandey, J. P., & Tiwari, A. (2014). Socio-religious importance of plants in Rewa region of Madhya Pradesh. *Int. J. Instit. Pharm. Lif. Scie. 4*(3), 17–21.

Panghal, M., Arya, V., Yadav, S., Kumar, S., & Yadav, J. P. (2010). Indigenous knowledge of medicinal plants used by Saperas community of Khetawas, Jhajjar District, Haryana, India, *J. Ethnobio. Ethnomedi. 6*(4), 1–11.

Panigrahi, G., & Murti, S. K. (1989–1999). Flora Bilaspur district, *M. P.* Bot. Sur. India, Kolkata, Vol. 1–2.

Parul & Vashistha, B. D. (2015). An Ethnobotanical study of plains of Yamuna Nagar District, Haryana, India. *Int. J. Innov. Res. in Sci., Engi. & Tech. 4*(1), 18600–18607.

Patel, S., Tiwari, S., Pisalkar, P. S., Mishra, N. K., Naik, R. K., & Khokhar, D. (2015). Indigenous processing of Tikhur (*Curcuma angustifolia* Roxb.) for the extraction of starch in Baster, Chhattisgarh. *Indian J. Nat. Prod. Reso. 6*(3), 213–220.

Prasad, L., Jalaj, R., Pandey, S., & Kumar, A. (2013). Few indigenous traditional fishing method of Faizabad district of eastern Uttar Pradesh, India *Indian J. Trad. Knowl. 12*(1), 116–122.

Rahangdale, C. P., Patley, R. K., & Yadav, K. C. (2014). Phytodiversity of ethnomedicinal plants in Sacred Groves and its traditional uses in Kabirdham district of Chhattisgarh. *Indian Forester 140* (1), 86–92.

Rai, R., & Nath, V. (2006). Socio-economic and livelihood pattern of ethnoc group Baiga in Achanakmar sal reserve forest in Bilaspur Chhattisgarh. *J. Trop. Forestry 22*, 62–70.

Rai, R., Nath, V., & Shukla, PK. (2004). Plants in Magico-religious beliefs of Baiga tribe in Central India. *J. Trop. Forestry. 20*, 39–50.

Raju, R. A. (1998). Wild plants of Indian Sub continent and their economic use. CBS Publishers.

Ramsankar, B. (2001). The "Soharai" festival of "Mundas' in Purulia. *Ethnobotany 13*, 140–141.

Sachan, K., & Kapoor, V. P. (2007). Optimization of extraction and dyeing conditions for traditional turmeric dye. *Indian J. Trad. Knowl. 6*(2), 270–278.

Sahu, C. (2008). Cultural identity of Jharkhand. *Jharkhand J. Dev. Manag. Stud. 1*(1), 139–145.

Sahu, P. K., Kumari, A., Sao, S., Singh, M., & Pandey, P. (2013). Sacred plants and their Ethno-botanical importance in Central India: A mini review. *Int. J. Pharm. & Life Sci.* *4*(8), 2910–2914.

Sandya, K., & Ahirwar, R. K. (2015). Diversity of medicinal plants and conservation by the tribes of Jaisinghnagar forest area, District Shahdol, Madhya Pradesh, India. *Intern. J. Sci. and Res. 4*(4), 509–512.

Saxena, H. O. (1986). Observations on the ethnobotany of Madhya Pradesh. *Bull. Bot. Surv. India 28*, 149–156.

Shil, S., Choudhury, M. D., & Das, S. (2014). Indigenous knowledge of medicinal plants used by the Reang tribe of Tripura of India. *J. Ethnopharmacol. 152*(1), 135–141.

Shukla, G., & Chakravarty, S. (2012). Ethnomedicinal plants use of Chilapatta reserved forest in West Bengal. *Indian forester. 138*(12), 1116–1124.

Siddiqui, M. O., & Dixit, S. N. (1975). Some noteworthy plant species from Gorakhpur. *J. Bombay Nat. Hist. Soc. 72,* 620–621.

Sikarwar, R. L. S. (1994). Wild edible plants of Morena district, Madhya Pradesh. *Vanyajati 42*(4), 31–35.

Sikarwar, R L S. (2002). Mahua [*Madhuca longifolia* (Koen.) Macbride] – A paradise tree for the tribals of Madhya Pradesh. *Indian J. Trad. Knowl. 1*(1), 87- 92.

Singh, A., Satanker, N., Kushwaha, M., Disoriya, R., & Gupta, A. K. (2013). Ethno-botany and uses of non-graminaceous forage species of Chitrakoot region of Madhya Pradesh. *Indian J. Nat. Prod. Reso. 4*(4), 425–431.

Singh, K. K., & Maheshwari, J. K. (1985). Forest in the life and economy of the tribals of Varanasi district, U. P. *J. Econ. Taxon. Bot. 6,* 109–116.

Singh, K. K., Prakash, A., & Palvi, S. K. (1999). Observations on some energy plants among the tribals of Madhya Pradesh. *J. Econ. Taxon. Bot. 23*(2), 291–296.

Singh, K. K., Saha, S., & Maheshwari, J. K. (1985). Ethnobotany of *Helicteres isora* L. in Kheri district Uttar Pradesh. *J. Econ. Taxon. Bot. 7,* 487–492.

Singh, L. Kasture, J., Singh, U. S., & Shaw, S. S. (2011). Ethnobotanical Practices of Tribals in Achanakmar Amarkantak Biosphere Reserve. *Indian forester. 137*(6), 767–777.

Singh, L. R. (2014). Food security through Wild Leafy Vegetables in Chotanagpur Plateau, Jharkhand. *Int. J. Res.Envir. Sci. Tech. 4(*4), 114–118.

Singh, R. V. (2001). Colouring plants-An innovative media to spread the message of conservation. *Down to Earth,* 20 September 2001. pp. 25–31.

Singh, U., & Bharti, A. K. (2015). Ethnobotanical study of plants of Raigarh area, Chhattisgarh, India. *Int. Res. J. Biol.Sci. 4*(6), 36–43.

Singh, V., & Singh R. V. (2002). Healthy hues. *Down to Earth* 11, 25–31.

Siva, R. (2007). Status of natural dyes and dye-yielding plants in India. *Curr. Sci. 92*(7): 916–924.

Srivastava, D. K., & Varma, S. K. (1981). An ethnobotanical study of Santhal Pargana, Bihar. *Indian Forester 107*, 30–41.

Srivastava, S. K., Tewari, J. P., & Shukla, D. S. (2008). A folk dye from leaves and stem of *Jatropha curcas* L. used by Tharu tribes of Devipatan division. *Indian J. Trad. Knowl. 7*(1), 77–78.

Swarup, V. (1998). Ornamental horticulture. Macmillan India Limited, New Delhi.

Tarafder, C. R. (1986). Ethnobotany of Chotanagpur (Bihar). *Folklore 27*, 119–124.

Thomas, B., Rajendran, A, Aravindhan, V., & Maharajan. M. (2011). Wild ornamental chasmophytic plants for rockery. *J. Mod. Biol. Tech. 1* (3), 20–21.

Tiwari, S. C., & Bharat, A. (2008). Natural dye-yielding plants and indigenous knowledge of dye preparation in Achanakmar-Amarkantak Biosphere Reserve, Central India. *Indian J. Nat. Prod. Reso. 7*(1), 82–87.

Tripathy, B. K., Panda, T., & Mohanty, R. B. (2014). Traditional artifacts from Bena grass [*Chrysopogon zizanioides* (L.) Roberty] (Poaceae) in Jaipur district of Odisha, India. *Indian J. Trad. Know. 13*(4), 771–777.

Upadhyay, R., & Choudhary M. S. (2012). Study of some common plants for natural dyes. *Int. J. Pharma. Res. & Bio-Sci. 1*(5), 309–316.

Upadhyay, R., & Choudhary, M. S. (2014). Tree Barks as a source of Natural Dyes from The forests of Madhya Pradesh. *Global J. Biosci. Biotech., 3*(1), 97–99.

Wagh, V.V & Jain, A. K. (2010a). Ethnomedicinal observation among Bheel and Bhilala tribe of Jhabua district, Madhya Pradesh, India. *Ethnobotanical Leaflets 14,* 715–720.

Wagh, V. V., & Jain, A. K. (2010b). Traditional herbal remedies among Bheel and Bhilala tribes of Jhabua district Madhya Pradesh. *Int. J. Bio. Tech. 1*(2), 20–24.

Yadav, S. S., & Bhandoria, M. S. (2013). Ethnobotanical exploration in Mahendragarh district of Haryana (India). *J. Med. Plant. Res. 7*(18), 1263–1271.

CHAPTER 9

A REVIEW ON ETHNOBOTANY OF HEPATOPROTECTIVE PLANTS OF INDIA

MADDI RAMAIAH

Department of Pharmacognosy, Hindu College of Pharmacy, Guntur – 522002, A.P., India, E-mail: rampharma83@gmail.com

CONTENTS

ABSTRACT

In recent times medicinal plant have received much needed attention as sources of bioactive substances used to treat wide variety of diseases and disorders of major body organs including liver as a hepatoprotective and antioxidants. Ethnobotany is a study of how people of particular cultures and regions make use of the plants in their local environments. Liver is the heaviest gland of the body and plays the major role in metabolic activities and bio-chemical conversions. Hepatic disease is a basic collective term of conditions, diseases, and infections that affect the cells, tissues structures, or functions of the liver. The present study is an attempt to report the ethnobotany of hepatoprotective plants of India, utilizing among the different tribal culture in India by using scientific studies. This may be useful to researchers who are working in the area hepatopharmacology and therapeutics.

9.1   INTRODUCTION

The simplest definition of ethnobotany is provided by the word itself: *ethno* (people) and *botany* (science of plants). In essence, it is a study of how people of particular cultures and regions make use of the plants in their local environments. These uses can include as food, medicine, fuel, shelter, and in many cultures, in religious ceremonies. In 1895, during a lecture in Philadelphia, a botanist named John Harshberger, provided the first definition of *ethnobotany* as the study of *how* native tribes used plants for food, shelter, or clothing (Harshberger, 1999). One of the best-known modern ethnobotanists was Richard Evans Schulte identified the field of ethnobotany as an interdisciplinary field, combining botany, anthropology, economics, ethics, history, chemistry, and many other areas of study.

Ethnobotanists need to be prepared to ask the following questions:

1. What are the fundamental ideas and conceptions of people living in a particular region about the plant life surrounding them?
2. What effect does a given environment have on the lives, customs, religion, thoughts, and everyday practical affairs of the people studied?
3. In what ways do the people make use of the local plants for food, medicine, material culture, and ceremonial purposes?
4. How much knowledge do the people have of the parts, functions, and activities of plants?
5. How are plant names categorized in the language of the people studied, and what can the study of these names reveal about the culture of the people?

The modern system of medicine still lack in providing suitable medicament for a large number of disease conditions, in spite of tremendous advances made in the discovery of new compounds. A few of these diseases are hepatic disorders, viral infections, AIDS, rheumatic diseases (Mohammad, 1994), etc. The available therapeutic agents only bring about symptomatic relief without any influence on the curative process, thus causing the risk of relapse and the danger of untoward effects. A large number of populations suffer due to various reasons from hepatic diseases of unknown origin. The development of antihepatotoxic drugs being a major thrust area has drawn the attention of workers in the field of natural product research.

India is a vast country with greatest emporia of plant wealth and represents a colorful mosaic of about 563 tribal communities which possess considerable knowledge regarding use of plants for livelihood, healthcare and other proposes through their long association with forestry inheritance and experiences (Bala et al., 2011).

Liver has a pivotal role in the maintenance of normal physiological process through its multiple and diverse functions, such as metabolism, secretion, storage and detoxification of variety of drugs. The bile secreted by the liver has, among other things, an important role in digestion. Liver diseases are the most serious ailments.

Treating liver diseases with botanical drugs has a long tradition, but evidence for efficacy is sparse. In the absence of reliable liver protective drugs in modern medicine, in India, a number of medical plants and their formulations are used to cure hepatic disorders in traditional systems of medicine (Stickel and Schuppan, 2007). Many folklore remedies from plant origin have long been used for the treatment of liver diseases. Hundreds of plants have been examined for use in a variety of liver disorders but only a few are well researched. Nearly 170 phyto constituents from 110 plants have been claimed to possess liver protecting activity. In India, more than 93 medical plants are used in different combinations in the preparation of 40-patented herbal formulations (Doreswamy and Sharma, 1995; Handa et al., 1986; Girish et al., 2009).

Several plants reported as hepatoprotective against hepatotoxicity in animals during the last decade and the polyherbal formulations have been proven to have hepatoprotective action against chemically induced liver damage in experimental animals (Sharma et al., 1991).

Apart from use of plants as a hepatoprotective, some of the herbals cause of adverse hepatic reactions has been published recently. Several herbals have been identified as a cause of acute and chronic hepatitis, cholestasis, drug-induced autoimmunity, vascular lesions, and even hepatic failure. So proper care must be taken for identification and selection for use of a hepatoprotective (Stickel et al., 2005).

9.1.1 LIVER DISORDERS

Liver is exposed to a variety of genobiotics and therapeutic agents due to inadequately controlled environmental pollution and expanding therapeutic uses of potent drugs. Thus, the disorders associated with this organ are numerous and varied. The following are some of the liver diseases that are commonly observed.

1. Necrosis
2. Cirrhosis
3. Hepatitis: may be of viral, toxic or deficiency type
4. Hepatic failure: Acute or Chronic
5. Chemical/drug induced hepatotoxicity: Generally may be hepatitis, jaundice and carcinogenesis.
6. Liver disorders due to impaired metabolic function, generally the disorders associated with fat (liposis) and bilirubin (jaundice) metabolism are very commonly seen.
   a. Disorders associated with fat metabolism: Fatty liver
   b. Disorders associated with bilirubin metabolism: Jaundice which may be of different types based upon mechanism of action and etiology. Hemolytic/pre-hepatic jaundice or obstructive (post-hepatic/cholestatic) jaundice or hepatogenous/hepatic jaundice/cholestasis. In these conditions there occurs unconjugated hyperbilirubinaemia.

Hereditary jaundice or pure cholestasis: Gilbert's syndrome, Dubin-Johnson syndrome, Crigler-Najjar syndrome, Rotor syndrome are some of the hereditary jaundice types usually observed. Gilbert's syndrome and Crigler-Najjar syndrome are examples of hereditary non-hemolytic unconjugated hyper bilirubinemia, whereas Dubin-Johnson syndrome and Rotor's syndrome are conditions with hereditary conjugated hyperbilirubinemia.

9.1.2 LIVER FUNCTION TESTS

When the liver is in a diseased state, one or more but not all of its functions are impaired. There can be no test for liver functions as a whole. The various liver function tests (LFTS) are tests of derangements of individual functions of liver, since many tests give similar abnormal results in a particular liver disease; it may be possible to extend a conclusion drawn from a single test.

The liver biopsy results may not be comparable with the LFTS since many functional changes are not mirrored by obvious structural changes in the liver cells (Praful and Drashan, 1996). Thus a battery of liver function tests is employed for accurate diagnosis, to assess the severity of the damage, to judge the prognosis and to evaluate therapy. These tests are described below in relation to major liver functions.

1. Tests for manufacture and excretion of bile (Mohan, 2005).
2. Serum enzyme assay include Alkaline phosphatase, Transaminases: SGOT (AST) and SGPT (ALT), γ-Glutamyl transpeptidase (γ-GT), 5-Nucleotidase and lactic dehydrogenase.
3. Immunologic tests.

9.1.3 PLANTS AS A HEPATOPROTECTIVE RESOURCE AND THEIR RECENT PERSPECTIVES

Use of herbal medicine for the treatment of various ailments dated back to thousands of years. Plants have emerged as a great source of pharmaceutical products. There has been increasing scientific and industrial interest in ethnobotanical medicine during the past few decades. Very often the synthetic drugs are associated with adverse effects. Thus, the plant-derived bioactive compounds have drawn the main attention as a source of complementary and alternative medicine. The use of alternative medicines for the treatment of liver diseases has a long history and medicinal plants and their derivatives are extensively used around the globe for this purpose. Huge interest of the scientific and pharmaceutical community over the therapeutic use of plant-based materials used in various ethnobotanical practices have led to purification and characterization of various bioactive compounds, which have proven to be hepatoprotective. A literature search in a traditional Oriental Medicine Database identified a number of herbal mixtures that are supposed to treat liver diseases (Stickel and Schuppan, 2007).

Herbal formulations have often been found to work better in a synergistic manner than working alone, e.g., LIV 52. The oldest mentions of hepatoprotective plant, Milk Thistle (*Silybum marianum*) is found in the Bible (Genesis 3:18). Silymarin, silybin, silydianin and silychristen were isolated from the plant, which has proven to be beneficial in liver related disorders (Tables 9.1 and 9.2).

TABLE 9.1 List of Some Important Hepatoprotecactive Medicinal Plants Mentioned in Ayurveda

Name of the plant	Part used	Family
Achillea millefolium L.	Whole plant	Asteracceae
Andrographis paniculata (Burm.f.) Wall. ex Nees	Whole plant	Acanthaceae
Aphanamixis polystachya (Wall.) R. N. Parker	Bark	Meliaceae
Apium graveolens L.	Seed	Umbelliferae
Asteracantha longifolia Nees	Leaf, root, seed	Acanthaceae
Berberis lycium Royle	Leaf	Berberidaceae
Bryonia alba L.	Root	Cucurbitaceae
Canavalia ensiformis (L.) DC.	Root	Leguminosae
Cichorium intybus L.	Whole plant	Asteracceae
Cosmostigma racemosa Roxb.	Root and bark	Asclepiadaceae
Delphinium zalil L.	Whole plant	Ranunculaceae
Euphorbia neriifolia L.	Fruit	Euphorbiaceae
Ficus carica L.	Fruit	Moraceae
Ficus heterophylla L. f.	Root	Moraceae
Fumaria officinalis L.	Whole plant	Fumariaceae
Fumaria parviflora Lamarck.	Whole plant	Fumariaceae
Garcinia indica (L.) Robs.	Fruit	Guttiferae
Gentiana kurroo Royle	Root	Gentianaceae
Gymnema sylvestre (Retz.) R.Br.ex Schult.	Leaf	Asclepiadaceae
Hedyotis corymbosa (L.) Lam.	Leaf	Rubiaceae
Hemidesmus indicus R.Br.	Root	Asclepiadaceae
Hygrophila spinosa T. Anders.	Leaf, root, stem, seed	Acanthaceae
Hyssopus officinalis L.	Whole plant	Labiatae
Luffa echinata Roxb.	Fruit and seed	Cucurbitaceae
Lycopersicon esculentum L.	Fruit	Solanceae
Mentha longifolia (L.) Huds.	Leaf	Labiatae
Myristica fragrans Houtt.	Seed	Myristicaeae
Nelumbo nucifera Gaertn.	Flower	Nymphaceae
Peonia emodi Wall. ex Royle	Tuber	Ranunculaceae
Phyllanthus niruri L.	Whole plant	Euphorbiaceae
Pinus roxburghii Sarg.	Oil	Pinaceae

TABLE 9.1 (Continued)

Name of the plant	Part used	Family
Prunus armeniaca L.	Fruit	Rosaceae
Pyrenthrum indicum DC.	Flower	Asteracceae
Rumex crispus L.	Root	Polygonaceae
Swertia chirata (Wall.) C. B. Clarke	Whole plant	Gentianaceae
Taraxacum officinale F. H. Wigg	Root	Asteracceae
Tinospora cordifolia (Willd.) Hook. f.	Stem	Menispermaceae
Trigonella foenum-graecum L.	Seed	Leguminosae
Vitex negundo L.	Whole plant	Verbenaceae
Zingiber officinale Roxb.	Rhizome	Zingiberaceae

TABLE 9.2 List of Some Commercially Available Polyherbal Hepatoprotective Tablets

S. No	Brand name	Composition	Manufacturer
1.	Arosi	Silymarin, L-ornithine-L-aspartate	Avincare
2.	Hepa	Silymarin, Calcium Pantothenate, Choline Bitartrate, Coenzyme Q 10, Ferrous Fumarate, Folic Acid, Inositol, L-Carnitine, L-Glutalthione, L-Ornithine, Pyridoxine, Riboflavin, Sodium Selenate, Vit B1, Vit B12, Vit C, Vit D3, Vit E, Zinc, N-Acetyl Cysteine	Venus Remedies
3.	Hepanit tablet	Silymarin, L-ornithine-L-Aspartate	Mac Organics
4.	Liv 52	Himsra (*Capparis spinosa*), Kasani (*Cichorium intybus*), Mandura bhasma, Kakamachi (*Solanum nigrum*), Arjuna (*Terminalia arjuna*), Kasamarda (*Cassia occidentalis*), Biranjasipha *(Achillea millefolium)*, Jhavuka (*Tamarix gallica*)	Himalaya drug company
	Liv 52 DS	Himsra (*Capparis spinosa*), Kasani (*Cichorium intybus*), Mandura bhasma, Arjuna (*Terminalia arjuna*), Kasamarda (*Cassia occidentalis*), Biranjasipha (*Achillea millefolium*), Jhavuka (*Tamarix gallica*), Kakamachi (*Solanum nigrum*)	Himalaya drug company
5.	Liveril	Silymarin, Calcium Pantothenate, Choline Bitartrate, Coenzyme Q 10, Folic Acid, Inositol, Iron, L-Carnitine, L-Glutalthione, L-Ornithine, Selenium, N-Acetyl Cysteine, Vit A, Vit B1, Vit B12, Vit B2, Vit B6, Vit C, Vit D3, Zinc	Meyer Organics Pvt. Ltd.

TABLE 9.2   (Continued)

S. No	Brand name	Composition	Manufacturer
6.	Livomap tablets	Punarnava (*Boerhavia diffusa*), Nimb (*Melia azadirachta*), Tikta patola (*Trichosanthes cucumerina*), Shunthi (*Zingiber officinale*), Katuki (*Picrorhiza kurroa*), Guduchi (*Tinospora cordifolia*), Devdaru (*Cedrus deodara*), Haritaki (*Terminalia chebula*), Varuna (*Crataeva religiosa*), Shigru (*Moringa oleifera*), Daruharidra (*Berberis aristata*), Afsantin (*Artemisia absinthium*), Sharapunkha (*Tephrosia purpurea*), Bhumiamalaki (*Phyllanthus niruri*)	Maharishi Ayurveda
7.	Livomyn tablet	*Andrographis paniculata, Phyllanthus niruri, Triphala, Cichorium intybus, Boerhaavia diffusa, Amoora rohituka, Eclipta alba, Adhatoda vasica, Zingiber officinale, Berberis aristata, Tinospora cordifolia, Tephrosia pupurea, Fumaria officinalis, Embelia ribes, Coriandrum sativum, Aloe barbadensis, Picrorrhiza kurroa*	Charak Pharmaceuticals Pvt Ltd
8.	Livotrit Tablets	Arogyvardhini Rasa, Mandur Bhasma, *Boerhavia diffusa, Eclipta alba, Andrographis paniculata*	Zandu Ayuveda
9.	Nitlor Plus	Silymarin, L-ornithine-L-aspartate	BMW Pharmaco India Pvt Ltd
10.	Nitoliv	Silymarin, L-ornithine-L-aspartate	Biomax Biotechnics
11.	Shamliv	Silymarin, Lecithin	Shamsri Pharma Pvt. Ltd
12.	SILYBON-70	Silymarin	Micro labs
13.	Silymarin Plus	Vitamin C, Vitamin E, Silymarin, Inositol, Choline	Source naturals

9.2   ETHNOBOTANY OF HEPATOPROTECTIVE PLANTS

Humans have long understood the medicinal properties of plants and have imbued trees, plants and flowers with spiritual properties. Herbs are used in many religions – such as in Christianity (myrrh (*Commiphora myrrha*), ague root (*Aletris farinosa*) and frankincense (*Boswellia spp.*). In Hinduism a form of Basil called Tulsi is worshipped as a goddess for its medicinal value since the Vedic times. Many Hindus have a Tulsi plant in front of their houses. In India, early Vedic texts describe the energies within plants and their use as medicine. The Rig Veda describes plants and their actions. The Atharva Veda mentions the therapeutic uses of plant medicines in greater detail. Charaka Samhita and Sushruta Samhita, the two classic Ayurvedic texts classified all medicinal substances into three groups: vegetable, animal and mineral origin. Astanga Hrdaya and Astanga Samgraha deal with Ayurveda material medica.

The present review is an attempt to report the ethnobotany of hepatoprotective plants of India, utilizing among the different tribal culture in India. We reviewed the different scientific studies regarding the ethnobotany of plants published in reputed journals, books, thesis and reports and finally literature was prepared. We do not claim to have included all the existing tribal communities of India information about ethnobotanical use of medicinal plants but we rather focused on information easily accessible to students and researchers. A list was produced showing the name(s), part of the plant used, name of the tribal community, local name of the plant, uses and/or mode of use. The precession of botanical identification in this review depends on that from original sources.

9.2.1 NOTE

Readers, please note that all material provided herein is for information only and may not be construed as medical advice. Readers should consult appropriate health professionals on any matter relating to their health and well-being. The effectiveness of the flowers and the leaves are at their best, when they are requested and plucked from the plants after they are fully bloomed on the plant. It is preferred not to offer the flowers plucked well in advance, in a bud condition to avoid the ill effects (Table 9.3).

9.3 CONCLUSION

Many Indian plants described above have been traditionally used individually or in combination for the treatment of variety of liver diseases and disorders. This review hopefully will help to find out the effective and safe plant for the treatment of liver disorders. However, the active ingredients in the herbal formulations are not well defined. It is therefore important to know

TABLE 9.3 Ethnobotanical Literature Review of Indian Medicinal Plants Used for Treatment of Various Liver Disease/Disorders

S. No.	Plant name	Part used	Tribe/Local community	Local names (State)	Uses/Mode of use
1	*Aerva lanata* (L.) Juss.	Roots	Local people	Kali-Bui (Shekhawari region, Rajasthan)	Root extract is given to patients of liver congestion and jaundice (Rishikesh and Ashwani, 2012)
2	*Abrus precatorius* L.	Roots	Mahadevkoli, Thakur and Katkari	(Konkan Region, Maharashtra)	The decoction of roots has been used for jaundice (Naikade and Meshram, 2014)
3	*Achyranthes aspera* L.	Leaf, root, whole plant	Local medical practitioners/ healers	Puthkanda (Shivalik Hills of Northwest Himalaya)	Decoction of root is used for jaundice (Devi et al., 2014)
4	*Achyranthes aspera* L.	Leaves	Traditional healers	Uttareni (Parnasala Sacred Grove Area Eastern Ghats of Khammam District, Telangana)	Jaundice: Tender leaves along with the tender leaves of *Careya arborea*, *Mimosa pudica* and *Ziziphus mauritiana* are crushed to paste and the paste along with cow milk is administered for 7 days (Srinivasa et al., 2015)
5	*Arbus precatorius* L.	Seeds	Local healers	Gunja/Ratti (Arunachal Pradesh)	Seeds are used in cough, cold and colic complaints, also used for jaundice and hemoglobin uric bile (Rama et al., 2012)
6	*Aconitum rotundifolium* Kar. & Kir.	Whole plant	Larje, Amchis	Bonkar, Pongtha, Vashi (Lahaul-spiti region Western Himalaya)	Whole plant juice extract is taken orally with equal volume of water to cure jaundice (Koushalya, 2012)
7	*Adhatoda vasica* Nees	Flower, Leaves	Local peoples of Taindol	Adusa (Taindol, Jhansi, Uttar Pradesh)	Leaves extract used for jaundice (Jitin, 2013)
8	*Adhatoda zeylanica* Nees	Leaf	Local healers	Vasaka (Arunachal Pradesh)	Fresh leaf juice is administered for one month for jaundice (Rama et al., 2012)

S. No.	Plant name	Part used	Tribe/Local community	Local names (State)	Uses/Mode of use
9	*Adiantum lunulatum* Burm.f.	Leaves	Yenadi	Hamsa padi (Andhra Pradesh)	Leaves are grinded to get juice and used in jaundice (Bala et al., 2011)
10	*Adiantum venustum* D.Don.	Whole plant	Gujjars, Bhoris Bakerwals	Kakbai (Bandipora district, Jammu and Kashmir)	Dried fronds are crushed to obtain powder. Powder is added to a glass of water and kept as such overnight. The extract is given next day early in the morning for the treatment of cough, jaundice and stomach ailments (Parvaiz et al., 2013)
11	*Aegle marmelos* (L.) Correa	Fruit	Local peoples of Taindol	Bel (Taindol, Jhansi, Uttar Pradesh)	Used for jaundice (Jitin, 2013)
12	*Aegle marmelos* (L.) Corr.	Leaf	Malayali tribes	Vilvam (Shevaroy Hills, Tamil Nadu)	The leaf juice 50 ml mixed with cow's milk used to cure jaundice (Alagesaboopathi, 2015)
13	*Aegle marmelos* (L.) Corr.	Leaf	Mahadevkoli, Thakur and Katkari	Bel (Konkan Region, Maharashtra)	Leaf powder is given along with goat milk for jaundice (Naikade and Meshram, 2014)
14	*Agave americana* L.	Leaves	Local peoples of Taindol	Kantala (Taindol, Jhansi, Uttar Pradesh)	Used for jaundice (Jitin, 2013)
15	*Agave americana* L.	Pulp	Local medical practitioners/ healers	Ramban (Shivalik Hills of Northwest Himalaya)	The pulp of plant is taken orally as blood purifier, constipation, acidity, as liver tonic (Devi et al., 2014)
16	*Alangium salvifolium* (L.f.) Wang.	Root, twig, bark	Local traditional healers	Ankula (Nawarangpur District, Odisha)	Root bark paste (5 gm) mixed with 7 black peppers (*Piper nigrum*) is administered twice a day for 10 days against hepatitis (Dhal et al., 2014)

TABLE 9.3 (Continued)

S. No.	Plant name	Part used	Tribe/Local community	Local names (State)	Uses/Mode of use
17	*Aloe barbadensis* L.	Leaf	Local healers	Ghritkumari (Arunachal Pradesh)	Fresh leaf juice is administered for one month (Rama et al., 2012)
18	*Alstonia scholaris* R. Br.	Bark	Local healers	Saptaparna/ Sitan gachha (Arunachal Pradesh)	Pieces of bark are worn in a garland for curing jaundice (Rama et al., 2012)
19	*Alternanthera pungens* L.	Whole plant	Local people	Katua Shak (Amarkantak region, Madhya Pradesh)	Used for liver and spleen complaints (Achuta et al., 2010)
20	*Alternanthera sessilis* (L.) R.Br.	Leaves	Irula tribals	Ponaganikerai (Kalavai village, Vellore district, Tamil Nadu)	Used in jaundice (Natarajan et al., 2013)
21	*Alternanthera sessilis* (L.) R.Br.	Whole plant	Fisherman community	Ponnanganni (Pudhukkottai District, Tamil Nadu)	The whole plant is used to treat diarrhea, skin disease, dyspepsia, hemorrhoids, liver (Rameshkumr and Ramakritinan, 2013)
22	*Amaranthus spinosus* L.	Roots	Kondh, Sabra, Naik tribes	Kanta Marish (Khordha forest division of Khordha District, Odisha)	A handful of dried roots made into fine powder. 3–5 g powder is given twice a day with sufficient water to treat jaundice (Mukesh et al., 2014)
23	*Amomum subulatum* Roxb.	Seed	Ethnic and rural people in Eastern Sikkim Himalayan Region	Elaichi (Sikkim)	Liver tonic (Trishna et al., 2012)
24	*Amomum subulatum* Roxb.	Seeds	Hakims, priests, tribal people	Elaichi (Eastern Sikkim Himalaya region)	Stomachic, heart and liver tonic (Trishna et al., 2012)

S. No.	Plant name	Part used	Tribe/Local community	Local names (State)	Uses/Mode of use
25	*Andrographis affinis* Nees	Leaf	Malayali tribes	Kodikkurundhu, Keeripparandai (Shevaroy Hills, Tamil Nadu)	Leaf paste mixed with cow's milk used in liver ailments. Leaf extract 50 ml. with mixed with Buffalo curd given internally a day for one week to cure jaundice (Alagesaboopathi, 2015)
26	*Andrographis alata* (Vahl) Nees	Leaf	Malayali tribes	Periyanangai (Shevaroy Hills, Tamil Nadu)	Leaf extract 25 ml. mixed with hot water is given orally twice a day for seven to 10 days in jaundice (Alagesaboopathi, 2015)
27	*Andrographis echioides* (L.) Nees	Whole plant	Chenchu	Dontarala aku (Nallamalais, Andhra Pradesh)	Whole plant water extract used to cure liver disease (Sabjan et al., 2014)
28	*Andrographis lineata* Nees	Root	Malayali tribes	Periyanangai (Shevaroy Hills, Tamil Nadu)	Root paste (25 gram) given along with cow's milk for four to six days to treat enlargement of liver. Leaf juice 50 ml mixed with cow's milk and taken orally twice a day for 5 days in liver diseases (Alagesaboopathi, 2015)
29	*Andrographis macrobotrys* Nees	Leaf, root	Malayali tribes	—	Fresh leaf juice is given orally thrice a day for one week to treat liver disorders. The root powder mixed with goat's milk and taken orally to treat jaundice (Alagesaboopathi, 2015)
30	*Andrographis neesiana* Wight	Leaf	Malayali tribes	—	Fresh leaf ground with water and the paste is given orally to cure jaundice (Alagesaboopathi, 2015)

TABLE 9.3 (Continued)

S. No.	Plant name	Part used	Tribe/Local community	Local names (State)	Uses/Mode of use
31	*Andrographis ovata* C. B. Clarke	Leaf	Malayali tribes	—	Leaf extract (25 ml) mixed with common salt and *Piper nigrum* (black pepper) three times a day for 7 days to treat liver ailments (Alagesaboopathi, 2015)
32	*Andrographis paniculata* (Brum.f.) Nees	Leaf	Koch, Meich, Rava, Munda, Santhal, Garo, Oraon	(Coochbehar district, West Bengal)	Leaf extract to treat jaundice (Tanmay et al., 2014)
33	*Andrographis paniculata* (Brum.f.) Nees	Whole plant	Local tribe	Kalmegh; Chirata (Oraon) (Jalpaiguri district, West Bengal)	Whole plant or leaf extract used as liver tonic (Bose et al., 2015)
34	*Andrographis paniculata* (Brum.f.) Nees	Leaf, root	Chenchu	Nelavemu (Nallamalais, Andhra Pradesh)	Leaf or root extract filtered and administered for liver diseases. This Plant leaf extract especially used by chenchu tribals in Bairulety tribal hamlet to get rid of hangover symptoms (Sabjan et al., 2014)
35	*Andrographis paniculata* (Burm.f.) Nees	Leaves	Yerukala, Yanadi, Sugali	Nelavemu (Sheshachala hill range of Kadapa District, Andhra Pradesh)	Decoction of leaves cures jaundice (Rajagopal et al., 2011)
36	*Andrographis paniculata* (Burm.f.) Nees	Whole plant	Malayali tribes	Siriyanangai, Periyanangai, Nilavembu (Shevaroy Hills, Tamil Nadu)	The decoction of the whole plant mixed with goat's milk is given two times a day for seven to 10 days for jaundice and live complaints (Alagesaboopathi, 2015)

S. No.	Plant name	Part used	Tribe/Local community	Local names (State)	Uses/Mode of use
37	*Andrographis paniculata* (Burm. f.) Nees	Leaves	Local healers	Kalmegha/Chirata	Leaves and young twigs are smashed and made paste; 20–30 gms paste taken three times daily after meal for 2–3 weeks to cure (Rama et al., 2012)
38	*Andrographis serpyllifolia* Wight	Leaf	Malayali tribes	Kaatuppooraankodi, Siyankodi, Thutuppoondu (Shevaroy Hills, Tamil Nadu)	Decoction of the leaf juice 50 ml mixed with cow's milk and drink to treat liver related stomach pain (Alagesaboopathi, 2015)
39	*Ardisia paniculata* Roxb.	Roots	Local healers	—	Root in combination with those of *Smilax ovalifolia* and *Bridelia tomentosa* are crushed and boiled in water and drunk @ 1 cup (100 ml) twice daily for jaundice (Rama et al., 2012)
40	*Areca catechu* L.	Young inflorescence	Kani, Kurumar, Kurumbar, Paniyan (Adakkamaram).	Adakkamaram (Kerala)	The pounded mass of young inflorescence about 5–10 g mixed in goat's milk is given orally twice daily for 7 days to treat jaundice (Asha and Pushpangadan, 2002)
41	*Arenga wightii* Griff.	Young inflorescence, fruit husk	Paniyan (Ayasthingu), Kani (Kudappana)	Ayasthingu, Kudappana (Kerala)	Fresh toddy obtained from the young inflorescence is given internally for jaundice; expressed juice of the fruit husk is also given to treat jaundice (Asha and Pushpangadan, 2002)
42	*Argemone mexicana* L.	Latex, root	Gond, Kol, Baiga, Panica, Khairwar, Manjhi, Mawasi, Agaria	Bharbhanda, kateli (Rewa dist, Madhya Pradesh)	Latex used in jaundice (Achuta et al.. 2010)

TABLE 9.3 (Continued)

S. No.	Plant name	Part used	Tribe/Local community	Local names (State)	Uses/Mode of use
43	*Argemone mexicana* L.	Latex, stems, shoots	Local medical practitioners/ healers	Barbhand (Shivalik Hills of Northwest Himalaya)	Latex is applied for skin problems (ringworm infection). Stems extraction is used in diabetes and that of leaves is used for jaundice, malaria and fever (Devi et al., 2014)
44	*Argemone mexicana* L.	Whole plant	Local people and traditional healers	Pillibhutti, satyanasi (Ambala district, Haryana)	Plant juice is used orally; 2–3 spoons daily for one week to cure jaundice (Vashistha and Kaur, 2013).
45	*Argemone mexicana* L.	Leaf	Local healers	Bhant/Satyanasi (Arunachal Pradesh)	Decoction of leaf is used in jaundice (Rama et al., 2012)
46	*Argemone mexicana* L.	Seeds	Malayali tribes	Perammathandu (Shevaroy Hills, Tamil Nadu)	Seed powder is taken with hot water internally twice a day for one week to treat jaundice (Alagesaboopathi, 2015)
47	*Aristolochia indica* L.	Roots	Kani (Kattukka Mooli), Kurumbar (Garudakodi), Muthuvan (Cheriya arayan)	Kattukkamooli, Garudakodi, Cheriya arayen (Kerala)	Paste of tender roots along with cow's milk administered internally for 5 days to treat jaundice (Asha and Pushpangadan, 2002)
48	*Asparagus racemosus* Willd.	Root	Mahadevkoli, Thakur and Katkari	(Konkan Region, Maharashtra)	Decoction obtained from the root has been used to cure jaundice (Naikade and Meshram, 2014)
49	*Asparagus racemosus* Willd.	Tuberous root	Ethnic and Rural people, Hakims, priests	Kurilo (Sikkim)	Decoction used as for diabetes, jaundice, urinary disorder (Trishna et al., 2012)

S. No.	Plant name	Part used	Tribe/Local community	Local names (State)	Uses/Mode of use
50	*Asparagus racemosus* Willd	Roots	Kani (Shathavai), Kurumbar, Mannan (Kilavari)	Sathavari, Kilavari (Kerala)	Expressed juice of fresh root (20–30 ml) given internally twice daily for 7 days to treat jaundice. Roots roasted and taken on an empty stomach in the morning to treat lever disorders (Asha and Pushpangadan, 2002)
51	*Asparagus recemosus* Willd.	Roots	Local tribe	Sharanoi (Kumaun region, Uttarakhand)	Dried root powder is used to cure liver disorders (Gangwar et al., 2010)
52	*Asplenium adiantoides* C. Chr.	Whole plant	Local healers	—	Plant decoction is used in jaundice (Rama et al., 2012)
53	*Asteracantha longifolia* Nees	Leaves, seeds	Local people	Talimkhana (Ajara Tahsil, Kolhapur District, Maharashtra)	Used in jaundice (Sadale and Karadge, 2013)
54	*Asteracantha longifolia* Nees	Whole plant	Local healers	Talmakhana (Arunachal Pradesh)	Plant extract and decoction of leaves is used in jaundice (Rama et al., 2012)
55	*Averrhoa carambola* L.	Fruit	Local healers	Kamrakh (Arunachal Pradesh)	3–4 slices of the fruit are taken for jaundice or juice of crushed fruit is taken orally for jaundice @ 1/2–1 cup (50 ml-100 ml) 3 times daily (Rama et al., 2012)
56	*Azadirachta indica* A. Juss.	Bark	Malayali tribes	Vembu (Shevaroy Hills, Tamil Nadu)	Decoction of the bark mixed with sugar is given internally in jaundice (Alagesaɔoopathi, 2015)
57	*Azadirachta indica* A. Juss.	Leaves	Mahadevkoli, Thakur and Katkari	(Konkan Region, Maharashtra)	Young leaves are fried with salt and powder given with milk (Naikade and Meshram, 2014)

TABLE 9.3 (Continued)

S. No.	Plant name	Part used	Tribe/Local community	Local names (State)	Uses/Mode of use
58	*Baliospermum montanum* (Willd.) Muell	Root	Kondareddis of Khammam dist., Telangana	Konda amudam (Andhra Pradesh, Telangana)	Root decoction (3 teaspoons) administered daily once a week (Reddy et al., 2008)
59	*Bauhinia acuminata* L.	Bark	Gond, Kol, Baiga, Panica, Khairwar, Manjhi, Mawasi, Agaria	Sivamalli (Rewa dist., Madhya Pradesh)	Root bark decoction is used for treating inflammation of liver (Achuta et al., 2010)
60	*Begonia laciniata* Roxb.	Roots	Yenadi	Hooirjo (West Bengal), Teisu (Nagaland)	A decoction of the root is given for liver diseases and fever (Ganapathy et al., 2013)
61	*Benincasa hispida* (Thunb.) Cogn.	Fruit	Local healers	Petha/ Chaulkumhra (Arunachal Pradesh)	Boiled extract of fruit is given in stomach ulcer and jaundice (Rama et al., 2012)
62	*Benincasa hispida* (Thunb.) Cogn.	Fruit	Local communities	Torbot (Manipur)	About a glass of extracted and filter juice is mixed with 2–3 teaspoon of sugar candy and a spoon of honey for treatment of jaundice patient, thrice daily (Rajesh et al., 2012)
63	*Berberis aristata* DC.	Root, Bark	Ethnic and Rural people in Eastern Sikkim Himalayan Region	Chutro (Sikkim)	Used in jaundice, malaria, fever and Diarrhea. It is also used externally to cure eye disease (Trishna et al., 2012)
64	*Berberis aristata* DC.	Root, stem	Local healers	Daru Haridra (Arunachal Pradesh)	Root and stem decoction is taken orally for jaundice (Rama et al., 2012)

S. No.	Plant name	Part used	Tribe/Local community	Local names (State)	Uses/Mode of use
65	*Berberis asiatica* DC.	Bark, stem, wood and fruits	Local tribe	Rasanjana, Daruhaldi, Kilmora (Kumaun region, Uttarakhand)	The roots are used for curing diabetes and jaundice (Gangwar et al., 2010)
66	*Bergenia stracheyi* (Hook. f. & Thoms.) Engl.	Rhizome and leaves	Local elderly people, hermits, shepherds, Vaids, Gujjars and Gaddies	Shamlot (Tribal Pangi Valley, Chamba, Himachal Pradesh)	Rhizome paste is given internally to cure jaundice (Bhupendar et al., 2014)
67	*Bixa orellana* L.	Leaves	Local healers	Sinduriya (Arunachal Pradesh)	Leaves are useful in jaundice (Rama et al., 2012)
68	*Boerhavia diffusa* L.	Root	Local healers	Punarnava (Arunachal Pradesh)	Root is used in various ways for jaundice (Rama et al., 2012)
69	*Boerhavia diffusa* L.	Leaf, root	Gond, Kol, Baiga, Panica, Khairwar, Manjhi, Mawasi, Agaria	Santan, Santh, Gadapurena, Punarnava (Rewa dist, Madhya Pradesh)	Root decoction used in treatment of jaundice (Achuta et al., 2010)
70	*Boerhavia diffusa* L.	Roots	Chenchu	Atuka mamidaku (Nallamalais, Andhra Pradesh)	Root paste mixed with water used in liver diseases (Sabjan et al., 2014)
71	*Boerhavia diffusa* L.	Root, leaf	Local medical practitioners/ healers	(Shivalik Hills of Northwest Himalaya)	The decoction of leaves is used for jaundice (Devi et al., 2014)
72	*Boerhavia diffusa* L.	Root	Malayali tribes	Mukkurattai (Shevaroy Hills, Tamil Nadu)	The root powder mixed with cow's milk is used in jaundice (Alagesaboopathi, 2015)

TABLE 9.3 (Continued)

S. No.	Plant name	Part used	Tribe/Local community	Local names (State)	Uses/Mode of use
73	*Boerhavia diffusa* L.	Whole plant	Mahadevkoli, Thakur and Katkari	(Konkan Region, Maharashtra)	Fresh whole plant material is boiled in water along with sugar, half cup of the decoction is given to the patient thrice a day for 2–3 weeks (Naikade and Meshram, 2014)
74	*Boerhavia erecta* L.	Root	Fisherman community	Padarmookirattai (Pudhukkottai District, Tamil Nadu)	Used to treat jaundice, enlarged spleen and gonorrhea (Ramesh Kumar and Ramakritinan, 2013)
75	*Brassica juncea*L.	Seeds	Mahadevkoli, Thakur and Katkari	(Konkan Region, Maharashtra)	Alum 40 gm (White mineral salt) + Brassica seed 3 g made into the paste and eaten along with fruit of banana twice a day for jaundice (Naikade and Meshram, 2014)
76	*Bridelia monoica* (Lour.) Mec.	Root	Local healers	Karagnalia (Arunachal Pradesh)	The root in combination with the root of *Smilax ovalifolia* and *Ardisia paniculata* are rubbed on grindstone and the paste is collected in a cup of water. The mixture is boiled and taken orally for jaundice (Rama et al., 2012)
77	*Bridelia montana* (Roxb.) Willd.	Bark	Sugali, Yerukala, Yanadi	Sankumanu (Salugu Panchayati of Paderu Mandalam, Visakhapatnam, District, Andhra Pradesh)	Used in jaundice (Padal et al., 2012)
78	*Bridelia stipularis* (L.) Bl.	Leaves	Local healers	—	Leaves are used for jaundice (Rama et al., 2012)

S. No.	Plant name	Part used	Tribe/Local community	Local names (State)	Uses/Mode of use
79	*Butea monosperma* (Lam.) Taub	Gum	Chenchus, Lambadi and Yerukulas	Moduga (Prakasam District, Andhra Pradesh)	A red astringent gum or resin from stem is administered in the treatment of Jaundice and diarrhea (Mohan and Murthy, 1992)
80	*Caesalpinia bonduc* L.	Root bark	Local tribe	Kalaachikottai (Athinadu Pachamalai Hills Of Eastern Ghats In Tamil Nadu)	Root bark, with garlic, pepper, kasakása cure jaundice (Anandkumar et al., 2014)
81	*Cajanus cajan* (L.) Millsp.	Leaf	Koch, Meich, Rava, Munda, Santhal, Garo, Oraon	Coochbehar district, West Bengal	Leaf decoction for jaundice (Tanmay et al., 2014)
82	*Cajanus cajan* (L.) Millsp.	Leaf	Local tribe	Arhar; Tauri kalai (Koch): Jehu (Garo); Kokhleng (Mech) (Jalpaiguridistrict, West Bengal)	Leaf decoction beneficial for jaundice (Bose et al., 2015)
83	*Calotropis gigantea* (L.) Dryand	Leaves	Yanadi	Tellajilledu (Kavali, Nellore district, Andhra Pradesh)	Leaves tied with black thread and tied around neck for jaundice patients (Swapna, 2015)
84	*Canavalia gladiata* (Jacq) DC.	Fruits	Local villagers	Abai, Ghevada (Kolhapur district, Maharashtra)	The root is ground in cow urine and administered internally for consecutive days is said to cure enlargement of liver (Jadhav et al., 2011)
85	*Canavalia gladiata* (Jacq.) DC.	Roots	Chenchu	Adavi thamba (Nallamalais, Andhra Pradesh)	Root paste (20 grams) given along with rice gravel for 2 to 3 days to cure enlargement of liver (Sabjan et al., 2014)

TABLE 9.3 (Continued)

S. No.	Plant name	Part used	Tribe/Local community	Local names (State)	Uses/Mode of use
86	*Capparis spinosa* L.	Shoots	Larje, Amchis	Martokpa, Rutoka (Lahaul-spiti region Western Himalaya)	Green shoots are cut and dried in shade and powdered. The powder is taken orally twice a day to cure liver pain in jaundice (Koushalya, 2012)
87	*Capsicum annuum* L.	Fruit	Local healers	Mircha (Arunachal Pradesh)	10–15 gms splitted fruit without seed are kept in 100–150 ml. water for 3–4 hours and after removing the fruit, water is taken against orally for jaundice (Rama et al., 2012)
88	*Carica opaca* Stapf. ex Haines	Leaf, root	Local medical practitioners/ healers	Garnu (Shivalik Hills of Northwest Himalaya)	Leaf and root decoction is used against asthma, jaundice (Devi et al., 2014)
89	*Carthamus tinctorius* L.	Fruit	Local healers	—	Fruit juice is used for cure of jaundice (Rama et al., 2012)
90	*Casearia graveolens* Dalz.	Roots, Leaves	Gond, Halba and Kawar	Arni (Darekasa hill range, Gondia district, Maharashtra)	Root and leaves extract mixed with warm water and given bath to child in jaundice (Chandra Kumar et al., 2015)
91	*Cassia fistula* L.	Bark, leaves	Yenadi,	Rela (Andhra Pradesh)	Used in treatment of jaundice in the form of powder (Bala et al., 2011)
92	*Cassia fistula* L.	Flowers	Malayali tribes	Konnai	Powdered flower is used to cure liver ailments (Alagesaboopathi, 2015)
93	*Cassia fistula* L.	Fruit	Herbal practitioners	Rela-kayalu, Relarala, Reylu, Suvarnam (Vizianagaram District, Andhra Pradesh)	One spoon of fruit pulp is administered with sugarcane juice to cure jaundice (Padal et al., 2013)

S. No.	Plant name	Part used	Tribe/Local community	Local names (State)	Uses/Mode of use
94	*Cassia fistula* L.	Fruit	Sugali, Yerukala, Yanadi	Rela (Salugu Panchayati of Paderu Mandalam, Visakhapatnam, District, Andhra Pradesh)	Used in jaundice (Padal et al., 2013b)
95	*Cassia italica* Lam.	Leaves, fruits	Tribal people Kailasagirikona Forest, Chittor dist., Andhra Pradesh	—	About 5–10 g paste is prepared from leaves and fruits are administered orally for a month to recover from jaundice (Pratap et al., 2009)
96	*Cassia occidentalis* L.	Leaf	Traditional healers	Kasintha (Parnasala Sacred Grove area Eastern Ghats of Khammam District, Telangana)	Jaundice: 10 spoonfuls of leaf juice mixed with buttermilk is given thrice a day for 7 days (Srinivasa et al., 2015)
97	*Cassia tora* L.	Leaves, seeds	Fisherman community	Oosithagarai (Pudhukkottai District, Tamil Nadu)	Skin diseases, dandruff, constipation, cough, hepatitis, fever, and hemorrhoids (Ramesh Kumar and Ramakritinan, 2013)
98	*Centella asiatica* (L.) Urb.	Leaves	Koragar (Kudagon), Kurumbar (Kumara), Kani (Varambil), Hill Pulayar Kodakan)	Kudagan, Kumara, Varambil, Kodakan (Kerala)	Expressed juice of fresh leaves (20–30 ml) given internally twice daily for 7 days to treat jaundice (Asha and Pushpangadan, 2002)

TABLE 9.3 (Continued)

S. No.	Plant name	Part used	Tribe/Local community	Local names (State)	Uses/Mode of use
99	*Ceratopteris siliquosa* (L) Copel.	Whole plant	Muthuvan and Kani (Shirunagal)	Shirunagal (Kerala)	Decoction of the whole plant taken internally (25–30 ml) twice daily to treat jaundice and other ailments (Asha and Pushpangadan, 2002)
100	*Chenopodium album* L.	Whole plant, root, fruit	Local medical practitioners/healers	Bhadhu (Shivalik Hills of Northwest Himalaya)	Decoction of roots is effective against jaundice (Devi et al., 2014)
101	*Chenopodium album* L.	Whole plant	Local people and traditional healers	Bathu sag (Ambala district, Haryana)	Root is used in jaundice and urinary problems (Vashistha and Kaur, 2013)
102	*Chloroxylon sweitenia* DC.	Root bark	Local tribe	Vambarai (Athinadu Pachamalai Hills of Eastern Ghats in Tamil Nadu)	Root bark is used for jaundice cure (Anandkumar et al., 2014)
103	*Cichorium intybus* L.	Leaves	Gujjars, Bhoris Bakerwals	Kasni/Wari Hundh (Bandipora district, Jammu and Kashmir)	Leaves are cooked and given to fresh mothers to cure body weakness, loosening of joints, body muscular pains, frequent bleeding, as appetizer and liver tonic (Jadhav et al., 2011)
104	*Cichorium intybus* L.	Seeds, root and leaves	Local tribe	Kasni (Kumaun region, Uttarakhand)	Herb is taken internally to cure liver disorders, spleen problems (Gangwar et al., 2010)
105	*Cissampelos pareira* L.	Root	Local healers	Ambashtha/Patha (Arunachal Pradesh)	Root is placed in water for overnight and the extract is taken orally for jaundice (Rama et al., 2012)

S. No.	Plant name	Part used	Tribe/Local community	Local names (State)	Uses/Mode of use
106	*Cissmpelos pareira* L.	Leaves	Chenchus, Lambadi and Yerukulas	—	The juice obtained from crushing the leaves is mixed with black pepper and is given to cure jaundice (Mohan and Murthy, 1992)
107	*Citrullus colocynthis* (L.) Schrad	Roots	Chenchus, Lambadi and Yerukulas	Chedupuccha (Prakasam District, Andhra Pradesh)	Crushed roots and fruits are boiled with water and the decoction is given for drinking to cure jaundice and urinary diseases (Mohan and Murthy, 1992)
108	*Citrus aurantifolia* (Christm.) Wingle, *C. medica* L., *C. reticulata, C. sinensis* (L.) Osbeck	Fruit	Local healers	—	Fruit administered during jaundice (Rama et al., 2012)
109	*Citrus medica* salib.	Fruit, leaves	Local people of Taindol	Bara nimbu (Taindol, Jhansi, Uttar Pradesh)	Used for jaundice (Jitin, 2013)
110	*Clerodendrum indicum* (L.) O.Kuntze	Root	Local healers	Vanabhenda (Arunachal Pradesh)	Root is soaked in water for overnight and fresh extract is taken orally for 7–15 days for jaundice (Rama et al., 2012)
111	*Clome viscosa* L.	Leaf	Malayali tribes	Naaivelai (Shevaroy Hills, Tamil Nadu)	Fresh leaf juice mixed with hot water is used in jaundice (Alagesaboopathi, 2015)
112	*Coccinia grandis* (L.) Voigt	Leaf	Koal, Panika, Bhuriya, Kharvar, Gaund	Kularu (Renukot forest division, Sonbhadra, Uttar Pradesh)	Fresh leaf juice is used in the treatment of jaundice (Anurag et al., 2012)
113	*Coccinia grandis* (L.) Voigt	Leaves	Irula tribals	Kovai (Kalavai village, Vellore district, Tamil Nadu)	Used in jaundice (Natarajan et al., 2013)

TABLE 9.3 (Continued)

S. No.	Plant name	Part used	Tribe/Local community	Local names (State)	Uses/Mode of use
114	*Cocculus hirsutus* (L.) Diels	Root, whole plant	Local traditional healers	Dahdahiya (Nawarangpur District, Odisha)	Root paste is taken thrice a day for liver dysfunction (Dhal et al., 2014)
115	*Coptis teeta* Wall.		Local healers	Mishmi teeta (Arunachal Pradesh)	Root is soaked in water for overnight and fresh extract is taken orally for 7–15 days for jaundice (Rama et al., 2012)
116	*Cordia dichotoma* Forst. f.	Leaves	Sugali, Yerukala, Yanadi	Banka nakkeri (Salugu Panchayati of Paderu Mandalam, Visakhapatnam, District, Andhra Pradesh)	Used in jaundice (Padal et al., 2012)
117	*Cordyceps sinensis*	Whole plant	Ethnic and Rural People	Yarcha gombuk (Eastern Sikkim Himalaya region)	Rejuvenates liver and heart (Trishna et al., 2012)
118	*Coriandrum sativum* L	Seeds, leaf	Local healers	Dhanyaka/Dhaniya (Arunachal Pradesh)	Leaf and fruits are taken orally during jaundice (Rama et al., 2012)
119	*Coriandrum sativum* L.	Seeds	Gujjars, Bhoris Bakerwals	Daniwaal (Bandipora district, Jammu and Kashmir)	Seed decoction is given to cure jaundice, drying of mouth and headache (Parvaiz et al., 2013)
120	*Costus speciostus* (Koenig ex Retz.) Sm.	Rhizome	Local healers	Kebuk/Keun (Arunachal Pradesh)	Fresh juice of rhizome is taken orally (Rama et al., 2012)
121	*Crepis flexuosa*	Whole plant	Larje, Amchis	Homa sili (Lahaul-spiti region Western Himalaya)	Fresh juice of the plant is mixed with equal amount of water is taken regularly once a day to cure jaundice until cured (Koushalya, 2012)

S. No.	Plant name	Part used	Tribe/Local community	Local names (State)	Uses/Mode of use
122	*Crinum amoenum* Roxb. ex Ker.-Gawler	Root	Local tribe	Astachatur (Rava) (Jalpaiguridistrict, West Bengal)	Root used to treat jaundice and diarrhea (Bose et al., 2015)
123	*Cryptolepis buchanani* Roem. and Schult.	Latex	Chenchus, Lambadi and Yerukulas	Adavipalathiga (Prakasam District, Andhra Pradesh)	The milky latex mixed with water is given to cure dysentery and hepatic troubles (Mohan and Murthy, 1992)
124	*Cucumis sativus* L.	Fruit	Local healers	Kheera (Arunachal Pradesh)	Fresh fruit is administered during jaundice (Rama et al., 2012)
125	*Cuminum cyminum* L.	Fruit	Kani and Kurumbar (Jera) Kani, aniyan, Irular, Kurumar, Kurichar (Jeeragam)	Jera Jeerakom, Cheerakam (Kerala)	Decoction of the fruit is given in jauncice to purify blood and as diuretic (Asha and Pushpangadan, 2002)
126	*Curculigo orchioides* Gaertn.	Roots	Gondand Madiya	Kali-musli (Markanda Forest Range of Gadchiroli District, Maharashtra, India)	Used for Jaundice (Pankaj et al., 2015)
127	*Curculigo orchioides* Gaertn.	Tubers	Local villagers	Kali-musali (Kolhapur district, Maharashtra)	Tubers cut and shade dry, in that add equal amount of sugar and one glass milk mix well to make thick mucilage, this mixture is used in asthma, jaundice and diarrhea (Jadhav et al., 2011)

TABLE 9.3 (Continued)

S. No.	Plant name	Part used	Tribe/Local community	Local names (State)	Uses/Mode of use
128	*Curculigo orchioides* Gaertn.	Rhizome	Local healers	Kali Musali (Arunachal Pradesh)	Rhizome is prescribed in piles, jaundice, asthma, diarrhea and gonorrhea, considered demulcent, tonic; used as poultice for itches and skin diseases (Rama et al., 2012)
129	*Curculigo orchioides* Gaertn.	Root	Kondareddis of Khammam dist.	Adavi taadi, Naela tadi (Andhra Pradesh, Telangana)	Tuberous root paste (12 g) administered daily twice for three days (Reddy et al., 2008)
130	*Curcuma domestica* Valeton	Rhizome	Kani, Kurichar, Paniyan Kattunaikan, Koragar, Hill Pulayar (Manjal)	Manjal (Kerala)	Pounded mass of fresh rhizome, equal to the size of the fruit of *Emblica officinalis* is mixed with calcium hydroxide (equal to the size of the seed of *Adenanthera pavonia* L.), diluted with water and kept overnight in an airtight bottle. Three ounces of the liquid portion from this preparation is given on an empty stomach in the morning for 7 days to treat jaundice and other chronic liver disorders. Daily bath is recommended when this drug is administered (Asha and Pushpangadan, 2002)
131	*Curcuma longa* L.	Whole plant	Local people of Taindol	Haldi (Taindol, Jhansi, Uttar Pradesh)	Used for jaundice (Jitin, 2013)
132	*Curcuma longa* L.	Rhizome	Local healers	Haridra/Haldi (Arunachal Pradesh)	40–50 gms rhizome pounded and made extract; extract is mixed with fruits of *Piper longum* L. and taken daily for 20–25 days during jaundice (Rama et al., 2012)

S. No.	Plant name	Part used	Tribe/Local community	Local names (State)	Uses/Mode of use
133	*Curcuma longa* L.	Rhizome	Mahadevkoli, Thakur and Katkari	Konkan Region, Maharashtra	Paste of rhizome is mixed with cow milk and taken once day for 12–13 days (Naikade and Meshram, 2014)
134	*Cuscuta epithymum* (L.) L.	Whole plant	Yenadi	Sitammapogunalu (Andhra Pradesh, Telangana)	Extract was used in urinary, spleen and liver disorders (Ganapathy et al., 2013)
135	*Cuscuta reflexa* Roxb	Whole plant	Local healers	Amarvela (Arunachal Pradesh)	Plant juice is taken for 6–7 days for jaundice (Rama et al., 2012)
136	*Cuscuta reflexa* Roxb.	Whole plant	Local tribe	Swarnalata; Alokzori (Oraon) (Jalpaiguridistrict, West Bengal)	Whole plant juice used to treat jaundice (Bose et al., 2015)
137	*Cydonia oblonga* Mill.	Seeds, fruits and flowers	Gujjars, Bhoris, Bakerwals	Bumchuont (Bandipora district, Jammu and Kashmir)	The seeds also form an important ingredient of a combination of different herbs such as seeds of *Cucumis sativus, Malva neglecta, Foeniculum vulgare,* fruits of *Ziziphus jujuba,* leaves and flowers of *Arnebia benthamii* and fronds of *Adiantum capillus-veneris*. This combination is locally called as "Sharbeth." The composite decoction of "Sherbeth" is given to cure jaundice, cough, cold, chronic constipation, fever and as a good blood purifier (Parvaiz et al., 2013)
138	*Decalepis hamiltonii* Wt. & Arn	Tuberous roots	Nakkala, Suygali, Yenadi, Yerukala.	Maredu kommulu (Chittor, Andhra Pradesh)	The root bark is powdered and used in a dose of 3–5 g, 2 times a day to stop jaundice (Jyothi et al., 2011)

TABLE 9.3 (Continued)

S. No.	Plant name	Part used	Tribe/Local community	Local names (State)	Uses/Mode of use
139	*Deeringia amaranthoides* (Lamarck) Merrill	Root	Local tribe	Chhorachurisag (Oraon) (Jalpaiguridistrict, West Bengal)	Root is used to treat jaundice (Bose et al., 2015)
140	*Dendrobium ovatum* (L.) Krenzl.	Whole plant	Yenadi	Nagli (Maharashtra)	Juice of fresh plant used for stomachic, carminative, antispasmodic, laxative, liver tonic (Ganapathy et al., 2013)
141	*Desmodium biflorum* L.		Kurichar and Kurumbar, Hill Pularyar (Cherupullari) Kani (Nilampullari)	Cheupulladri, Nilampullari (Kerala)	The whole plant is pounded well and 10 g of this is mixed with goat's milk, given on an empty stomach in the morning and evening for five days to treat jaundice and other liver complaints (Asha and Pushpangadan, 2002)
142	*Desmodium laxiflorum* DC.		Local healers	—	50–100 gms roots are crushed and boiled in water used for jaundice (Rama et al., 2012)
143	*Desmostachya bipinnata* (L.) Stapf		Local healers	Kusha (Arunachal Pradesh)	50 ml. extract mixing with powder of three fruits of *Piper longum* is taken daily for 10–15 days to cure jaundice (Rama et al., 2012)
144	*Dioscorea bulbifera* (L)	Tubers, bulbs	Local people	Kadu karanda (Ajara Tahsil, Kolhapur District, Maharashtra)	Used in jaundice (Sadale and Karadge, 2013)
145	*Eclipta alba* (L.) Hassk.	Leaves	Malayali tribes	Majalkarisalankanni (Shevaroy Hills, Tamil Nadu)	Decoction of leaves mixed with hot water used in liver disorders (Alagesaboopathi, 2015)

S. No.	Plant name	Part used	Tribe/Local community	Local names (State)	Uses/Mode of use
146	*Eclipta prostrata* (L.) L.	Leaves	Chenchu	Gunta-galijaraku (Rudrakod Sacred Grove, Nallamalais, Andhra Pradesh)	Leaves are squeezed in the first to get leaf sap and a teaspoonful of sap is given in orally, thrice in a day for 15 days to cure jaundice (Rao and Sunitha, 2011)
147	*Ecipta prostrata* (L.) L.	Leaves	Malayali tribes	Karisalankanni (Shevaroy Hills, Tamil Nadu)	Leaf juice mixed with cow's milk is given internally twice a day for one week to cure jaundice (Alagesaboopathi, 2015)
148	*Eclipta prostrata* (L.) L.	Whole plant	Local healers	Bhringaraja (Arunachal Pradesh)	20–30 gms paste of the whole plant mixed with salt is taken once daily for 15–20 days to cure jaundice (Rama et al., 2012)
149	*Eclipta prostrata* (L.) L.	Whole plant	Local people	Karisalanganni (Ariyalur District, Tamil Nadu)	The powder of *Eclipta prostrata, Leucas aspera* and *Phyllanthus niruri* are mixed with butter milk and taken orally to cure jaundice (Satishpandiyan et al., 2014)
150	*Eclipta prostrata* (L.) L.	Whole plant	Irula tribals	Manjal (Kalavai village, Vellore district, Tamil Nadu)	Used in jaundice (Natarajan et al., 2013)
151	*Elaeagnus caudata* Schlecht	Stem bark	Local healers	–	200–250 gms stem bark and fruit of the species are pounded and boiled in water; 100 ml. extract mixed with *Piper longum* is taken daily for 2–3 weeks to cure jaundice and other liver troubles (Rama et al., 2012)
152	*Elettaria cardamomum* Maton	Seed	Many Kerala tribes	Kerala	Seed extract is used for jaundice

TABLE 9.3 (Continued)

S. No.	Plant name	Part used	Tribe/Local community	Local names (State)	Uses/Mode of use
153	*Emblica officinalis* Gaertn.	Bark, fruits	Local tribe	Aonla, Aonwala (Kumaun region, Uttarakhand)	Bark decoction is used for treating diarrhea, dysentery, Cholera and jaundice (Gangwar et al., 2010).
154	*Emblica officinalis* Gaertn.	Stems	Local healers	Amlaki (Arunachal Pradesh)	100- 150 g pith of young branches are boiled in cow milk and made extract. 100 g of extract is taken one time daily in the early morning for 20 days to get relief from jaundice (Rama et al., 2012)
155	*Emblica officinalis* Geartn.	Fruit	Malayali tribes	Nelli (Shevaroy Hills, Tamil Nadu)	Fruit is consumed orally to control jaundice (Alagesaboopathi, 2015)
156	*Erythrina variegata* L.	Stem bark	Local healers	Mura (Arunachal Pradesh)	Stem bark is used for jaundice. (Rama et al., 2012)
157	*Erythroxylum monogynum* Roxb.	Leaf	Chenchu	Devadari, Dadiri (Rudrakod Sacred Grove, Nallamalais, Andhra Pradesh)	Leaf juice is administered for jaundice (Rao and Sunitha, 2011)
158	*Euphorbia geniculata* Ortega	Aerial parts	Gondand Madiya	Bada dudhi (Markanda Forest Range of Gadchiroli District, Maharashtra, India)	Used for Jaundice (Pankaj et al., 2015)
159	*Euphorbia hirta* L.	Whole Plant	Local peoples of Taindol	Dudhi (Taindol, Jhansi, Uttar Pradesh)	Used for jaundice (Jitin, 2013)

S. No.	Plant name	Part used	Tribe/Local community	Local names (State)	Uses/Mode of use
160	*Euphorbia hirta* L.	Leaf, latex	Herbalists, farmers, spiritualist	Dudhi (Yamuna Nagar district of Haryana)	Leaf juice is used to treat jaundice, fever, fungal infection, and syphilis and body nodes (Parul and Vashistha, 2015)
161	*Euphorbia hirta* L.	Whole plant	Local medical practitioners/healers	Doddak (Shivalik Hills of Northwest Himalaya)	Decoction of the whole plant is used in cough, asthma, bronchitis, jaundice and digestive problems (Devi et al., 2014)
162	*Exacum tetragonum* Roxb.	Whole plant	Local healers	Chireta (Arunachal Pradesh)	Plant juice or decoction is used thrice a day for jaundice (Rama et al., 2012)
163	*Ficus glomerata* Roxb	Fruit	Many Kerala tribes	Kerala	Used in jaundice
164	*Ficus racemosa* L.	Roots, leaves	Kani, Kurichar, Mannan (Athi)	Kerala	Tender roots and leaves are powdered to make a paste and 5–6 gm taken daily 5 days to treat jaundice. Tender roots powdered well and boiled in Goat's milk and given for liver complaints (Asha and Pushpangadan, 2002)
165	*Ficus religiosa* L.	Seed, latex, bark	local peoples of Taindol	Pipal (Taindol, Jhansi, Uttar Pradesh)	Used for jaundice (Jitin, 2013)
166	*Ficus semicordata* Buch.-Ham.	Leaf	Local healers	Padhotado (Arunachal Pradesh)	Leaf decoction in combination with that of *Bytmeria pilosa* Roxb. and *Phyllanthus fraternus* and the bark of *Callicarpa arborea* is taken orally for get relief from jaundice (Rama et al., 2012)

TABLE 9.3 (Continued)

S. No.	Plant name	Part used	Tribe/Local community	Local names (State)	Uses/Mode of use
167	*Flacourtia indica* Burm.f	Leaf, fruit	Gond, Kol, Baiga, Panica, Khairwar, Manjhi, Mawasi, Agaria	Rakatsonk, kateyya (Rewa dist, Madhya Pradesh)	Fruit juice is given in the morning in liver problems (Achuta et al., 2010)
168	*Foeniculum vulgare* Mill.	Whole plant	Gujjars, Bhoris Bakerwals	Bodiyaan (Bandipora district, Jammu and Kashmir)	Used in jaundice (Parvaiz et al., 2013)
169	*Fumaria indica* (Haussk.) Pugsely	Whole plant	Local medical practitioners/ healers	Pitpapar (Shivalik Hills of Northwest Himalaya)	Powder of whole plant is used for jaundice (Devi et al., 2014)
170	*Garcinia pedunculata* Roxb.	Fruits	Local healers	Thekera (Arunachal Pradesh)	Young fruits are prescribed in jaundice besides use as stimulant, emetic diuretic pulmonary and renal troubles (Rama et al., 2012)
171	*Gardenia jasminoides* Ellis	Whole plant	Local healers	Gandharaja (Arunachal Pradesh)	Plant is considered an indigenous medicine for cough, rheumatism, anemia and jaundice (Rama et al., 2012)
172	*Garuga pinnata* Collebr.	Bark	Local tribe	Jum, Tinn, Kharpat, Nil bhadi; Rosuni (Rava) (Jalpaiguridistrict, West Bengal)	Bark is used in jaundice (Bose et al., 2015)
173	*Gentiana moorcroftiana* G. Don	Whole plant	Larje, Amchis	Santik (Lahaul-spiti region Western Himalaya)	Fresh plant parts are crushed and the juice obtained is taken in empty stomach to cure jaundice (Koushalya, 2012)

S. No.	Plant name	Part used	Tribe/Local community	Local names (State)	Uses/Mode of use
174	*Gentiana tubiflora* (Wall. ex G.Don) Grisebach	Whole plant	Larje, Amchis	Chatik (Lahaul-spiti region Western Himalaya)	Juice of the whole plant is mixed with equal quantity of water and about half glass is taken orally during morning hours with empty stomach to cure jaundice (Koushalya, 2012)
175	*Geranium pratense* L.	Whole plant	Larje, Amchis	Podh-Lo, Tapan (Lahaul-spiti region Western Himalaya)	About one spoon of plant powder is taken orally with water to cure jaundice (Koushalya, 2012)
176	*Glycosmis arborea* (Roxb.) DC.	Root	Local tribe	Ashshewra (Jalpaiguri district, West Bengal)	Root powder used to treat fever, hepatopathy, eczema, skin diseases, wounds, liver disorder (Bose et al., 2015)
177	*Glycosmis pentaphylla* (Retz.) Correa	Whole plant	Local healers	—	Boiled extract of plant is administered in jaundice (Rama et al., 2012)
178	*Hedychium spicatum* Buch. Ham ex. Smith	Rhizome	Local tribe	Kapoor kachri (Kumaun region, Uttarakhand)	The powder of root is useful in the treatment of liver complaints (Gangwar et al., 2010)
179	*Hedyotis auricularia* L.	Whole plant	Local healers	—	Decoction of plant is used for 20–25 days or till complete relief (Rama et al., 2012)
180	*Helicteres isora* L.	Whole plant	Local tribe	—	Whole plant used to treat jaundice (Bose et al., 2015)
181	*Helminthostachya zeylanica* Hook.	Rhizome	Local tribe	Dinshabalindo (Meich); Nagdhup (Rava) (Jalpaiguri district, West Bengal)	Rhizome used to treat jaundice (Bose et al., 2015)
182	*Hemidesmus indicus* (L.) R.Br.	Roots	Mahadevkoli, Thakur and Katkari	Konkan Region, Maharashtra	Root powder given along with honey once a day for jaundice (Naikade and Meshram, 2014)

TABLE 9.3 (Continued)

S. No.	Plant name	Part used	Tribe/Local community	Local names (State)	Uses/Mode of use
183	Herabalformula-2	Leaves	Mahadevkoli, Thakur and Katkari	Konkan Region, Maharashtra	*Eclipta alba* L., *Phyllanthus amarus* Schum & Thonn. *Leucas aspera* (Willd.) Link. Leaves of above three plants are ground and extract is given for jaundice (Naikade and Meshram, 2014).
184	Herabalformula-3	Stems, fruits	Mahadevkoli, Thakur and Katkari	Konkan Region, Maharashtra	*Musa paradisiaca* L., *Lablab purpureus* L. Interior stem portion and fruits legume plants are prepared as a vegetable curry and given along with diet for jaundice (Naikade and Meshram, 2014)
185	Herabalformula-4	Leaf	Mahadevkoli, Thakur and Katkari	Konkan Region, Maharashtra	*Cynodon dactylon* (L.) Pers., *Phyllanthus amarus* Schum. & Thonn. Leaf extracts of above two plants are mixed and given with water (Naikade and Meshram, 2014)
186	Herbal formula-1	—	Mahadevkoli, Thakur and Katkari	Konkan Region, Maharashtra	*Cynodon dactylon, Phyllanthus amarus* and *Piper nigram.* Mixture of three plants leaves, fruits made into a juice and given to patient to cure the jaundice (Naikade and Meshram, 2014)
187	*Hibisucs lampas* Cav.	Roots	Kani, Kurumbar, Kuruchar (kolukatta)	Kolukatta (Kerala)	Expressed juice of the fresh roots (10–15 ml) is administered internally for 7 days for treating jaundice (Asha and Pushpangadan, 2002)
188	*Hippophae tibetana* Schltdl.	Berries	Larje, Amchis	Chha-Tuan (Lahaul-spiti region Western Himalaya)	Dried berries are crushed and boiled in water and decoction obtained is taken as a tea to cure jaundice (Koushalya, 2012)

S. No.	Plant name	Part used	Tribe/Local community	Local names (State)	Uses/Mode of use
189	*Homalomena aromatica* Schott	Rhizome	Local healers	Sugandhamantri (Arunachal Pradesh)	Rhizome juice is taken orally in jaundice and other liver complaints (Rama et al., 2012)
190	*Houttuynia cordata* Thunb.	Leaves	Local healers	Masundhuri (Arunachal Pradesh)	3–4 fresh leaves are eaten twice daily in case of jaundice. It is also used as condiment. Plant extract is also used for stomach complaints and jaundice (Rama et al., 2012)
191	*Hoya pendula* R. Br.	Root	Sugali, Yerukala, Yanadi	Pala thiga (Salugu Panchayati Of Paderu Mandalam, Visakhapatnam, district, Andhra Pradesh)	Used in jaundice (Padal et al., 2012)
192	*Hygrophila salicifolia* Nees	Whole plant	Local healers	Talmakhana, Minjonjo (Arunachal Pradesh)	Pounded whole plant is taken orally for stomach complaints and jaundice (Rama et al., 2012)
193	*Impatiens henslowiana* Arn.	Flowers	Kani, Mannan (Manja thetchi), Kurichar, Paniyan, Muthuvan (Nedimoorkhan)	Perumthumba (Kerala)	Expressed juice of flowers and tender leaves is used as nasal drops in severe conditions of jaundice (Asha and Pushpangadan, 2002)
194	*Inula cappa* (Buch-Ham.ex D. Don) DC.	Leaves	Local healers	–	The leaves are crushed with those of *Plantago asiatica* and *Lobelia angulata* and the juice is taken orally 10 ml. twice a day for jaundice (Rama et al., 2012)
195	*Ipomoea aquatica* Forsk.	Arial parts	Local healers	Kalmi (Arunachal Pradesh)	Arial part is taken orally for jaundice cure (Rama et al., 2012)

TABLE 9.3 (Continued)

S. No.	Plant name	Part used	Tribe/Local community	Local names (State)	Uses/Mode of use
196	*Ixora coccinea* L.	Roots	Many Kerala tribes (perumthumba)	Manjathetchi, Nedimoorkhan (Kerala)	Powdered mass of fresh roots (10–15 gm) in cold water are given for 7 days (thrice daily) to treat jaundice (Asha and Pushpangadan, 2002)
197	*Jatropha curcas* L.	Fruits, seeds, leaves	Local peoples of taindol	Taindol, Jhansi, Uttar Pradesh	Used for jaundice (Jitin, 2013)
198	*Lagenaria siceraria* (Mol.) Standl.	Whole plant	Local healers	Lao, Lauki (Arunachal Pradesh)	Plant decoction mixed with sugar is taken as protective agent during jaundice (Rama et al., 2012)
199	*Lagerstroemia speciosa* (L.) Pers.	Root bark	Local healers	Ajar (Arunachal Pradesh)	Decoction of root bark is given in jaundice and enlargement of spleen (Rama et al., 2012)
200	*Lawsonia inermis* L.	Stem bark	Local healers	Menhadi/ Jetuka (Arunachal Pradesh)	Stem bark is pounded in water and taken in jaundice and spleen enlargement (Rama et al., 2012)
201	*Lawsonia inermis*	Roots	Godaba	Manjuati (Semiliguda block, Koraput district, Odisha)	The root is crushed and taken with water of raw rice to cure jaundice (Raut et al., 2013)
202	*Leucas aspera* (Willd.) Link	Leaf	Malayali tribes	Thumbai (Shevaroy Hills, Tamil Nadu)	Fresh leaf juice is taken with water orally thrice a day for five days to liver diseases (Alagesaboopathi, 2015)
203	*Leucas aspera [Syn.: L. plukenetii* (Roth) Spreng.]	Whole plant	Local healers	Tumakusir (Arunachal Pradesh)	30–40 gms paste of the whole plant is taken three times in meal for 2–3 weeks to cure jaundice (Rama et al., 2012)

S. No.	Plant name	Part used	Tribe/Local community	Local names (State)	Uses/Mode of use
204	*Leucas aspera [Syn.: L. plukenetii* (Roth) Spreng.]	Leaf	Koch, Meich, Rava, Munda, Santhal, Garo, Oraon	(Coochbehar district, West Bengal)	Leaf juice used in jaundice (Tanmay et al., 2014)
205	*Leucas aspera* (Willd.) Link	Leaf	Mahadevkoli, Thakur and Katkari	(Konkan Region, Maharashtra)	Leaf paste applied on head to cure Jaundice (Naikade and Meshram, 2014)
206	*Limonia acidissima* L.	Leaves, fruits	Local communities	Vilam pazham (Jawadhu hills, Thiruvannamalai district of Tamil Nadu)	The fruit is much used in India as a liver and cardiac tonic (Ranganathan et al., 2012)
207	*Litsea monopetala* (Roxb.) Pers.	Stem bark	Local healers	Rogu (Arunachal Pradesh)	The stem bark is taken 2 times a day for one week orally to cure jaundice associated with hepatitis as follows: The bark is ground with the bark of *Vitex peduncularis*, 3 leaves of *Piper betel*, 4 clones of *Allium sativum*, 2–3 grains of *Piper nigrum* (Golmarich) and added 2 spoonfuls (10 gms) of sugar. The paste is made into pills and taken orally (Rama et al., 2012)
208	*Ludwigia adscendens* (L.) Hara	Young twigs	Local healers	Talijuri (Arunachal Pradesh)	200–300 gms of young twigs are smashed and boiled in water; 100 ml extract is taken orally for 15–20 days to cure jaundice (Rama et al., 2012)
209	*Mahonia nepalensis* DC.	Root	Local healers	—	Root juice is pounded and 2 teaspoons juice is given in jaundice and other liver disorders (Rama et al., 2012)

TABLE 9.3 (Continued)

S. No.	Plant name	Part used	Tribe/Local community	Local names (State)	Uses/Mode of use
210	*Malus domestica* Borkh.	Fruits	Gujjars, Bhoris Bakerwals	Maharaji Treil (Bandipora district, Jammu and Kashmir)	Fruits are harvested and stored at some warm place for 15–20 days so as to ripe completely. Ripe fruits are eaten to cure dyspepsia, diabetes, jaundice, urinary problems, loss of appetite and to remove phlegm from the chest, quench the thirst and dissolve the body fats (Jadhav et al., 2011)
211	*Manilkara hexandra* (Roxb.) Dubard.	Fruits	Gond and Madiya	Khirani (Markanda Forest Range of Gadchiroli district, Maharashtra, India)	Used for Jaundice (Pankaj et al., 2015)
212	*Maytenus emarginata* (Willd.) Ding Hou	Root bark, leaves	Herbal practitioners	Danti, Pedda chintu (Vizianagaram district, Andhra Pradesh)	10 to 15 leaves with sugar cube taken orally two times for 7 days to cure jaundice (Padal et al., 2013)
213	*Melothria purpusilla* (Blume) Cogn.	Whole plant	Local communities	Lamthabi (Manipur)	Vegetative parts of this plant is boiled with sugar candy in water and given in patient of Jaundice, Kidney infection (Rajesh et al., 2012)
214	*Momordica charantia*L.	Fruits	Mahadevkoli, Thakur and Katkari	Konkan Region, Maharashtra	Fruits are dried and fried given with normal diet for jaundice (Naikade and Meshram, 2014).
215	*Momordica charantia* L.	Fruit, leaf	Local medical practitioners/ healers	Karela (Shivalik Hills of Northwest Himalaya)	Fruits extract is used for diabetes, jaundice and cholera (Devi et al., 2014)

S. No.	Plant name	Part used	Tribe/Local community	Local names (State)	Uses/Mode of use
216	*Momordica charantia* L.		Local healers	Karela (Arunachal Pradesh)	The leaves are boiled with that of *Benincasa hispida* (Mainpawl) in the proportion of 5:100 gms, and the extract is taken orally against jaundice (Rama et al., 2012)
217	*Momordica charantia* L.	Leaves	Yenadi	Kakara (Andhra Pradesh)	Leaves juice used in jaundice (Bala et al., 2011)
218	*Momordica subangulata* Bl.	Fruits	Muthuvan Kurichar, Kani and Paniyan (kattupaval)	Kattupaval (Kerala)	Fresh juice of tender fruits used for jaundice (Asha and Pushpangadan, 2002)
219	*Morinda pubescens* Smith	Stem bark	Chenchu	Maddi (Rudrakod Sacred Grove, Nallamalais, Andhra Pradesh)	Fresh stem bark is crushed and stained in a glass of water throughout the night. The infusion is given orally in the morning for 7 days for jaundice (Rao and Sunitha, 2011)
220	*Moringa concanensis* Nimmo ex Dalz. & Gibson	Leaves	Gond tribe	Munaga (Adilabad district, Andhra Pradesh)	Leaves are boiled along with pulses and taken as food in anemia and jaundice (Murthy, 2012)
221	*Moringa concanensis* Nimmo ex Dalz. & Gibs.	Leaf	Local tribes	Jangli saragavo (R. D. F. Poshina forest range of Sabarkantha district, North Gujarat)	Half teaspoonful paste is prepared from the leaves and applied over the surface of body for a week to relief from jaundice (Patel and Patel, 2013)
222	*Moringa oleifera* Lam	Stem bark	Many Kerala tribes	Kerala	Used in jaundice
223	*Morus alba* L.		Local healers	Tuda (Arunachal Pradesh)	Root decoction is taken orally for 10–12 days for cure of jaundice (Rama et al., 2012)

TABLE 9.3 (Continued)

S. No.	Plant name	Part used	Tribe/Local community	Local names (State)	Uses/Mode of use
224	*Murdannia japonica* (Thunb.) Faden	Root	Local tribe	–	Root used to treat jaundice (Bose et al., 2015)
225	*Murraya koenigii* L.	Fruits, leaves, seeds	Irula tribals	Karuveppilai Kalavai village, Vellore district, Tamil Nadu)	Used in jaundice (Natarajan et al., 2013)
226	*Musa paradisiaca* L.	Stem	Mahadevkoli, Thakur and Katkari	Konkan Region, Maharashtra	Interior stem portion is dried and powder is given with honey for jaundice (Naikade and Meshram, 2014)
227	*Mussaenda frondosa* L.	Leaves	Local healers	–	Juice of fresh leaves is good for jaundice. Root decoction is taken orally for 10–12 days. 100 gms. pounded roots are boiled in water and made extract, 100 ml. extract is orally taken daily for 15 days to reduce jaundice (Rama et al., 2012)
228	*Myristica fragrans* Houtt.	Fruit	Many Kerala tribes	Kerala	Used to cure jaundice
229	*Naregamia alata.* Wight & Arn.	Whole plant	Kurumbar, Kani, Paniyan (Nilachar) Mannan, Kurumar, Muthuvan (Nilavepu)	Nilachara, Nilavepu (Kerala)	Whole plant extract is used for jaundice (Asha and Pushpangadan, 2002)

S. No.	Plant name	Part used	Tribe/Local community	Local names (State)	Uses/Mode of use
230	*Oldenlandia corymbosa* L.	Whole plant	Traditional healers	Vermela–vemu (Parnasala sacred grove area, Eastern Ghats of Khammam District, Telangana)	The fresh Plant extract is given in jaundice and other liver complaints. The decoction is given in low fever with gastric problems (Srinivasa et al., 2015)
231	*Oroxylum indicum* (L.) Vent.	Bark	Local healers	Shyonaka/ Bhatgila (Arunachal Pradesh)	Half kg. crushed bark is boiled in water and 100 ml. extract is taken thrice daily for 2–3 weeks to cure jaundice (Rama et al., 2012)
232	*Oroxylum indicum* Vent.	Fruit, seed, bark	Local tribe	Sona, Kanaidingi; Hatipanjara, Totala (Oraon); Dagduya (Munda); Jamblaophang (Rava); Kharukhandai (Meich) (Jalpaiguri district, West Bengal)	Paste of hydrated fruit or seed or bark applied in stomach pain, chest pain, used as appetizer, and for jaundice (Bose et al., 2015)
233	*Oxalis corniculata* L.	Leaf	Gaddi tribe	Amblu (Bharmor, Himachal Pradesh)	Leaf juice is useful in liver problems (Dutt et al., 2011)
234	*Peltigera canina* Willd.	Whole plant	Local healers	–	Plant juice is recommended for cure of jaundice and other liver disorders (Rama et al., 2012)
235	*Pergularia daemia* (Forrshk.) Chiov.	Leaf	Malayali tribes	Vealiparuthi (Shevaroy Hills, Tamil Nadu)	Leaf juice is mixed with cow's milk and drink to treat jaundice and liver problems (Alagesaboopathi, 2015)
236	*Phyllanthes amaras* Schum. & Thon.	Whole plant	Local healers	Bhumyamlaki (Arunachal Pradesh)	Plant juice as well as powder of dried plant is taken orally with water for 10–12 days (Rama et al., 2012)

TABLE 9.3 (Continued)

S. No.	Plant name	Part used	Tribe/Local community	Local names (State)	Uses/Mode of use
237	*Phyllanthus amarus* Schum. & Thonn.	Fruit	Traditional healers	Nelausiri (Parnasala sacred grove area, Eastern Ghats of Khammam district, Telangana)	Jaundice: Plant paste mixed with curd 3 spoonfuls is given orally twice a day for 7 days (Srinivasa et al., 2015)
238	*Phyllanthus amarus* Schum. & Thon.	Entire plant	Gond tribe	Nela usiri (Adilabad district, Andhra Pradesh)	Entire plant powder along with pepper powder is administered for jaundice (Murthy, 2012)
239	*Phyllanthus amarus* Schum. & Thonn.	Leaves	Chenchu	Nela usirika (Nallamalais, Andhra Pradesh)	Leaves mixed with curd given orally for jaundice-3 spoonfuls twice a day for 7 days (Sabjan et al., 2014)
240	*Phyllanthus amarus* Schum. & Thonn.	Root, fruit	Local people	Keela nelli (Ariyalur District, Tamil Nadu)	Roots and fruits are crushed and mixed with goat's milk. The mixture is taken orally to cure jaundice and liver problems (Satishpandiyan et al., 2014)
241	*Phyllanthus amarus* Schum. & Thonn.	Whole plant	Malayali tribes	Kellanelli (Shevaroy Hills, Tamil Nadu)	Decoction of the whole plant mixed with cow's milk and taken two times a day for one week to manage jaundice and liver complaints (Alagesaboopathi, 2015)
242	*Phyllanthus amarus* Schum. & Thonn.	Whole plant	Yandi	Nela vusiri (Chittor dist, Andhra Pradesh)	Oral administration of whole plant paste, about 10 g once daily for one week, cures jaundice (Ganesh and Sundarsanam, 2013)
243	*Phyllanthus amarus* Schum. & Thonn.	Leaf	Mahadevkoli, Thakur and Katkari	Konkan Region, Maharashtra	Leaf juice 2 spoonfuls + cow milk is given early in the morning for jaundice (Naikade and Meshram, 2014).

S. No.	Plant name	Part used	Tribe/Local community	Local names (State)	Uses/Mode of use
244	*Phyllanthus emblica* L.	Fruits, roots	Local tribe	Amlaki, Amlab (Jalpaiguridistrict, West Bengal)	Fresh fruit and root paste used to treat jaundice (Bose et al., 2015)
245	*Phyllanthus emblica*L.	Fruits	Mahadevkoli, Thakur and Katkari	Konkan Region, Maharashtra	Dried fruit and seeds of *Punica granatum* L. are grounded together along with sugar and made into powder, two-three teaspoons of the powder are dissolved in one cup of water and taken orally twice a day for three weeks (Naikade and Meshram, 2014)
246	*Phyllanthus fraternus* Webst.	Whole plant	Many Kerala tribes	Kerala	Used to cure liver problems
257	*Phyllanthus niruri* L.	Whole plant	Yenadi	Nela usiri (Andhra Pradesh)	Juice used in jaundice (Bala et al., 2011)
248	*Physalis minima* L.	Leaf	Chenchu	Buda busara (Nallamalais, Andhra Pradesh)	Leaf extract (15–20 ml) mixed with buffalo curd or sheep milk (100 ml) taken daily two times for 21 days to cure jaundice (Sabjan et al., 2014)
249	*Physalis minima* L.	Leaves	Traditional healers	Budima (Parnasala sacred grove area, Eastern Ghats of Khammam district, Telangana)	Leaves used to cure jaundice (Srinivasa et al., 2015)
250	*Picrorhiza kurrooa*Benth.	Root	Local healers	Katuki (Arunachal Pradesh)	Root extract/ decoction is used in jaundice for 12 days (Rama et al., 2012)

TABLE 9.3 (Continued)

S. No.	Plant name	Part used	Tribe/Local community	Local names (State)	Uses/Mode of use
251	*Piper nigrum* L.	Fruits	Rural people	Jalook (Ahoms, Chutiyas and Deuris villages of Chenijan, Jorhat district, Assam)	Fruits are crushed and mixed with two teaspoonful leaf extracts of Bel, *Aegle marmelos* Corr. (Rutaceae) is given to take orally thrice daily (Sonia et al., 2011)
252	*Polygala arvensis* Wlld.	Whole plant	Malayali tribes	Milakunanka (Shevaroy Hills, Tamil Nadu)	Decoction of the whole plant is given internally twice a day for one week to cure liver disorders (Alagesaboopathi, 2015)
253	*Polygala glabra* Willd.	Stem	Local healers	Maradu (Arunachal Pradesh)	Stem decoction is used in jaundice and related disorders (Rama et al., 2012)
254	*Polygonum glabrum* Wild	Root	Local healers	Bihagni, Bhilamgori (Arunachal Pradesh)	Root juice is given in jaundice and other disorders (Rama et al., 2012)
255	*Polygonum tortuosum* D. Don	Arial parts	Larje, Amchis	Nayalo (Lahaul-spiti region Western Himalaya)	Powder obtained from aerial parts is consumed orally with water to cure jaundice (Koushalya, 2012)
256	*Psidium guajava* L.	Fruit	Local healers	Madhurika, Amrud (Arunachal Pradesh)	Juice from one fruit, 1/4-liter goat milk and root of *Sida cordifolia* are mixed together thoroughly. The preparation is administered orally. 3 doses is sufficient and will result in disappearance of symptoms like clear urine and removal of yellowness from the eyes of the patients (Rama et al., 2012)
257	*Pterocarpus marsupium* Roxb.	Bark	Malayali tribes	Vengai (Shevaroy Hills, Tamil Nadu)	Decoction of the bark is taken as liver tonic (Alagesaboopathi, 2015)

S. No.	Plant name	Part used	Tribe/Local community	Local names (State)	Uses/Mode of use
258	*Pterocarpus santalinus* L.	Bark	Malayali tribes	Semmaram, Chandana vengai (Shevaroy Hills, Tamil Nadu)	Decoction of the bark is given orally twice a day for 5 days in jaundice (Alagesaboopathi, 2015)
259	*Raphanus sativus* L	Leaf, root	Local medical practitioners/healers	Muli (Shivalik Hills of Northwest Himalaya)	Extract of leaves and root is used for jaundice (Devi et al., 2014)
260	*Raphanus sativus* L.	Roots, leaves	Local people	Muli (Solan, Himachal Pradesh)	Young leaves taken for curing jaundice (Mamata and Sood, 2013)
261	*Raphanus sativus* L.	Leaf, root	Gond, Kol, Baiga, Panica, Khairwar, Manjhi, Mawasi, Agaria	Mooli (Rewa dist, Madhya Pradesh)	Juice of roots and leaves is given for curing jaundice (Achuta et al., 2010)
262	*Ricinus communis* L.	Leaves	Yenadi	Amudam (Andhra Pradesh)	Leaves paste used in jaundice (Bala et al., 2011)
263	*Ricinus communis* L.	Leaf	Gond, Halba and Kawar	Erandi (Darekasa hill range, Gondia district, Maharashtra)	Juice is prepared and given orally to cure jaundice (Chandra Kumar et al., 2015)
264	*Ricinus cummunis* L.	Leaves	Yerukala, Yanadi, Sugali	Patcha amudam (Sheshachala hill range of Kadapa District, Andhra Pradesh)	Decoction of leaves cures jaundice (Rajagopal et al., 2011)
265	*Rosa webbiana* Wall. ex Royle	Fruit	Larje, Amchis	T-siya, Seva, Shanab, susli (Lahaul-spiti region Western Himalaya)	Ripened fruit juice is taken with water to cure jaundice and impotency (Koushalya, 2012)

TABLE 9.3 (Continued)

S. No.	Plant name	Part used	Tribe/Local community	Local names (State)	Uses/Mode of use
266	*Rubia cordifolia* L.	Tubers	Konda dora tribe	Mangala katthi (Visakhapatnam district, Andhra Pradesh)	Tuber ground into paste with that of *Mirabilis jalapa* and made into pills. One pill each is administered daily thrice with water on empty stomach for jaundice (Padal et al., 2013a)
267	*Rumex nepalensis* Spreng.	Roots	Hakims, priests, tribal people	Halhaley (Eastern Sikkim Himalaya region)	Dried or fresh extract used orally in hepatitis, loss of hair, also plant used as dyes (Trishna et al., 2012)
268	*Saccharaum officinarum* L.	Whole plant	Koal, Panika, Bhuriya, Kharvar, Gaund	Ganna (Renukot forest division, Sonbhadra, Uttar Pradesh)	Used in jaundice (Anurag et al., 2012)
269	*Saccharum officinarum* L.	Stem	Local medical practitioners/ healers	Ganna (Shivalik Hills of Northwest Himalaya)	Extract of stem is used for constipation and jaundice (Devi et al., 2014)
270	*Sapindus emarginatus* Vahl	Leaves	Yerukala, Yanadi, Sugali	Kunkudu (Sheshachala hill range of Kadapa District, Andhra Pradesh)	Extraction of leaves cures jaundice (Rajagopal et al., 2011)
271	*Scorzonera divaricata* Turcz.	Leaves, shoots	Larje, Amchis	Thunpu (Lahaul-spiti region Western Himalaya)	Decoction of leaves and shoots prepared at low temperature is prescribed orally to cure jaundice (Koushalya, 2012)
272	*Sesbania grandiflora* (L.) Pers.	Leaf	Koch, Meich, Rava, Munda, Santhal, Garo, Oraon	Coochbehar district, West Bengal	Extract of leaves used in jaundice (Tanmay et al., 2014)

S. No.	Plant name	Part used	Tribe/Local community	Local names (State)	Uses/Mode of use
273	*Solanum nigrum* L.	Fruits	Bhil, Meena, Sahariya	Makoe (Baran district from Rajasthan)	Fruits are eaten thrice in a day in jaundice (Rishikesh and Ashwani, 2012)
274	*Solanum nigrum* L		Local healers	Kakamachi (Arunachal Pradesh)	Plant juice is administered with juice of equal quantity of *Phyllanthus amarus* and *Aloe barbadensis* for jaundice (Rama et al., 2012)
275	*Solanum nigrum* L.	Leaf	Malayali tribes	Manathaka (Shevaroy Hills, Tamil Nadu)	The leaf juice mixed with *Piper nigrum* (black pepper) and drink to treat liver ailments (Alagesaboopathi, 2015)
276	*Solanum nigrum* L.	Whole herb	Local tribe	Makoi (Kumaun region, Uttarakhand)	Decoction of leaves is used for liver (Gangwar et al., 2010)
277	*Solanum surattense* Burm.f.	Root bark	Traditional healers	Errivanga (Parnasala sacred grove area, Eastern Ghats of Khammam district, Telangana)	Jaundice: Root bark pound with stem bark of *Moringa oleifera*. 3 g of the paste had given orally once a day for 6 days (Srinivasa et al., 2015)
278	*Solanum xanthocarpum* Schrad. and Wendl.	Whole plant	Local people	Kantakari (Taindol, Jhansi, Uttar Pradesh)	Used for jaundice (Jitin, 2013)
279	*Solena heterophylla* Lour.	Roots	Local healers	Amtamoola (Arunachal Pradesh)	Fresh roots inhaled reduce jaundice; also the fresh roots are cut into pieces and tied with root of *Plumbago indica* and rhizome of *Curcuma longa* then worn around neck for 15 to 20 days to reduce jaundice (Rama et al., 2012)
280	*Sonchus oleraceus* L.	Leaves	Local people and traditional healers	Bakri booti (Ambala district, Haryana)	Plant extract mixed with clove is taken orally to cure liver diseases, particularly enlarged liver (Vashistha Kaur, 2013)

TABLE 9.3 (Continued)

S. No.	Plant name	Part used	Tribe/Local community	Local names (State)	Uses/Mode of use
281	*Spinaceae oleracea* L.	Seeds	Local healers	Palangshak (Arunachal Pradesh)	Seeds boiled in water and extract is taken for 7- 10 days (Rama et al., 2012)
282	*Swertia chirata* Buch.-Ham. ex Wall.	Whole plant	Local tribe	Bhucharitta, Kariyata, Chirata (Kumaun region, Uttarakhand)	Flowers, stem and roots are used in asthma, jaundice and anemia (Gangwar et al., 2010)
283	*Swertia chirata*Buch.-Ham. ex Wall.		Local healers	Chireta (Arunachal Pradesh)	Plant juice or decoction after boiling in water for 3–4 hours is taken orally one teaspoon full, thrice a day for 7–8 days. (Rama et al., 2012)
284	*Symplocos racemosa* Roxb	Stem, bark	Local traditional healers	Ludho (Nawarangpur District, Odisha)	Stembark decoction with honey (3:2) is given to children below 10 years against liver complaints (Dhal et al., 2014)
285	*Tabernaemontana divaricata* (L.) R. Br.	Roots	Local healers	—	100–200 gms roots are crushed and boiled in water; extract is taken three times to cure jaundice (Rama et al., 2012)
286	*Tabernaemontana divaricata* (L.) R.Br.	Root	Mahadevkoli, Thakur and Katkari	Konkan Region, Maharashtra	Root powder is boiled in water and the extract is given thrice a day for two-three weeks (Naikade and Meshram, 2014)
287	*Tamarindus indica* L.	Fruit, bark	Local people	Imli (Taindol, Jhansi, Uttar Pradesh)	Used for jaundice (Jitin, 2013)
288	*Taraxacum officinale* Wigg.	Root, rhizome	Local medical practitioners/healers	Dulal, Dudhi (Shivalik Hills of Northwest Himalaya)	Decoction of rhizome is used for jaundice (Devi et al., 2014)

S. No.	Plant name	Part used	Tribe/Local community	Local names (State)	Uses/Mode of use
289	*Tephrosia purpurea* L.	Roots	Chenchu	Vempalaku (Nallamalais, Andhra Pradesh)	Root extract (10 ml) mixed with a pinch of salt for Liver related stomach pain (Sabjan et al., 2014)
290	*Tephrosia purpurea* L.	Whole plant, roots	Gond, Kol, Baiga, Panica, Khairwar, Manjhi, Mawasi, Agaria	Sarpokha (Rewa dist, Madhya Pradesh)	Decoction is given in liver disorders (Achuta et al., 2010)
291	*Terminalia catappa* L.	Bark	Fisherman community	Nattu Vathamaram (Pudhukkottai District, Tamil Nadu)	It is used against liver diseases (Ramesh Kumar and Ramakritinan, 2013)
292	*Terminalia chebula* Retz.	Fruit	Many Kerala tribes	Kerala	Used to cure jaundice
293	*Terminalia chebula* Retz.	Fruit, bark	Yerukala, Yanadi, Sugali	Karakkaya (Sheshachala hill range of Kadapa district, Andhra Pradesh)	The decoction of bark cures fractures, ulcers, asthma, cough and jaundice (Rajagopal et al., 2011)
294	*Thespesia populnea* Corr.	Leaves	Herbal practitioners	Gangaravi, Gangaraya, Gangarenu, Gangirana, Muniganga-ravi (Vizianagaram district, Andhra Pradesh)	Extract of 3–4 fleshy leaves ground with an equal quantity of cow milk. This mixture is taken on empty stomach early in the morning for 7 days, it is effective remedy for jaundice (Padal et al., 2013)
295	*Thymus serpyllum* L.	Leaves	Local tribe	Ajwain (Kumaun region, Uttarakhand)	The herb is given in weak vision, complaints of liver and stomach, suppression of urine and menstruation (Gangwar et al., 2010)

TABLE 9.3 (Continued)

S. No.	Plant name	Part used	Tribe/Local community	Local names (State)	Uses/Mode of use
296	*Tinospora cordifolia* (Willd.) Miers	Whole plant	Mahadevkoli, Thakur and Katkari	Konkan Region, Maharashtra	Infusions of whole plant along with sugar juice are given to patient (Naikade and Meshram, 2014)
297	*Tinospora cordifolia* (Willd.) Miers	Stems	Local healers	Amrita, Guduchi (Arunachal Pradesh)	Fresh stem juice 10 ml. or plant soaked in water for overnight taken twice a day for 7 days (Rama et al., 2012)
298	*Tinospora cordifolia* (Willd.) Miers	Fruit	Yanadi	–	Dry fruits powder, about 5–10 g once daily for one month, with honey is administered orally to relieve from jaundice and burning micturition (Ganesh and Sudarsanam, 2013)
299	*Tinospora cordifolia* (Willd.) Miers	Leaf	Fisherman community	Seendhil (Pudhukkottai District, Tamil Nadu)	Leaf extract is taken orally with equal quantity of honey daily in the morning for jaundice until cure (Ramesh Kumar and Ramakritinan, 2013)
300	*Trianthema portulacastrum* L.	Leaf	Gond, Kol, Baiga, Panica, Khairwar, Manjhi, Mawasi, Agaria	Patharchatta (Rewa dist, Madhya Pradesh)	Leaf juice is given in jaundice (Achuta et al., 2010)
301	*Tribulus terrestris* L.	Leaf	Yanadi	–	Leaf paste, about 10–15 g once daily for a week, is administered orally to relieve from jaundice (Ganesh and Sudarsanam, 2013)
302	*Tridax procumbens* (L.) L.	Whole plant	Traditional healers	Gaddichamanthi (Parnasala Sacred Grove Area, Khammam District, Telangana)	Jaundice: Plant paste with jaggery is administered in doses of 2 spoonfuls per day for 7 days (Srinivasa et al., 2015).

S. No.	Plant name	Part used	Tribe/Local community	Local names (State)	Uses/Mode of use
303	*Trigonella emodi* Benth.	Flowers, leaves	Larje, Amchis	Rebuksu (Lahaul-spiti region Western Himalaya)	Flowers and leaves are dried and ground to prepare powder. One spoon of powder is taken twice a day for one week to cure jaundice (Koushalya, 2012)
304	*Valeriana wallichii* DC.	Whole herb	Local tribe	Samoy (Kumaun region, Uttarakhand)	Hepatoprotective (Gangwar et al., 2010)
305	*Vanda tessellata* (Roxb.) Hook.	Root	Local traditional healers	Malang (Nawarangpur District, Odisha)	Root decoction with common salt (2:1) is administered twice a day for 7 days against hepatitis (Dhal et al., 2014)
306	*Vitex negundo* L.		Malayali tribes	Notchi (Shevaroy Hills, Tamil Nadu)	Decoction of the flowers is given internally twice a day for five days in liver complaints (Alagesaboopathi, 2015)
307	*Woodfordia fruticosa* (L.) Kurz	Dried flower	Ethnic and Rural People	Dhayeroo (Sikkim)	For piles, liver complaints Bark for gastric trouble (Trishna et al., 2012)
308	*Woodfordia fruticosa* (L.) Kurz	Bark, flowers	Hakims, priests, tribal people	Dhayeroo (Eastern Sikkim Himalaya region)	Dried flower for piles, liver complaints Bark for gastric trouble (Trishna et al., 2012)
309	*Xeromphis spinosa* (Thunb.) Keay	Stems, roots	Local healers	—	Decoction of 200–300 gms. of stem and root bark of the species is mixed with *Piper longum* powder and taken thrice daily for 2–3 weeks to relieve jaundice (Rama et al., 2012)
310	*Zizphus jujuba* Mill.	Fruits	Local people	Ber (Taindol, Jhansi, Uttar Pradesh)	Used for treatment of liver disorders (Jitin, 2013)

the active component and their molecular interactions, which will help to analyze the therapeutic efficacy of the product so that it can be standardized and commercialized by pharma companies. Pre-clinical and clinical investigation of traditional medicinal plants for hepatoprotective activity may provide valuable leads for the development of safe and effective drugs.

KEYWORDS

- **ethnobotany**
- **hepatoprotective**
- **liver**
- **liver disorders**

REFERENCES

Achuta, N. S., Sharad, S., & Rawat, A. K. S. (2010). An ethnobotanical study of medicinal plants of Rewa district of Madhya Pradesh. *Indian J. Trad. Knowl., 9*(1), 191–202.

Alagesaboopathi, C. (2015). Medicinal plants used for the treatment of liver diseases by Malayali tribes in Shevaroy hills, Salem district, Tamil Nadu, India. *World J. Pharmaceut. Res., 4*(4), 816–828.

Anandakumar, D., Rathinakumar, S. S., & Prabakaran, G. (2014). Ethnobotanical survey of Athinadu Pachamalai hills of Eastern Ghats in Tamil Nadu, South India. *Int. J. Adv. Interdis. Res., 1*(4), 7–11.

Anurag, S., Singh, G. S., & Singh, P. K. (2012). Medico-ethnobotanical inventory of Renukot forest division, Sonbhadra, Uttar Pradesh, India. *Indian J. Nat. Prod. Resour., 3*(3), 448–457.

Asha, V. V., & Pushpangadan, P. (2002). Hepatoprotective plants used by the tribals of Wynadu, Malappuram and Palghat districts of Kerala, India. *Anc. Sci. of life, 22*(1), 1–8.

Bala, K. M., Mythili, S., Sravan, K. K., Ravinder, B., Murali, T., & Mahender, T. (2011). Ethnobotanical survey of medicinal plants in Khammam district, Andhra Pradesh, India. *Int. J. Appl. Boil. Pharma. Tech., 2*(4), 366–370.

Bose, D., Roy, J. G., Mahapatra, S. D., Datta, T., Mahapatra, S. D., & Biswas, H. (2015). Medicinal plants used by tribals in Jalpaiguri district, West Bengal, India. *J. Med. Plants Stud., 3*(3), 15–21.

Chandrakumar, P., Praveenkumar, N., & Sushama, N. (2015). Ethnobotanical studies on the medicinal plants of Darekasa hill range of Gondia district, Maharashtra, India. *Int. J Res. Plant Sci, 5*(1), 10–16.

Devi, U., Sharma, P., & Rana, J. C. (2014). Assessment of ethnomedicinal plants in Shivalik hills of Northwest Himalaya, India. *Am. J. Ethnomed., 1*(4), 186–205.

Dhal, N. K., Panda, S. S., & Muduli, S. D. (2014). Ethnobotanical studies in Nawarangpur district, Odisha, India. *Am. J. Phytomed. Clin. Therapeu., 2*(2), 257–276.

Doreswamy, R., & Sharma, D. (1995). Plant drugs for liver disorders management. *Indian drugs,* 32, 139–144.

Dutt, B., Sharma, S. S., Sharma, K. R., Gupta, A., & Singh, H. (2011). Ethnobotanical survey of plants used by gaddi tribe of Bharmor area in Himachal Pradesh. *ENVIS Bulletin: Himalayan Ecology, 19,* 22–27.

Ganapathy, S., Ramaiah, M., Sarala, S., & Babu, P. M. (2013). Ethnobotanical literature survey of three Indian medicinal plants for hepatoprotective activity. *Int. J. Res. Ayur. Pharm., 4*(3), 378–381.

Ganesh, P., & Sudarsanam, G. (2013). Ethnomedicinal plants used by Yanadi tribes in Seshachalam biosphere reserve forest of Chittoor district, Andhra Pradesh India. *Int. J. Pharm. & Life Sci., 4*(11), 3073–3079.

Gangwar, K. K., Deepali & Gangwar, R. S. (2010). Ethnomedicinal plant diversity in Kumaun Himalaya of Uttarakhand, India. *Nature and Science, 8*(5), 66–78.

Girish, C., Koner, B. C., Jayanthi, S., Rao, K. R., Rajesh, B., & Pradhan, S. C. (2009). Hepatoprotective activity of six polyherbal formulations in paracetamol induced liver toxicity in mice. *Indian J. Med. Res., 129*(5), 569–578.

Handa, S. S., Sharma, A., & Chakraborty, K. K. (1986). Natural products and plants as liver protecting drugs. *Fitoterapia, 57,* 307–351.

Harshberger, J. (1999). "What is Ethnobotany and why is it important?" Ethnobotany at Fort Lewis College. Available online at http://anthro. fortlewis.edu/ethnobotany/ethno2. htm. Accessed on 07 June 2015.

Jadhav, V. D., Mahadkar, S. D., & Valvi, S. R. (2011). Documentation and ethnobotanical survey of wild edible plants from Kolhapur district. *Recent Res.Sci. & Tech., 3*(12), 58–63.

Jitin, R. (2013). An ethnobotanical study of medicinal plants in Taindol Village, district, Jhansi, region of Bundelkhand, Uttar Pradesh, India. *J. Med. Plants Stud., 1*(5), 59–71.

Jyothi, B., Prasad, G. P., Sudarsanam, G., Sitaram, B., & Vasudha, K. (2011). Ethnobotanical investigation of underground plant parts form Chittoor district, Andhra Pradesh, India. *Life Sci. Leafl., 18,* 695–699.

Koushalya, N. S. (2012). Traditional knowledge on ethnobotanical uses of plant biodiversity: A detailed study from the Indian Western Himalaya. *Biodiv. Res. Conserv., 28,* 63–77.

Mamta, S., & Sood, S. K. (2013). Ethnobotanical Survey for wild plants of district Solan, Himachal Pradesh, India. *Int. J. Environ. Biol., 3*(3), 87–95.

Mohammad, A. (1994). Textbook of Pharmacognosy. Publ. SBS publishers, New Delhi. 8–14.

Mohan, H.(2005).Text book of pathology, (5th Ed). Jayapee Brother's Medical, New Delhi. 610–613.

Mohan, R. K., & Murthy, P. V. B. (1992). Plants used in traditional medicine by tribals of Prakasam district, Andhra Pradesh. *Anc. Sci. Life, 11*(3), 176–181.

Mukesh, K., Tariq, A. B., Hussaini1, S. A., Kishore, K., Hakimuddin, K., Aminuddin, & Samiulla, L. (2014). Ethnomedicines in the Khordha forest division of Khordha district, Odisha, India. *Int. J. Curr. Microbiol. App. Sci., 3*(1), 274–280.

Murthy, E. N. (2012). Ethnomedicinal plants used by Gonds of Adilabad district, Andhra Pradesh, India. *Int. J. Pharm. & Life Sci., 3*(10), 2034–2043.

Naikade, S. M., & Meshram, M. R. (2014). Ethno-medicinal plants used for jaundice from Konkan region, Maharashtra, India. *Int. J. Pharm. Sci. Inven., 3*(12), 39–41.

Natarajan, A., Leelavinodh, K. S., Jayavelu, A., Devi, K., & Senthil, K. B. (2013). A study on ethnomedicinal plants of Kalavai, Vellore district, Tamil Nadu, India. *J. App. Pharm Sci., 3*(1), 99–102.

Padal, S. B., Butchi, R. J., & Chandrasekhar, P. (2013a). Traditional knowledge of Konda Dora tribes, Visakhapatnam district, Andhra Pradesh, India. *IOSR J. Pharm., 3*(1), 22–28.

Padal, S. B., Chandrasekhar, P., & Vijakumar, Y. (2013b). Traditional uses of plants by the tribal communities of Salugu Panchayati of Paderu Mandalam, Visakhapatnam district, Andhra Pradesh, India. *Int. J. Comp. Eng. Res., 3*(5), 98–103.

Padal, S. B., Venkaiah, M., Chandrasekhar, P. & Vijayakumar, Y. (2013). Traditional phytotherapy of Vizianagaram district, Andhra Pradesh, India. *IOSR J. Pharm., 3*(6), 41–50.

Pankaj, R., Chavhan & Aparna, S. M. (2015). Ethnobotanical survey of Markanda forest range of Gadchiroli district, Maharashtra, India. *Brit. J. Res., 2*(1), 055–062.

Parul, B., & Vashistha, D. (2015). An Ethnobotanical study of plains of Yamuna Nagar district, Haryana, India. *Int. J. Inn. Res. Sci., Eng. & Tech., 4*(1), 18600–18607.

Parvaiz, A. L., Ajay, K. B., & Fayaz, A. B. (2013). A study of some locally available herbal medicines for the treatment of various ailments in Bandipora district of J&K, India. *Int. J. Pharm. Bio. Sci., 4*(2), 440–453.

Patel, H. R., & Patel, R. S. (2013). Ethnobotanical plants used by the tribes of R. D. F. Poshina forest range of Sabarkantha district, North Gujarat, India. *Int. J. Sci. & Res. Pub., 3*(2), 1–8.

Praful, B. G., & Darshan P. G. (1996). Text book of medical laboratory technology. Publ. Bhalani Publishing House, Bombay. p. 186.

Pratap, G. P., Prasad, G. P., & Sudarsanam, G. (2009). Ethnomedical studies in Kilasagirikona Forest Range of Chittor district, Andhra Pradesh. *Anc. Sci. Life, 29*(2), 40–45.

Rajagopal, R. S., Madhusudhana, R. A., Philomina, N. S., & Yasodamma, N. (2011). Ethnobotanical survey of sheshachala hill range of Kadapa district, Andhra Pradesh, India. *Indian J. Fund. & Appl. Life Sci., 1*(4), 324–329

Rajesh, S. Y., Onita, D, C. H., Santosh, K. S., & Abujamand, D. C. (2012). Study on the Ethnomedicinal system of Manipur. *Int. J. Pharm. & Biol. Arch., 3*(3), 587–591.

Rama, S., Rawat, M. S., Deb, S., & Sharma, B. K. (2012). Jaundice and its traditional cure in Arunachal Pradesh. *J. Pharm. & Sci. Inn., 1*(3), 93–97.

Rameshkumar, S., & Ramakritinan, C. M. (2013). Floristic survey of traditional herbal medicinal plants for treatments of various diseases from coastal diversity in Pudhukkottai district, Tamil Nadu, India. *J. Coastal Life Med., 1*(3), 225–232.

Ranganathan, R., Vijayalakshmi, R., & Parameswari, P. (2012). Ethnomedicinal survey of Jawadhu hills in Tamil Nadu. *Asian J. Pharm. Clin. Res., 5*(2), 45–49.

Rao, B. R. P., & Sunitha, S. (2011). Medicinal plant resources of Rudrakod Sacred Grove in Nallamalais, Andhra Pradesh, India. *J. Biodiversity, 2*(2), 75–89.

Raut, S., Raut, S., Sen, S. K., Satpathy, S., & Pattnaik, D. (2013). An Ethnobotanical survey of medicinal plants in Semiliguda of Koraput district, Odisha, India. *Res. J. Recent Sci., 2*(8), 20–30.

Reddy, K. N., Reddy, C. S., & Raju, V. S. (2008). Ethnomedicinal Observations among the Kondareddis of Khammam District, Andhra Pradesh, India. *Ethnobot. Leaflets, 12*, 916–926.

Rishikesh, M., & Ashwani, K. (2012). Ethnobotanical survey of medicinal plants from Baran-district of Rajasthan, India. *J. Ethnobiol. Trad. Med., 117*, 199–203.

Sabjan, G., Sudarsanam, G., Dharaneeshwara R. D., & Muralidhara Rao, D. (2014). Ethnobotanical crude drugs used in treatment of liver diseases by Chenchu tribes in Nallamalais, Andhra Pradesh, India. *Am. J. Ethnomed., 1*(3), 115–121.

Sadale, A. N., & Karadge, B. A. (2013). Survey on Ethno-Medicinal plants of Ajara Tahsil, district Kolhapur, Maharashtra, India. *Trend. Life sci., 2*(1), 21–25.

Sathishpandiyan, S., Prathap, S., Vivek, P., Chandran, M., Bharathiraja, B., Yuvaraj, D. & Smila, K. H. (2014). Ethnobotanical study of medicinal plants used by local people in Ariyalur district, Tamil Nadu, India. *Int. J. Chem Tech Res., 6*(9), 4276–4284.

Sharma, A., Shing, R. T., Sehgal, V., & Handa, S. S. (1991). Antihepatotoxic activity of some plants used in herbal formulations. *Fitoterapia, 62*(2), 131–138.

Sonia, S. G. S., Neeliand, S. S., & Jalapure. (2011). Ethnobotanical survey of some of the herbs used in Jorhat district, Assam. *Int J Curr Pharm Res., 3*(4), 53–54.

Srinivasa, R. D., Shanmukha, R. V., Prayaga, M. P., Narasimha, R. G. M., & Venkateswara, R. Y. (2015). Some ethnomedicinal plants of Parnasala Sacred Grove Area Eastern Ghats of Khammam district, Telangana, India. *J. Pharm. Sci. & Res., 7*(4), 210–218.

Stickel, F., Patsenker, E., & Schuppan, D. (2005). Herbal hepatotoxicity. *J. Hepatol., 43,* 901–910.

Stickel, F., & Schuppan, D. (2007). Herbal medicine in the treatment of liver diseases. *Dig. Liver Dis., 39*, 293–304.

Swapna, B. (2015). An ethnobotanical survey of plants used by Yanadi tribe of Kavali, Nellore district, Andhra Pradesh, India. *J Sci. Innov. Res., 4*(1), 22–26.

Tanmay, D., Amal, K. P., & Santanu, G. D. (2014). Medicinal plants used by tribal population of Coochbehar district, West Bengal, India-an ethnobotanical survey. *Asian Pac. J. Trop. Biomed. 4*(1), S478–S482.

Trishna, D., Shanti, B. M., Dipankar, S., & Shivani, A. (2012). Ethnobotanical survey of medicinal plants used by ethnic and rural people in Eastern Sikkim Himalayan Region. *African J. Basic & Appl. Sci., 4*(1), 16–20.

Vashistha, B. D., & Kaur, M. (2013). Floristic and ethnobotanical survey of Ambala district, Haryana. *Int. J. Pharm. Bio. Sci., 4*(2), 353–360.

CHAPTER 10

ETHNOMEDICINE FOR SKIN DISEASES IN INDIA

J. KOTESWARA RAO, J. SUNEETHA, R. RATNA MANJULA, and T. V. V. SEETHARAMI REDDI

Department of Botany, Andhra University, Visakhapatnam – 530003, India, E-mail: reddytvvs@rediffmail.com

CONTENTS

ABSTRACT

The review deals with 583 species of plants covering 419 genera and 135 families used by different tribes living throughout India for curing a variety of skin ailments. They are used in curing 34 skin diseases under 904 ethnomedicinal practices. *Azadirachta indica* and *Jatropha curcas* cures a maximum of 9 skin ailments each followed by *Abutilon indicum, Argemone mexicana, Callicarpa macrophylla, Datura metel, Lawsonia inermis* each curing 7 ailments and others. As many as 27 plants cure 5–9 skin diseases. Of the 904 practices 789 involve single plant only and the rest 115 involve a combination of one to seven plants. Detailed phytochemical and pharmacological studies are needed to determine the effective constituents and characteristic biological activity.

10.1   INTRODUCTION

Primitive man gets much enthusiasm towards green plants and started to examine the property of using plants by trial and error and obtained different beneficial properties. Later he became enriched with the knowledge of many useful and harmful plants. His enriched knowledge has been transferred from one generation to another without any written documents. Much of the information about the medicinal value of plants has come from the knowledge gathered by the aboriginals since the tribes mostly inhabit forested and hilly terrains. The familiarity with plant species producing medicines, essential oils and insecticides dates back to the beginning of civilization. They must have tested the plant resources of the earth and the knowledge thus gained must have been retained by their ancestors.

In India the scheduled tribe population is 84,326,240 constituting about 8.2% of country total population of 1,026,103,289 belonging to over 550 tribal communities and 227 ethnic groups. They inhabit about 5000-forested villages and about 15% of countries geographical area is occupied by them. They are spread over varied geographic and climatic zones of the country. The skin is the largest in the human body. It is the first line of defense against hostile environment, being waterproof and germ- proof. The inflammatory skin disorders include a range of rashes and lesions that cause irritation and inflame the skin, viral skin problems and fungal infections.

10.2   REVIEW OF LITERATURE

Sinha et al. (1996) dealt with 27 ethnomedicinal plants belonging to 24 genera and 19 families of Andaman and Nicobar Islands for curing various skin disorders among the tribes *Jarawas, Sentinelese, Great Andamanese, Onges, Nicobarese* and the *Shompens.* Nayak et al. (2004) provided information on the use of plant crude drugs for various diseases used by *Paraja, Kandha,*

Kutia kandha, Tekeria and *Jhadia* tribes of Kalahandi district, Orissa. Ayyanar and Ignacimuthu (2005) reported the use of 14 plants to cure skin diseases by the *Kani* ethnic group in Tirunelveli hills of Tamil Nadu. Jeeva et al. (2007) described 30 plant species belonging to 29 genera and 22 families used in the treatment of skin diseases in south Travancore, southern peninsular India. Bhattacharjee and Chatterjee (2007) reported 10 plant species used for skin diseases by the *Oraon* tribe of North 24 Paraganas district of West Bengal. The study of Wayanad district, Kerala by Nisha and Sivadsan (2007) resulted in 62 species of plants used by tribes (*Paniyas, Kurumas, Adiyars, Kurichyas, Kattanaikans, Kadans*) for curing various skin diseases. Lal and Singh (2008) communicated information on 18 plant species belonging to 14 families used by the *Lahaula* and *Bhotia* communities for curing different skin disorders in Himachal Pradesh. Prashantkumar and Vidyasagar (2008) described 26 plant species of 25 genera belonging to 16 families used by the *Kadukuruba* and *Lambani* folk of Bidar district, Karnataka for the treatment of skin diseases. Das et al. (2008) studied medicinal plants used by *Jaintia, Riang, Chorai, Hrangkhol, Mizo, Vaiphei paite, Karbi, Naga* and *Kuki* tribes of Cachar district, Assam for curing various ailments. Silja et al. (2008) revealed the use of 136 plant species for traditional medicinal purposes by the *Mullu kuruma* tribe of Wayanad district, Kerala. Kingston et al. (2009) reported 30 plant species belonging to 29 genera and 22 families for treating skin diseases in Kanyakumari district of Tamil Nadu. Raju et al. (2010) recorded 104 species of plants belonging to 91 genera and 49 families used by the *Konda reddi* tribe of Andhra Pradesh for curing various skin diseases. Shukla et al. (2010) studied 166 plant species used against different diseases including skin disorders by the tribes *Gond, Kol, Baiga, Punica, Khairwar, Manjhi, Mawasi* and *Agaria* of Rewa district, Madhya Pradesh. Shah et al. (2011) reported 54 grandmas' prescriptions for skin ailments in Valsad district, Gujarat. Manjula and Reddi (2015) reported 77 species of plants covering 68 genera and 37 families used for treating scabies and skin diseases by the *Koya, Lambada, Gond/Naikpod, Yerukula, Nayak* and *Konda reddi* tribes of Khammam district of Andhra Pradesh. A study on ethnomedicinal plants used for skin diseases by the *Bagata, Gadaba, Khond, Konda dora, Konda kammara, Kotia, Mali, Muka dora, Porja* and *Valmiki* tribes of Visakhapatnam district, Andhra Pradesh resulted in 51 species of plants covering 49 genera and 39 families (Babu and Reddi, 2015). Suneetha and Reddi (2015) dealt with 67 species of plants covering 59 genera and 36 families used for curing skin diseases by the *Konda reddi, Konda dora,*

Koya dora, Konda kammara, Konda kapu, Manne dora and *Valmiki* tribes of East Godavari district, Andhra Pradesh. The present review provides folk prescriptions of botanical origin for treating some of the cutaneous elements like acne, alopecia, allergy, anemia, blisters, boils, bruises, burns, carbuncle, cuts, dandruff, eczema, excoriation, heel cracks, herpes, itches, leprosy, leucoderma, pimples, psoriasis, scabies, scurvy, tumors, ulcers, whitlow and wounds.

10.3 ENUMERATION

The plants are arranged in a tabular form in alphabetical order with botanical name followed by practice and reference (*see* Table 10.1).

10.4 RESULTS AND DISCUSSION

The present review deals with 583 species of plants covering 419 genera and 135 families used by different tribes for curing a variety of skin diseases ranging from acne to wounds. Fabaceae is the dominant family with 40 species followed by Euphorbiaceae (31 spp.); Asteraceae (24 spp.); Apocynaceeeae, Caesalpiniaceae, Mimosaceae (each 22 spp.); Lamiaceae (18 spp.); Verbenaceae, Poaceae (17 spp. each); Malvaceae, Rubiaceae (16 spp. each); Solanaceae (15 spp.); Cucurbitaceae, Acanthaceae (14 spp. each); Convolvulaceae (13 spp.); Rutaceae (11 spp.); Moraceae, Asclepiadaceae (9 spp. each); Arecaceae, Zingiberaceae (8 spp. each); Amaranathaceae, Liliaceae, Combretaceae, Menispermaceae (each 7 spp.); Anacardiaceae, Sterculiaceae (each 6 spp.) and others ranging from 1–5 species. The plants have been presented with their botanical name, method of preparation, dosage and mode of administration. Five hundred and eighty three (583) plants are used in curing 34 skin diseases under 904 ethnomedicinal practices. *Azadirachta indica* and *Jatropha curcas* cures a maximum of 9 skin ailments each followed by *Abutilon indicum, Argemone mexicana, Callicarpa macrophylla, Datura metel, Lawsonia inermis* each curing 7 ailments; *Acacia catechu, Aegle marmelos, Aloe barbadensis, Cassia alata, Curcuma longa, Hibiscus rosa-sinensis, Mallotus philippensis, Ocimum tenuiflorum, Zanthoxylum armatum* each curing 6 diseases; *Ageratum conyzoides, Cassia fistula, Cocos nucifera, Datura innoxia, Eclipta prostrata, Heliotropium indicum, Holoptelea integrifolia, Leonotis nepetaefolia, Psoralea corylifolia,*

TABLE 10.1 Ethnomedicinal Plants Used for Skin Diseases

S. No.	Name of the plant	Practice	Reference
1	*Abelmoschus manihot* (L.) Medik.	Eczema: Root paste mixed with a pinch of turmeric is applied on the affected areas daily twice till cure.	Suneetha, 2007
2	*A. moschatus* (L.) Medik.	Skin diseases: Seed paste mixed with coconut oil is applied on the affected areas daily twice for 3 days.	Raju et al., 2010
3	*Abrus precatorius* L.	Dandruff: Leaf paste is applied externally before going to head bath till cure.	Naidu, 2003; Rao, 2010; Suneetha, 2007
		Leucoderma: White variety leaf paste is applied on the white patches and exposed to sunlight daily for one hour for 30 days.	
		Ringworm: Leaf paste of red variety is applied on the affected areas daily twice for 2 days.	
4	*Abutilon indicum* (L.) Sweet	Burns: A paste made of the seed and *Plumbago* root is applied as a stimulant dressing.	Prashantkumar and Vidyasagar, 2008; Singh and Maheshwari, 1989; Naidu, 2003
		Boils, abscesses, scabies, carbuncles and itches: Ash of whole plant is applied on affected areas.	
		Leprosy: A spoonful of root paste mixed with a spoonful of root paste of *Amaranthus spinosus* is administered daily twice for 21 days.	
5	*Acacia auriculiformis* A. Cunn. ex Benth.	Eradication of lice: Stem bark paste is applied to the scalp.	Babu, 2007
6	*A. caesia* (L.) Willd.	Common skin diseases: Leaf paste mixed with a pinch of turmeric is applied on the affected areas daily twice for 2 days.	Suneetha and Reddi, 2015

TABLE 10.1 (Continued)

S. No.	Name of the plant	Practice	Reference
7	*A. catechu* (L.f.) Willd.	Leprosy: The wood is used as an ingredient of concoction for curing.	Babu, 2007; Suneetha and Reddi, 2015; Nisha and Sivadasan, 2007; Naidu, 2003; Babu and Reddi, 2015
		Common skin diseases: Stem bark paste is applied on the affected areas daily once till cure.	
		Cuts and wounds: Stem bark paste is applied on the affected areas.	
		Leprosy: Stem bark decoction is administered thrice a day for one month.	
		Scabies: Leaf paste mixed with a pinch of turmeric is applied on the affected areas daily twice for 2 days.	
8	*A. chundra* (Roxb. ex Rott.) Willd.	Wounds: The gum extracted from the bark is used as astringent.	Suneetha and Reddi, 2015; Babu, 2007; Sharma et al., 2003
		Leprosy: Bark ground with leaf bases of neem and the paste is applied on ulcers.	
		Common skin diseases: Stem bark decoction is taken orally in 2–3 spoonfuls thrice a day till cure.	
9	*A. concinna* DC.	Hair growth, splitting, falling and dandruff: Dry powder of the pod is used.	Shukla et al., 2010
10	*A. indica* L.	Ringworm: Leaf paste with lime juice is given in infection.	Shukla et al., 2010; Suneetha and Reddi, 2015
		Eczema: Handful of leaves, a garlic bulb and a few pepper seeds are ground. The paste is made into pills of peanut seed size and administered daily twice for 7 days.	
11	*A. leucophloea* (Roxb.) Willd.	Leucoderma: Leaf paste is applied on the affected areas twice daily till cure.	Swamy, 2009
12	*A. nilotica* (L.) Del.	Eczema: Twenty-five g of the bark is boiled in one liter of water and the vapors are allowed to foment on the affected parts.	Suneetha, 2007; Babu, 2007
		Heel cracks: Gum paste is applied on the cracks.	

S. No.	Name of the plant	Practice	Reference
13	*A. polycantha* Willd.	Leprosy: Twenty ml of oil extracted from seeds is given twice a day for 30 days.	Raju, 2009
14	*A. racemosa* Baill.	Skin diseases: Powder of root and leaf along with seeds of *Strychnos nux-vomica* mixed with cold water is applied externally on affected places for 40 days.	Ayyanar and Ignacimuthu, 2005
15	*A. sinuata* (Lour.) Merr.	Dandruff: Fruit paste is applied to head 1 h before bath twice a week for 2–3 weeks. Ringworm: 2 spoonfuls of dried fruit ash mixed with a spoonful of sesame oil is applied on the affected areas daily twice for 5 days.	Silja et al., 2008; Swamy, 2009
16	*A. torta* (Roxb.) Craib	Wounds: Stem bark paste mixed with a pinch of alum is applied on the affected areas daily twice for 2 days.	Manjula, 2011
17	*Acalypha ciliata* Forssk.	Common skin diseases: Leaves are ground with turmeric and the paste is applied on the affected areas.	Suneetha, 2007
18	*Acanthospermum hispidum* DC.	Ringworm: Whole plant paste mixed with castor oil is applied on the affected parts till cure.	Rao, 2009; Shukla et al., 2010
19	*Acanthus leucostachys* Wallich	Skin diseases: Leaf paste is applied on the affected parts. Cuts and wounds: Leaf paste is applied externally.	Das et al., 2008
20	*Achyranthes aspera* L.	Eczema and psoriasis: A paste of leaf is made with turmeric and smeared on the infected areas as an effective remedy. Eczema and psoriasis: Root paste is applied on the affected area till cure. Psoriasis: Ten g of leaves are ground with 5 g of dried zinger and made into paste. This paste is applied on the affected areas daily twice till cure.	Nisha and Sivadasan, 2007; Prashantkumar and Vidyasagar, 2008; Babu, 2007

TABLE 10.1 (Continued)

S. No.	Name of the plant	Practice	Reference
21	*Acorus calamus* L.	Eczema: Pounded rhizomes along with *Curcuma aromatica* rhizomes and *Azadirachta indica* leaves are applied on the affected parts twice a day for one week.	Jeeva et al., 2007
22	*Adansonia digitata* L.	Skin disease: Leaf paste mixed with turmeric powder is applied on the affected areas till cure.	Manjula and Reddi, 2015
23	*Adenostemma lavenia* (L.) Kunt.	Cuts and wounds: Leaf paste is externally used.	Sinha et al., 1996
24	*Adhatoda vasica* Nees	Scabies: A bath using leaves along with tender leaves of *Vitex negundo*, boiled in water is done to get rid of them.	Shetty et al., 2015
25	*A. zeylanica* Medik.	Scabies: Leaf paste mixed with a pinch of turmeric powder is applied on the affected areas daily twice for 3 days.	Raju, 2009
26	*Adiantum philippense* L.	Allergy: Root paste mixed with 50 ml of water is administered orally daily twice till cure. Herpes: The rhizome paste is also applied on affected parts.	Babu, 2007; Naidu, 2003
27	*Aegle marmelos* (L.) Corr.	Burn injuries: Crushed leaf paste is applied on them. Scurvy: Leaf juice is mildly heated and administered in half cup once a day till cure. Scabies and other skin diseases: Fruits crushed with seeds of *Strychnos nux-vomica* and *Pongamia pinnata*, boiled with coconut oil is applied on the affected parts twice a day till recovery occurs. Itching: Leaves with those of *Cassia fistula* made into paste is applied on the affected parts to cure itching and irritations caused by rashes. Leucoderma: Leaf paste is applied on the affected areas once a day for 15 days.	Nayak et al., 2004; Manjula and Reddi, 2015; Jeeva et al., 2007; Arya et al., 2015; Babu, 2007

S. No.	Name of the plant	Practice	Reference
28	*Aeschynomene indica* L.	Cuts and wounds: Leaf paste is applied on the affected areas once a day till cure.	Babu, 2007
29	*Aganosma caryophyllata* (Roxb. ex Sims) G.Don	Cuts and wounds: The cottony wool in the fruit is used to stop bleeding from cuts and wounds.	Naidu, 2003
30	*Agave cantula* Roxb.	Bruises: Leaf juice is applied on the affected areas daily twice till cure.	Naidu, 2003
31	*Ageratum conyzoides* L.	Cuts and sores: Crushed leaf paste is applied till cure.	Changkija, 1999; Das et al., 2008; Naidu, 2003; Shukla et al., 2010
		Cuts and wounds: Leaf juice is used in fresh cuts and wounds.	
		Leprous sores and other skin diseases: A hot poultice of the leaves and stem is applied on sores.	
		Leprosy: Fomented leaves and stems are used.	
32	*Ailanthus excelsa* Roxb.	Common skin disease: Stem bark paste is applied on affected areas till cure.	Swamy, 2009
33	*Alangium salvifolium* (L. f.) Wang.	Skin diseases: Root bark paste is applied on the affected areas till cure.	Babu and Reddi, 2015
34	*Albizia amara* Boivin	Dandruff: Leaf paste mixed with fruit juice of *Sapindus emarginatus* (1:10) is used as hair wash.	Babu, 2007; Manjula, 2011; Ayyanar and Ignacimuthu, 2005
		Excoriation: Leaf paste is applied on the affected areas.	
		Skin diseases: Paste of leaf and root bark with root bark of *Jasminum angustifolium* and rhizome of *Cyperus rotundus* is heated with neem oil and applied externally on affected places for 10 days.	
35	*A. chinensis* (Osbeck) Merr.	Scabies: Leaf juice is applied on the affected areas daily twice for 2 days.	Babu and Reddi, 2015
36	*A. lebbeck* (L.) Willd.	Skin diseases: Stem bark powder mixed with turmeric is applied all over the body.	Manjula and Reddi, 2015

TABLE 10.1 (Continued)

S. No.	Name of the plant	Practice	Reference
37	*A. odoratissima* (L. f.) Benth.	Ringworm: Stem bark paste mixed with a pinch of cheese is applied on the affected areas daily twice for 3 days.	Suneetha, 2007; Babu, 2007
		Leprosy: Stem bark paste is applied on the affected areas daily twice for 21 days.	
38	*Allium cepa* L.	Suppurating wounds: They are swabbed with sufficient amount of onion juice for complete cure.	Naidu, 2003
39	*A. sativum* L.	Infected area of the skin: Bulb paste is applied on for 2–3 days	Prashantkumar and Vidyasagar, 2008
40	*Alnus nepalensis* Don	Cuts and wounds: Paste from the leaves is applied on affected parts.	Changkija, 1999
41	*Aloe barbadensis* Mill.	Dandruff: Leaf pulp is used for the preparation of hair oil for dandruff.	Silja et al., 2008; Bhattacharjee and Chatterjee, 2007; Prashantkumar and Vidyasagar, 2008; Arya et al., 2015; Naidu, 2003
		Sunburn and skin irritation: Leaf decoction has a moisturizing affect on the skin and is a common remedy.	
		Acne: Leaf juice has antibacterial, antiviral, antifungal and wound healing properties; reduces scars, lesions and useful in acne vulgaris.	
		Rough patches on the skin: Mix thoroughly two cups of organic salt, one cup of *Aloe vera*, one cup of coconut oil and two tablespoon of honey. Apply this to the rough patches on the skin. Continued application will soften these patches.	
		Carbuncle: Leaves with those of *Dillenia indica*, *Marsilea quadrifolia* and *Oxalis corniculata* taken in equal quantities are ground. Paste mixed with a pinch of turmeric and camphor is applied on the affected areas daily once for 5 days.	

S. No.	Name of the plant	Practice	Reference
42	*Alstonia scholaris* (L.) R.Br.	Dandruff and itching of the scalp: The whole leaf paste is applied on the scalp and after 1 h the paste is washed thoroughly.	Naidu, 2003; Das et al., 2008
43	*A. venenata* Br.	Allergy and abscess: Latex is applied on the affected areas.	Rao, 2010
		Scabies and boils: Dried stem bark powder mixed with coconut oil is applied on the affected parts.	
44	*Alternanthera sessilis* (L.) R. Br. ex DC.	Psoriasis: A spoonful of whole plant paste with a spoonful of leaf paste of *Andrographis paniculata* mixed in a glass of water is administered daily once for 21 days.	Suneetha, 2007
45	*Alysicarpus monilifer* (L.) DC.	Boils and blisters: Paste of the plant is applied on affected areas.	Naidu, 2003
46	*A. vaginalis* (L.) DC.	Blisters and boils: Whole plant paste is applied on affected areas till cure.	Swamy, 2009
47	*Amaranthus spinosus* L.	Herpes: Root paste is heated then applied on the affected areas, particularly in between fingers of the legs daily once for 2 days.	Babu, 2007; Suneetha and Reddi, 2015
		Scabies: Root paste is applied on the affected areas daily twice till cure.	
48	*A. viridis* L.	Pimples: Dried fruit ash mixed with water is applied on the pimples twice a day for 3 days.	Suneetha, 2007
49	*Ammania baccifera* L.	Cancerous ulcers: Leaf paste is applied intermittently along with the latex of *Argemone mexicana*.	Naidu, 2003; Shah et al., 2011; Chetty et al., 1998: Suneetha, 2007
		Eczema: Leaf paste is applied over the affected skin.	
		Skin eruptions: Whole plant is burnt and the ashes are mixed in sesame oil is applied.	
		Blisters on the breast: Whole plant paste is applied on affected areas daily once for 3 days.	

TABLE 10.1 (Continued)

S. No.	Name of the plant	Practice	Reference
50	*Ampelocissus latifolia* (Vahl) Planch.	Common skin diseases: Branches made into poultice is applied on affected areas till cure.	Swamy, 2009; Suneetha, 2007
		Ringworm: Root paste is applied on the affected areas daily twice for 3 days.	
51	*A. tomentosa* Planch.	Boils and wounds: Leaf paste is applied on the affected areas daily twice for 3 days.	Suneetha, 2007
52	*Anacardium occidentale* L.	Leprosy: Powdered bark mixed with honey is taken orally continuously for 6 months.	Jeeva et al., 2007; Manjula and Reddi, 2015; Arya et al., 2015; Raju et al., 2010; Suneetha, 2007
		Skin diseases: Seed coat is made into ash and mixed with coconut oil and applied on the affected areas.	
		Skin diseases: The outer seed coat of mature cashew nut is burned and the latex obtained is applied to the affected part on the skin.	
		Scabies: Fruit paste is applied on the affected areas daily twice for 3 days.	
		Ringworm: Gum from stem bark is applied on the affected areas daily once for 3 days.	
53	*Andrographis paniculata* (Burm. *f.*) Wall.	Common skin diseases: Five or 6 spoonful of dry leaf powder mixed in a bucket of hot water is used for bath only once.	Babu and Reddi, 2015; Raju et al., 2010; Suneetha, 2007
		Scabies: Leaf paste is applied on the affected areas daily twice for 3 days.	
		Leprosy: Leaves and whole plant of *Byttneria herbacea* are taken in equal quantities and ground. 2 spoonfuls of paste is administered daily twice for 5 days. Meanwhile paste is applied on the affected areas daily twice for 5 days.	

S. No.	Name of the plant	Practice	Reference
54	*Anisochilus carnosus* (L.f.) Wall.	Fissures and cracks: Leaf paste mixed with turmeric is applied to the toes and feet.	Naidu, 2003
55	*Anisomeles indica* (L.) Ktze.	Scabies: Leaf juice mixed with castor oil is applied on the affected areas once a day till cure.	Rao, 2010
56	*A. malabarica* (L.) R. Br. ex Sims.	Excoriation: Leaf juice is applied on the affected areas till cure.	Manjula, 2011
57	*Annona reticulata* L.	Blisters: Flower paste mixed with turmeric powder is applied on the affected areas.	Manjula and Reddi, 2015; Babu and Reddi, 2015
		Scabies: Leaf paste is applied on the affected areas once a day till cure.	
58	*A. squamosa* L.	Eczema: Leaf paste mixed with coconut oil is applied to the affected areas.	Manjula, 2011; Manjula and Reddi, 2015; Manjula and Reddi, 2015; Swamy, 2009
		Common skin diseases: Stem bark or leaf or seed powder mixed with groundnut oil is applied on affected parts till cure.	
		Scabies: Leaf paste is applied on the affected parts.	
		Dandruff: Stem bark ground with neem leaves is applied to head 1 hour before bath twice or thrice a week.	
59	*Anodendron paniculatum* DC.	Leprosy: Tuber paste is applied on the affected parts and administered orally with water twice a day for 2 days.	Rao, 2010
60	*Anthocephalus chinensis* (Lamk.) Rich. ex Walp.	Tumors: Leaf paste is applied on the affected areas.	Manjula, 2011
61	*Antigonon leptopus* Hook. & Arn.	Cuts and wounds: Leaves with those of *Euphorbia ligularia* are taken in equal quantities and ground. Paste mixed with a pinch of turmeric is applied on the affected areas daily once for 2 days.	Suneetha, 2007

TABLE 10.1 (Continued)

S. No.	Name of the plant	Practice	Reference
62	*Aquilegia fragrans* Benth.	Boils: Flower powder paste in ghee is used for massage once a day for 10–15 days.	Lal and Singh, 2008
63	*Areca catechu* L.	Burns and sprains: Young leaves and young fruits are used for the preparation of oil used in the treatment.	Silja et al., 2008
64	*Argemone mexicana* L.	Leprous skin: Leaf juice mixed with *Acorus calamus* rhizome paste is applied on the affected parts.	Naidu, 2003; Naidu, 2003; Das et al., 2008; Prashantkumar and Vidyasagar, 2008; Jeeva et al., 2007; Bhattacharjee and Chatterjee, 2007; Nayak et al., 2004
		Psoriasis: Whole plant with leaves of *Coldenia procumbens* and rhizome of *Curcuma longa* in equal proportions are taken and burnt into ash. It is mixed with coconut oil and the paste is applied continuously for about a month.	
		Cuts and wounds: Root juice is used in fresh cuts and wounds.	
		Common skin diseases: Whole plant paste is applied on affected parts of the skin.	
		Skin diseases: Pounded seeds along with rhizomes of *Curcuma aromatica* and *Acorus calamus* made into paste are applied on all types of skin diseases.	
		Skin diseases: The root paste mixed with sugar (4:1) is taken orally with water.	
		Ringworm, scabies and skin diseases: The leaf juice mixed with common salt and water (1:1) and applied locally.	
65	*Argyreia nervosa* (Burm. f.) Boj.	Carbuncles, boils and blisters: Leaves are dipped in castor oil and mildly heated and applied on the affected areas.	Naidu, 2003
66	*Arisaema tortuosum* (Wall.) Schott. (Figure 10.1)	Scabies: Tuber paste is applied on the affected parts till cure.	Rao, 2010

FIGURE 10.1.   (1) *Arisaema tortuosum,* (2) *Aristolochia indica,* (3) *Asclepias curassavica,* (4) *Asparagus racemosus,* (5) *Bauhinia variegata,* (6) *Breynia retusa.*

TABLE 10.1 (Continued)

S. No.	Name of the plant	Practice	Reference
67	*Aristolochia bracteolata* Lamk.	Boils and eczema: Leaf paste is applied on the affected portion.	Silja et al., 2008; Manjula and Reddi, 2015
		Skin diseases: Leaf paste is applied on the affected parts once a day till cure.	
68	*A. indica* L.	Common skin diseases: Root paste mixed with 2 or 3 drops of water is applied on the affected areas daily once till cure.	Babu and Reddi, 2015; Manjula, 2011; Manjula, 2011
		Leprosy: 2 spoonfuls of root powder mixed with one spoon of black pepper powder is administered along with a cup of water twice a day for 30 days.	
		Leucoderma: One spoonful of root powder mixed with one spoon of honey is administered twice a day for 7 days.	
69	*Artemisia maritima* L.	Skin diseases: Fresh root juice is applied on skin to cure boils.	Lal and Singh, 2008
70	*A. nilagirica* (C. B. Clarke) Pamp.	Leprosy and skin diseases: Leaf juice is used externally.	Silja et al., 2008
71	*Artocarpus heterophyllus* Lamk.	Wounds: Stem latex is applied on the affected areas daily twice for 2 days.	Suneeetha, 2007
72	*A. hirsutus* Lam.	Pimples and cracks on the skin: Infusion of bark is applied.	Silja et al., 2008
73	*Asclepias curassavica* L.	Skin diseases: Latex is applied on the affected areas till cure.	Rao, 2012
74	*Asparagus racemosus* Willd.	Skin diseases: Tubers along with leaves of *Plumbago indica* made into paste and applied.	Kingston et al., 2009; Suneetha, 2007
		Wounds: Tuberous root paste is applied on the affected areas daily twice for 2 days.	

S. No.	Name of the plant	Practice	Reference
75	*Atalantia monophylla* (L.) Corr.	Scabies: Root bark paste is applied on the affected areas daily twice till cure.	Raju et al., 2010; Babu and Reddi, 2015
		Scabies: Fruit juice is applied on the affected areas twice a day for 3 days meanwhile 50 ml of fruit juice is administered orally.	
76	*Azadirachta indica* Juss.	Scabies: Leaf paste is applied to the affected areas about 2–3 times a day till cure. Leaf decoction bath is taken daily for 15 days.	Prashantkumar and Vidyasagar, 2008; Shukla et al., 2010; Kumar and Chauhan, 2007; Jeeva et al., 2007; Swamy, 2009; Suneetha, 2007; Suneetha, 2007; Nayak et al., 2004
		Skin diseases: Leaf decoction is applied over the affected parts.	
		Skin diseases: Fruit and leaf juice cures various skin diseases.	
		Dandruff: Flowers are boiled in gingelly oil and applied on the head.	
		Eczema, ringworm, scabies: Apply seed oil on affected areas.	
		Ringworm: Leaf paste mixed with a pinch of turmeric is applied on the affected areas daily twice for 3 days.	
		Leprosy: Stem bark with that of *Bauhinia variegata* and stem of *Tinospora cordifolia* taken in equal quantities are ground. 2 spoonfuls of paste mixed in a glass of water is administered daily twice for 15 days.	
		Wounds: Leaf decoction is applied locally.	
77	*Bacopa monnieri* (L.) Penn.	Skin diseases: Whole plant juice is applied on the affected areas once a day till cure.	Manjula and Reddi, 2015
78	*Balanites aegyptiaca* (L.) Del.	Leucoderma: Fruit pulp is applied on the affected areas.	Manjula, 2011
79	*Baliospermum montanum* Muell.-Arg.	Common skin diseases: Leaves with those of *Eclipta prostrata* and *Aloe vera* taken in equal quantities are ground. Paste is applied on the affected areas daily twice for 3 days.	Suneetha and Reddi, 2015

TABLE 10.1 (Continued)

S. No.	Name of the plant	Practice	Reference
80	*Bambusa arundinacea* Ait.	Leucoderma: Leaf paste mixed with a pinch of turmeric is applied on the affected areas daily twice for 3 days.	Suneetha, 2007
81	*Barleria cristata* L.	Burns: Roots with stem of *Santalum album* taken in equal quantities are ground. Paste is applied on the affected areas daily twice for 3 days.	Suneetha, 2007
82	*B. prionitis* L.	Common skin diseases: Leaf paste is applied on the affected areas once a day till cure.	Babu and Reddi, 2015; Manjula, 2011
		Ringworm and scabies: Leaf paste is applied on the affected areas till cure.	
83	*Basella alba* L. var. *rubra* (L.) Stew.	Wounds: Leaf paste is applied on the affected areas daily once till cure.	Suneetha, 2007; Babu, 2007
		Burns and cuts: Leaf paste is applied on the affected areas daily once till cure.	
84	*Bauhinia acuminata* L.	Skin diseases: Leaf paste is applied externally on the affected areas.	Silja et al., 2008
85	*B. purpurea* L.	Skin diseases: Bark is crushed and the juice extract is drunk to prevent skin diseases.	Changkija, 1999
86	*B. racemosa* Lamk.	Skin diseases: Leaf paste is applied on the affected areas.	Manjula and Reddi, 2015
87	*B. vahlii* (Wt.& Arn.) Benth.	Scabies: Wash the affected areas with root decoction twice a day till cure.	Manjula and Reddi, 2015; Suneetha, 2007
88	*B. variegata* L.	Pimples: Leaf paste mixed with a pinch of turmeric is applied on the affected areas daily twice for 3 days.	Babu, 2007
		Leprosy: A spoonful of stem bark paste is administered on empty stomach daily once for 15 days.	

S. No.	Name of the plant	Practice	Reference
89	*Benincasa hispida* (Thunb.) Cogn.	Dandruff: Fruit paste is applied on the head once a day before taking head bath.	Manjula, 2011
90	*Bergia ammannioides* Roxb.	Scabies: Whole plant infusion is applied on the affected areas.	Chetty et al., 1998
91	*Beta vulgaris* L.	Burns and cuts: Leaf paste is applied on the affected areas daily twice for 2 days.	Suneetha, 2007
92	*Biophytum nervifolium* Thw.	Skin diseases and ulcers: Leaf paste along with garlic is applied externally.	Naidu, 2003
93	*B. sensitivum* (L.) DC.	Wounds: Seed paste is applied on the affected areas once a day till cure.	Babu, 2007
94	*Blechnum orientale* L.	Cuts and wounds: Rhizome and fronds are used in fresh cuts and wounds.	Das et al., 2008
95	*Blumea mollis* (Don) Merr.	Tumors: Whole plant paste is applied on the affected areas. Bruises: Plant poultice is applied on affected areas.	Manjula, 2011; Swarny, 2009
96	*Blyxa octandra* Planch. ex Thw.	Scabies: Ribbon shaped leaves with those of *Tridax procumbens* are made into paste with turmeric powder and applied on the affected areas.	Chetty et al., 1998
97	*Boehmeria platyphylla* Don	Carbuncle: Root paste is applied on the affected areas daily twice till cure.	Suneetha, 2007

TABLE 10.1 (Continued)

S. No.	Name of the plant	Practice	Reference
98	Boerhavia chinensis (L.) Asch. & Schw.	Anemia: Tuber with root bark of *Maerua oblongifolia*, roots of *Withania somnifera* and *Gmelina asiatica* taken in equal quantities are ground. 2 spoonfuls of paste mixed in a glass of cow milk is administered daily twice for 15 days.	Suneetha, 2007; Suneetha, 2007
		Bruises and cuts: Tubers with roots of *Withania somnifera* and leaves of *Tridax procumbens* are taken in equal quantities and ground. A spoonful of paste is administered daily twice for 3 days.	
99	B. diffusa L.	Anemia: Leaf extract is used internally in the treatment.	Silja et al., 2008; Rao, 2010
		Burns: Leaf and root paste is applied on the burnt areas.	
100	Bombax ceiba L.	Boils and sores: Flower paste is applied on the affected parts once a day till cure.	Suneetha, 2007; Manjula, 2011; Babu, 2007
		Pimples: Leaf paste is applied on the face.	
		Boils: Root paste of young plant is applied on the boils twice a day for 2 days.	
101	Borassus flabellifer L.	Cuts and wounds: Leaf paste is applied on the affected parts.	Manjula, 2011; Naidu, 2003
		Burns: The dried male inflorescence made into ash and mixed with coconut oil is applied.	
102	Boswellia serrata Colebr.	Eruptions: The resin formed by hardened latex is used in the preparation of ointment for sores and external eruptions.	Naidu, 2003
103	Brassica juncea (L.) Czern.	Common skin diseases: Seed oil is applied externally.	Swamy, 2009
104	B. nigra (L.) Koch.	Heel cracks: A handful of seeds with 10 seeds of *Datura metel* are taken and ground. Then mildly heated paste is applied on the cracks twice a day till cure.	Suneetha, 2008

S. No.	Name of the plant	Practice	Reference
105	*Breynia retusa* (Dennst.) Alston	Skin diseases: Wash the affected areas with whole plant decoction till cure.	Manjula and Reddi, 2015
106	*Bridelia scandens* (Roxb.) Willd.	Ringworm and athlete's foot: The leaves are used for cleaning the sores on the scalp of the children. The leaves are made into a fine paste and applied.	Nisha and Sivadasan, 2007; Nisha and Sivadasan, 2007
		Psoriasis: The leaf is an ingredient of medicated oil, which is used against psoriasis.	
107	*Bryophyllum spathulata* Salisb.	Burns: Paste of the plant is applied on burns.	Changkija, 1999
108	*Buchanania axillaris* (Desv.) Ramam.	Common skin diseases: Kernel oil is applied on the affected areas twice a day for 3 days.	Babu and Reddi, 2015
109	*B. lanzan* Spr.	Burns: Decoction of bark (100 ml) is taken internally once a day for 3 days. Seed paste is applied till cure.	Prashantkumar and Vidyasagar, 2008
110	*Butea monosperma* (Lamk.) Taub.	Ringworm: Seed paste is applied on the affected parts.	Raju et al., 2010; Bhattacharjee and Chatterjee, 2007
		Eczema: Root paste is applied externally on weeping eczema.	
		Bruises: Flower paste is applied on the affected areas.	
111	*Byttneria herbacea* Roxb.	Cuts and wounds: Leaf paste mixed with a pinch of turmeric is applied on the affected areas daily twice for 2 days.	Suneetha, 2007
112	*Caesalpinia bonduc* (L.) Roxb.	Skin infection: The leaf paste in coconut oil is used for skin infection.	Sinha et al., 1996
113	*C. pulcherrima* (L.) Sw.	Sores in mouth: Half glass of leaf decoction with a spoonful of salt is used for gargling to cure tongue sores.	Suneetha, 2007
114	*C. sappan* L.	Skin diseases: Stem bark paste is applied on the affected areas.	Manjula and Reddi, 2015
115	*Cadaba fruticosa* (L.) Druce	Boils: Leaf paste is applied on the affected areas twice a day till cure.	Babu, 2007

TABLE 10.1 (Continued)

S. No.	Name of the plant	Practice	Reference
116	Cedrus deodara (Roxb.) Loud.	Boils: Heartwood oil is applied on skin.	Lal and Singh, 2008
117	Cajanus cajan (L.) Millsp.	Sores in mouth: 2 spoonfuls of leaf decoction with a pinch of salt is used to gargle twice a day for 2 days.	Suneetha, 2007
118	Calamus rotang L.	Ringworm: 2 spoonfuls of leaf paste mixed with a pinch of turmeric is applied on the affected areas daily twice for 3 days.	Suneetha, 2007
119	C. viminalis Willd.	Skin diseases: Rachis and petiole are burnt and the ash mixed with coconut oil is applied till cure.	Naidu, 2003
120	Callicarpa macrophylla Lamk.	Cuts, wounds and ulcers: Leaf paste in coconut oil is used locally. Acne, pimples, blemishes and other allergic skin patches: Dry fruit powder is blended in creams to treat them.	Sinha et al., 1996; Sharma et al., 2003
121	Calophyllum inophyllum L.	Scabies: Fruit pulp is applied on the affected areas daily twice till cure. Leprosy: 2 spoonfuls of fruit paste is administered daily once for 15 days.	Babu and Reddi, 2015; Suneetha, 2007
122	Calotropis gigantea (L.) R.Br. ex Ait.	Leucoderma: Latex is placed in a steel vessel and exposed to sunlight for 21 days and the dried powder is given in the size of a rice grain for 20 days. Leprosy: Root bark powder soaked in buttermilk is applied on dark patches for 21 days. Leaf paste is applied on swelling portion till it cures. Latex is applied. Common skin diseases: Leaf paste mixed with a pinch of cheese is applied on the affected areas daily twice till cure. Leprosy: Latex is applied on affected areas. Ringworm: The latex is used to cure.	Naidu, 2003; Prashantkumar and Vidyasagar, 2008; Suneetha and Reddi, 2015; Prashantkumar and Vidyasagar, 2008

S. No.	Name of the plant	Practice	Reference
123	*C. procera* Br.	Whitlow: Leaf paste mixed with a pinch of turmeric powder is applied on the affected areas daily twice for 5 days.	Raju et al., 2010; Suneetha, 2007
		Leprosy: Flowers and leaves are taken in equal quantities and ground.	
		2 spoonfuls of paste mixed with a pinch of ghee is administered daily twice for 21 days.	
124	*Calycopteris floribunda* Lamk.	Ulcer wounds and boils: Root bark paste is applied on the affected areas.	Naidu, 2003; Manjula, 2011
		Boils: Root paste is applied on the affected areas once a day till cure.	
125	*Canna orientalis* Rosc.	Ringworm: Tuber paste is applied on the affected areas.	Rao, 2010
126	*Canscora diffusa* (Vahl) Br.	Ulcers, wounds and boils: The whole plant ground with garlic and the paste is applied.	Naidu, 2003
127	*Capparis zeylanica* L. (Figure 10.2)	Boils: Leaf paste is applied on affected areas till cure.	Swamy, 2009
128	*Capsicum annuum* L.	Wounds: Fruit powder is applied on the affected areas with oil cloth.	Manjula, 2011
129	*Caralluma umbellata* Haw.	Scabies: Stem paste is applied on the affected areas.	Manjula and Reddi, 2015
130	*Cardiospermum halicacabum* L.	Dandruff: Paste of whole plant is used as poultice.	Silja et al., 2008; Manjula and Reddi, 2015; Swamy, 2009
		Scabies: Leaf paste is applied on the affected areas.	
		Ringworm: Tender fruit pulp is applied externally till cure.	
131	*Careya arborea* Roxb.	Leucoderma: 2 spoonfuls of stem bark paste mixed with a spoonful of seed oil of *Azadirachta indica* is applied on the affected areas once a day till cure.	Suneetha, 2007

FIGURE 10.2.   (7) *Capparis zeylanica*, (8) *Couropita guinensis*, (9) *Entada rheedi*, (10) *Mirabilis jalapa*, (11) *Pueraria tuberosa*, (12) *Rauvolfia tetraphylla*.

TABLE 10.1 (Continued)

S. No.	Name of the plant	Practice	Reference
132	*Carica papaya* L.	Skin diseases and ring worm: The latex is applied on affected areas.	Changkija, 1999; Babu and Reddi, 2015
		Common skin diseases and scabies: Whole plant latex is applied on the affected areas externally for 10 minutes then washed with Dettol or lifebuoy soap twice a day till cure.	
133	*Carissa carandas* L.	Scabies: Root paste is applied on the affected areas daily twice for 3 days.	Suneetha and Reddi, 2015; Suneetha, 2007
		Ringworm: Fruit paste mixed with a pinch of turmeric is applied on the affected areas daily twice for 3 days.	
134	*C. spinarum* L.	Wounds: Root paste is applied on affected areas till cure.	Swamy, 2009
135	*Carum carvi* L.	Skin infection: Powdered seeds are roasted in *desi* ghee. The preparation is applied on skin for a week.	Lal and Singh, 2008
136	*Caryota urens* L.	Dandruff: Nut powder made into paste is applied to the head and bath is taken after 1 h.	Naidu, 2003
137	*Cascabela thevetia* (L.) Lipp.	Eczema: Milky latex of leaf is applied on the affected part for one hour twice a day till cure.	Babu and Reddi, 2015
138	*Casearia elliptica* Willd.	Ringworm: Stem bark juice is applied on affected parts.	Shukla et al., 2010
139	*Cassia absus* L.	Ringworm: Seed paste is applied on the affected areas.	Manjula, 2011
140	*C. alata* L.	Ringworm, eczema, black heads and fungal diseases: The leaves are recommended for use. The leaf juice with juice of lime is used for curing ringworm infection.	Sinha et al., 1996; Nisha and Sivadasan, 2007; Suneetha and Reddi, 2015
		Common skin diseases: Fresh leaf paste mixed with a pinch of turmeric is applied on the affected areas once a day till cure.	
		Leprosy: 2 spoonfuls of dried leaf powder mixed with a spoonful of honey is administered daily twice for 21 days. Meanwhile fresh leaf paste is applied on the affected areas daily once for 21 days.	

TABLE 10.1 (Continued)

S. No.	Name of the plant	Practice	Reference
141	C. auriculata L.	Skin diseases: Paste of dried leaves with vinegar is applied on affected areas.	Jeeva et al., 2007
142	C. fistula L.	Scabies: Flower paste is applied on the affected areas till cure.	Manjula and Reddi, 2015; Prashantkumar and Vidyasagar, 2008; Arya et al., 2015; Naidu, 2003; Naidu, 2003
		Scabies: Root bark decoction is taken internally twice a day for 15–20 days. Leaf paste is applied on the affected areas till it cures.	
		Skin diseases: Flowers and leaves are made into a decoction and drunk for relief.	
		Eczema and leprosy: Leaves are ground with rice washed water and the paste is applied on the affected areas twice a day till cure.	
		Leucoderma: One teaspoonful of bark powder is taken early in the morning with water for about a fortnight.	
143	C. occidentalis L.	Boils and wounds: Pounded leaves paste mixed with coconut oil is used externally.	Sinha et al., 1996; Babu and Reddi, 2015; Das et al., 2008; Shah et al., 2011
		Skin diseases: Seed and leaf paste is applied externally.	
		Leucoderma: Leaf paste mixed with sesame oil is applied on the affected areas daily twice till cure.	
		Ringworm: Seeds are roasted, crushed and soaked in curdled milk for 3 days; curd is rubbed on the patches.	
144	C. siamea Lamk.	Scabies: Roots with leaves of Oxalis corniculata are taken in equal quantities and ground. Paste mixed with a pinch of lime is applied on the affected areas daily once for 3 days.	Suneetha and Reddi, 2015
145	C. sophera L.	Ringworm: Root paste is applied on the affected areas daily twice for 3 days.	Suneetha, 2007

S. No.	Name of the plant	Practice	Reference
146	C. tora L.	Skin diseases: Leaves and seeds are applied on the affected areas.	Das et al., 2008; Swamy, 2009; Manjula and Reddi, 2015; Suneetha, 2007
		Ringworm: Roots rubbed on stone with lime juice and the poultice is applied on affected areas till cure.	
		Skin diseases: Leaf paste is applied on the affected areas.	
		Ringworm: Root paste mixed with 5 drops of lemon juice is applied on the affected areas daily once till cure.	
147	Cassytha filiformis L.	Scabies: Whole plant paste mixed with turmeric powder is applied on the affected areas till cure.	Manjula and Reddi, 2015
148	Catharanthus roseus (L.) Don	Leucoderma: Plant latex is placed in an earthenware pot and exposed to sunlight for 21 days. The dried powder in doses of rice grain size with honey is administered for 21 days.	Naidu, 2003
149	Catunaregam spinosa (Thunb.) Tirv.	Bruises: Stem bark paste is warmed and applied externally as a stimulant.	Swamy, 2009
150	Cayratia auriculata Gamble	Wounds, ulcers and boils: Leaves ground with turmeric powder and the paste is applied externally for children for chicken pox. The paste is also applied on affected parts.	Naidu, 2003
151	C. pedata (Lamk.) Juss.ex Gagnep.	Itches and scabies: Tuberous root paste is applied on the affected areas daily once till cure.	Babu and Reddi, 2015; Suneetha, 2007
		Black patches on the face: 2 spoonfuls of tuberous root paste mixed with a spoonful of curd and lemon juice is applied on the affected areas daily twice for 3 days.	
152	Cedrus deodara (Roxb.) Loud.	Skin itching: Heartwood oil is used for massage.	Lal and Singh, 2008
153	Ceiba pentandra (L.) Gaertn.	Cuts: Stem bark is ground with that of Mangifera indica. Paste is applied on the affected areas and bandaged.	Babu, 2007

TABLE 10.1 (Continued)

S. No.	Name of the plant	Practice	Reference
154	*Celastrus paniculatus* Willd.	Leucoderma and scabies: Seed oil mixed with a pinch of turmeric is applied on the affected areas daily twice till cure.	Suneetha, 2007
155	*Celosia argentea* L.	Wounds: Seed paste is applied on the affected parts till cure.	Prashantkumar and Vidyasagar, 2008; Babu, 2007
		Sores and wounds: Leaf paste with turmeric is applied on the wounds once a day till cure.	
156	*Centella asiatica* (L.) Urban	Leprosy and anemia: Two or three spoons of shade dried powder of the leaves along with powder made from the pepper seed is given early in the morning with a glass of cow milk for about a month.	Naidu, 2003; Babu and Reddi, 2015; Suneetha, 2007
		Common skin diseases: Leaf paste is applied on the affected areas once a day till cure.	
		Leprosy: Hundred ml of leaf decoction is administered twice a day for 3 months.	
157	*Claoxylon indicum* Hassk.	Cuts and wounds: Leaf paste is applied locally.	Sinha et al., 1996
158	*Chloris barbata* Sw.	Skin diseases: Whole plant decoction mixed with seed powder of cardamom is administered in 20 ml dose once a day for 5 days. Take bath with this decoction.	Manjula and Reddi, 2015
159	*Chromolaena odorata* (L.) King & Robin.	Cuts and wounds: Whole plant is ground into paste in castor oil and applied on affected parts.	Naidu, 2003
160	*Cicer arietinum* L.	Black patches on the face: Seed powder with gum of *Terminalia alata* and turmeric are taken in equal quantities and applied on the affected areas daily once for 10 days and also used as bath powder.	Suneetha, 2007
161	*Cinnamomum camphora* (L.) Nees & Eberm.	Ringworm and scabies: Leaf oil with turmeric (2:1) is applied on the affected areas daily twice for 7 days.	Hari Babu, 2015

S. No.	Name of the plant	Practice	Reference
162	C. verum Presl.	Psoriasis: Bark is one of the ingredients of medicated oil which is used against psoriasis.	Nisha and Sivadasan, 2007
163	Cipadessa baccifera (Roth) Miq.	Eczema and ring worm: 2 spoonfuls of leaf extract with a spoonful of Jatropha gossypifolia seed oil and a spoonful of turmeric is applied on the affected areas thrice a day till cure.	Naidu, 2003
164	Cissampelos pareira L.	Eczema, itches and wounds: Leaf juice is applied on the affected areas once a day for half an hour till cure.	Babu, 2007
165	Cissus repens Lamk.	Burns and wounds: Paste of the tuberous roots is applied.	Naidu, 2003
166	Citrullus colocynthis (L.) Schr.	Scabies and skin diseases: Leaf juice mixed with a pinch of turmeric powder is applied on the affected parts till cure.	Manjula and Reddi, 2015; Manjula, 2011
		Warts: Root is pasted with cow urine and applied on the affected parts for 5 days.	
167	Citrus aurantifolia (Chr.) Sw.	Scabies: Dry leaf ash mixed with coconut oil is applied on the affected areas daily twice for 3 days.	Suneetha and Reddi, 2015; Naidu, 2003
		Whitlow: Turmeric powder and a pinch of salt is inserted into the fruit by making a hole and later on the affected finger is inserted in the hole for 3–5 days.	
168	C. limon (L.) Burm.	Whitlow: A pinch of salt and turmeric is inserted into the fruit by making a small hole and kept around the affected finger.	Suneetha, 2007
169	C. medica L.	Eczema: A pinch of salt and a spoonful of turmeric is added to about 100 ml of leaf juice and applied to the entire body and exposed to sunlight for an hour and later on full bath is taken.	Naidu, 2003
170	C. sinensis (L.) Osbeck	Scabies: 2 spoonfuls of leaf paste mixed with a spoonful of ash of whole plant of Eclipta prostrata is applied on the affected areas daily once till cure.	Suneetha and Reddi, 2015

TABLE 10.1 (Continued)

S. No.	Name of the plant	Practice	Reference
171	*Cleistanthus collinus* (Roxb.) Beth. ex Hook.*f.*	Ringworm: Root bark paste in doses of 10 g in lukewarm water is taken once a day for 15 days and the paste is simultaneously applied locally.	Naidu, 2003
172	*Clematis gouriana* Roxb.	Herpes: Leaves are given orally.	Kumar and Chauhan, 2007
173	*Cleome chelidonii* L.*f.*	Itching and tumors: Leaf paste is applied on the affected areas.	Manjula, 2011
174	*C. viscosa* L.	Ringworm: Leaves with those of *Phyllanthus emblica* and *Leucas aspera* taken in equal quantities are ground with a pinch of salt. Paste is applied on the affected areas daily once for one hour for 3 days.	Suneetha, 2007
175	*Clerodendrum inerme* (L.) Gaertn.	Skin diseases: Paste of leaf juice mixed with bee wax, resins of *Vateria indica* and *Nigella sativa* seeds kept in hot water bath, cooled and applied.	Jeeva et al., 2007
176	*C. paniculatum* L.	Skin diseases: Leaf plant paste is applied on the affected parts.	Manjula and Reddi, 2015
177	*C. phlomidis* L.*f.*	Ringworm: Leaf paste with a pinch of turmeric is applied on the affected areas daily twice for 3 days.	Suneetha, 2008
178	*Clitoria ternatea* L.	Scabies: Leaf juice is given orally twice a day for 6 days.	Jeeva et al., 2007; Manjula, 2011
		Leucoderma: Whole plant paste is applied on the affected areas till cure.	
179	*Coccinia grandis* (L.) Voigt	Leprosy and psoriasis: Paste of leaves and roots are used externally.	Silja et al., 2008; Rao, 2010; Babu and Reddi, 2015
		Ringworm: Leaf paste is applied on the affected areas.	
		Scabies: Leaf paste is applied on the affected areas twice a day till cure.	
180	*Cocculus hirsutus* (L.) Diels	Eczema: Leaf juice is applied on the affected areas twice a day till cure.	Babu, 2007

S. No.	Name of the plant	Practice	Reference
181	*Cochlospermum religiosum* (L.) Alston	Tumors: Leaf paste is applied on the affected areas.	Manjula, 2011; Swamy, 2009
		Dandruff: Leaf paste is applied to head, 1-hour before bath twice a week for 3-weeks.	
182	*Cocos nucifera* L.	Skin diseases, ringworm, eczema and boils: The coconut oil is used for preparing different kinds of medicinal oils. The upper portion of tender nut is made into a paste and mixed with rice-washed water. Then it is cooled and allowed to ferment. It is used as a poultice against the skin diseases.	Nisha and Sivadasan, 2007; Suneetha and Reddi, 2015; Babu, 2007
		Common skin diseases: Endocarp of fruit ash mixed with a pinch of turmeric is applied on the affected areas daily once till cure.	
		Leucoderma: Mildly heated epicarp of fruit ground with shade dried leaves of *Sansevieria roxburghiana* and paste is applied on the affected areas along with honey by quill of a cock once a day till cure (Figure 10.3).	
183	*Coldenia procumbens* L.	Psoriasis: The leaves of *Coldenia procumbens* along with whole plant of *Argemone mexicana* and rhizome of *Curcuma longa* in equal proportions are taken and burnt to ash. It is mixed in coconut oil and applied continuously for about 30 days.	Naidu, 2003; Manjula and Reddi, 2015; Swamy, 2009; Swamy, 2009
		Scabies: Leaves are roasted on fire and mixed with coconut oil is applied on the affected parts.	
		Leucoderma: Leaves powdered with those of *Argemone mexicana* and rhizome of *Curcuma longa*, mixed with coconut oil and applied on patches, then exposed to morning sunlight for 30–40 minutes for about 3 months or till cure.	
		Psoriasis: Leaves are ground with those of *Argemone mexicana* and turmeric, mixed with coconut oil and applied externally till cure.	

FIGURE 10.3. (13) *Sansieveria roxburghiana,* (14) *Solanum nigrum,* (15) *Terminalia bellirica,* (16) *Terminalia chebula,* (17) *Thysanolaena maxima,* (18) *Wrightia arborea*

TABLE 10.1 (Continued)

S. No.	Name of the plant	Practice	Reference
184	*Coleus amboinicus* Lour.	Eczema: Leaves and jaggery are taken in equal quantities and ground. Paste made into pills of peanut seed size. Two pills are administered daily twice for 2 days.	Suneetha, 2007
185	*Colocasia esculenta* (L.) Schott	Skin diseases: Juice of leaves and rhizome along with roots of *Toddalia asiatica* and root bark of *Crataeva adansonii* is mixed with gingelly oil to prepare syrup and this is applied externally for 21 days.	Ayyanar and Ignacimuthu, 2005
186	*Combretum roxburghii* Spr.	Boils, blisters and skin diseases: Warm leaf paste is applied on affected parts.	Naidu, 2003
187	*Commelina benghalensis* L.	Leprosy: Stem pieces are ground into soft mass and applied externally on ulcers.	Swamy, 2009
188	*Commiphora mukul* Jacq.	Tumors: Gum boiled in water is applied on the affected areas once a day till cure.	Manjula, 2011
189	*Corallocarpus epigaeus* Hook.*f*.	Leprosy: Tubers boiled in coconut oil are applied on the affected parts continuously for 6 months.	Jeeva et al., 2007
190	*Corchorus olitorius* L.	Ringworm: Leaf paste with a pinch of turmeric is applied on the affected areas daily twice till cure.	Hari Babu, 2008
191	*Cordia dichotoma* Forst.*f*.	Ringworm: Seed powder mixed with coconut oil is applied on the affected areas till cure.	Manjula, 2011
192	*Costus speciosus* (Koen.) Sm.	Scabies: Whole plant paste mixed with turmeric powder is applied on the affected parts once a day till cure.	Manjula and Reddi. 2015
193	*Couroupita guianensis* Aubl.	Skin diseases: Leaf paste is applied on the affected parts.	Manjula and Reddi. 2015
194	*Crateva magna* (Lour.) DC.	Whitlow: Stem bark paste is applied externally till cure.	Swamy, 2009

TABLE 10.1 (Continued)

S. No.	Name of the plant	Practice	Reference
195	*Crinum asiaticum* L.	Common skin diseases: Leaf paste with turmeric is applied on the affected areas once a day till cure.	Babu and Reddi, 2015
196	*Crotalaria juncea* L.	Cuts and wounds: Leaf paste mixed with a pinch of turmeric is applied on the affected areas daily twice for 2 days.	Suneetha, 2007
197	*C. laburnifolia* L.	Allergy: Five spoonful of leaf juice is administered orally to get relief.	Naidu, 2003
198	*C. prostrata* Rottl. ex Willd.	Scabies: Juice of the leaves and tender branches administered in 2 spoonfuls twice a day for about 10 days.	Swamy, 2009
199	*C. retusa* L.	Scabies: Leaf paste is applied on the affected areas daily twice till cure.	Babu and Reddi, 2015
200	*C. verrucosa* L.	Scabies: Leaf paste mixed with a pinch of cheese is applied on the affected areas daily twice for 3 days.	Suneetha and Reddi, 2015; Manjula and Reddi, 2015; Manjula, 2011
		Skin diseases: Leaf paste is applied on the affected areas.	
		Herpes: Leaf paste is applied on the affected areas.	
201	*Croton bonplandianum* Baill.	Scabies: Latex of whole plant mixed with a pinch of turmeric is applied on the affected areas daily twice for 3 days.	Raju et al., 2010; Manjula, 2011
		Ring worm: Leaf paste mixed with lemon juice is applied.	
202	*Cryptolepis buchananii* Roem. & Schult.	Leucoderma: Root paste is applied on the affected areas. Ten g of it is administered orally once a day till cure.	Rao, 2010; Suneetha, 2007
		Tumors: 2 spoonfuls of stem bark paste mixed with a pinch of asafoetida and jaggery is administered daily twice for 5 days. Meanwhile latex is massaged on the affected areas daily once for 5 days.	
203	*C. grandiflora* R.Br.	Cuts and wounds: Root paste is applied on the affected parts.	Manjula, 2011
204	*Cucumis melo* L.	Ringworm: Tender fruit pulp is applied externally.	Naidu, 2003; Swamy, 2009
		Eczema: Tender fruit pulp with seed poultice is applied externally till cure.	

S. No.	Name of the plant	Practice	Reference
205	*Curculigo orchioides* Gaertn.	Herpes: Paste of the corm is applied on the affected areas daily twice till cure.	Raju, 2009; Babu, 2007
		Leprosy, leucoderma and psoriasis: Ten g of tuberous root ground along with 2 spoonfuls of honey is administered daily twice till cure.	
206	*Curcuma amada* Roxb.	Common skin diseases: Rhizome paste is applied on the affected areas daily twice till cure.	Manjula and Reddi, 2015; Sharma et al., 2003
		Acne and blemishes: Dry rhizome powder is used in face packs to cure.	
207	*C. aromatica* Sal.	Skin diseases: Leaf paste is applied twice a day on the affected parts.	Manjula and Reddi, 2015
208	*C.caesia* Roxb.	Scabies: Rhizome paste mixed with a cup of water is administered once a day for 40 days.	Manjula and Reddi, 2015; Manjula and Reddi, 2015; Manjula, 2011
		Molting of skin in children: Half g of rhizome powder mixed with a cup of water is taken once a day till cure.	
		Leprosy: One g of rhizome powder mixed with a cup of water is administered once a day for 40 days.	
209	*C. longa* L.	Enhancing the color: The rhizome is used.	Sharma et al., 2003; Prashantkumar and Vidyasagar, 2008; Babu and Reddi, 2015; Naidu, 2003; Suneetha and Reddi, 2015
		Septic and wounds: Rhizome paste is used externally.	
		Common skin diseases: Leaf paste and lemon juice (3:1) is applied on the affected areas before going to bed till cure.	
		Psoriasis: Rhizome, whole plant of *Argemone mexicana* with the leaves of *Coldenia procumbens* in equal proportions are taken and burnt to ash. The resulting ash is mixed in coconut oil and made into a paste. It is applied continuously for about a month.	
		Scabies: Rhizome paste is applied on the affected areas once a day for 3 days.	

TABLE 10.1 (Continued)

S. No.	Name of the plant	Practice	Reference
210	C. pseudomontana Grah.	Anasarca: Mildly heated tuberous root paste is applied allover the body daily twice for 3 days.	Suneetha, 2007; Suneetha, 2007
		Leprosy: Tuberous roots and leaves of *Andrographis paniculata* are taken in equal quantities and ground. 2 spoonfuls of paste is administered daily twice for 21 days. Meanwhile paste is applied on the affected areas daily once for 21 days.	
211	*Cuscuta reflexa* Roxb.	Dandruff: Paste of the stems and branches is applied on head for removing it.	Kumar and Chauhan, 2007
212	*Cycas circinalis* L.	Burns: Stem bark decoction is applied on the affected areas once a day till cure.	Manjula, 2011
213	*Cyclea peltata* (Lamk.) Hook.f. & Thoms.	Skin diseases: The root is used as an ingredient of concoction.	Nisha and Sivadasan, 2007
214	*Cymbopogon flexuosus* (Steud.) Wats.	Dried skin: Three spoonful of whole plant decoction is administered twice a day till cure.	Manjula and Reddi, 2015; Suneetha, 2007
		Leprosy: 2 spoonfuls of leaf paste mixed with a spoonful of leaf paste of *Andrographis paniculata* is administered daily twice for 15 days. Meanwhile paste is applied on the affected areas daily once for 15 days.	
215	*C.martini* (Roxb.) Wats.	Ringworm: Whole plant paste mixed with a pinch of lime is applied on the affected areas daily once till cure.	Suneetha, 2007; Babu, 2007
		Leprosy: Leaf paste is applied on the affected areas daily once till cure.	

S. No.	Name of the plant	Practice	Reference
216	*Cynodon dactylon* (L.) Pers.	Common skin diseases: Whole plant with leaves of *Tridax procumbens* taken in equal quantities is dried and powdered. 2 spoonfuls of powder mixed in a glass of water is administered daily twice for 3 days.	Suneetha and Reddi, 2015; Jeeva et al., 2007; Manjula and Reddi, 2015; Manjula, 2011
		Skin diseases: Pounded leaves boiled in coconut oil is applied.	
		Scabies: Tuber paste mixed with a pinch of turmeric and coconut oil is applied on the affected areas.	
		Herpes and wounds: Whole plant paste is applied on the affected areas once a day till cure.	
217	*Cynometra ramiflora* L.	Skin infections, leprosy and scabies: Seed paste and seed oil are used externally.	Sinha et al., 1996
218	*Cyperus pangorei* Rottb.	Boils: Root paste is applied externally till cure.	Swamy, 2009
219	*C. rotundus* L.	Wounds, sores and ulcers: Paste of tuber and 2–3 betel leaves is applied for 3–4 days on the affected parts.	Prashantkumar and Vidyasagar, 2008
220	*Dalbergia lanceolaria* L. f.	Wounds: Mildly heated leaf paste mixed with a pinch of camphor is applied on the affected areas daily once for 2 days.	Suneetha, 2007;
221	*D. latifolia* Roxb.	Leucoderma: Stem bark powder is applied on the affected parts and a teaspoonful powder is administered internally once a day for about a month.	Naidu, 2003; Naidu. 2003; Suneetha and Reddi, 2015
		Leprosy: The bark or the leaf extract is administered in doses of 1 spoonful once a day for 3 months.	
		Common skin diseases: Stem bark with leaves of *Momordica charantia* is applied on the affected areas daily once before bath for about 5 days.	
222	*D. sissoo* Roxb.	Scabies: Stem bark paste mixed with a pinch of turmeric is applied on the affected areas daily twice for 2 days.	Suneetha and Reddi, 2015

TABLE 10.1 (Continued)

S. No.	Name of the plant	Practice	Reference
223	*Datura innoxia* Mill.	Boils and blisters: Paste of leaves along with salt is applied on affected areas.	Singh and Maheshwari, 1989; Naidu, 2003; Suneetha, 2007; Babu, 2007
		Reddish rashes: The seeds are pound and mixed with cow ghee and applied externally.	
		Alopecia: Leaf juice is applied on the scalp for an hour before washing the hair. This is continued for 10 days.	
		Leprosy: A tender leaf is eaten twice a day for 10 days.	
224	*D. metel* L.	Scabies: Leaf and nut paste is applied on the affected parts.	Babu and Reddi, 2015; Rao, 2010; Babu, 2007; Swamy, 2009; Shah et al., 2011
		Alopecia and bald head: Leaf juice is applied on the scalp twice a day for one week. Carbuncle and tumors: Unripened fruit paste is applied on the affected areas for an hour then wounds are washed with Dettol soap twice a day for 3 or 4 days.	
		Dandruff: Flower juice is applied to head by mixing with hair oil.	
		Eczema: Juice of crushed leaves is applied.	
225	*D. stramonium* L.	Scabies: Paste of equally mixed flowers of the plant and *Tylophora indica* leaves is applied on the affected part for 11 days.	Prashantkumar and Vidyasagar, 2008; Das et al., 2008
		Skin diseases: Seeds are taken once daily as the remedy.	
226	*Dendrocalamus strictus* Nees	Cuts and wounds: Tender leaf paste is applied on the affected areas daily once for 3 days.	Suneetha, 2007
227	*Dendrophthoe falcata* (L. f.) Etting	Cuts and wounds: Leaf paste is applied on the affected areas till cure.	Manjula, 2011; Rao, 2010
		Tumors: Bark paste along with 50 ml of water is administered twice a day till cure.	

S. No.	Name of the plant	Practice	Reference
228	*Desmodium gangeticum* (L.) DC.	Ringworm: Seed paste mixed with a pinch of cheese is applied on the affected areas daily once for 3 days.	Suneetha, 2007
229	*D. triflorum* DC.	Skin diseases: Dried leaves are ground with those of *Vicoa indica* and few grains of black pepper and the paste is applied on affected parts.	Naidu, 2003
230	*Dichrostachys cinerea* (L.) Wt. & Arn.	Common skin diseases: Root paste is applied externally till cure.	Swamy, 2009; Swamy, 2009
		Skin allergy: Root paste given in 2 spoonfuls with a glass of water once or twice a day.	
231	*Didymosperma nana* Wendl. & Drude.	Skin diseases: Fruit scraps are applied on affected areas.	Changkija, 1999
232	*Digera muricata* (L.) Mart.	Wounds: Leaf paste with a pinch of turmeric is applied on the affected areas daily twice for 2 days.	Babu, 2007
233	*Dillenia indica* L.	Lice and dandruff: Fruit is toxic but fruit juice is used as hair wash.	Naidu, 2003; Babu, 2007
		Dandruff and scabies: Fruit pulp is applied on the scalp for half an hour and then bath with soap nut once a day for 4 days.	
234	*Dioscorea bulbifera* L.	Skin diseases: The tuber is tied to the body till cure.	Kumar and Chauhan, 2007
235	*D. oppositifolia* L.	Cuts and wounds: Leaf paste mixed with a pinch of turmeric is applied on the affected areas daily twice for 2 days.	Suneetha, 2007
236	*D. pentaphylla* L.	Heel cracks: Leaf paste is applied on the cracks twice a day till cure.	Babu, 2007
237	*Diospyros melanoxylon* Roxb.	Anemia: Stem bark extract is given to pregnant women in 2 spoonfuls with a glass of rice cooked water once in morning for about 2 months from 6th or 7th month of pregnancy.	Swamy, 2009
238	*D. montana* Roxb.	Anasarca: Stem bark is crushed into paste and the paste in doses of 2 spoonfuls along with boiled rice is taken once a day for 5 days.	Naidu, 2003

TABLE 10.1 (Continued)

S. No.	Name of the plant	Practice	Reference
239	*Diplocyclos palmatus* (L.) Jeffrey	Common skin diseases: Whole plant paste mixed with a pinch of cheese is applied on the affected areas daily twice for 2 days.	Suneetha and Reddi, 2015
240	*Dodonaea viscosa* (L.) Jacq.	Wounds: Leaf paste is applied once a day for 3 days.	Babu, 2007
241	*Duranta repens* L.	Abscess: Leaf paste mixed with coconut oil is applied on the affected parts.	Manjula, 2011
242	*Echinochloa frumentacea* Link	Burns: Root paste is applied on the affected areas daily once for 2 days.	Suneetha, 2007
243	*Eclipta prostrata* L.	Premature graying of hair and dandruff: Leaf juice is applied on head. After 1 h full bath is taken. This can be repeated once in a week for few weeks.	Naidu, 2003; Suneetha, 2007; Suneetha, 2007; Suneetha, 2007
		Black patches on face: Whole plant and that of *Marsilea quadrifolia* is taken in equal quantities and ground. Paste is applied on the affected areas daily once till cure.	
		Leprosy: Leaves with those of *Cajanus cajan* taken in equal quantities are ground. 2 spoonfuls of paste is administered daily twice for 15 days. Meanwhile paste mixed with sesame oil is applied on the affected areas daily once for 15 days.	
		Leucoderma: Leaves and seeds of *Psoralea corylifolia* taken in equal quantities are ground with jaggery. Paste incubated along with copper coins for one day. Paste is applied on the affected areas daily twice for 5 days.	
244	*Elaeocarpus ganitrus* Roxb.	Skin diseases: Fruit paste is applied on the affected areas once a day till cure.	Manjula and Reddi, 2015

S. No.	Name of the plant	Practice	Reference
245	*Elephantopus scaber* L.	Ringworm and eczema: The leaves are ground into a paste with leaves of *Vernonia cineria* and turmeric and applied externally till cure.	Nisha and Sivadasan, 2007; Naidu, 2003
		Eczema and ringworm: Dried root paste mixed with mustard oil is applied.	
246	*Eleusine coracana* (L.) Gaertn.	Common skin diseases: Seed paste mixed with a pinch of turmeric is applied on the affected areas daily once for 3 days.	Suneetha and Reddi, 2015
247	*Elytraria acaulis* (L.f.) Lind.	Anasarca: Decoction of the tuberous roots is administered in doses of 2 spoonfuls twice a day for weekdays.	Naidu, 2003
248	*Emilia sonchifolia* (L.) DC.	Breast abscess: Leaves are crushed and applied.	Naidu, 2003; Suneetna, 2007; Swamy, 2009
		Cuts and wounds: Leaves with those of *Tridax procumbens* and *Aloe vera* are taken in equal quantities and ground with lime. Paste is applied on the affected areas daily twice for 2 days.	
		Ear sores: Leaf juice is poured in 2–3 drops thrice a day for 3–4 days or till cure.	
249	*Entada rheedii* Spr.	Skin diseases: Seed paste in water is applied externally.	Naidu, 2003
250	*Ervatamia heyneana* Cooke	Skin diseases: Fresh stem bark extract boiled with *Curcuma* and cumin is applied.	Shetty et al., 2015
251	*Erythrina variegata* L.	Cuts and wounds: Stem bark is ground with a pinch of turmeric powder. The paste is applied on the affected areas and bandaged once only.	Raju, 2009
252	*Erythroxylon monogynum* Roxb.	Skin diseases: Leaf paste with leaves of *Cassia senna* and *Jatropha curcas* mixed with groundnut flour is applied externally on the whole body before taking bath for 30 days.	Ayyanar and Ignacimuthu, 2005

TABLE 10.1 (Continued)

S. No.	Name of the plant	Practice	Reference
253	*Eucalyptus globulus* Labil.	Common skin diseases: Leaf oil is applied on the affected areas daily twice till cure.	Suneetha and Reddi, 2015; Manjula, 2011
		Cuts and wounds: Affected areas are washed with leaf decoction once a day till cure.	
254	*Eulophia nuda* Lindl.	Heel cracks: Tuberous root paste is applied on the heel cracks twice a day till cure.	Babu, 2007
255	*Euphorbia antiquorum* L.	Skin diseases: Stem juice is applied on the affected areas.	Manjula and Reddi, 2015; Manjula, 2011
		Warts: Latex is applied on the affected areas once a day till cure.	
256	*E. atoto* Forst.f.	Skin infection, cuts and wounds: Leaf paste is applied locally.	Sinha et al., 1996
257	*E. hirta* L.	Skin diseases: Whole plant is roasted on fire, powdered and mixed with coconut oil is applied on the affected areas till cure.	Manjula and Reddi, 2015; Suneetha, 2007; Swamy, 2009
		Warts: Latex of whole plant mixed with a pinch of turmeric is applied on the affected areas daily once till cure.	
		Lip cracks: Latex is applied externally till cure.	
258	*E. ligularia* Roxb.	Leprosy: 2 spoonfuls of stem paste mixed with a spoonful of honey is administered daily twice for 21 days.	Suneetha, 2007
259	*E. nerifolia* L.	Heel cracks: Latex mixed with gingelly oil is applied on the affected areas.	Manjula, 2011
260	*E. nivulia* Buch.-Ham.	Warts: Latex is applied on the warts once a day till cure.	Babu, 2007
261	*E. tirucalli* L.	Warts: Latex is applied on the affected areas daily once till cure.	Babu, 2007; Manjula, 2011
		Heel cracks: Latex is applied on the cracks once a day till cure.	
262	*Evolvulus alsinoides* L.	Scabies and black spots on skin: Root paste is applied on the affected areas.	Rao, 2010

S. No.	Name of the plant	Practice	Reference
263	*E. nummularius* L.	Boils on head: Whole plant paste is applied on the affected areas of the head for half an hour twice a day for 2 days.	Babu, 2007
264	*Ficus benghalensis* L.	Fissures in the feet, boils and blisters: Latex is applied continuously till they disappear.	Naidu, 2003; Manjula, 2011; Swamy, 2009
		Heel cracks: Latex is applied on the affected areas.	
		Leprosy: Tender prop root decoction is given in 2 spoonfuls twice a day for 3 months.	
265	*F. carica* L.	Anaemia: Presoaked fruits are eaten early in the morning without brushing the teeth.	Manjula, 2011; Shah et al., 2011
		Warts: Fresh juice of barely ripe figs is applied on the affected areas.	
266	*F. hispida* L.f.	Leucoderma: Decoction of stem bark given twice a day for one month.	Singh and Maheshwari, 1989; Naidu, 2003; Suneetha, 2007
		Leucoderma: Leaf juice is rubbed over the patches.	
		Psoriasis: Fruit paste is applied on the affected areas daily twice till cure.	
267	*F. microcarpa* L.f.	Wounds: Root bark paste is applied over them.	Suneetha, 2007; Manjula, 2011; Babu, 2007
		Bruises and wounds: Aerial root paste is applied on the affected areas daily twice for 2 days.	
		Carbuncle and tumors: Latex is applied on the affected areas once a day till cure.	
268	*F. racemosa* L.	Cuts and wounds: Bark paste is applied on the affected parts.	Rao, 2010; Swamy, 2009
		Wounds: Stem bark decoction is used in washing.	

TABLE 10.1 (Continued)

S. No.	Name of the plant	Practice	Reference
269	*F. religiosa* L.	Cuts: Stem bark paste is applied on the cuts and bandaged once a day till cure. Ulcers: Stem bark paste is applied on affected areas.	Babu, 2007; Swamy, 2009
270	*Flacourtia indica* (Burm. *f.*) Merr.	Leprosy: 2 spoonfuls of fruit paste mixed with a spoonful of honey is administered daily twice for 21 days. Fruit paste is also applied on the affected areas daily once for 21 days.	Suneetha, 2007
271	*Foeniculum vulgare* Mill.	Eczema: Fruit paste is applied on the affected areas daily twice for 3 days.	Suneetha, 2007
272	*Fraxinus xanthoxyloides* (Wall. ex G. Don) DC.	Wounds and injuries: Stem bark decoction is taken once a day for one week.	Lal and Singh, 2008
273	*Galium aparine* L.	Wounds and cuts: Shade dried aerial parts powder is applied to stop bleeding and for quick healing.	Lal and Singh, 2008
274	*Gardenia latifolia* Ait.	Ringworm: Stem bark paste mixed with a pinch of turmeric is applied on the affected areas daily once for 2 days. Scabies: 2 spoonfuls of stem bark paste mixed with a pinch of turmeric is applied on the affected areas daily once for 3 days.	Suneetha, 2007; Babu and Reddi, 2015
275	*Garuga pinnata* Roxb.	Burns and wounds: Leaf juice is applied on them.	Naidu, 2003
276	*Girardinia diversifolia* (Link) Friis	Scabies: Leaf paste along with pepper powder mixed with *Pongamia pinnata* oil is applied on the affected parts.	Rao, 2010
277	*Gloriosa superba* L.	Chicken pox and leprosy: Tuber paste is applied on the body. Leprosy: Tuber paste is applied on the affected parts daily twice for 30 days.	Ayyanar and Ignacimuthu, 2005; Suneetha, 2007
278	*Glycosmis arborea* (Roxb.) DC.	Anemia: 2 spoonfuls of root bark powder mixed with a spoonful of honey is administered daily twice for 5 days.	Suneetha, 2007

S. No.	Name of the plant	Practice	Reference
279	*Glycyrrhiza glabra* L.	Leucoderma: Paste of stem along with *Withania somnifera* roots is applied on the affected parts continuously for one year to cure leucoderma and other skin diseases.	Jeeva et al., 2007
280	*Glyricidia sepium* (Jacq.) Kunth ex Walp.	Skin diseases: Leaf paste is applied on the affected parts till cure.	Manjula and Reddi, 2015
281	*Gmelina arborea* Roxb.	Boils: Stem bark and leaves of *Dichrostachys cinerea* are ground. Paste is applied on the boils daily twice for 3 days.	Raju, 2009
282	*G. asiatica* L.	Dandruff: The ripe fruit paste is applied to the scalp 1 h before bath for 4 weeks.	Naidu, 2003
283	*Gossypium arboreum* L.	Ringworm: Twenty g of leaf paste mixed with a pinch of turmeric is applied on the affected areas daily twice till cure.	Suneetha, 2007
284	*Grewia asiatica* Mast.	Skin diseases: Fruit paste is applied on the affected parts.	Manjula and Reddi, 2015
285	*G. flavescens* Juss.	Skin diseases: Washed with root decoction twice a day on affected areas.	Manjula and Reddi, 2015
286	*G. tiliaefolia* Vahl	Lice and dandruff: Leaf paste is used as hair wash.	Naidu, 2003
287	*Gyrocarpus asiaticus* Willd.	Bruises: Stem bark juice is applied on the affected parts.	Manjula, 2011
288	*Haldina cordifolia* (Roxb.) Rids.	Tumors and wounds: Stem bark paste is applied on the affected areas once a day till cure.	Manjula, 2011
289	*Hedyotis corymbosa* (L.) Lamk.	Burning sensation: Whole plant paste is applied on the affected areas.	Manjula, 2011
290	*Helianthus annuus* L.	Common skin diseases: Flower paste is applied on the affected areas daily twice for 3 days.	Babu and Reddi, 2015

TABLE 10.1 (Continued)

S. No.	Name of the plant	Practice	Reference
291	*Helicteres isora* L.	Cuts and wounds: Root paste along with bark paste of *Litsea deccanensis* and *Holarrhena antidysenterica* is bandaged around the affected parts.	Rao, 2010; Babu, 2007
		Wounds: Warm stem bark paste is applied on the affected areas once a day till cure.	
292	*Heliotropium indicum* L.	Skin diseases: Ten ml of whole plant decoction mixed with one spoon of honey is administered once a day for 7 days.	Manjula and Reddi, 2015; Das et al., 2008; Swamy, 2009
		Cuts and wounds: Fresh leaf extract is applied externally.	
		Sores and ulcers: Leaf poultice is applied on affected areas till cure	
293	*H. supinum* L.	Cuts and wounds: Root paste is applied on affected areas till cure.	Swamy, 2009
294	*Hemidesmus indicus* (L.) Br.	Eczema: The roots are ground with water and made into a paste. It is used as poultice.	Nisha and Sivadasan, 2007; Suneetha and Reddi, 2015; Babu and Reddi, 2015; Babu and Reddi, 2015; Suneetha, 2007
		Common skin diseases: Root decoction is applied on the affected areas once a day till cure.	
		Common skin diseases: Roots with stem bark of *Gmelina arborea* and unripe fruits of *Solanum surattense* taken in equal quantities are ground with 4 pepper seeds and paste is applied on the affected areas twice a day till cure.	
		Scabies: Leaves or roots mixed with 5 seeds of pepper are ground. Paste is applied on the affected areas with cock quill till cure.	
		Psoriasis: Roots and leaves of *Andrographis paniculata*, roots and leaves of *Calotropis procera* and rhizome of *Acorus calamus* are taken in equal quantities and powdered. This powder mixed with a spoonful of honey and cow ghee and made into pills of soap nut seed size. Two pills are administered daily twice for 21 days.	

S. No.	Name of the plant	Practice	Reference
295	*Hernandia peltata* Meissn.	Cuts, wounds and ulcers: Leaf paste is applied locally.	Sinha et al., 1996
296	*Hibiscus cannabinus* L.	Anemia: A handful of leaves are ground with a pinch of salt. A spoonful of paste is administered daily twice for 15 days.	Raju, 2009
297	*H.rosa-sinensis* L.	Alopecia and dandruff: Leaf paste is applied on the scalp daily continuously for 7 days.	Naidu, 2003; Shetty et al., 2015; Manjula, 2011; Changkija, 1999
		Scabies: Different dishes prepared from leaves and petals are provided in the patient's diet.	
		Alopecia and bald head: Flowers are rubbed at the affected areas of the head till cure.	
298	*Holarrhena antidysenterica* Wall.	Skin diseases and ringworm: Latex is applied till cure.	Manjula, 2011; Babu, 2007
		Leprosy: Seed paste is applied on the affected areas.	
		Ringworm: Stem bark with that of *Strychnos nux-vomica* is taken in equal quantities and ground. Paste is applied on the affected areas twice a day for 3 days.	
299	*Holoptelea integrifolia* (Roxb.) Planch.	Rheumatic swellings, ulcers and scabies: The juice of the boiled bark is applied.	Naidu, 2003; Kumar and Chauhan, 2007; Shukla et al., 2010
		Leucoderma: The paste of the bark is made into pills of the size of pea and two pills are taken twice a day till cure.	
		Ringworm: Leaves are made into paste and applied over the body.	
300	*Holostemma ada-kodien* Schult.	Ringworm: Leaf paste with turmeric (2:1) is applied on the affected areas daily twice for 7 days.	Manjula, 2011

TABLE 10.1 (Continued)

S. No.	Name of the plant	Practice	Reference
301	*Homonoia riparia* Lour.	Lice and dandruff: Leaf paste is applied on head. Dandruff: Leaf paste is applied to head 1 h before going to bath once in 3 days for about a month.	Naidu, 2003; Swamy, 2009
302	*Hydnocarpus kurzii* (King) Warb.	Leprosy: Paste made from the seeds is applied.	Changkija, 1999
303	*Hydrolea zeylanica* (L.) Vahl	Scabies: Equal amounts of *Hydrolea zeylanica*, *Aristolochia bracteolata* and *Tridax procumbens* leaves are made into paste. It is applied on the affected areas.	Chetty et al., 1998
304	*Hygrophila auriculata* (Schum.) Heine.	Skin diseases: Dried leaf powder mixed with castor oil is applied on the affected parts. Sores on scalp: The powder of whole plant dried under sunlight is applied on affected parts.	Jeeva et al., 2007; Rao, 2010
305	*Hymenodictyon orixense* (Roxb.) Mabb.	Cuts and wounds: Stem paste is applied on the affected areas.	Rao, 2010
306	*Hyptis suaveolens* (L.) Poit.	Sores and fungal infection: Leaf paste is applied on them. Skin diseases: Young twigs and leaves are used.	Prashantkumar and Vidyasagar, 2008; Das et al., 2008
307	*Hyssopus officinalis* L.	Skin for eruptions: The paste prepared from flowers and leaves are applied on affected areas.	Lal and Singh, 2008;
308	*Ichnocarpus frutescens* (L.) Br.	Common skin diseases: Root paste mixed with a pinch of lime is applied on the affected areas daily once for 3 days. Eczema: Root paste is applied on the affected areas once a day till cure.	Suneetha and Reddi, 2015; Babu, 2007

S. No.	Name of the plant	Practice	Reference
309	*Impatiens balsamina* L.	Burns and cuts: Tender shoot and flower paste is applied on the affected areas daily twice till cure.	Babu, 2007
310	*Indigofera aspalathoides* Vahl	Leprosy: Powdered barks mixed with coconut oil are applied on the affected parts continuously for 6 months.	Jeeva et al., 2007
311	*I.linifolia* (L.f.) Retz.	Leprosy: Whole plant paste is applied at the affected portion till it cures.	Prashantkumar and Vidyasagar, 2008
312	*I.tinctoria* L.	Ringworm: Whole plant paste is applied on the affected areas daily twice for 3 days. Scabies: Leaf paste mixed with a pinch of turmeric is applied on the affected areas daily twice till cure.	Suneetha, 2008; Babu and Reddi, 2015
313	*Indoneesiella echioides* (L.) Sreem.	Ringworm: Root paste is applied on the affected areas daily twice for 3 days.	Suneetha, 2007; Babu, 2007
		Eczema and scabies: Whole plant paste is applied on the affected areas once a day till cure.	
314	*Inula racemosa* Hook.f.	Boils: Paste prepared from root powder mixed with fine butter is applied twice a day for a week in skin eruptions. One spoon of the decoction is taken daily as cure for boils.	Lal and Singh, 2008; Lal and Singh, 2008
		Blisters: Paste prepared from the root is applied for 7 days.	
315	*Ipomoea aquatica* Forssk.	Eczema: Leaf paste with turmeric is applied on the affected parts twice a day for 3 days.	Suneetha, 2007
316	*I. carnea* (Mart. ex Choisy) Austin.	Boil: Leaves applied with mustard oil are tied on a boil to make it ripen fast.	Naidu, 2003; Das et al., 2008
		Cuts and wounds: Fresh milky juice of the plant is useful.	
317	*I. eriocarpa* Br.	Scabies: Root paste mixed with coconut oil is applied on the affected parts.	Rao, 2010
318	*I. hederifolia* L.	Burns: Root and flower paste is applied on the affected parts.	Rao, 2010

TABLE 10.1 (Continued)

S. No.	Name of the plant	Practice	Reference
319	*I. obscura* (L.) Ker-Gawl.	Pimples and sores: Whole plant paste is applied on the affected areas daily once for 5 days.	Suneetha, 2007
320	*I. pes-tigridis* L.	Scabies: Leaf paste is applied on the affected areas daily twice for 3 days.	Suneetha and Reddi, 2015
321	*I. quamoclit* L.	Carbuncle: Leaf paste is applied on the affected parts twice a day till cure.	Manjula, 2011
322	*Ixora coccinea* L.	Eczema: Leaf paste is applied on the affected areas daily twice till cure.	Suneetha, 2008
323	*I. pavetta* Andr.	Anemia: Stem bark decoction given in 1 spoonful twice a day for about 2 months.	Swamy, 2009
324	*Jasminum angustifolium* (L.) Willd.	Ringworm: Leaf juice mixed with lemon juice is applied on the affected areas once a day till cure. Scabies: Whole plant paste is applied on the affected areas once a day till cure.	Manjula, 2011; Suneetha and Reddi, 2015
325	*J. auriculatum* Vahl	Ringworm: Root paste mixed with a pinch of turmeric is applied on the affected areas daily twice for 3 days.	Suneetha, 2007
326	*J. grandiflorum* L.	Scabies: Root paste mixed with a pinch of turmeric is applied on the affected areas daily twice till cure. Common skin diseases: Root paste is applied on the affected areas twice a day till cure. Leprosy: 2 spoonfuls of flower paste mixed with a spoonful of honey is administered daily twice for 21 days. The paste is applied on the affected areas daily twice for 21 days.	Suneetha and Reddi, 2015; Babu and Reddi, 2015; Babu, 2007
327	*J. officinale* L.	Skin diseases: Flowers mixed with oil is applied on the body till cure.	Arya et al., 2015

S. No.	Name of the plant	Practice	Reference
328	*J. sambac* (L.) Ait.	Ringworm: Leaf paste mixed with a pinch of turmeric is applied on the affected areas daily twice for 3 days.	Suneetha, 2007; Babu, 2007
		Leprosy: Leaf paste with flower paste of *Jasminum grandiflorum* (1:1) is applied on the affected areas once a day till cure.	
329	*Jatropha curcas* L.	Lymph gland swelling, measles, burns and eczema: Latex is applied on them.	Naidu, 2003; Silja et al., 2007; Raju et al., 2010; Manjula and Reddi, 2015; Babu, 2007
		Wounds, eczema and scabies: Fresh stem latex is applied.	
		Skin diseases: Seed oil is applied on the affected areas once a day till cure.	
		Leprosy: 2 spoonfuls of root powder mixed with jaggery is administered daily twice for 40 days.	
		Dandruff and eczema: Seeds are crushed and boiled in water. The oil formed as a thin layer is collected with a quill of cock and applied to scalp and hair twice a day till cure.	
330	*J. gossypifolia* L.	Leprosy: Roots are ground. Paste mixed with a pinch of turmeric powder is applied on the affected areas daily once for 15 days.	Raju, 2009
331	*Juniperus communis* L.	Boils: Heartwood oil is used for massage.	Lal and Singh, 2008; Lal and Singh, 2008
		Skin itching and infections: Regular massage of heartwood oil is considered effective against them.	
332	*Justicia adhatoda* L.	Psoriasis: The leaf powder is boiled with sesame oil and applied on the affected parts.	Naidu, 2003; Suneetha and Reddi, 2015
		Scabies: Leaf paste mixed with a pinch of turmeric is applied on the affected areas daily twice for 3 days.	

TABLE 10.1 (Continued)

S. No.	Name of the plant	Practice	Reference
333	*J. glauca* Rottl.	Cuts: Leaf paste mixed with a pinch of turmeric is applied on the affected areas once a day till cure.	Suneetha, 2007
334	*J. gendarussa* Burm.	Leucoderma: Leaves ground with those of *Lawsonia inermis* and the paste is applied on patches and exposed to morning sunlight for 30–45 minutes for about 2 months.	Swamy, 2009
335	*Kalanchoe pinnata* (Lamk.) Pers.	Tumors: Oil is applied on the leaf, warmed and pasted on the affected areas. Carbuncle: Leaf paste with castor oil is applied on the affected areas once a day till cure.	Manjula, 2011; Babu, 2007
336	*Kydia calycina* Roxb.	Skin diseases: Stem bark paste or leaf paste is applied externally. Dandruff: Stem bark paste is used to wash hair thrice a week for three weeks.	Naidu, 2003; Babu, 2007
337	*Lablab purpureus*(L.) Sw.	Ringworm and scabies: Leaf juice is applied on the affected areas. Skin diseases: Stems are roasted, powdered and mixed with coconut oil is applied on the affected areas.	Manjula, 2011; Manjula and Reddi, 2015
338	*Lagenaria siceraria* Standl.	Burns: Leaf juice is applied on them.	Naidu, 2003
339	*Lannea coromandelica* (Houtt.) Merr.	Cuts and burns: Paste of stem bark is applied on them. Dandruff: Seed infusion is used for head bath to get relief.	Singh and Maheshwari, 1989; Naidu, 2003
340	*Lantana camara* L.	Cuts and wounds: Leaf paste is applied for 7–8 days. Cuts and wounds: Bruised leaves are used in fresh cuts and wounds. Skin diseases: Leaf juice is applied on the affected areas.	Prashantkumar and Vidyasagar, 2008; Das et al., 2007; Manjula and Reddi, 2015

S. No.	Name of the plant	Practice	Reference
341	*Lawsonia inermis* L.	Leucoderma: 25 g of leaves with 10 pepper seeds are ground into a paste and made into pills. Two pills are administered thrice a day with rice washed water for about 40 days.	Naidu, 2003; Prashantkumar and Vidyasagar, 2008; Manjula and Reddi, 2015; Arya et al., 2015; Swamy, 2009; Swamy, 2009
		Cuts and wounds: Leaf paste is applied on them for 7–8 days.	
		Scabies and skin diseases: Leaf paste is applied on the affected parts.	
		Skin diseases: Leaves of henna applied to infected skin portions.	
		Leprosy: Bark decoction is administered in 2 spoonfuls twice a day for about 3 months.	
		Psoriasis: Stem bark decoction is administered orally in 2 spoonfuls twice a day for 2–3 months. Also the paste mixed with castor oil is applied externally on affected areas.	
342	*Leea indica* (Burm. *f.*) Merr.	Cuts, wounds and sores: Leaf paste is used on affected areas.	Sinha et al., 1996
343	*Leonotis nepetaefolia* Br.	Skin diseases: Plant paste is applied on the affected parts.	Nayak et al., 2004; Manjula, 2011; Babu, 2007
		Cuts and wounds: Flowers with leaves of *Tridax procumbens, Holarrhena pubescens, Ventilago denticulata, Santalum album,* bark of *Boswellia serrata, Ficus racemosa* and roots of *Sarcostemma secamone* are grind in equal quantities to fine paste. Paste is applied with honey on the affected parts till cure.	
		Eczema and ringworm: Dried flower and seed powder with mustard oil is applied on the affected areas daily till cure.	
344	*Lepidogathis cristata* Willd.	Dandruff: Whole plant paste mixed with gingelly oil is applied on the head before going to bath.	Manjula, 2011

TABLE 10.1 (Continued)

S. No.	Name of the plant	Practice	Reference
345	*Leptadenia reticulata* (Retz.) Wt. & Arn.	Eczema and scabies: Leaf paste is applied on the affected areas daily twice till cure.	Babu, 2007; Swamy, 2009
346	*Leucas aspera* (Willd.) Link.	Ringworm: Leaf paste is applied on affected areas till cure.	Shetty et al., 2015; Suneetha, 2007
		Scabies: Leaf extract is applied on the affected areas.	
347	*L. biflora* (Vahl) R.Br.	Psoriasis: Leaf paste mixed with a pinch of camphor is applied on the affected areas daily twice for 3 days.	Ayyanar and Ignacimuthu, 2005
		Skin diseases: Paste of whole plant is mixed with coconut oil and applied externally on affected places for 14 days to cure.	
348	*L. cephalotes* (Roth) Spr.	Scabies: Leaf paste mixed with a pinch of camphor is applied on the affected areas daily once for 2 days.	Suneetha and Reddi, 2015
349	*Limonia acidissima* L.	Scabies: Stem bark paste is applied on the affected areas once a day till cure.	Manjula and Reddi, 2015
350	*Lindelofia longiflora* (Benth.) Baill.	Scabies: Leaves are burnt to obtain ash and the ash is applied on cuts and wounds to check bleeding and for quick healing.	Lal and Singh, 2008
351	*Lindernia crustacea* (L.) F. Muel.	Ringworm: Plant poultice is applied on affected areas.	Swamy, 2009; Swamy, 2009
		Sores: Plant poultice is applied externally.	
352	*Lippia geminata* Kunth	Cuts and wounds: Fresh leaf juice is useful against them.	Das et al., 2008
353	*L. javanica* (Burm. *f.*) Spr.	Burns: Dried leaf powder is applied on the affected areas once a day till cure.	Babu, 2007
354	*Litsea deccanensis* Gamble	Scabies: Bark paste is applied on the affected parts.	Rao, 2010
355	*L. glutinosa* (Lour.) Robins.	Bruises, sprains and wounds: Stem bark paste is applied over them.	Naidu, 2003; Swamy, 2009
		Wounds: Bark paste is applied externally till cure.	

S. No.	Name of the plant	Practice	Reference
356	*Ludwigia parviflora* Roxb.	Vitiligo: Leaf paste mixed with that of *Clerodendrum viscosum* is applied on the affected parts during night and morning after bath.	Rao, 2010
357	*Luffa acutangula* (L.) Roxb.	Skin diseases: Seed paste is applied on the affected areas.	Manjula and Reddi, 2015
358	*L. cylindrica* (L.) Roem.	Common skin diseases: Seed paste mixed with a pinch of camphor is applied on the affected areas daily twice till cure.	Suneetha and Reddi, 2015; Silja et al., 2008
		Leprosy: Leaf paste is externally applied.	
359	*Lygodium flexuosum* (L.) Sw.	Anemia: One pinch of root powder is taken orally once a day for 15–21 days.	Naidu, 2003
360	*Macaranga peltata* (Roxb.) Muell.-Arg.	Ringworm: Root paste mixed with nut soup of *Terminalia chebula* and bark paste of *Litsea deccanensis* is applied on the affected areas.	Rao, 2010; Babu, 2007
		Eczema, ringworm and skin diseases: Root paste is applied on the affected areas once a day till cure.	
361	*Madhuca longifolia* (Koen.) Mac Br.	Sores on the scalp: Seed oil is applied once a day till cure.	Manjula and Reddi, 2015; Babu and Reddi, 2015; Jeeva et al., 2007; Manjula, 2011; Manjula, 2011
		Common skin diseases: Stem bark paste is applied on the affected areas once a day till cure.	
		Skin diseases: Pounded seeds with *Ocimum tenuiflorum* leaf extract are applied on the affected parts.	
		Allergy: Seed oil mixed with 100 g of leaf powder of *Datura innoxia* and 30 g of black pepper powder is taken in doses of 5 g once a day for 2 days.	
		Sores on the scalp and dandruff: Seed oil is applied once a day till cure.	

TABLE 10.1 (Continued)

S. No.	Name of the plant	Practice	Reference
362	*Mallotus philippensis* Muell.-Arg.	Boils and blisters: Red glandular hair powder along with ghee is applied locally to cure.	Nayak et al., 2004; Naidu, 2003; Suneetha, 2007; Suneetha and Reddi, 2015; Babu, 2007
		Eruptions of leprous skin: Fruit paste or oil in sesamum oil is applied.	
		Ringworm: Fruit paste mixed with a pinch of turmeric and honey is applied on the affected areas daily twice till cure.	
		Scabies: Root paste mixed with *Pongamia pinnata* oil is applied on the affected areas once a day till cure.	
		Psoriasis: 2 spoonfuls of fruit paste with a spoonful of castor oil is applied on the affected areas daily once for one month.	
363	*Malvastrum coromandelianum* Garcke	Wounds: The juice of leaves is externally applied for cooling and quick healing.	Kumar and Chauhan, 2007
364	*Mangifera indica* L.	Scabies: Gum resin mixed with coconut oil is applied on the affected part for 7–8 days.	Prashantkumar and Vidyasagar, 2008; Suneetha, 2007
		Heel cracks: Stem bark gum mixed with coconut oil is applied on the affected areas daily once for 3 days.	
365	*Manihot esculenta* Crantz	Cuts and wounds: Tuberous root paste mixed with a pinch of turmeric and camphor is applied on the affected areas daily twice for 2 days.	Babu, 2007
366	*Marsilea quadrifolia* L.	Allergy: Whole plant paste mixed with castor oil is anointed allover the body till cure.	Rao, 2010; Suneetha, 2007
		Ringworm: Leaves with those of *Justicia adhatoda*, *Eclipta prostrata* and *Sida cordata* taken in equal quantities are ground. Paste is applied on the affected areas daily twice till cure.	

S. No.	Name of the plant	Practice	Reference
367	*Martynia annua* L.	Wounds: Seed oil mixed with a pinch of turmeric is applied on the affected areas daily once for 3 days.	Suneetha, 2007
368	*Maytenus senegalensis* (Lamk.) Exell.	Dandruff: Leaf paste is used as hair wash thrice a week for one month.	Babu, 2007; Swamy, 2009
		Sores: Leaf paste is applied externally till cure.	
369	*Medicago falcata* L.	Wounds and injuries: The aerial parts paste is applied on them for healing purposes.	Lal and Singh, 2008
370	*Melastoma malabathricum* L.	Scabies: Leaf paste is applied on the affected areas daily twice for 3 days.	Suneetha and Reddi, 2015; Babu, 2007
		Carbuncle: Leaf paste is applied on the carbuncle once a day till cure.	
371	*Melia azedarach* L.	Eczema: Roots and leaves are taken in equal quantities and ground. The paste is applied on the affected areas daily twice for 3 days.	Raju, 2009; Babu, 2007; Swamy, 2009
		Psoriasis: Leaves with those of *Azadirachta indica*, *Cassia occidentalis* and whole plant of *Oxalis corniculata* are taken in equal quantities and ground. Paste with coconut oil or castor oil is applied on the affected areas once a day for a month.	
		Leprosy: Dried fruit powder given in 2 spoonfuls twice a day for about 2 months.	
372	*Memecylon edule* Roxb.	Pimples and skin diseases: Leaf juice is applied on the affected areas.	Manjula and Reddi, 2015
373	*M. umbellatum* Burm.f.	Anasarca: Poultice of bark with black pepper and turmeric is applied once a day for about 3 days.	Naidu, 2003
374	*Mentha spicata* L.	Common skin diseases: Leaf juice is applied on the affected areas daily twice for 5 days.	Suneetha and Reddi, 2015
375	*Merremia aegyptia* (L.) Urban.	Skin diseases: Leaves mixed with those of *Acalypha indica* are ground and applied on the affected areas.	Manjula and Reddi, 2015

TABLE 10.1 (Continued)

S. No.	Name of the plant	Practice	Reference
376	*M. emarginata* (Burm. *f.*) Hall.*f.*	Wounds and boils: Whole plant paste is applied externally.	Shukla et al., 2010
377	*Michelia champaca* L.	Eczema, ring worm and skin diseases: A spoonful of seed oil mixed with a spoonful of honey is applied on affected areas once a day till cure.	Babu, 2007
378	*Mimosa pudica* L.	Skin diseases: Whole plant is heated in oil and applying the resulting concoction on the body will cure.	Arya et al., 2015; Arya et al., 2015; Nisha and Sivadasan, 2007; Swamy, 2009
		Allergy: Grounded leaf juice is taken orally to get rid of various skin allergies.	
		Eczema: The tender leaves are made into a paste and applied on the sores.	
		Psoriasis: The medicated oil is prepared using roots of *Mimosa pudica* along with leaves of *Hibiscus rosa-sinensis* and the roots of *Sida rhombifolia*, which is very effective.	
379	*Mimusops elengi* L.	Scabies: Latex is applied on the affected areas daily once for 3 days.	Raju et al., 2010
380	*Mirabilis jalapa* L.	Allergy: Leaf paste mixed with turmeric is applied on the affected areas daily twice for 3 days.	Rao, 2010; Swamy, 2009
		Blisters and boils: Root paste is applied externally till cure.	
381	*Mollugo pentaphylla* L.	Wounds, scabies and skin diseases: Whole dried plant is crushed and the paste is applied locally.	Nayak et al., 2004; Suneetha, 2007
		Ringworm: Whole plant ground with seeds of *Canavalia virosa* is applied on the affected areas daily twice for 3 days.	
382	*M. nudicaulis* Lamk.	Common skin diseases: Plant juice is applied externally for prevention of itching.	Swamy, 2009

S. No.	Name of the plant	Practice	Reference
383	*Momordica charantia* L.	Psoriasis: Water extract of the fruit is taken orally in the morning daily till cure.	Shah et al., 2011
384	*M. dioica* Roxb.ex Willd.	Dandruff: Tuber paste is applied on head.	Naidu, 2003; Babu, 2007
		Burns and cuts: Leaf paste is applied on the affected areas daily twice for 2 days.	
385	*Morinda pubescens* Sm.	Scabies: Leaf paste is applied on the affected areas till cure.	Manjula and Reddi, 2015; Suneetha, 2007
		Wounds: Leaf paste is applied on the affected areas daily twice for 2 days.	
386	*Moringa oleifera* Lam.	Herpes: Stem bark is boiled and applied on the affected areas.	Manjula, 2011; Manjula, 2011; Manjula, 2011
		Ring worm: Fine paste of root is applied on the affected areas once a day till cure.	
		Whitlow: Gum is applied around the affected areas.	
387	*Mucuna pruriens* (L.) DC.	Allergy: Roots with black pepper seeds are ground and applied on the affected areas.	Rao, 2010
388	*Mukia maderaspatana* (L.) Cogn.	Dandruff: Leaf paste is applied to head 1 h before bath twice or thrice a week.	Swamy, 2009
389	*Murraya koenigii* (L.) Spr.	Heel cracks: Leaves ground along with those of *Lawsonia inermis* is applied on the cracked areas.	Manjula, 2011; Swamy, 2009
		Dandruff: Leaf paste applied to head 1 h before bath for 2–3 times a week	
390	*Musa paradisiacal* L.	Boils: Fruit boiled and ground with a pinch of turmeric and salt is applied on the boils and bandaged.	Suneetha, 2007
391	*Mussaenda frondosa* L.	Wounds: Paste made from leaves and flowers is applied.	Changkija, 1999; Changkija, 1999
		Leprosy: Root decoction is given orally.	

TABLE 10.1 (Continued)

S. No.	Name of the plant	Practice	Reference
392	*Myristica fragrans* Houtt	Cuts, skin diseases and wounds: Fruit paste is applied on the affected areas.	Manjula, 2011
393	*Naravelia zeylanica* (L.) DC.	Leprosy: Root paste with leaf paste of *Achyranthes aspera* (1:1) is applied on the affected areas once a day for 21 days.	Babu, 2007; Nisha and Sivadasan, 2007
		Ringworm: The entire leaf is steamed and it is placed on the portion affected.	
394	*Nelumbo nucifera* Gaertn.	Skin diseases: Root paste is applied on the affected areas till cure.	Manjula and Reddi, 2015; Babu, 2007
		Ringworm: A spoonful of root powder is administered daily once for 5 days.	
395	*Nerium indicum* Mill.	Scabies: Stem paste mixed with 5 ml of mustard oil is applied on the affected areas till cure.	Suneetha and Reddi, 2015
396	*N. oleander* L.	Scabies: Stem paste mixed with mustard oil is applied on the affected areas till cure.	Raju, 2009
397	*Nicotiana tabacum* L.	Common skin diseases: Leaf paste is applied on the affected areas daily twice for 3 days.	Suneetha and Reddi, 2015; Manjula, 2011; Babu, 2007
		Septic wounds: Take equal quantities of leaf along with that of *Justicia adhatoda* and boiled in castor oil is applied on the affected areas till cure.	
		Sores: Leaf paste with that of *Securinega leucopyrus* (1:1) is applied on the sores to expel worms once a day for 3 days.	
398	*Nyctanthes arbor–tristis* L.	Dandruff: Seed paste is applied to the scalp and head bath is taken after 1 h.	Naidu, 2003; Suneetha, 2007
		Alopecia: 2 spoonfuls of seed powder mixed with a spoonful of honey is administered daily twice for 15 days. Meanwhile the paste is applied on the scaly patches.	

S. No.	Name of the plant	Practice	Reference
399	*Nymphaea nouchali* Burm.f.	Wounds: Rhizome paste is applied on the affected areas daily twice for 3 days.	Suneetha, 2007
400	*Ocimum americanum* L.	Skin disease: Leaf paste is applied on the affected areas once a day till cure.	Manjula and Reddi, 2015
401	*O. basilicum* L.	Ringworm: Leaf paste mixed with a pinch of cheese is applied on the affected areas daily twice for 3 days.	Suneetha, 2007; Marjula and Reddi, 2015
		Scabies and swellings: Leaf paste is applied on the affected areas.	
402	*O. gratissimum* L.	Eczema: Leaf paste mixed with a pinch of turmeric (3:1) is applied on the affected areas daily twice for 3 days.	Suneetha, 2007
403	*O.sanctum* L.	Ringworm: Leaf paste is applied on the skin.	Prashantkumar and Vidyasagar, 2008
404	*O. tenuiflorum* L.	Dandruff: Leaf juice is anointed on the head.	Naidu, 2003; Shetty et al., 2015; Suneetha, 2007; Suneetha, 2007; Manjula, 2011; Swamy, 2009
		Scabies: Leaves are used to cure them.	
		Eczema: Leaf paste mixed with a pinch of camphor is applied on the affected areas daily twice till cure.	
		Pimples: Leaf paste is applied on the pimples daily once for 3 days.	
		Ring worm: Leaf paste mixed with salt is applied on the affected areas.	
		Itching and ringworm: Leaf sap is rubbed over skin to get relief.	
405	*Operculina turpethum* (L.) Silva Manso	Allergy: Leaf paste is applied on the affected areas.	Shukla et al., 2010
406	*Opilia amentacea* Roxb.	Tumors: Phylloclade paste is applied once day till cure.	Babu and Reddi, 2015

TABLE 10.1 (Continued)

S. No.	Name of the plant	Practice	Reference
407	*Opuntia dillenii* (Ker-Gawl.) Haw.	Carbuncle: Phyllode paste is ground along with 5 or 6 seeds of *Piper nigrum* and paste is applied on the affected areas twice a day till cure.	Rao, 2010; Babu, 2007
		Dandruff: Fruit paste is applied to the scalp.	
408	*Oroxylum indicum* (L.) Vent.	Wounds: Stem bark paste is applied on the affected areas till cure.	Manjula, 2011
409	*Orthosiphon rubicundus* Benth.	Ringworm: Leaf paste is applied on affected areas.	Swamy, 2009
410	*Oryza sativa* L.	Abscess and tumors: Boiled seed paste is applied on the affected areas.	Manjula, 2011
411	*Oxalis corniculata* L.	Ringworm: Leaf paste mixed with a pinch of turmeric is applied on the affected areas daily twice till cure.	Suneetha, 2007
412	*Pandanus fascicularis* Lamk.	Common skin diseases: Leaf paste is applied on the affected areas daily twice for 5 days.	Suneetha and Reddi, 2015; Babu and Reddi, 2015
		Common skin diseases: Flower paste is applied on the affected areas daily once till cure.	
413	*Panicum sumatrense* Roth ex Roem. & Schult.	Carbuncle: Root paste is applied on the affected areas daily twice till cure.	Suneetha, 2008
414	*Paracalyx scariosus* (Roxb.) Ali	Boils and blisters: Leaf paste is applied on them.	Naidu, 2003
415	*Parkinsonia aculeata* L.	Skin diseases: Leaf decoction is administered orally in 2 spoonfuls twice a day till cure.	Manjula and Reddi, 2015
416	*Passiflora foetida* L.	Cuts and wounds: Dried leaf powder mixed with that of *Woodfordia fruticosa* is applied on the affected parts.	Rao, 2010
417	*Pavetta indica* L.	Common skin diseases: Leaf paste is applied on itchings.	Swamy, 2009

S. No.	Name of the plant	Practice	Reference
418	*Pavonia odorata* Willd.	Anasarca: Leaf paste is applied on the affected areas daily once for 3 days.	Babu, 2007
419	*P. zeylanica* (L.) Cav.	Ringworm: Whole plant paste mixed with a pinch of turmeric is applied on the affected areas daily once till cure.	Suneetha 2007
420	*Pedilanthus tithymaloides* (L.) Poit.	Psoriasis and warts: Latex is applied on the affected areas.	Manjula, 2011
421	*Peltophorum pterocarpum* (DC.) Baker	Sores: Stem bark paste is applied on the affected areas daily once till cure.	Suneetha, 2007
422	*Peperomia tetraphylla* (Forst. *f*.) Hook. & Arn.	Sores on the scalp: Plant paste is applied on affected parts.	Rao, 2010
423	*Pergularia daemia* (Forsk.) Chiov.	Scabies: Fruits and leaves are made into paste in mustard oil and the oil is applied locally thrice a day for about 7 days to cure.	Naidu, 2003; Manjula, 2011
		Tumors: Leaf paste mixed with coconut oil is mildly heated and applied on the affected areas.	
424	*Phyla nodiflora* (L.) Green	Dandruff: Leaf juice mixed and boiled with equal volume of gingelly oil is applied twice a week on head.	Jeeva et al., 2007
425	*Phyllanthus amarus* Schum. & Thonn.	Skin diseases: Plant extract with rhizome paste of turmeric is used internally.	Silja et al., 2007; Manjula and Reddi, 2015; Raju et al., 2010
		Anemia and dandruff: Leaf juice is used.	
		Scabies: Leaf paste is applied on the affected areas once a day till cure.	
426	*P. emblica* L.	Ringworm: Bark is made into a paste and mixed with curd and applied all over the body and washed after 3 h, which is very effective for curing infection.	Nisha and Sivadasan, 2007

TABLE 10.1 (Continued)

S. No.	Name of the plant	Practice	Reference
427	P. reticulatus Poir.	Anasarca: The stem bark extract is taken in doses of 5 spoonful once a day for 3 days.	Naidu, 2003; Swamy, 2009
		Burns, cuts and sores: Leaf paste is applied on affected areas.	
428	Physalis minima L.	Wounds: Leaf juice is applied on the affected areas once a day till cure.	Manjula, 2011
429	Piper betle L.	Warts: Leaf paste mixed with lime is applied on the affected areas once a day till cure.	Manjula, 2011
430	P. longum L.	Wounds: Long pepper ground with fronds of Hemionitis arifolia and the paste is applied.	Naidu, 2003
431	P. nigrum L.	Scabies: Seed powder mixed with coconut oil is applied on the affected areas.	Manjula and Reddi, 2015; Arya et al., 2015; Manjula, 2011
		Skin rashes: Pepper is infused into heated oil, when cooled is effective in the treatment of skin rashes.	
		Ringworm and scabies: Seed powder mixed with coconut oil is applied on the affected areas.	
432	Pinus roxburghii Sarg.	Tumors: Gum mixed with coconut oil is applied on the affected areas.	Manjula, 2011
433	Pistia stratiotes L.	Burning sensation: Leaf juice is applied on the affected areas.	Manjula, 2011
434	Plantago major L.	Boils: Slightly warmed leaves applied with mustard oil are kept on them overnight to burst it out. It is believed that it reduces both swelling and pain.	Lal and Singh, 2007
435	Plumbago auriculata Lamk.	Warts: Root paste is applied over them.	Naidu, 2003

S. No.	Name of the plant	Practice	Reference
436	*P. indica* L.	Ringworm: Roots with leaves of *Eclipta prostrata, Marsilea quadrifolia* taken in equal quantities are ground. Paste mixed with a pinch of camphor is applied on the affected areas daily twice for 3 days.	Suneetha, 2007
437	*P. rosea* L.	Ringworm and scabies: Roots with leaves of *Eclipta prostrata, Marsilea quadrifolia* are taken in equal quantities and ground. Paste mixed with a pinch of camphor is applied on the affected areas daily twice for 3 days.	Suneetha, 2007; Manjula, 2011
		Leucoderma: Root decoction mixed with one spoon of honey is administered once a day for 15 days.	
438	*P. zeylanica* L.	Scabies: Leaf paste is applied on the affected areas once a day till cure.	Manjula and Reddi, 2015; Swamy, 2009; Raju, 2009
		Skin diseases: Roots along with stem barks of *Holarrhena pubescens, Azadirachta indica, Cassia fistula, Pterocarpus marsupium* and *Alstonia scholaris* taken in equal quantities are powdered. 2 spoonfuls of powder is boiled in four glasses of water till one glass of decoction remains. This is made into two parts and taken in the morning and evening with half spoonful of cow ghee and one spoonful of honey.	
		Leprosy: Root paste is applied on the affected areas daily once for 10 days.	
439	*Plumeria alba* L.	Scabies and ulcers: Latex along with lemon juice is applied on the affected areas once a day till cure.	Babu, 2007
340	*P. rubra* L.	Skin diseases: A piece of cloth is soaked in the latex and dried. The dried cloth is burnt to ash and the ash mixed with coconut oil is applied warm externally on the affected parts.	Naidu, 2003
441	*Polyalthia cerasoides* (Roxb.) Bedd.	Bruises: Stem bark paste is applied on affected areas.	Swamy, 2009

TABLE 10.1 (Continued)

S. No.	Name of the plant	Practice	Reference
442	P. longifolia Benth.& Hk.f.	Common skin diseases: Leaf paste mixed with a pinch of turmeric and cheese is applied on the affected areas daily once for 3 days.	Suneetha and Reddi, 2015
443	Polygonum glabrum Willd.	Tumors: Whole plant paste is applied on the affected parts till cure.	Manjula, 2011
444	Pongamia pinnata (L.) Pierre	Leucorrhoea: Stem bark crushed with that of Alangium salvifolium, Syzygium cumini and Terminalia arjuna and the extract mixed with little jaggery is given in doses of 3 spoonfuls once a day for 9 days.	Naidu, 2003; Bhattacharjee and Chatterjee, 2007
		Common skin diseases: Seed oil is applied externally till cure.	
445	Portulaca oleracea L.	Common skin diseases: Whole plant with leaves of Psidium guajava and tuberous roots of Asparagus racemosus and roots of Rauvolfia tetraphylla taken in equal quantities are ground. 2 spoonfuls of paste mixed in a glass of water is administered daily twice for 15 days.	Suneetha and Reddi, 2015; Suneetha, 2007
		Scurvy: Whole plant with leaves of Psidium guajava and tuberous roots of Asparagus racemosus and roots of Rauvolfia tetraphylla are taken in equal quantities and ground. 2 spoonfuls of paste mixed in a glass of water is administered daily twice for 15 days.	
446	P. quadrifida L.	Common skin diseases, burns and scalds: Leaf paste is applied externally till cure.	Swamy, 2009
447	Pouzolzia cymosa Wight	Skin diseases: Leaf decoction along with leaves of Naravelia zeylanica and Cassia senna is heated with water and used to take bath for 30 days.	Ayyanar and Ignacimuthu, 2005
448	P. zeylanica (L.) Benn.	Boils: Leaf paste mixed with a pinch of turmeric is applied on the affected areas till cure.	Suneetha, 2007
449	Premna latifolia Roxb.	Allergy: Half a glass of decoction of the bark is taken once a day.	Kumar and Chauhan, 2007

S. No.	Name of the plant	Practice	Reference
450	*Prosopis chilensis* (Mol.) St.	Wounds: Handful of leaves crushed with 2 spoonfuls of kerosene is applied on the affected areas daily twice for 2 days.	Suneetha, 2007
451	*P. cineraria* (L.) Druce	Wounds: Stem bark, fruits and leaves are taken in equal quantities and ground. Paste is applied on the affected areas daily once for 3 days.	Raju et al., 2010
452	*P. juliflora* (Sw.) DC.	Cuts and wounds: Leaf paste mixed with kerosene is applied on the affected areas daily twice for 2 days.	Singh and Maheshwari, 1989; Manjula, 2011
		Tumors: Leaf paste is applied on the affected areas once a day till cure.	
453	*Psidium guajava* L.	Burns: Paste of stem bark is applied.	Singh and Maheshwari, 1989; Manjula, 2011
		Scurvy: Leaf decoction heated mildly is taken orally twice a day till cure.	
454	*Psoralea corylifolia* L.	Scabies: Whole plant paste is applied on the affected areas till cure.	Manjula and Reddi, 2015; Suneetha and Reddi, 2015; Suneetha, 2007; Manjula, 2011; Babu, 2007
		Common skin diseases: Seed paste mixed with a pinch of turmeric and sesame oil is applied on the affected areas daily twice till cure.	
		Leprosy: Seeds and seeds of *Sesamum indicum* are taken in equal quantities are ground. A spoonful of powder mixed with half spoon of honey is administered daily twice for 10 days. Meanwhile paste is applied on the affected areas daily once for 10 days.	
		Leucoderma: Seeds and fruits of *Phyllanthus emblica* and leaves of *Andrographis paniculata* taken in equal quantities are ground. Paste is dried and powdered. A spoonful of powder mixed in a glass of goat milk is administered daily twice for 15 days.	
		Ring worm: Whole plant paste is applied on the affected areas till cure.	
455	*Pteris vittata* L.	Cuts and wounds: Juice of fronds is applied on the affected areas daily once for 3 days.	Suneetha, 2007

TABLE 10.1   (Continued)

S. No.	Name of the plant	Practice	Reference
456	*Pterocarpus marsupium* Roxb.	Cracks of lips: Gum obtained from bark is applied over them.	Nayak et al., 2004; Manjula, 2011
		Leucoderma: Fifty ml of stem bark decoction is applied on the affected areas once a day till cure.	
457	*P. santalinus* L.f.	Leprosy: One spoonful of stem bark paste mixed with a cup of water is administered twice a day till cure.	Manjula, 2011
458	*Pterospermum xylocarpum* (Gaertn.) Sant. & Wagh.	Common skin diseases, eczema, ring worm and scabies: 50 g of stem bark paste mixed in a bucket of water is bathed twice a day for 3 days.	Babu, 2007
459	*Pueraria tuberosa* (Willd.) DC.	Cuts and wounds: Tuber paste is administered twice a day till cure and the paste is also applied on the affected parts.	Rao, 2010
460	*Pupalia lappacea* (L.) Juss.	Cuts and wounds: Three spoonful of leaf powder mixed with a spoonful of coconut oil is applied on the affected areas daily once for 2 days.	Suneetha, 2007; Swamy, 2009
		Leprosy: Fruit paste with coconut oil is applied on ulcers.	
461	*Putranjiva roxburghii* Wall.	Skin allergy: Dried fruits beaded in a garland, is put on the neck to cure.	Shukla et al., 2010; Manjula, 2011
		Boils: Seed paste is applied on the affected areas.	
462	*Quisqualis indica* L.	Wounds: Leaf paste mixed with a pinch of powder of coffee is applied on the affected areas daily twice for 2 days.	Suneetha, 2007; Manjula, 2011
		Boils: Flower paste is applied on the affected areas till cure.	
463	*Rauvolfia serpentina* (L.) Benth. ex Kurz	Psoriasis: The leaves are made into a paste and applied on sores found on the body. A paste is made with boiled stem and leaves and used as poultice.	Nisha and Sivadasan, 2007; Manjula, 2011
		Leucoderma: Root paste along with those of *Eclipta alba*, *Hemidesmus indicus* and leaf paste of *Withania somnifera* is administered in doses of 2 spoonfuls twice a day till cure.	

S. No.	Name of the plant	Practice	Reference
464	*R. tetraphylla* L.	Skin diseases: Root paste is applied on the affected areas till cure.	Manjula and Reddi, 2015
465	*Rheum ustrale* Don	Boils and injuries: Root decoction mixed with ghee or oil is applied for 2–3 days.	Lal and Singh, 2008
466	*Richardia scabra* L.	Skin diseases: Leaf paste along with leaves of *Euphorbia hirta*, *Wrightia tinctoria*, *Toddalia asiatica* and *Clitoria ternatea* combined with the coconut oil is applied externally till cure.	Ayyanar and Ignac·muthu, 2005
467	*Richardia scabra* L.	Boils: Leaves are pasted on the affected areas till cure.	Manjula, 2011
468	*Rorippa indica* (L.) Hiern	Ringworm: Leaf paste is applied externally.	Swamy, 2009
469	*Rubia cordifolia* L.	Ringworm and leucoderma: Paste made from plant is applied on affected areas.	Changkija, 1999; Sharma et al., 2003
		Dark spots: Dry roots and stem powder rubbed with honey is recommended as a cure.	
470	*Rubus ellipticus* Sm.	Heel cracks: Root paste is applied on the cracks before bed till cure.	Babu, 2007
471	*Ruta graveolens* L.	Tumors: Leaves are warmed with castor oil and tied on the affected areas.	Manjula, 2011
472	*Saccharum officinarum* L.	Anemia: Fifty ml of stem juice mixed with one spoonful of amla powder and honey is administered once a day for 30 days.	Manjula, 2011
473	*S. spontaneum* L.	Common skin diseases: Root paste mixed with a pinch of turmeric is applied on the affected areas daily twice for 2 days.	Suneetha and Redd., 2015
474	*Samanea saman* (Jacq.) Merr.	Scabies: Leaf paste is applied on the affected areas.	Manjula and Reddi, 2015; Manjula, 2011
		Wounds: Leaf paste is applied on the affected areas.	
475	*Sanseviera roxburghiana* Schult.f.	Ringworm and skin diseases: Leaf paste is applied on the affected areas.	Manjula, 2011

TABLE 10.1 (Continued)

S. No.	Name of the plant	Practice	Reference
476	*Santalum album* L.	Common skin diseases: Seed oil is applied for all kinds of skin diseases.	Swamy, 2009; Suneetha, 2007
		Wounds: Stem paste mixed with a pinch of salt is applied on the affected areas daily twice for 2 days.	
477	*Sapindus emarginata* Vahl	Dandruff: Fruit juice is used as hair wash to kill lice and to treat dandruff.	Naidu, 2003
478	*Saraca indica* L.	Scabies: Dried flowers boiled with coconut oil are applied on the affected parts with the help of cock feather thrice a day till it is cured as a remedy.	Jeeva et al., 2007
479	*Sarcostemma secamone* (L.) Bennet.	Boils and wounds: Leaves are boiled in water and made into paste and applied.	Babu and Reddi, 2015
480	*Sauropus quadrangularis* (Willd.) Muell.- Arg.	Common skin diseases: Leaf paste is applied on the affected areas till cure.	Suneetha and Reddi, 2015; Suneetha, 2007
		Sores: Leaf paste is applied on the affected areas till cure.	
481	*Saussurea costus* (Falc.) Lipsch.	Boils and blisters: Root paste is applied directly on them.	Lal and Singh, 2008; Lal and Singh, 2008
		Leprosy: Root paste is applied externally for 7–8 days.	
482	*Schleichera oleosa* (Lour.) Oken	Scabies: Seed oil is applied on the affected areas once a day till cure.	Suneetha and Reddi, 2015; Babu and Reddi, 2015
		Common skin diseases and scabies: Dry seeds are crushed and boiled in water. The oil formed as a thin layer is collected and applied on the affected parts twice a day till cure.	
483	*Scilla hyacinthina* (Roth) Macb.	Carbuncle: Bulb paste is applied on the affected areas once a day for 3 days. Leprosy: Tuber paste is applied on ulcers.	Babu, 2007; Swamy, 2009

S. No.	Name of the plant	Practice	Reference
484	*Scleropyrum pentandrum* (Dennst.) Mabb.	Skin diseases: Paste of stem bark and leaf is applied externally till cure.	Ayyanar and Ignacimuthu, 2005
485	*Scutellaria discolor* Colebr.	Cuts and wounds: Fresh leaf juice is used against them.	Das et al., 2008
486	*S. violacea* Heyne	Skin diseases: Leaf paste is heated and mixed with castor oil and applied externally for 3 days.	Ayyanar and Ignacimuthu, 2005
487	*Securinega virosa* (Roxb. ex Willd.) Pax & Hoffm.	Scabies: Leaves with those of *Marsilea quadrifolia* and *Azadirachta indica* taken in equal quantities are made into powder. Powder mixed with coconut oil is applied on the affected areas daily twice for 3 days.	Suneetha and Reddi, 2011
488	*Semecarpus anacardium* L. f.	Burning wounds: Stem bark powder mixed with coconut oil is applied on the affected areas till cure.	Manjula, 2011
489	*Senecio cappa* D. Don	Skin diseases and boils: Paste of the plant is applied over them.	Changkija, 1999
490	*Sesamum indicum* L.	Alopecia: Plant paste is applied to the head and after one hour bath is taken.	Manjula, 2011
491	*Sesbania grandiflora* Pers.	Itching: Fresh bark paste is applied on affected areas till cure. Skin diseases: Grounded leaves mixed with coconut oil are applied on the body for relief.	Manjula, 2011; Arya et al., 2015
492	*S. sesban* (L.) Merr.	Skin disease: Leaf paste is applied on the affected parts twice a day for 15 days.	Manjula and Reddi, 2015
493	*Shorea robusta* Gaertn. f.	Cracks of feet: Gum is useful in healing.	Kumar and Chauhan, 2007
494	*Sida acuta* Burm. f.	Herpes: Leaf paste is applied on the affected areas till cure.	Manjula, 2011
495	*S. cordata* (Burm. f.) Bross.	Cuts and wounds: Leaf paste mixed with a pinch of turmeric powder is applied on the affected areas daily once for 2 days.	Raju, 2009

TABLE 10.1 (Continued)

S. No.	Name of the plant	Practice	Reference
496	S.rhombifolia L.	Ringworm: The entire plant is crushed and applied on the infected areas.	Nisha and Sivadasan, 2007
497	Smilax zeylanica L.	Pimples: Stem paste mixed with honey is applied on the affected areas. Sores and ulcers: Leaf paste is applied on affected areas till cure.	Manjula, 2011; Swamy, 2009
498	Solanum americanum Mill.	Cuts: Leaf juice is applied on the affected areas daily twice till cure.	Suneetha, 2007
499	S. anguivi Lamk.	Scabies: Fried seed paste mixed with a pinch of turmeric is applied on the affected areas daily once for 3 days.	Suneetha and Reddi, 2015; Babu and Reddi, 2015;
		Scabies: Three or 4 fruits are ground with ant eggs. Paste is applied on the affected areas twice a day till cure.	Bhattacharjee and Chatterjee, 2007
		Skin diseases: Green fruit paste is applied on affected parts.	
500	S. melongena L.	Heel cracks: Root paste is applied on the affected areas daily once for 2 days.	Suneetha, 2007
501	S. nigrum L.	Common skin diseases: Leaf paste is applied on the affected areas once a day till cure.	Babu and Reddi, 2015
502	S. surattense Burm.f.	Alopecia and itching: Fruit paste is applied on the affected areas once a day till cure.	Manjula, 2011
503	S. virginianum L.	Pimples: Dried seed boiled with water is applied on them.	Nayak et al., 2004
504	S. torvum Sw.	Heel cracks: Root paste is applied on the affected areas daily once for 2 days.	Suneetha, 2007; Manjula, 2011
		Cuts and wounds: Leaf paste is applied on the affected areas once a day for 5 days.	

S. No.	Name of the plant	Practice	Reference
505	*S. tuberosum* L.	Burns: Tuber paste mixed with a pinch of turmeric is applied on the affected areas daily once for 3 days.	Babu, 2007
506	*Solena heterophylla* Lour.	Cuts on the tongue: Tuberous root or root stock is given till cure.	Naidu, 2003
507	*Sorghum vulgare* (L.) Pers.	Tumors: Seed paste mixed with leaf of *Coccinia grandis* is applied on the affected areas till cure.	Manjula, 2011
508	*Soymida febrifuga* (Roxb.) Juss.	Leprosy and common skin diseases: A spoonful of leaf extract mixed in a glass of hot water is administered daily once on empty stomach for 21 days.	Babu, 2007
509	*Spathodea campanulata* P.Beauv	Skin diseases: Stem bark paste is applied on the affected areas once a day till cure.	Manjula and Reddi. 2015
510	*Sphaeranthus indicus* L.	Leucoderma: Paste of whole plant along with seed paste of *Psoralea corylifolia* and neem oil is applied for two to three months.	Singh and Maheshwari, 1989; Manjula and Reddi, 2015
		Ring worm and skin diseases: Whole plant juice is applied with a pinch of salt once a day till cure.	
511	*Stachytarpheta jamaicensis*(L.) Vahl	Bruises and sprains: Leaves are crushed, boiled and then applied over them.	Manjula, 2011
512	*S. urticaefolia* (Salisb.) Sims.	Cuts and wounds: Leaves ground with those of *Leucas cephalotes* and the paste is applied on affected parts.	Naidu, 2003
513	*Stephania hernandifolia* Walp.	Burns: Plant paste is applied on affected parts.	Changkija, 1999;
514	*Sterculia foetida* L.	Ringworm and scabies: 2 spoonfuls of seed oil mixed with a spoonful of turmeric powder is applied on the affected areas daily once for 4 days.	Hari Babu and Reddi, 2008
515	*S. urens* Roxb.	Heel cracks: Gum is applied on the cracks.	Rao, 2010

TABLE 10.1 (Continued)

S. No.	Name of the plant	Practice	Reference
516	*Streblus asper* Lour.	Scabies: Leaf paste mixed with turmeric is applied on the affected areas daily twice for 3 days.	Suneetha and Reddi, 2015; Hari Babu and Reddi, 2008
		Eczema and scabies: A handful of leaves are ground along with 5 pepper seeds and the paste is applied on the affected areas twice a day till cure.	
517	*Strychnos nux-vomica* L.	Leucoderma: Stem paste mixed with *Schleichera oleosa* seed oil (3:2) is applied.	Bhattacharjee and Chatterjee, 2007; Suneetha and Reddi, 2015; Swamy, 2011
		Scabies: Root bark with stem bark of *Cassia fistula* and seed gum of *Mangifera indica* taken in equal quantities are powdered. 2 spoonfuls of powder mixed with a spoonful of mildly heated sunflower oil is applied on the affected areas daily once till cure.	
		Common skin diseases: The leaves are boiled in water along with those of *Alangium salvifolium*, *Azadirachta indica* and *Leucas aspera* and then used for bathing and said to be effective. Leaves ground with leaves of *Vitex negundo* and the paste with gingelly oil are applied externally till cure.	
518	*S. potatorum* L. f.	Common skin diseases: Seed paste is applied on the affected areas daily twice for 5 days.	Raju et al., 2010
519	*Swietenia mahogani* (L.) Jacq.	Cuts and wounds: Stem bark paste mixed with a pinch of turmeric powder is applied on the affected areas till cure.	Manjula, 2011

S. No.	Name of the plant	Practice	Reference
520	*Syzygium cumini* L.	Skin diseases: Leaf paste is applied on the affected areas.	Manjula and Reddi, 2015; Swamy, 2009; Nisha and Sivadasan, 2007; Babu, 2007
		Common skin diseases: Stem bark extract is applied on the affected areas.	
		Leprosy: Decoction of bark in combination with decoction of the bark of *Tinospora cordifolia*, pod of *Terminalia chebula* and wood of *Acacia catechu* is taken orally.	
		Whitlow: Stem bark and seeds of *Cleome viscosa* is thoroughly pasted. This paste is cooked in urine of the patient till it becomes semi-solid. It is applied on the affected areas by using quill of a hen.	
521	*Tabernaemontana divaricata* (L.) Br. ex Roem. & Schult.	Common skin diseases: Roots are ground with turmeric and black pepper and applied on the affected areas till cure.	Manjula and Reddi. 2015; Arya et al., 2015
		Skin diseases: Leaf juice is applied on the affected areas.	
522	*T. heyneana* Wall.	Skin diseases: Powder of leaf and stem bark along with the stem bark of *Ficus benghalensis*, *Madhuca longifolia*, *Strychnos nux-vomica* and leaves of *Evolvulus alsinoides* is heated with coconut oil and applied externally.	Ayyanar and Ignacimuthu, 2005
523	*Tagetes erecta* L.	Skin disease: Leaf paste is applied on the affected areas till cure.	Manjula and Reddi, 2015
524	*Tamarindus indica* L.	Anemia: Extract of old tamarind fruits mixed with old jaggery in 1:2 proportion is given to women.	Naidu, 2003; Manjula and Reddi, 2015; Suneetha, 2007; Manjula, 2011
		Itching and scabies: Tender leaf paste is applied on the affected areas.	
		Boils: Leaf paste mixed with a pinch of camphor and turmeric is applied on the affected areas once a day till cure.	
		Warts: Leaf juice is applied on the affected areas.	

TABLE 10.1 (Continued)

S. No.	Name of the plant	Practice	Reference
525	*Taraxacum officinale* Weber	Boils and blisters: Flower decoction is taken in equal proportion with water for 2 days to cure.	Lal and Singh, 2008
526	*Tarenna asiatica* (L.) Ktze. ex Schum.	Boils: Dried fruits are burnt and the ash mixed with coconut oil is applied externally till cure.	Swamy, 2009
527	*Tectaria coadunate* (Wall.) Chr.	Pimples: Rhizome paste used as facial cream once a day for 3 days.	Babu, 2007
528	*Tectona grandis* L.f.	Dandruff: Flower paste with castor oil (3:1) is used as hair tonic.	Babu, 2007; Raju et al., 2010
		Scabies: Leaf paste mixed with a pinch of camphor is applied on the affected areas daily twice for 5 days.	
529	*Tephrosia purpurea* Pers.	Skin diseases: Leaf decoction is used for bath everyday for 10–15 days to cure.	Prashantkumar and Vidyasagar, 2008
530	*T. villosa* (L.) Pers.	Eczema: Crushed leaves are taken internally once a day early in the morning for 51 days.	Shetty et al., 2015; Swamy, 2009
		Leucoderma: Root bark ground with seeds of *Piper nigrum* and the paste dissolved in water is administered in ½ glass once in morning for about 8 weeks with a gap of 2 days between weeks.	
531	*Terminalia arjuna* (Roxb. ex DC.) Wt. & Arn.	Boils and blisters: Bark paste is applied allover the body and 50 mg of bark paste is given orally once a day till cure.	Rao, 2010; Manjula, 2011
		Pimples: Fine paste of stem bark is applied on the affected areas once a day till cure.	
532	*T. bellirica* (Gaertn.) Roxb.	Common skin diseases: Seed paste mixed with a pinch of sesame oil is applied on the affected areas till cure.	Suneetha and Reddi, 2015

S. No.	Name of the plant	Practice	Reference
533	*T. chebula* Retz.	Burns and skin diseases: Fruit juice is applied on the affected areas.	Manjula and Reddi, 2015; Babu and Reddi, 2015; Nayak et al., 2004
		Common skin diseases: Paste made out of sun dried fruit is applied on the affected areas once a day till cure.	
		Skin diseases, eczema and cuts: Fruit paste is applied locally to cure.	
534	*T. coriacea* (Roxb.) Wt & Arn.	Black patches on the skin: Powder prepared by mixing the gum extracted from the bark, turmeric, and Bengal gram seeds in 1:2:2 proportions is used as bath powder to remove them.	Naidu, 2003; Manjula and Reddi, 2015
		Common skin diseases: Leaf paste mixed with a pinch of cheese and 2 drops of castor oil is applied on the affected areas daily once for 2 days.	
535	*Themeda cymbarica* (Roxb.) Hack.	Wounds: 2 spoonfuls of root paste mixed with a pinch of mustard oil is applied on the affected areas daily twice for 2 days.	Suneetha, 2007
536	*Thespesia lampas* (Cav.) Dalz. & Gibs.	Scabies: Leaf paste mixed with a pinch of turmeric is applied on the affected areas daily once for 2 days.	Suneetha and Reddi, 2015
537	*T. populnea* (L.)Sol. ex Correa.	Scabies, eczema, ringworm and psoriasis: The bark, leaves, flowers and fruits are said to be useful to control them. Tender leaves are tied over the affected areas.	Nisha and Sivadasan, 2007; Suneetha and Reddi, 2015; Suneetha, 2007; Suneetha, 2007
		Scabies: Leaf paste is applied on the affected areas daily twice till cure.	
		Psoriasis: Bark decoction taken in 2 spoonfuls twice a day for 15 days, also the paste is applied externally on affected areas.	
		Ringworm: Bark decoction given in 1 spoonful twice a day for 5–7 days or till cure.	
538	*Thuja orientalis* L.	Pimples: Twenty g of leaf extract mixed with half spoonful of turmeric is used as facial cream daily twice for a week.	Babu, 2007

TABLE 10.1 (Continued)

S. No.	Name of the plant	Practice	Reference
539	*Thunbergia fragrans* Roxb.	Cuts and wounds: Leaves with those of *Euphorbia ligularia* taken in equal quantities are ground. Paste mixed with a pinch of turmeric is applied on the affected areas daily twice for 2 days.	Suneetha, 2007
540	*Thysanolaena maxima* Ktze.	Boils: Root paste is applied on the affected areas.	Manjula, 2011; Rao, 2010
		Cracks on tongue: Water droplets present in the axils of leaves are applied on the tongue once a day till cure.	
541	*Tiliacora acuminata* (Lamk.) Miers	Itches: Whole plant paste with seed oil of *Pongamia pinnata* is applied on the affected areas twice a day till cure.	Babu, 2007
542	*Tinospora cordifolia* (Willd.) Miers	Leucoderma: Tuber and aerial root paste is applied.	Naidu, 2003; Suneetha, 2007; Changkija, 1999
		Leprosy: Roots with tuberous roots of *Dioscorea alata*, leaves of *Eclipta prostrata* and *Andrographis paniculata* taken in equal quantities are ground. 2 spoonfuls of paste mixed in a glass of cow milk is administered daily once for 10 days. Meanwhile paste mixed with a pinch of turmeric is applied on the affected areas daily once for 10 days.	
		Leucoderma: Five spoonful of dried root powder mixed with a spoonful of castor oil is applied on the affected areas once a day for 21 days.	
543	*T. sinensis* (Lour.) Merr.	Burns: Plant paste is applied on affected areas.	Naidu, 2003
544	*Tragia involucrata* L.	Anemia: Whole plant decoction is administered once a day for 7–10 days.	Manjula, 2011
545	*Tribulus terrestris* L.	Skin diseases: Leaf paste is applied on the affected parts for 3–4 days.	Prashantkumar and Vidyasagar, 2008

S. No.	Name of the plant	Practice	Reference
546	*Trichosanthes tricuspidata* Lour.	Ringworm: Leaf paste is applied externally till cure.	Swamy, 2009
547	*Tridax procumbens* L.	Cuts and wounds: Leaf juice mixed with a pinch of lime is applied on the affected areas twice a day.	Prashantkumar and Vidyasagar, 2008; Nayak et al., 2004
		Ringworm: Leaves are applied externally.	
548	*Trigonella foenum–graecum* L.	Boils: Seed paste is applied on the affected areas daily once for 3 days.	Raju, 2009
549	*Triticum aestivum* L.	Psoriasis: Root paste mixed with fruit paste of *Piper longum* (1:1) is applied on the affected areas daily twice for 5 days.	Suneetha, 2007
550	*Triumfetta rhomboidea* Jacq.	Eczema and scabies: Leaf paste is applied on the affected areas daily twice till cure.	Babu, 2007; Manjula, 2011
		Leprosy: Twenty ml of flower decoction is taken twice a day till cure.	
551	*T. rotundifolia* Lam.	Leprosy: Leaves and flowers are ground and made into paste and applied on ulcers.	Swamy, 2009
552	*Typha angustata* Bory & Chaub.	Scabies: Leaf ash mixed with castor oil (5:1) is applied on the affected areas daily twice for 2 days.	Suneetha and Reddi, 2015
553	*Urena lobata* L.	Dandruff and premature fall of hair: Leaf paste is applied on head, hair wash is done after 1 h. The same treatment is repeated for every 3 days for about 15 days to cure.	Naidu, 2003; Ayyanar and Ignacimuthu, 2005
		Skin diseases: Leaf powder along with leaves of *Jasminum flexile* is mixed with water and taken internally for 20 days.	
554	*Urginea indica* (Roxb.) Kunth.	Boils: Paste of the bulb is applied on the affected areas daily twice for 3 days.	Raju, 2009
555	*Vanilla fragrans* Ames	Scabies: Stem and leaf paste is applied on the affected parts.	Rao, 2010

TABLE 10.1 (Continued)

S. No.	Name of the plant	Practice	Reference
556	*Ventilago denticulata* Willd.	Cuts and wounds: Leaf is pestled with that of *Tridax procumbens* and *Holarrhena pubescens* and applied on the affected areas.	Manjula, 2011
557	*Verbascum thapsus* L.	Boils: Paste of leaf and flower powder in mustard oil is applied.	Lal and Singh, 2008
558	*Vernonia albicans* DC.	Leucoderma: A spoonful of seed powder mixed with 2 black pepper fruits is administered once a day for about a month.	Naidu, 2003; Nisha and Sivadasan, 2007; Manjula and Reddi, 2015; Babu, 2007
		Ringworm and eczema: The leaf juice is used in the treatment.	
		Scabies and wounds: Leaf paste is applied on the affected areas once a day till cure.	
		Leucoderma: A spoonful of seed powder mixed with half spoon of seed powder of *Piper nigrum* is administered daily once. Meanwhile seed paste prepared with cow urine is applied on the affected areas daily once till cure.	
559	*Vetiveria zizanioides* (L.) Nash	Common skin diseases: Leaf oil is applied on the affected areas daily twice till cure.	Suneetha and Reddi, 2015; Suneetha, 2007
		Allergy: 2 spoonfuls of root paste mixed in a glass of water is administered daily twice for 3 days.	
560	*Vicoa indica* (L.) DC.	Skin diseases: Dried leaves are ground with those of *Desmodium triflorum* and grains of black pepper and the paste is applied on affected areas.	Naidu, 2003
561	*Vigna trilobata* (L.) Verd.	Wounds: Leaf paste is applied on the affected areas daily twice for 2 days.	Suneetha, 2007
562	*Viscum orientale* Willd.	Cuts and wounds: Leaf paste is applied on affected areas till cure.	Swamy, 2009
563	*Vitex negundo* L.	Scabies: Tender leaves along with *Cuminum* and *Adhatoda* leaves, clove and *Myristica* boiled in coconut oil are applied.	Shetty et al., 2015

S. No.	Name of the plant	Practice	Reference
564	*V. quinata*(Lour.) Williams	Wounds: Stem bark powder is mildly heated and applied on the wounds once a day till cure.	Babu, 2007
565	*Waltheria indica* L.	Wounds and ulcers: Plant powder is applied externally for drying and healing.	Naidu, 2003
566	*Wattakaka volubilis* (L.f.) Stapf.	Blisters and boils: Leaf paste is applied externally till cure.	Swamy, 2009
567	*Wedelia biflora* (L.) DC.	Cuts and wounds: Leaves mixed with lime are used.	Sinha et al., 1996
568	*W. chinensis* (Osb.) Merr.	Boils and warts: Whole plant powder mixed with turmeric powder is applied on the affected areas once a day till cure.	Manjula, 2011
569	*Wendlandia tinctoria* (Roxb.) DC.	Cramps: Stem bark paste is applied on the affected parts.	Manjula, 2011
570	*Withania somnifera* (L.) Dunal	Common skin diseases: Leaf paste is applied on the affected areas daily twice till cure.	Raju et al., 2010
571	*Woodfordia fruticosa* (L.) Kurz	Leprosy: 2 spoonfuls of stem bark paste is administered daily once for 15 days early in the morning before breakfast. Meanwhile flower paste is applied on the affected areas daily twice for 10 days. Burns and scabies: Fine flower powder is applied on the affected areas once a day till cure.	Suneetha, 2007; Manjula, 2011
572	*Wrightia arborea* (Dennst.) Mabb.	Skin diseases: Stem bark paste is applied on the affected parts.	Manjula and Reddi, 2015
573	*W. tinctoria* Br.	Skin diseases, ringworm and leprosy: Bark is made into paste and applied. Psoriasis: Pounded leaves mixed with coconut oil are applied on the affected parts.	Changkija, 1999; Jeeva et al., 2007

TABLE 10.1 (Continued)

S. No.	Name of the plant	Practice	Reference
574	*Xanthium pungens* L.	Herpes: Roots with leaves of *Azadirachta indica* and *Marsilea quadrifolia* taken in equal quantities are ground. Paste is applied on the affected areas twice a day till cure	Suneetha, 2007
575	*X. strumarium* L.	Herpes: The leaf powder in doses of 2 g with water is administered twice a day for 5 days.	Naidu, 2003; Manjula and Reddi, 2015
		Scabies: Nut juice is applied on the affected areas till cure.	
576	*Zanthoxylum armatum* DC.	Boils, dandruff, eczema, ringworm, scabies and skin diseases: 50 g of leaf paste ground along with 10 seeds of *Piper nigrum* and 20 g of crustose lichen. Paste is applied on the affected areas once a day till cure.	Babu, 2007
577	*Zea mays* L.	Eczema: Male flowers paste is applied on the affected areas daily twice for 2 days.	Suneetha, 2007
578	*Zingiber officinale* Rosc.	Allergy: One spoonful of rhizome juice mixed with one spoonful of honey is administered once a day for 3–5 days.	Manjula, 2011
579	*Z. zerumbet* (L.) Sm.	Leprosy: Rhizome and seeds of *Piper nigrum* taken in equal quantities are ground. A spoonful of paste is administered daily twice for 3 days.	Manjula, 2009
580	*Ziziphus mauritiana* Lamk.	Whitlow: Leaf paste is applied on the affected areas daily twice for 3 days.	Raju, 2009
581	*Z. oenoplia* (L.) Mill.	Eczema: Stem bark powder mixed with ghee is applied externally till cure.	Swamy, 2009; Suneetha, 2007
		Blisters: Root paste is applied on the affected areas till cure.	

S. No.	Name of the plant	Practice	Reference
582	Z. xylopyrus (Retz.) Willd.	Skin diseases: Bark is ground with that of *Strychnos nux-vomica* and whole plant of *Andrographis paniculata* and the paste is applied on the affected areas till cure.	Swamy, 2009; Suneetha, 2007
		Leucoderma: Stem bark with those of *Bauhinia recemosa, Madhuca indica* and *Strychnos nux-vomica* and leaves of *Andrographis paniculata* taken in equal quantities are ground. Paste is made into pills of peanut seed size and dried. Two pills are administered daily once for 20 days.	
583	*Zornia gibbosa* Span.	Scabies: Leaves with those of *Marsilea quadrifolia* taken in equal quantities are ground. Paste is applied on the affected areas daily twice for 3 days.	Suneetha and Reddi, 2015

Tamarindus indica, Vernonia albicans each curing 5 ailments and others ranging from one to four. As many as 27 plants are used in curing 5–9 skin diseases. Of the 904 practices 789 involve single plant only and the rest 115 involve a combination of one to seven plants in addition to the original one; 67 practices involve one plant, 29 involve two plants, 14 involve three plants, 3 involve four plants, and 1 involve five and seven plants each. They are administered either in the form of ash, curry, decoction, infusion, juice, latex, mucilaginous jelly or gum, oil, paste, poultice, powder, pulp or sap along with either alum, ant eggs, bee wax, butter, butter milk, camphor, castor oil, cheese, coconut oil, cold water, cow ghee, cow milk, cow urine, egg albumen, feathers of peacock, garlic, goat milk, honey, hot water, jatropha seed oil, lime juice, mustard oil, palm oil, pepper, pongamia oil, salt, sesame oil, sugar candy, turmeric or wine. Large number of plants or their parts are used singly to treat the skin diseases given a clue of their use in combination with other suitable plants and the resultant experiences can be exploited in future practices by the vaidhyas for effective cure of the skin diseases.

Most common route of administration was external application. Common methods of obtaining medicinal plants were from the wild and cultivation. The collection, identification and documentation of ethnomedicinal data on biological resources are inevitable steps for bio-prospecting. The green wave all over the world is pushing the knowledge of primitive societies on the ascending spirals. These will straightway lead remarkable discoveries from the world of plant-based ethnomedicines. Folklore medicine in India has developed through the knowledge passed on orally from generation to generation. The earlier medicine men were reluctant to part with their knowledge on account of superstitions and other related factors but this has changed gradually in recent times. They freely discuss and exchange their views and ideas with them. Such cooperation is helpful in documenting the ethnic skills and has added many novelties to the medicinal wealth of India. After the advent of synthetic drugs, the attention towards herbal cure eclipsed for a time during the revolution of these allopathic medicines. Again people around the world became aware and attracted towards herbal therapy.

The same plant may not be used for curing the same skin disease in different parts of India but it may be used to cure some other skin ailment. Sometimes even if it is used for the same disease the plant part may be varying. The present study represents a contribution to the existing knowledge of folk remedies that are in current practice for treatment of skin diseases, which happens to be the most common ailment amongst tribal population because of their unhygienic habitat. Once the efficacy of these herbal drugs in treating skin diseases is scientifically established, the popularization of these remedies can be recommended in Indian healthcare system for wider application, since these plants are well within the reach of the tribal masses.

10.5 CONCLUSION

The detailed phytochemical and pharmacological studies are however required to determine the effective constituents and characteristic biological activity of these potential medicinal plants in the field of dermatology. Skin being delicate and sensitive part of the human body its tolerance to herbal applications varies from individual to individual and as such these applications must be used with utmost care.

KEYWORDS

- **burns**
- **eczema**
- **leprosy**
- **psoriasis**
- **ringworm**
- **scabies**
- **skin diseases**

REFERENCES

Arya, S., Adarsh, G., Anand, S. K., Kumar, N. S., & Santoshkumar, R. (2015). Selected medicinally and economically important plants growing in Aruvippuram area of Thiru-vananthapuram district, Kerala, India. *J. Trop. Med. Plants 15,* 19–31.

Ayyanar, M., & Ignacimuthu, S. (2005). Medicinal plants used by the tribals of Tirunelveli hills, Tamil Nadu to treat poisonous bites and skin diseases. *Indian J. Trad. Knowl. 4,* 229–236.

Bhattacharjee, A., & Chatterjee, S. (2007). Medicinal plants used in skin disease in Deganga West Bengal. *Indian J. Trad. Knowl. 6*, 358–359.

Changkija, S. (1999). Folk medicinal plants of the Nagas in India. *Asian Folklore Studies 58*, 205–230.

Chetty, K. M., Chetty, M. L., Sudhakar, A., & Ramesh, C. (1998). Ethno-medicobotany of some aquatic Angiospermae in Chittoor district of Andhra Pradesh, India. *Fitoterapia 69*, 7–12.

Das, A. K., Dutta, B. K., & Sharma, G. D. (2008). Medicinal plants used by different tribes of Cachar district, Assam. *Indian J. Trad. Knowl. 7*, 446–454.

Hari Babu, M. (2007). Ethnomedicine from Visakhapatnam district, Andhra Pradesh, India. PhD Thesis, Andhra University, Visakhapatnam.

Hari Babu, M., & Reddi, T. V. V. S. (2015). Ethnomedicinal plants against skin diseases in Visakhapatnam district of Andhra Pradesh. *J. Non-Timber Forest Products 22*, 167–171.

Jeeva, G. M., Jeeva, S., & Kingston, C. (2007). Traditional treatment of skin diseases in South Travancore, southern peninsular India. *Indian J. Trad. Knowl. 6*, 498–501.

Kingston, C., Jeeva, S., Jeeva, G. M., Kiruba, S., Mishra, B. P., & Kannan, D. (2009). Indigenous knowledge of using medicinal plants in treating skin diseases in Kanyakumari district, Southern India. *Indian J. Trad. Knowl. 8*, 196–200.

Kumar, N., & Chauhan, N. S. (2007). Ethnobotanical studies of Nahan area, district Sirmour, Himachal Pradesh. *J. Non-Timber Forest Products 14*, 307–312.

Lal, B., & Singh, K. N. (2008). Indigenous herbal remedies used to cure skin disorders by the natives of Lahaul-Spiti in Himachal Pradesh. *Indian J. Trad. Knowl. 7*, 237–241.

Manjula, R. R. (2011). Ethnobotany of Khammam district, Andhra Pradesh. PhD Thesis, Andhra University, Visakhapatnam.

Manjula R. R., & Reddi T. V. V. S. (2015). Ethnomedicine for scabies and skin diseases in Khammam district of Andhra Pradesh. *Medicinal Plant Res. 5*, 1–6.

Naidu, B. V. A. R. (2003). Ethnomedicine from Srikakulam district, Andhra Pradesh, India. PhD Thesis, Andhra University, Visakhapatnam.

Nayak, S., Behera, S. K., & Misra, M. K. (2004). Ethno-medico-botanical survey of Kalahandi district of Orissa. *Indian J. Trad. Knowl. 3*, 72–79.

Nisha V. M., & Sivadasan, M. (2007). Ethnodermatologically significant plants used by traditional healers of Wayanad district, Kerala. *Ethnobotany 19*, 55–61

Prashantkumar, P., & Vidyasagar, G. M. (2008). Traditional knowledge on medicinal plants used for the treatment of skin diseases in Bidar district, Karnataka. *Indian J. Trad. Knowl. 7*, 273–276.

Raju, M. P. (2009). Ethnobotany of the *Konda Reddis*. PhD Thesis, Andhra University, Visakhapatnam.

Raju, M. P., Prasanthi, S., & Reddi, T. V. V. S. (2010). Ethnotherapeutic management of skin diseases among the Konda Reddis of Andhra Pradesh. *J. Econ. Taxon. Bot. 34*, 456–465.

Rao, J. K. (2010). Ethnobotany of primitive tribal groups of Visakhapatnam district, Andhra Pradesh. PhD Thesis, Andhra University, Visakhapatnam.

Shah, B., Seth, F., & Parabia, M. (2011). Documenting Grandmas' prescriptions for skin ailments in Valsad district, Gujarat. *Indian J. Trad. Knowl. 10*, 372–374.

Sharma, L., Agarwal, G., & Kumar, A. (2003). Medicinal plants for skin and hair care. *Indian J. Trad. Knowl. 2*, 62–68.

Shetty, R. G., Poojitha, K. G., & Sunanda, B. B. (2015). People's knowledge on medicinal plants in Mudigere Taluk, Karnataka, India. *J. Trop. Med. Plants 15*, 37–42.

Shukla, A. N., Srivastava, S., & Rawat, A. K. S. (2010). An ethnobotanical study of medicinal plants of Rewa district, Madhya Pradesh. *Indian J. Trad. Knowl. 9*, 191–202.

Silja, V. P., Samitha Varma, K., & Mohanan, K. V. (2008). Ethnomedicinal plant knowledge of the *Mullu kuruma* tribe of Wayanad district, Kerala. *Indian J. Trad. Knowl. 7*, 604–612.

Singh, K. K., & Maheshwari, J. K. (1989). Traditional herbal remedies among the *Tharus* of Bahraich District, U.P., India. *Ethnobotany 1,* 51–56.

Sinha, B. K., Maina, V., & Padhye, P. M. (1996). Ethno-medicinal plants of Bay islands for skin care. *J. Econ. Taxon Bot. Addl. Ser. 12,* 375–380.

Suneetha, J. (2007). Ethnobotany of East Godavari district, Andhra Pradesh, India. PhD Thesis, Andhra University, Visakhapatnam.

Suneetha, J., & Reddi, T. V. V. S. (2015). Ethnomedicine for skin diseases in East Godavari district of Andhra Pradesh. *Ethnobotany 27,* 79–85.

Swamy, N. S. (2009). Ethnobotanical knowledge from Adilabad district, Andhra Pradesh, India. PhD Thesis, Andhra University, Visakhapatnam.

CHAPTER 11

ETHNOBOTANY OF PLANT CONTRACEPTIVES

BALJOT KAUR

Stri Roga & Prasuti Tantra (Gyne & Obs), SKSS Ayurvedic Medical College, Sarabha, Ludhiana, Punjab, India, E-mail: bhbharaj@gmail.com

CONTENTS

ABSTRACT

Study of ethnobotany assumes great importance in enhancing our knowledge about the plants grown and used by the native/ethnic and the tribal communities, the diversity produced and assembled by them through generations of informal training for their own sustenance and different traditional technologies, means and methods adopted by them for conservation of that plant diversity. Herbs used by various Tribal communities in India are rich source of providing herbal contraceptives. These herbs are being evaluated scientifically also to develop safe contraceptives. Herbs having antifertility effect have been compiled in this review.

11.1 INTRODUCTION

Over growing population is one of the major threats in the developing countries, facing new challenges, with its inevitable consequences on all aspects of development (Ciganda and Laborde, 2003). Therefore, there is an urgent need to control population explosion, and to ensure better health for one and all. Efforts have been taken to tackle this serious problem by developing antifertility agents called contraceptive; those chemical substances that inhibit either the sperm production or sperm motility in males or prevent the formation of ovum and produce some changes in the endometrium, making it unsusceptible to a fertile ovum in females (Kaunitz and Benrubi, 1998).

Plants having such properties may have a role in rapid discharge of the fertilized ova from the fallopian tube, inhibition of implantation due to a interruption in estrogen- progesterone balance, fetal abortion due to lack of supply of nutrients to the uterus and the embryo, and also on the male by affecting sperm count, motility, and viability. In recent years, many workers have reported a lot of traditional plants used for antifertility purpose (Kumar and Mishra, 2011; Mitra and Mukherjee, 2009; Yadav et al., 2006). With the passage of time, tribal communities have developed a great deal of knowledge on the use of plants and plant products in curing various diseases, ailments (Maheshwari et al., 1986; Das et al., 2014).

Synthetic hormonal contraceptives cannot be used continuously because of their health related effects, like increase in blood transaminase and cholesterol levels, dyspepsia, headache, depression, tiredness, weight gain, hyper menorrhea and intermenorrheal hemorrhage and also disturb the metabolism of lipid, protein, carbohydrate, enzymes and vitamins (Noumi and Tchakonang, 2001). Therefore, scientists are on the hunt for newer alternatives, with lesser side effects, self-administrable, less expensive and with complete reversibility. Much of these properties are observed in drugs of natural plant origin. Many plants are reported to have fertility regulatory activity.

11.2 MECHANISM OF ACTION OF ANTIFERTILITY PLANTS (YADAV ET AL., 2014)

Plant drugs have been used since time immemorial for their effects upon sex hormones particularly for suppressing fertility, regularizing menstrual cycle, relieving dysmenorrhea, treating enlarged prostate, menopausal symptoms, breast pain and during and after childhood. Specific biological effects under the division of fertility regulating category are nonspecific contraceptive or antifertility effects, abortifacient, uterine stimulant and uterine relaxants, labor induction and labor inhibition, oxytocic and anti-oxytocic, estrogenic and anti-estrogenic, progestrogenic and antiprogesterogenic, ovulatory and anti-ovulatory, androgenic and anti-androgenic, spermicidal and anti-spermatogenic effects (Gediya et al., 2011; Goonasekera and Gunawardana, 1995). The site of action of antifertility agents in females consists of the hypothalamus, the anterior pituitary, the ovary, the oviduct, the uterus and the vagina. The Hypothalamus controls the action of the uterus via follicle stimulating hormone (FSH) and Luteinizing hormone (LH) releasing hormones. Antifertility agents may therefore exert their effort at this level either by disrupting hormonal function of the hypothalamus and/ or the pituitary, or by interrupting the neural pathway to the hypothalamus that control the liberation of gonadotrophin releasing hormones. Early researchers in the area of female fertility regulation focused their attention to phytestrogens following the recognition that excess ingestion of plants containing estrogenic compounds resulted in infertility in animals and humans (Pradhan et al., 2012; Gupta and Sharma, 2006; Ittiavirah and Habeeb, 2013).

11.3 PLANTS USED AS CONTRACEPTIVES

Abrus precatorius L. Family: Fabaceae (Figure 11.1)

About ½ teaspoon paste in tablet form twice a day in empty stomach for 3 days just after completion of menstrual period (Das et al., 2014).

About 5 g seeds boiled in 50 ml cow's milk, seed coat removed, powdered it and mix equal amount of turmeric and jaggery, make into small pills, 1 pill taken daily for 5 days starting on the fourth day of menstruation (Shrivastava et al., 2007).

One gram of seed powdered, mixed with milk and boiled for 30 minutes. 200 ml of this taken twice daily during menstrual cycle acts as contraceptive (Kumar and Mishra, 2011).

Powder taken orally with water during menstruation (Zingare, 2012).

The seeds kept in cow milk for a period of overnight and such milk soaked seed is given to woman in the morning at the end of menstruation cycle for preventing pregnancy (Prasad et al., 2014).

Acacia leucophloea Willd. Family: Leguminosae

Leaves and bark are used as contraceptive (Meena and Rao, 2010).

Acacia nilotica L Delile. Family: Mimosaceae

Powder of Babul flower given with hot milk on empty stomach for 5–7 days, is sufficient to prevent fertility (Ekka, 2012).

Achyranthes aspera L. Family: Amaranthaceae (Figure 11.2)

The roots are boiled and decoction is given (One tea cup thrice a day) after menstruation to induce sterility in women (Shah et al., 2009).

In Tripura 1 teaspoon paste of whole plant as tablet is given twice a day for 7 days on empty stomach (Das et al., 2014).

Roots of *Achyranthes aspera* and *Piper betle, Vitex negundo* in equal proportion, two teaspoons of dry ginger and 10 nos. of Golmarich (*Piper nigrum*) are taken and ground with water. The mixture was prepared in the form of pills. One pill is to be taken in empty stomach for three consecutive days after the last date of menstruation. It is very effective medicine for the tribal women to prevent pregnancy (Basak et al., 2016).

Roots of Apang (*Achyranthes aspera* L.), Krishna tulsi (*Ocimum americanum*), Iswari (*Aristolochia indica*), Singara (*Bauhinia purpurea*) and Saora tree (*Streblus asper* Lour.) are mixed and powdered. Two teaspoons of the powder with one teaspoonful of Rasamanik and Rasasindur are taken together for three days. It prevents pregnancy completely (Basak et al., 2016).

Roots of Chirchiti (*Achyranthes aspera*), Pan (*Piper betle*) and Sarpagandha (*Rauvolfia serpentina*), Banda of Begna (*Vitex negundo*), seeds of Methi (*Trigonella foenum-graecum*), Golmarich (*Piper nigrum*) are grinded together with water. The mixture was dried in the form of pills. One pill is to be taken in empty stomach for five consecutive days after the last date of menstruation. It is very effective medicine for the tribal women to prevent pregnancy (Basak et al., 2016).

Plant Decoction of the whole plant is taken internally (Sathiyaraj et al., 2012).

Aegle marmelos (L.) Correa Family: Rutaceae

The boiled leaves are eaten for contraceptive purpose (Sathiyaraj et al., 2012).

Albizia lebbeck (L.) Benth. Family: Leguminosae

The aqueous extract of bark is used (2 teaspoons daily for one week before menses) against conception in women (Shah et al., 2009).

Allium cepa L. Family: Alliaceae (Figure 11.3)

Bulb paste is prepared in combination with *Terminalia arjuna* (Roxb. ex DC.) Wight & Arn. fruit bark and *Allium sativum* L. bulb in tablet form. 2 tablets twice a day for 5 days (Das et al., 2014).

Allium sativum L. Family: Alliaceae

Bulb is used same as *Allium cepa* L. (Das et al., 2014).

Aloe barbadensis Mill. Family: Liliaceae (Figure 11.4)

Dried juice, fresh *Hibiscus rosa-sinensis* L. flower, latex of *Ferula asa-foetida* L. and dried powder of *Zingiber officinale* Roscoe rhizome are mixed in equal ratio (5 grams each) along with ½ teaspoon honey. 1 teaspoon of this mixture twice a day in empty stomach for 8 days (Das et al., 2014).

About 6–12 grams powder of leaves of Patha if taken during 4 days of menstrual cycle, is sufficient to prevent fertility for long time (Ekka, 2012).

Amaranthus spinosus L. (Figure 11.5) and *Amaranthus viridis* L. Family: Amaranthaceae

Fresh root (10 gram) is eaten by women before two days copulation. It works as an antifertility agent (Shah et al., 2009).

In Bastar, Madhya Pradesh, about 10–15 grams of the roots are powdered and mixed with 15–20 ml of rice water and kept overnight. This decoction water is regularly administered orally from third consecutive day after menstrual period, empty stomach early in the morning by women. It has been observed that root powder extract posses anti-ovulatory properties and inhibits enzymatic activity which avoids pregnancy (Rai and Nath, 2005).

Homogeneous crushed roots in rice water taken after menstruation cycle twice a day (Shrivastava et al., 2007).

It inhibits fusion of sperm and ovum (Kamboj and Dhawan, 1982).

Leaves are used orally (Zingare, 2012).

Homogeneously crushed Chaulai root in rice wash water if taken after menstrual cycle, women will not conceive (Ekka, 2012).

Andrographis paniculata (Burm.f.) Nees. Family: Acanthaceae

Leaves are used for contraceptive purpose (Sathiyaraj et al., 2012).

Anona squamosa L. Family: Annonaceae

2 to 5 unripe fruit taken raw with warm water act as contraceptive (Kumar and Mishra, 2011).

Seeds are taken orally with milk (Zingare, 2012).

Argemone mexicana L. Family: Papaveraceae

Satyanashi leaf (2 leaves) paste is taken with a glass of milk during menstruation cycle for 5 days is enough to prevent conception (Ekka, 2012).

Aristolochia bracteolata Lam. Family: Aristolochiaceae

Leaf extract is given for antifertility effects (Sathiyaraj et al., 2012).

Asparagus racemosus Willd. Family: Asparagaceae

Root paste of Satamuli (*Asparagus racemosus*) and Sarpagandha (*Rauvolfia serpentina*), fresh latex of Akanda (*Calotropis procera*), Rice water and soil of red water pond are macerated and taken orally during morning hours on every Saturday. It causes inhibition of fertilization and prevents pregnancy (Basak et al., 2016).

Satamuli plant (*Asparagus racemosus*), Ramdatan (root) (*Smilax zeylanica*), Ananta (root) (*Hemidesmus indicus*), Nilkanta (root) (*Polygala chinensis*) are ground with Golmorich (*Piper nigrum*) and form pills. Two pills are to be taken every day in empty stomach immediate after menstruation for fifteen days. It prevents pregnancy for six months without any side effects (Basak et al., 2016).

Azadirachta indica A. Juss. Family: Meliaceae

1 teaspoon of kernel oil, known as Neem oil, is taken after menses before copulation makes the women sterile. Sometimes oil is also inserted into vagina before copulation (Shah et al., 2009).

Seed powdered and given 3 grams daily (Shrivastava et al., 2007).

Seeds are crushed and the oil is expelled from local ghani available in tribal villages. The extracted oil is applied for 10–12 days in genital parts of men and womb (uterus) of tribal women 1–2 hours before sexual inter-course. This treatment is given to women right from 3rd day onwards of menses till the date or period pregnancy is to be avoided. The oil extracted from *Azadirachta indica* contains azadirachtin which possesses enzymatic activity for preventing conception. It has been observed that the oil acts as herbal contraceptives and avoids pregnancy (Rai and Nath, 2005).

Oil is applied locally (Zingare, 2012).

Leaf is used as a contraceptive (Sathiyaraj et al., 2012).

Balanites aegyptiaca (L.) Delile Family: Balanitaceae (Figure 11.6)

Fruit powder is used orally (Zingare, 2012).

Tablets are prepared from roots mixed with Hing (*Ferula asafoetida*), by adding *Piper betle* leaf juice are taken once with water for 9 days, soon after the menstruation, to avoid unwanted pregnancy (Vijigiri and Sharma, 2010).

Bambusa vulgaris Schrad. Family: Poaceae

Leaves extract is taken orally to reduce the sperm count (Sathiyaraj et al., 2012).

Berberis aristata DC. Family: Berberidaceae

Rasaut, Harra (*Terminalia chebula* Linn.), Amla (*Emblica officinalis* L.) (3 grams powder each), equal quantity of each if taken by women during menstrual cycle period, women will not conceive pregnancy (Ekka, 2012).

Bombax ceiba L. Family: Bombacaceae

Flower is boiled with black pepper (*Piper nigrum*). The mixture is orally taken to prevent pregnancy (Kumar and Mishra, 2011).

Borassus flabelifer L. Family: Palmae

Root is used as contraceptive and to induce sterility (Meena and Rao, 2010).

Bridelia retusa (L.) Spreng. Family: Euphorbiaceae

Bark is given orally to women to induce sterility and hence act as contraceptive (Meena and Rao, 2010).

Butea monosperma (Lam.) Taub. Family: Fabaceae

Inflorescence is dried and taken (3 grams daily for 15 days) possess contraceptive properties (Shah et al., 2009).

In Jharkhand, petals are boiled in cow milk for one hour. The mixture taken thrice-daily acts as contraceptive (Kumar and Mishra, 2011).

In Madhya Pradesh about 10–15 roasted seeds are powdered and taken every day on empty stomach (Shrivastava et al., 2007). The gum is taken orally with water for a week (Zingare, 2012).

About 5–6 grams of gum collected from bark of trees is mixed with cow-milk. This mixed milk is regularly administered orally to women, early in the morning empty stomach right from the first day after menses for consecutive three days. It is said that by drinking such milk some enzymatic activity takes place in women which prevents or inhibits conception. It has been observed that milk mixed with gum purifies blood and destroys pathogen, avoids pregnancy and acts as herbal contraceptives (Rai and Nath, 2005).

The fresh seeds (about 15 to 20) are to be taken every day on empty stomach by both males and females (Sinha and Nathawat, 1989). Seeds are also used as contraceptives (Dwivedi and Kaul, 2008).

Gum is used as the male herbal contraceptives by the tribals of Gujarat and Rajasthan (Billore and Audichiya, 1978).

Roasted Palas seed powder taken during 4 days of menstrual cycle, lead to permanent sterilization (Ekka, 2012).

Calotropis gigantea (L.) Dryand Family: Asclepiadaceae

Decoction of roots is taken internally (Sathiyaraj et al., 2012).

Carica papaya L. Family: Caricaceae

About 30 seeds powdered and to be taken with water regularly after the menses till the commencement of next menses (Shrivastava et al., 2007). ·

Prevents fusion of sperm with ovum by reducing the activity of sperms (Kamal et al., 2003).

Fresh or dried seeds paste is prepared. 2 teaspoons paste decoction taken every day after menstrual period till commencement of next menstrual period (Das et al., 2014).

Peel the bark of male plant root and sizeable fruit *Xylopia aethiopica*, little oil of *Elais guineensis* and cook with cat fish, then eat. To restore fertility, carry out the above but with female plant (Kumar and Mishra, 2011).

About 50 seeds are to be eaten with water by both male and female regularly. It is to be taken early morning, empty stomach (Sinha and Nathawat, 1989). Seed powder is also used as contraceptive (Dwivedi and Kaul, 2008).

About 50 seeds are to be eaten with water by both male and female regularly. It is to be taken early morning, empty stomach (Sinha and Nathawat, 1989).

Leaf juice is used for contraceptive purpose (Sathiyaraj et al., 2012).

Cassia alata L. Family: Caesalpiniaceae

10 grams fresh root is mixed with 10 *Cynodon dactylon* (L.) Pers., 5 fresh *Piper betle* L. leaves, and 10 grams fresh *Ricinus communis* L. root. Paste decoction is prepared. 5 ml of this is taken daily in empty stomach for 7 days after completion of menstruation cycle (Das et al., 2014).

Cassia auriculata L. Family: Caesalpiniaceae

Flowers are crushed and mixed with water and taken orally (Sathiyaraj et al., 2012).

Cheilocostus specious (J. Koen.) C. Specht (Syn.: *Costus speciosus* (J. Koen.) Sm. Family: Zingiberaceae (Figure 11.7)

The paste of rhizome is used orally (Choudhury et al., 2011).

Cissampelos pareira L. Family: Menispermaceae

Juice of tender leaves is taken orally (Sathiyaraj et al., 2012).

Citrus aurantifolia (Christm.) Swingle Family: Rutaceae

A paste is made by crushing leaves of Kagji (*Citrus aurantifolia*), Barundaru (*Crateva nurvala*), and leaves of Vilati tulsi (*Hyptis suaveolens*). Pills are made with the paste. Three pills should be taken each day for three consecutive days from the first day of menstruation. It prevents pregnancy for that month (Basak et al., 2016).

Citrus limon (L.) Burm.f. Family: Rutaceae

Juice is used locally (Zingare, 2012).

Coriandrum sativum L. Family: Umbelliferae

Coriander powder if taken during menstruation period, it prevents conception (Ekka, 2012).

Crataeva nurvala Buch.-Ham. Family: Capparaceae

Powdered bark is regularly administered orally to tribal women, early in the morning empty stomach right from the first day after menses for three days. It is reported by tribals that this treatment not only purifies blood but also destroys pathogen, avoids pregnancy and acts as herbal contraceptives (Rai and Nath, 2005).

Pills are made from the roots of Barundaru (*Crataeva nurvula*) and Kanta saru (*Lasia spinosa*) plants. One pill with leaf bud mucilage of Kadam (*Haldina cordifolia*) should be taken for all the days of menstruation for checking conception. It prevents pregnancy permanently (Basak et al., 2016).

Bark powder of Varun, if taken in menstrual period, women will not be pregnant (Ekka, 2012).

Crotalaria juncea L. Family: Fabaceae

The flowers are boiled in water and filtered. The extract is taken orally (Sathiyaraj et al., 2012).

Curcuma longa L. Family: Zingiberaceae

Powder of turmeric (10–15 grams taken with water after menses for 5 days) induces sterility in women (Shah et al., 2009). 3 grams powder taken during menstruation cycle every day (Shrivastava et al., 2007).

In Bastar region of Chhattisgarh, about 8–10 pieces of rhizomes are washed, dried and powdered. 5–6 grams of powder is mixed with 5 grams of jaggery and is regularly administered orally for 4–5 days during menstrual period empty stomach early in the morning to women. The uses has been found to be very good herbal contraceptive (Rai and Nath, 2005).

In some villages it was found that Rhizomes of *Curcuma longa* L. (Haldi) are collected from forest by tribals, dried and crushed into fine powder. About 10–12 grams of haldi powder prepared from rhizome was mixed with fresh milk of cow and boiled. The milk on boiling when became warm was given 250 ml to women to drink early in the morning empty stomach right from the third day after menses for a period of about 10–12 days. It is reported by tribals that this treatment purifies blood and destroys pathogen. It has been found in tribal community that haldi powder is very good oral herbal contraceptive (Rai and Nath, 2005).

Haldi powder and old jaggery are taken in menstrual period, it prevents pregnancy if it is repeated many times it will not be needed (Ekka, 2012).

Haldi powder (3 grams) if taken during menstruation cycle till 6 days is enough to prevent conception (Ekka, 2012).

Decoction is taken by mouth before sexual intercourse (Sathiyaraj et al., 2012).

Cuscuta reflexa Roxb. Family: Cuscutaceae (Figure 11.8)

Seeds of the plant are given (3 grams after menses) to induce sterility in women (Shah et al., 2009).

Fresh plant paste is prepared separately and mixed with leaves paste of *Stephania japonica* (Thunb.) Miers in equal volume and tablets are prepared from it, one tablet contain 1/2 spoon paste. 2 tablets twice a day in empty stomach (Das et al., 2014).

Leaves are boiled and mashed. The extract is filtered with a piece of cloth and allowed to cool. The filtered extract is again boiled for 2–3 hours and allowed to cool. The extract is regularly administered orally to women, early in the morning on empty stomach right from the third day after menses for 21 days. It has been observed that this extract acts as anti-ovulatory and avoids pregnancy and makes tribal women permanently sterile (Rai and Nath, 2005).

Cynodon dactylon (L.) Pers. Family: Poaceae

Same as *Cassia alata* L. (Das et al., 2014).

Daucus carota L. Family: Apiaceae

Munda tribes takes its decoction to prevent pregnancy (Kumar and Mishra, 2011).

Seed powder is taken orally with liquor (Zingare, 2012).

Desmodium gangeticum DC. Family: Fabaceae

Whole plant 5 grams is powdered and taken during menstruation period daily (Shrivastava et al., 2007).

Panchang (whole plant) of Shalparni is to be taken during menstrual period, women become sterile (Ekka, 2012).

Dioscorea alata L. Family: Dioscoreaceae

Crushed tuber (Bajhkand) if taken one tuber daily during 4 days of menstruation; is one of the best sterilizers (for long life) (Ekka, 2012).

Dioscorea bulbifera L. Family: Dioscoreaceae

About 8–10 tubers are washed, dried and crushed. One tuber is regularly administered orally for 4–5 days during menstrual period empty stomach early in the morning to women. It has been found that tuber powder of climber acts as herbal contraceptive (Rai and Nath, 2005).

Tubers peeled out dried, roasted powdered about 10 grams for 5 days just after menses (Shrivastava et al., 2007).

Whole plant paste is prepared and mixed equally with *Ficus religiosa* L. leaves paste. Tablets are then prepared from this mixture and one tablet contains 1–2 teaspoons mixture and is taken twice a day till commencement of next menstrual cycle (Das et al., 2014).

Tuber is used as contraceptive (Meena and Rao, 2010).

Dioscorea deltoidea Wall. ex Kunth. Family: Dioscoreaceae

For birth control, the rhizome extract is taken before and after intercourse (Bhat et al., 2012).

Dioscorea floribunda M. Martens & Galeotti Family: Dioscoreaceae.

HO tribes use its tuber and root as contraceptive (Kumar and Mishra, 2011).

Embelia ribes Burm.f. Family: Myrsinaceae (Figure 11.9)

It is another plant with contraceptive properties mentioned in the ancient textbook of Ayurveda-the Yogaratnakar. It is being used today as a contraceptive by practitioners of the traditional systems of medicine in combination with *Piper longum* and borax (Chaudhary, 1993).

Embelia ribes alone was administered in a dose of 2 grams for five days followed by 1 gram daily for another 10 days. After observing the effect on 2051 cycles over four years, it was reported that the plant protected 95% of women from pregnancy (Tewari et al., 1976).

Ensete superbum (Roxb.) Cheesman. Family: Musaceae

Eating the inner part of the bud of this plant helps to prevent pregnancy (Prasad et al., 2014).

Eugenia jambolana Lam. Family: Myrtaceae

Flowers can be used to reduce sperm count (Sathiyaraj et al., 2012).

Euphorbia antiquorum L. Family: Euphorbiaceae

Ashes of dry small pieces of Sehud, if taken for 21 days with honey are one of the best for sterility (Ekka, 2012).

Euphorbia hirta L. Family: Euphorbiaceae

Leaf is crushed and mixed with hot water (Sathiyaraj et al., 2012).

Ferula asafoetida L. Family: Apiaceae

Same as *Aloe barbadensis* Mill. (Das et al., 2014).

The herb is considered useful in the treatment of several problems concerning women such as sterility, unwanted abortion, pre-mature labor, unusually painful, difficult and excessive menstruation and leucorrhoea (Mahendra and Bisht, 2012).

Ficus religiosa L. Family: Moraceae

5–10 receptacles are grinded with sugar and taken before one week of menses make the women sterile (Shah et al., 2009).

Leaves paste is mixed equally with the whole plant paste of *Dioscorea bulbifera* L. Tablets are then prepared from this mixture and one tablet contain 1–2 teaspoons mixture and is taken twice a day till commencement of next menstrual cycle (Das et al., 2014).

Seeds are dried and mixed with seeds of small tree *Embelia ribes* Burm. f. These seeds are roasted with suhag (Borax) and are orally administered for a period of 8–10 days empty stomach early in the morning to women after 5th day of menstrual period. It has been observed that the seeds when consumed act as herbal contraceptives, which avoids pregnancy (Rai and Nath, 2005).

Mixed powdered seed 5 grams with equal quantity of suhago (Borax) and taken for 4 days during menstruation period (Shrivastava et al., 2007).

Peepal, Baibedanga (*Embelia ribes*) and roasted suhaga (Borax); each mixed in equal quantity and taken for 4 days during menstruation period; is enough to prevent conception. (Ekka, 2012).

Foenicum vulgare Mill. Family: Apiaceae

Powder of fruits given with water (20 grams twice daily for 7 days) after menses induces sterility in women (Shah et al., 2009).

Gossypium herbaceum L. Family: Malvaceae

Decoction of fresh root is prepared and 1-teaspoon decoction is taken daily for 5 days (Das et al., 2014).

Hibiscus rosa-sinensis L. Family: Malvaceae

Flowers 5–8 are dried, crushed and mixed with honey and taken every morning on empty stomach (Shrivastava et al., 2007).

Roots are crushed and taken orally (Sathiyaraj et al., 2012).

Hyptis suaveolens (L.) Poit. Family: Lamiaceae

Decoction of leaf is taken orally (Sathiyaraj et al., 2012).

Indigofera glandulosa Willd. Family: Fabaceae

Extract of whole plant (one cup full) of Janglee neel is to be taken during menstrual period till 3 days is enough, women become sterile (Ekka, 2012).

Jasminum amplexicaule L. Family: Oleaceae

Flowers and buds of Chameli 3 grams if taken orally, women will not be pregnant; or one bud daily till 7 days if taken, women will not be pregnant (Ekka, 2012).

Lantana camara L. Family: Verbenaceae

Decoction of leaves is used orally (Sathiyaraj et al., 2012).

Lawsonia inermis L. Family: Lythraceae

Paste of fresh leaves is prepared and mixed with little amount of esabgul (*Plantago ovata*) powder. Tablets are then prepared from ½ teaspoon mixture, taken twice daily for 21 days from last menstrual period (Das et al., 2014).

Roots are soaked in water for 24 hours. The medicated water is taken orally (Zingare, 2012).

50 grams of juice of fresh leaves is to be taken regularly every day after the menses till the commencement of the next menses. Prolonged use can lead to permanent sterilization (Sinha and Nathawat, 1989).

Lygodium flexuosum (L) Sw. Family: Schizaeaceae

Leaves paste is prepared and mixed with paste of fresh *Moringa oleifera* Lam. root and in equal ratio. 1/2 teaspoons mixture is used to prepare 1 tablet and is taken daily on empty stomach after completion of menstrual cycle (Das et al., 2014).

Melia azedarach L. Family: Meliaceae

Bark is used for contraceptive purpose (Sathiyaraj et al., 2012).

Mentha arvensis L. Family: Lamiaceae

Whole plant is dried in shade and then made into a powder. Ten grams of powder is taken before intercourse for antifertility (Shah et al., 2009).

Leaves and branches are boiled and mashed. The extract is filtered with a piece of cloth and allowed to cool. This extract is administered orally to women two hours before performing sexual intercourse. It is said that this extract acts as antiovulatory and inhibits enzymatic activity which avoids pregnancy. The extract has been observed as an excellent herbal contraceptive popular among tribal localities in villages surveyed in Bastar region (Rai and Nath, 2005).

Mentha longifolia L. Family: Lamiaceae

Dried leaves are grinded and mixed and given (5–10 grams for a week) after menses to induce sterility (Shah et al., 2009).

Michelia champaca L. Family: Magnoliaceae

Stem bark powder taken orally with water for 7 days (Zingare, 2012).

Mimosa pudica L. Family: Mimosaceae

The decoctions of roots are used for the contraceptive purpose (Sathiyaraj et al., 2012).

Momordica charantia L. Family: Cucurbitaceae

Fresh seed paste is prepared and mixed with paste of *Stephania japonica* (Thunb.) Miers stem in equal volume. Tablets are then prepared from containing 1–2 teaspoons paste taken twice a day after completion of menstrual cycle to commencement of next menstrual cycle (Das et al., 2014).

The seed powder is mixed with water and the mixture is orally taken (Sathiyaraj et al., 2012).

Momordica dioica Roxb. ex Willd. Family: Cucurbitaceae

Root powder with jaggery is given orally (Zingare, 2012).

Moringa oleifera Lam. Family: Moringaceae

The leaf is used as contraceptive in females (Choudhury, 2011).

Leaves paste is prepared and mixed with paste of fresh root of *Moringa oleifera* Lam. and in equal ratio. ½ teaspoon mixture is used to prepare 1 tablet and is taken daily in empty stomach after completion of menstrual cycle (Das et al., 2014).

A paste is made by crushing bark of Sojna (*Moringa oleifera*) and flowers of Kagji (*Citrus aurantifolia*), talans (*Piper nigrum, Ferula foetida,*

Coriandrum sativum, Terminalia chebula). The dose is 4 pills per day taken orally at the date of onset of menstruation in empty stomach. It prevents pregnancy for that month (Basak et al., 2016).

Morus alba L. Family: Moraceae

Leaf juice is taken orally (Sathiyaraj et al., 2012).

Mucuna pruriens (L.) DC. Family: Fabaceae

Seed powder of 3 seeds given once daily for 3 days after menses (Shrivastava et al., 2007).

Seeds are used as oral contraceptives (Dwivedi and Kaul, 2008).

Musa balbisiana Colla Family: Musaceae

Tablets are prepared from fresh or dried seeds paste containing 5 grams paste and taken twice a day in empty stomach for 7 days (Das et al., 2014).

Musa paradisica L. Family: Musaceae

Juice of root is taken orally (Zingare, 2012).

Juice of root is used orally to stop conception (Meena and Rao, 2010).

Nelumbo nucifera Gaertn. Family: Nelumbonaceae

Flowers of Padma (*Nelumbo nucifera*) and Tal (*Borassus flabellifer*), roots of Pan (*Piper betle*) and Dhawai (*Woodfordia fruticosa*) are grinded with water. They are mixed with eggs of snail and are dried in the form of pills. Five pills are to be taken in empty stomach for three consecutive days after the last date of menstruation. It is very effective to prevent pregnancy (Basak et al., 2016).

Nerium oleander L. Family: Apocynaceae

The powder of leaves is mixed with water and taken orally (Sathiyaraj et al., 2012).

Ocimum sanctum L. Family: Lamiaceae

Leaves of *Ocimum sanctum*, buds of *Hibiscus rosa-sinensis*, flowers of *Butea monosperma* and root of Halud (*Curcuma domestica*) are ground together. One teaspoon of the mixture has to be taken at every morning in empty stomach for one week. This mixture will prevent pregnancy with normal sexual life (Basak et al., 2016).

Oroxylum indicum Vent. Family: Bignoniaceae

Paste of fresh stem bark is prepared and mixed with *Coccinia grandis* (L.) Voigt stem, fresh *Tacca laevis* Roxb. root, fresh *Lygodium flexuosum* (L.) Sw. leaves paste in equal ratio. Tablets are then prepared containing 1-teaspoon mixture and taken once a day in empty stomach for 7 days (just 2 days after menstrual period) (Das et al., 2014).

Parthenium hysterophorus L. Family: Asteraceae

Powder of whole plant of Gajar Ghass (2–3 gram), if taken during menstruation period, women will not be pregnant (Ekka, 2012).

Phlogacanthus thyrsiformis (Roxb. ex Hardw.) Mabb. Family: Acanthaceae

1 teaspoon leaves paste as tablet twice a day for 3–4 days in empty stomach (Das et al., 2014).

Phyllanthus amarus Schumach. & Thonn. Family: Euphorbiaceae

Leaf juice is taken orally (Sathiyaraj et al., 2012).

Phyllanthus emblica L. Family: Euphorbiaceae

A mixed paste of fruit along with *Terminalia chebula* Retz., *Phyllanthus emblica* L., *Terminalia bellerica* Roxb., are prepared. Rasanjan is added to this mixture. Tablets are prepared from this and one tablet contains 6 rati mixtures. 1 tab daily once for one month (Das et al., 2014).

Piper betle L. Family: Piperaceae

Roots of Pan (*Piper betle*) and Begna (*Vitex negundo*) and stem of Akanda (*Calotropis procera*) are macerated with a pinch of rock salt with the help of water. The mixture then mixed with small amount of mustard oil and ghee. The mixture should be taken on Saturday after puja. It prevents pregnancy for that month (Basak et al., 2016).

Roots of Pan (*Piper betle*), Nishinda (*Vitex negundo*) and Sarpagandha (*Rauvolfia serpentina*) are grinded together. Three teaspoonful of paste have to eat with water and common salt. After menstruation the drug has to be taken for three consecutive days at morning and in empty stomach. It prevents pregnancy for that month with no side effect (Basak et al., 2016).

About 2–4 grams of root of Pan (*Piper betle*) and 2–4 grams of Bamboo seeds (*Bambusa bambos* L.) are taken orally in the form of *Goli* (tablet) in empty stomach (Chandra et al., 2007).

Plumbago rosea L. Family: Plumbaginaceae

The fresh root is used in the vagina (Choudhury et al., 2011).

Plumbago zeylanica L. Family: Plumbaginaceae

About 5–6 pieces of roots weighing 10–12 grams are powdered and mixed with 10–15 ml. rice water and is kept overnight. This water is regularly administered orally for three consecutive days during menstrual period empty stomach early in the morning by tribal women. It has been observed in tribal community that root powder of the shrub causes sterility in women (Rai and Nath, 2005).

Chitrak root homogeneously mixed with rice wash water (6 ml Quath) if taken after menstrual cycle for 3 days, it is also one of the best for sterility (Ekka, 2012).

Plumeria rubra L. Family: Plumbaginaceae

About 10–12 grams of leaves are boiled and kept over-night in 250 ml of water. The leaves are crushed in the morning and extract is filtered. The filtered extract is regularly administered orally to women, early in the morning empty stomach right from the third day after menses for 15 days. It has been observed that this extract avoids pregnancy and acts as herbal contraceptive (Rai and Nath, 2005).

Leaves of Champa soaked overnight in water, in the morning these leaves crushed with this water and taken in menstrual period, women will not be pregnant during the time or number of year same as the number of leaves taken (Ekka, 2012).

Polygala chinensis L. Family: Polygalaceae

In a new mud pot, root of Nilkantha (*Polygala chinensis*), Ananta (*Hemidesmus indicus*), Kantikiari (*Solanum surattense*) and stem of Rakta chandan (*Anadenanthera pavonia*) and Swet chandan (*Santalum album*) is cooked in fire. Rasamanik, Rasasindur, Makaradhwaj, Golmarich (*Piper nigrum*), blood of black chicken is mixed. The Women will take as much as they can. The rest of the amount will be prepared in the form of pills. 3 pills in each day will be taken for one month. It prevents pregnancy permanently but if the women want to revive pregnancy they have to take different medicines from medicine men (Basak et al., 2016).

Prosopis cineraria (L.) Druce Family: Mimosaceae

Leaf juice is taken orally in morning (Sathiyaraj et al., 2012).

Punica granatum L. Family: Punicaceae

Decoction is used by "Oraon" tribe as anti-fertility agent (Kumar and Mishra, 2011).

Randia dumetorum L. Family: Rubiaceae

Powder of Fruits is mixed with milk and given twice-daily about 50 ml (Shrivastava et al., 2007).

Rauvolfia serpentina (L.) Benth. ex Kurz. Family: Apocynaceae

Roots of Sarpagandha (*Rauwolfia serpentina*) and Pan (*Piper betle*) are to be immersed in water for 7 days. The water should be mixed with juice of Ganda (*Tagetes patula*) and Tulsi (*Ocimum sanctum*) leaves. This mixture should be taken with small amount of Pipal powder (*Piper nigrum*) and Hincha (*Enhydra fluctuans*) on 5th day of menstruation. It prevents pregnancy for that month (Basak et al., 2016).

Ricinus communis L. Family: Euphorbiaceae

The seeds are grinded into powder form and 1 gram is given to women before intercourse as contraceptive (Shah et al., 2009).

Seeds are taken orally with water during menstruation (Zingare, 2012).

Seed oil is used locally (Zingare, 2012).

5–10 ml sap of seed in menstruation period once daily (Shrivastava et al., 2007).

Seeds are used as contraceptive (Dwivedi and Kaul, 2008).

Sap of Aundi seed in menstruation period if it is taken, prevents conception (Ekka, 2012).

Santalum album L. Family: Santalaceae

Chandan, Sarson (*Brassica compestris* Hook.f. & Thoms.) and Sugar mixed with 4–6 grams with rice wash water, are taken after menstrual cycle period, women become sterile (Ekka, 2012).

Saraca asoca (Roxb.) Willd. Family: Fabaceae

The bark of Asoka (*Saraca asoka*) tree washed thoroughly and sun-dried to reduce its moisture content. It is then grinded and sieved. The dry powder is pasted with floral parts of Jaba (*Hibiscus rosa-sinensis*). The medicine has to be taken every day in morning. It prevents pregnancy with no side effects (Basak et al., 2016).

Sesamum indicum L. Family: Pedaliaceae

Seeds oil is mixed with Lac of Ber (*Ziziphus jujuba*) and boiled. Taken orally (Zingare, 2012).

Solanum surattense Burm.f. Family: Solanaceae

Seed soaked in water and used to reduce sperm count (Sathiyaraj et al., 2012).

Soymida febrifuga (Roxb.) A. Juss. Family: Meliaceae

25 grams of Rohin (*Soymida febrifuga*) and Saora (*Streblus asper* Lour.) bark, 5 grams of white Akanda (*Calotropis procera*) flower, 10 grams of Golmarich (*Piper nigrum*), 10 grams of Darchini (*Laurus cinamonum*) and 10 grams of Pipul (*Piper longum*) was powdered and mixed to form pills. One pill has to be taken per day for 5 days from the last day of menstruation. It has to be continued for six months. It prevents pregnancy for one year. It has a long-term effects without any side effect (Basak et al., 2016).

Stephania japonica (Thunb.) Meirs Family: Menispermaceae

Leaves are used same as *Momordica charantia* L. (Das et al., 2014).

Syzygium aromaticum L. Family: Myrtaceae

One lounga daily till 21 days regularly is sufficient to prevent conception (Ekka, 2012).

Tacca laevis Roxb. Family: Taccaceae

Rhizomes are used same as *Oroxylum indicum* Vent. (Das et al., 2014).

Tamarindus indica L. Family: Fabaceae

The flowers with *Ferula asafoetida* are used for contraception (Choudhury et al., 2011).

Tecomella undulata (Sm.) Seem. Family: Bignoniaceae

Bark is chewed (Zingare, 2012).

Tectona grandis L.f. Family: Verbenaceae

The fruits and the young leaves (about 5 each) are to be eaten regularly with either milk or honey after the menses and till the commencement of next menses. It is to be taken early morning in empty stomach (Sinha and Nathawat, 1989).

Terminalia bellerica Roxb. Family: Combretaceae

Fruit is used same as *Phyllanthus emblica* L. (Das et al., 2014).

Terminalia chebula Retz. Family: Combretaceae

Fruit Bark is used same as *Phyllanthus emblica* L. (Das et al., 2014).

Tinospora cordifolia (L) Merr. Family: Menispermaceae

Guduchi powder (3 gram) of whole plant, if taken orally, women will not be pregnant (Ekka, 2012).

Vicoa indica Cass. Family: Asteraceae

Also called Banjhouri, the plant decoction is used for sterility in females. (Dwivedi and Kaul, 2008)

It is used for inducing sterility in the village of Sukhodeora in Bihar (Choudhary et al., 1988).

Vitex negundo L. Family: Verbenaceae

Banda of Boan (*Vitex negundo*), flowers of Palash (*Butea monosperma*) and Mushroom (*Psalliota campestris* L.) are crushed to form paste with all talans (*Myristica fragrans* Houtt./Jaiphal; *Piper longum* L./Pipul; *Piper nigum* L./Kali mirch). One pill is taken with one cup of Mahua (*Madhuca indica*) liquor early in the morning and in empty stomach for 7 consecutive days. This combination is an effective oral herbal contraceptive used among the tribal women (Basak et al., 2016).

Roots of Nishinda (*Vitex negundo*), banda of Ramdatan (*Smilax zeylanica*) and bark of Nim (*Azadirachta indica*) are pasted with Mahua (*Madhuca indica*) liquor. It prevents pregnancy (Basak et al., 2016).

A paste is made by crushing roots of Nishinda (*Vitex negundo*) and Shimul (*Bombax ceiba*), leaves of Bel (*Aegle marmelos*) with 10 Golmarich (*Piper nigrum*). The dose is 10 pills per day taken orally at the date of termination of menstruation in empty stomach. It prevents pregnancy for that month (Basak et al., 2016).

Zingiber officinale Roscoe Family: Zingiberaceae

Rhizome is used same as *Aloe barbadensis* Mill. (Das et al., 2014).

Ziziphus jujuba Mill. Family: Rhamnaceae

Lac of Ber mixed with 25–50 grams Til oil (*Sesamum indicum*) then boiled for some time, if taken during 4 days of menstrual cycle, women become sterile (Ekka, 2012).

Ziziphus oenoplia (L.) Mill. Family: Rhamnaceae

Root of Siakul (*Ziziphus oenoplia*), leaves of Hinga ara (*Enhydra fluctuans* Lour.), Banda of Arjun (*Terminalia arjuna* Roxb.) and Begna (*Vitex negundo*) are dried and smashed into pellets with the help of water. One pellet is taken with warm milk in empty stomach on the 5th day of menstruation and continued for three consecutive days during menstrual periods. This combination is very effective to prevent pregnancy (Basak et al., 2016).

Generally the tribal people take the medicine either with fresh cold drinking water or with country liquor, rice beer or with honey as advised by the medicine men. Internal medicine is prescribed to take in empty stomach in the morning, repeated at noon and again in the evening, according to necessity. Most of the tribal medicines are prepared in combination with some ingredients like long pepper/pipul (*Piper longum*), black pepper (*Piper nigrum*), darchini (*Laurus cinamonum*), elachi, etc., which are known as "Talan" (Basak et al., 2016).

11.4 CONCLUSION

It is observed that the dosages and duration of medicine generally depend on the intensity of the disease and age of patient. It is observed that tribals harvest that plant part, used for medicinal purpose at particular growth period or season, for example, before flowering and fruiting period, etc. presumably to obtain maximum concentration of the active principle. As tuberous plants remain in dormant phase and have a limited period for completing their life cycle, tribal preserve the tuber for various remedies, which is harvested in their particular period. Hence, the tribals have a specified way of collecting the herbs, preparing and applying the medicine. It is observed that single plant species or a combination of different plant species is used for curing various diseases (Choudhary et al., 2008).

Tribals are custodians of vast traditional knowledge and their wisdom with respect to plants is noteworthy. Their day-to-day need is totally dependent on forest and they obtain food, medicine, shelter and other materials from plants (Kumar and Mishra, 2011). Plants, since ancient times, have been used globally across varied cultures throughout the known civilizations as valuable and safe natural source of medicines and as agents of therapeutic, industrial and environmental utilities. The medical historians have recorded plants that could be used as contraceptives, emmenagogues, abortifacients and aphrodisiac (Kiritkar and Basu, 1975). The use of medicinal plants as decoction and infusion may be consistent with phyto-pharmacological effects. Traditional medicines are practiced worldwide for regulating fertility since ancient times. A large number of plants species have been screened for their antifertility efficacy. The recent review of Kumar et al. (2009) reported 577 plant species belonging to 122 families, having been used traditionally in antifertility regulatory agents in female. However, the search for an orally active, safe and effective plant preparation or compound is yet needed for fertility regulation due to incomplete inhibition of fertility or side effects.

KEYWORDS

- **antifertility**
- **ethnobotany**
- **ethnomedicine**
- **folklore medicine**
- **herbal contraceptives**
- **plant contraceptives**

REFERENCES

Basak, S., Banerjee, A., & Manna, C. K. (2016). Studies of some ethnomedicinal plants used by the Santal tribal people of the district Bankura, W. B., India, in controlling fertility ethnomedicine of the Santal tribal people. *Intern. J. Novel Research in Life Sci., 3*(1), 20–28.

Billore, K. V., & Audichiya, K. C. (1978). Some oral contraceptives – family planning tribal way. *J. Res. Ind. Med. Yoga and Homoeo, 13(2), 104–109.*

Chandra, R., Mahato, M., Mandal, S. C., Kumar, K., & Kumar, J. (2011). Ethnomedicinal formulations used by traditional herbal practitioners of Ranchi, Jharkhand. *Indian J. Trad. Knowl., 6*(4), 599–601.

Choudhary, K., Singh, M., & Pillai, U. (2008). Ethnobotanical survey of Rajasthan – An update. *American-Eurasian J. Bot., 1* (2), 38–45.

Choudhary, N., Mahanta, B., & Kalita, J. (2011). An ethnobotanical survey on medicinal plants used in reproductive health related disorders in Rangia Subdivision, Kamrup District, Assam. *Internation. J. Sci. Advanced Technol., 1(7),* 154–159.

Chaudhary, R. R. (1993). The quest for a herbal contraceptive. *The National Medical J. India, 6 (5),* 199–201.

Chaudhary, R. R., Mathur, V. S., & Sankarnarayanan, P.(1988). Clinical evaluation of plant contraceptives: Translating folklore into scientific application. In: Dhawan, B. N., Agarwal, K. K., Arora, R. B., Parmar, S. S. (eds.). *Pharmacology for health in Asia.* New Delhi: Allied Publishers. 432–436.

Ciganda, C., & Laborde, A. (2003). Herbal infusions used for induced abortion. *J. Toxicol. Clin. Toxicol. 41(3),* 235–239.

Das, B., Talukdar, A. D., & Choudhary, M. D. (2014). A few traditional medicinal plants used as antifertility agents by ethnic people of Tripura, India. *Intern. J. Pharm. Pharmaceut. Sci. 6*(3), 47–53.

Dwivedi, S., & Kaul, S. (2008). Ethnomedicinal uses of some plant species by ethnic and rural peoples of Indore district of Madhya Pradesh, India. *Pharmaceut. Reviews 6* (3).

Ekka, A. (2012). Some traditional medicine for anti-fertility used by the tribals in Chhattisgarh, India. *Intern. J. Biol., Pharmacy and Allied Sci. 1*(2), 108–112.

Gediya, S., Rbadiya, C., Soni, J. et al. (2011). Herbal plants used as contraceptives. *Intern. J. Curr. Pharmaceut. Rev. Res. 2* (1), 47–53.

Goonasekera, M., & Gunawardana, V. (1995). Pregnancy terminating effect of *Jatropha curcas* in rats. *J. Ethnopharmacol. 4*(7), 117–123.

Gupta, R., & Sharma, R. (2006). A review on medicinal plants exhibiting antifertility activity in males. *Nat. Prod. Radiance 5*(5), 389–410.

Ittiavirah, S., & Habeeb, R. (2013). Evaluation of spermicidal and antiandrogenic activities of aqueous extract of *Tinospora cordifolia* (Willd.) stem. *African J. Pharm. Pharmacol. 7*(34), 2392–2396.

Kamal, R., Gupta, R. S., & Lohiya, N. K. (2003). Plants for male fertility regulation. *Phytotherapy Res., 17*(6), 579–590.

Kamboj, V. P., & Dhawan, V. N. (1982). Research on plants for fertility Regulation in India. *J. Ethnopharmcol., 6,* 91–266.

Kaunitz, A. M., & Benrubi, G. I. (1998). The good news about hormonal contraception and gynaecologic cancer. *The female patients 23,* 43–51.

Kirtikar, K. R., & Basu, B. D. (1975). Indian Medicinal Plants. Bishen Singh Mahendra Pal Singh, Dehradun.

Kumar, D., & Mishra, P. K. (2011). Plant based contraceptive popular among tribals of Jharkhand. *Bioscience Discovery 2*(1), 11–14.

Kumar, M. N., Pai, N. B., Rao, T. S., & Goyal, N. (2009). Biology of sexual dysfunction. *Health and Allied Sciences 8*(1), 1–7.

Mahendra, P., & Bisht, S. (2012). *Ferula asafoetida:* Traditional uses and pharmacological activity. *Pharmacognosy review 6*(12), 141–146.

Maheshwari, J. K., Kalakoti, B. S., & Brijlal. (1986). Ethnomedicine of Bhil tribes of Jhabua district, Madhya Pradesh. *Anc. Sci. Life 5,* 255–261.

Meena, A. K., & Rao, M. M.(2010). Folk herbal medicines used by the Meena community in Rajasthan. *Asian J. Trad. Medicines, 5*(1), 19–31.

Mitra, S., & Mukherjee, S. K. (2009). Some abortifacient plants used by the tribal people of West Bengal. *Nat. Prod. Radiance 8*(2), 167–171.

Noumi, E., & Tchakonang, N. Y. C. (2001). Plants used as abortifacients in the Sangmelina region of southern Cameroon. *J. Ethnopharmacol. 76*(3), 263–268.

Pradhan, D. K., Mishra, M. R., Mishra, A., Panda, A. K., Behera, R. K., Jha, S. & Choudhury, S. (2012). A comprehensive review of plants used as contraceptives. *Int. J. Pharmaceut. Sci. Res. 4*(1), 148–155.

Prasad, A. G. D., Shyma, T. B., & Raghavendra, M. P. (2014). Traditional herbal remedies used for management of reproductive disorders in Wayanad district, Kerala, *Intern. J. Res. Pharmacy and Chemistry, 4*(2), 333–341.

Rai, R., & Nath, V. (2005). Some lesser known oral herbal contraceptives in folk claims as anti-fertility and fertility induced plants in Bastar region of Chhattisgarh. *J. Nat. Remedies, 5*(2), 153–159.

Sathiyaraj, K., Sivaraj, A., Thirumalai, T., & Senthilkumar, B. (2012). Ethnobotanical study of antifertility medicinal plants used by the local people in Kathiyavadi village, Vellore District, Tamil Nadu, India. *Asian Pacific J. Tropical Biomedicine*, S1285–S1288.

Shah, G. M., Khan, M. A., Ahmad, M., Zafar, M., & Khan, A. A. (2009). Observations on antifertility and abortifacient herbal drugs. *African J. Biotech. 8*(9), 1959–1964.

Shrivastava, S., Dwivedi, S., Dubey, D., & Kapoor, S. (2007). Traditional herbal remedies from Madhya Pradesh used as oral contraceptives – A field survey. *International J. Green Pharmacy. 1*(1), 18–22.

Sinha, R. K., & Nathawat, G. S. (1989). Anti-fertility effects of some plants used by the street herbal vendors for birth control. *Ancient Science of Life, 9*(2), 66–68.

Tewari, P. V., Sharma, S. K., & Basu, K. (1976). Clinical trial of an indigenous drug as an oral contraceptive. *J. Natl. Integrated Med. Assoc. 18,* 117–118.

Vijigiri, D., & Sharma, P. P. (2010). Traditional uses of plants in indigenous folklore of Nizamabad District, Andhra Pradesh, India. *Ethnobotanical Leaflets 14, 29–45.*

Yadav, J. P., Kumar, S., & Siwach, P. (2006). Folk medicine used in gynecological and other related problems by rural population of Haryana. *Indian J. Trad. Knowl. 5*(3), 323–326.

Zingare, A. K. (2012). Ethnomedicinal uses of plants among the Halba tribe of Gondia district of Maharashtra, India. *Bionano Frontier, 5*(2), 121–125.

A – *Achyranthus aspera* L.

B – *Albizia lebbeck* (L.) Benth.

C – *Allium cepa* L.

D – *Aloe barbadensis* Mill.

E – *Amaranthus spinosus* L.

F – *Asparagus racemosus* Willd.

PLATE 1

A – *Azadirachta indica* A.
Juss.

B – *Bambusa vulgaris*
Schrad.

C – *Coriandrum sativum*
L.

D – *Curcuma longa* L.

E – *Euphorbia hirta* L.

F – *Ficus religiosa* L.

G – *Hibiscus rosa–sinensis* L

PLATE 2

A – *Lantana camara* L.

B – *Melia azedarach* L.

C – *Mentha arvensis* L.

D – *Morus alba* L.

E – *Musa paradisica* L.

F – *Nelumbo nucifera* Gaertn.

G – *Nerium oleander* L.

H – *Ocimum sanctum* L.

I – *Plumbago zeylanica* L.

PLATE 3

A – *Ricinus communis* L.

B – *Solanum surattense* Burm.f.

C – *Vitex negundo* L.

PLATE 4

CHAPTER 12

ETHNOBOTANY OF THE NEEM TREE (*AZADIRACHTA INDICA* A. JUSS): A REVIEW

K. SRI RAMA MURTHY,[1] T. PULLAIAH,[2] BIR BAHADUR,[3] and K. V. KRISHNAMURTHY[4]

[1]*R&D Center for Conservation Biology and Plant Biotechnology, Shivashakti Biotechnologies Limited, S. R. Nagar, Hyderabad – 500038, Telangana, India, E-mail: drmurthy@gmail.com*

[2]*Department of Botany, Sri Krishnadevaraya University, Anantapur – 515003, Andhra Pradesh, India. E-mail: pullaiah.thammineni@gmail.com*

[3]*Department of Botany, Kakatiya University, Warangal – 505009, Telangana, India, E-mail: birbahadur5april@gmail.com*

[4]*Consultant, R&D, Sami Labs, Peenya Industrial Area, Bangalore – 560058, Karnataka, India. E-mail: kvkbdu@yahoo.co.in*

CONTENTS

ABSTRACT

Azadirachta indica, commonly known as neem, native of India-Myanmar and naturalized in most of tropical and subtropical countries has great value. The importance of the neem tree has been recognized by Indians since ancient times. Ethnobotanically neem is useful in medicine, veterinary medicine, as biopesticides and in many other ways. It contains many biologically active compounds including alkaloids, flavonoids, terpenoids, phenolic compounds, carotenoids, steroids and ketones. Biologically, the most active compounds in neem are the azadirachtins which are actually a mixture of seven isomeric compounds labeled as azadirachtin A-G of which azadirachtin E is the most effective. Other compounds that have a biological activity are salannin, volatile oils, meliantriol and nimbin.

12.1   INTRODUCTION

The neem tree, *Azadirachta indica* A. Juss. [=*Melia azadirachta* L., & *Melia indica* (A. Juss.) Brandis] is known as the Indian lilac or Margosa (Koul et al., 1990). Neem is native to the Indian subcontinent and was introduced into Africa, and is presently grown in many Asian countries, as well as tropical areas of the New World (Koul et al., 1990). Neem trees are fast growers, and in three years may grow to 20 feet in height from seed planting. It will grow where rainfall is only 18 inches per year and it thrives in areas of extreme heat up to 120 degrees Fahrenheit. Neem trees can live up to 200 years (Conrick, 2001). It can grow in very different types of soils but a soil pH value of 6.2 to 7.0 is the best (pH range 5.9 to 10). The main mode of reproduction is through seeds, although stem cuttings are also used in some places.

12.1.1 TAXONOMICAL POSITION

The taxonomic classification of neem is as follows: Kingdom: Plantae, Order: Rutales, Suborder: Rutineae, Family: Meliaceae, Subfamily: Melioideae, Tribe: Melieae, Genus: *Azadirachta*, Species: *Azadirachta indica.*

12.1.2 VERNACULAR NAMES

Arabic – Neeb, Azad-darakhul-hind, Shajarat Alnnim; Assamese – Neem; Bengali – Nim; English – Margosa, Neem Tree; French – Azadirac de l'Inde, margosier, margousier; German – Indischer zedrach, Grossblaettiger zedrach; Gujarati – Dhanujhada, Limba; Hausa – Darbejiya, Dogonyaro, Bedi; Hindi – Neem; Kannada – Bevu; Malayalam – Ariyaveppu; Manipuri – Neem; Marathi – Kadunimba; Myanmar – Burma – Tamar; Persian – Azad Darakthe hind, neeb, nib; Portuguese – Nimbo, Margosa, Amargoseira; Punjabi – Nimm; Sanskrit – Arishta (meaning "relieving sickness"), Pakvakrita, Nimbaka; Sinhala – Kohomba; Tamil – Veppai; Sengumaru; Telugu – Vepa; Thai – Sadao; tulu-besappu; Urdu – Neem. *See* https://en.wikipedia.org/wiki/Azadirachta_indica.

12.1.3 BOTANICAL DESCRIPTION

Semi-evergrecn trce with a wide trunk, which can grow up to 40–50 feet or more, with a straight trunk and long spreading branches forming a broad round crown, wood is red; bark rough, dark brown and with wide longitudinal fissures separated by flat ridges. The leaves are alternate, compound, imparipinnate, each comprising 5–15 leaflets. The leaflets are bright green, oblique at the base or slightly curved, coarsely toothed; with a pointed tip. Flowers are white, fragrant, in 10–20 cm long axillary panicles, mostly in the leaf axils. The sepals are ovate and about one cm long, while petals are sweet scented, white and oblanceolate. Fruits are ellipsoid drupes, glabrous, 12–20 mm long, green, turning yellow on ripening, aromatic with garlic like odor. Fruits alone contain latex. Fresh leaves and flowers come in March–April. Fruits mature between April and August depending upon geographical location.

12.1.4   ECOLOGY

The neem tree is noted for its drought tolerance. Normally it thrives in areas with sub arid to sub humid conditions, with an annual rainfall between 400 and 1200 mm. It can grow in regions with an annual rainfall below 400 mm, but in such cases it depends largely on ground water levels. Neem can grow in many different types of soil, but it thrives best on well-drained deep and sandy soils. It is a typical tropical to subtropical tree and exists at annual mean temperatures between 21–32°C. It can tolerate high to very high temperatures and does not tolerate temperature below 4°C. Neem is a life giving tree, especially for the dry coastal, southern districts of India. It is one of the very few shade giving trees that thrive in the drought prone areas. The trees are not at all delicate about the water quality and thrive on the merest trickle of water, whatever be the quality. In India it is very common to see neem trees used for shade lining the roads and avenues or in most people's back yards, parks, etc. In very dry areas the trees are planted in large tracts of land.

12.1.5   DISTRIBUTION

A native to Karnataka in India and Myanmar (Schmulterer, 1995), it grows in most parts of south East Asia and West Africa, and more recently Caribbean and south and Central America. In India it occurs naturally in Siwalik Hills, dry forests of Andhra Pradesh, Telangana, Tamil Nadu and Karnataka to an altitude of approximately 700 m. It is cultivated and frequently naturalized throughout the drier regions of tropical and subtropical India, Pakistan, Sri Lanka, Thailand and Indonesia. It is also grown and often naturalized in Peninsular Malaysia, Singapore, Philippines, Australia, Saudi Arabia, Tropical Africa, the Caribbean, Central and South America (Parotta, 2001).

12.1.6   WEED STATUS

Neem is considered a weed in many parts of the world/regions, including some parts of the Middle East, and most of Sub Saharan Africa including

West Africa. In Senegal it has been used as a malarial drug and Tanzania and other Indian Ocean states where in Kiswahili it is known as 'the panacea', literally 'the tree that cures as many as forty (diseases)'.

12.2   HISTORY OF NEEM USES

The neem tree's history dates back to antiquity, with indications that it was used in medical treatments about 4,500 years ago in India. There is evidence found from excavations at Harappa and Mohenjo-Daro in Northwestern and western India, in which several therapeutic compounds including neem leaves, were gathered in the ruins (Conrick, 2001). India's ancient books Charaka-Samhita (about 500 B.C.) and the Susruta Samhita (about 300 A.D.) mention neem in almost 100 entries for treating many diseases which affect human society (Conrick, 2001). In Sanskrit, the language of ancient Indian literature, neem is referred to as *Nimba*, which is derived from the term *Nimbati Swastyamdadati*, which means 'to give good health' (Randhawa, 1997).

Neem was used in every stage of life throughout India, and is still used today for its many beneficial qualities. Starting from birth, the Sarira Sthanam recommended that newborn infants be anointed with herbs and oil, laid on a silken sheet and fanned with neem tree branches. The child was given small doses of neem oil when ill, bathed with neem tea to treat cuts, rashes, and chicken pox. Neem twigs were used as toothbrushes to prevent gum diseases and tooth caries. Wedding ceremonies included neem leaves placed on the floor of the temple, and neem branches for fans. Neem oil was used in small lamps for lighting. Neem wood was used for cooking and for making the roof of the house. Grains and beans were stored in containers with neem leaves to keep out insects. At the time of death, neem branches were used to cover the body and neem wood was used in the funeral pyre (Conrick, 2001).

In the Indian book about healing plants for women, called "Touch Me, Touch-me-not: Women, Plants and Healing," the author describes the role of neem for the village folk of India. Shodini (1997) describes neem as an all-purpose medicine and as a tree used in some form of Goddess worship. Neem leaves were used in the primitive societies in India to exorcize the spirits of the dead. Branches of neem were placed in households because it was believed that the goddess lived in the branches and would guard the household against smallpox. Though smallpox is not as great a threat as it was, neem branches are still used even now for bathing scars, and is used as a ritual termination of an attack of chickenpox or measles even today. The neem tree was considered protective to women and children. Delivery chambers were fumigated with its burning bark. In some parts of India, to celebrate the New Year, neem leaves are mixed with other eatable ingredients symbolizing the sweet and sour experiences of the upcoming year (Shodini, 1997).

12.3 ETHNOBOTANY OF NEEM

12.3.1 ASSOCIATION WITH HINDU SOCIO-CULTURAL LIFE IN INDIA

Neem leaf or bark is considered an effective Pitta pacifier due to its bitter taste. Hence, it is traditionally recommended during early summer in Ayurveda (that is, the month of Chaitra as per the Hindu Calendar which usually falls in the month of March–April), and during Gudi Padva, which is the New Year in the state of Maharashtra, the ancient practice of drinking a small quantity of neem juice or paste on that day, before starting festivities, is found. As in many Hindu festivals and their association with some food to avoid negative side effects of the season or change of seasons, neem juice is associated with Gudi Padva to remind people to use it during that particular month or season to pacify summer pitta. In Tamil Nadu during the summer months of April to June, the Mariamman temple festival is celebrated, which is thousand year old tradition. The Neem leaves and flowers are the most important part of the Mariamman festival. The goddess Mariamman statue will be garlanded with Neem leaves and flowers. During most occasions of celebrations and weddings the people of Tamil Nadu adorn their surroundings with the Neem leaves and flowers as a form of decoration and also to ward off evil spirits and infections. In the eastern coastal state of Orissa the famous Jagannath temple idols are placed on a plate made of Neem heart wood along with some other essential oils and powders. This plate is replaced every seven/nine years. Telugu New years, Ugadi is famous during which *Pachadi* is prepared with jaggery, neem flowers, tamarind, etc. all over Telangana and Andhra Pradesh.

In many traditional Indian communities neem leaves are hung in bunches at the entrance to the house as a symbolic way to keep out infestations and evil spirit. Also, neem leaves are spread on the bed of patients suffering from Chicken and small pox. Neem bark or leaf decoctions are drunk by jaundice patients. Neem leaf soaked water is drunk every morning to keep the body to resist infections of various sorts. Similarly, this water is used to bathe young children. In some communities, it is customary for a bride to bathe in water soaked with neem leaves. Neem leaves are generally burnt to keep mosquitoes away; villagers often sleep under a neem tree for the same purpose.

The tree figures very largely in folk songs and lores of India. The tribal priest ties neem twigs around his waist to ward off evil spirits during his religious/ritual performance (Vartak and Ghate, 1990) (Figures 12.1–12.5).

12.3.2 ETHNOMEDICINAL USES

All parts of the tree have been used medicinally for centuries. It has been used in Ayurvedic, Siddha, Unani and folk medicine for more than 4000 years due to its medicinal properties (Table 12.1). The earliest Sanskrit medical writings refer to the benefits of Neem's fruits, seeds, oil, leaves, roots and

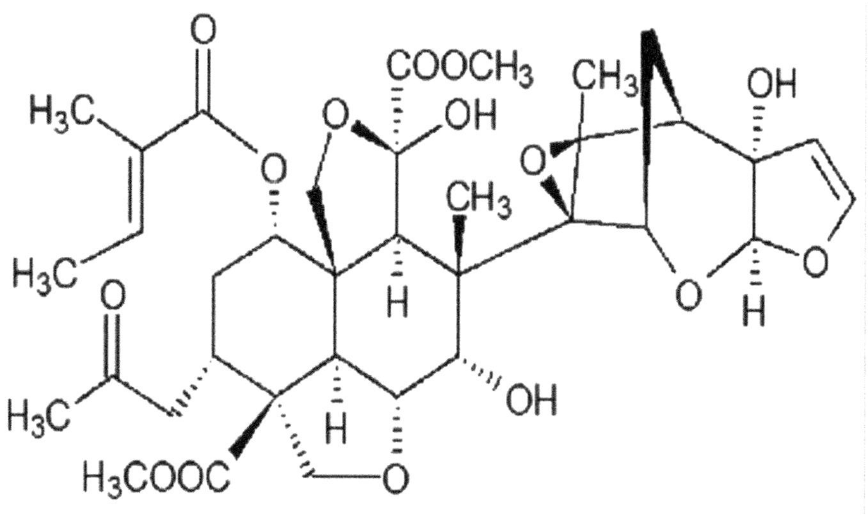

FIGURE 12.1 Chemical structure of the tetranortriterpenoid azadirachtin.

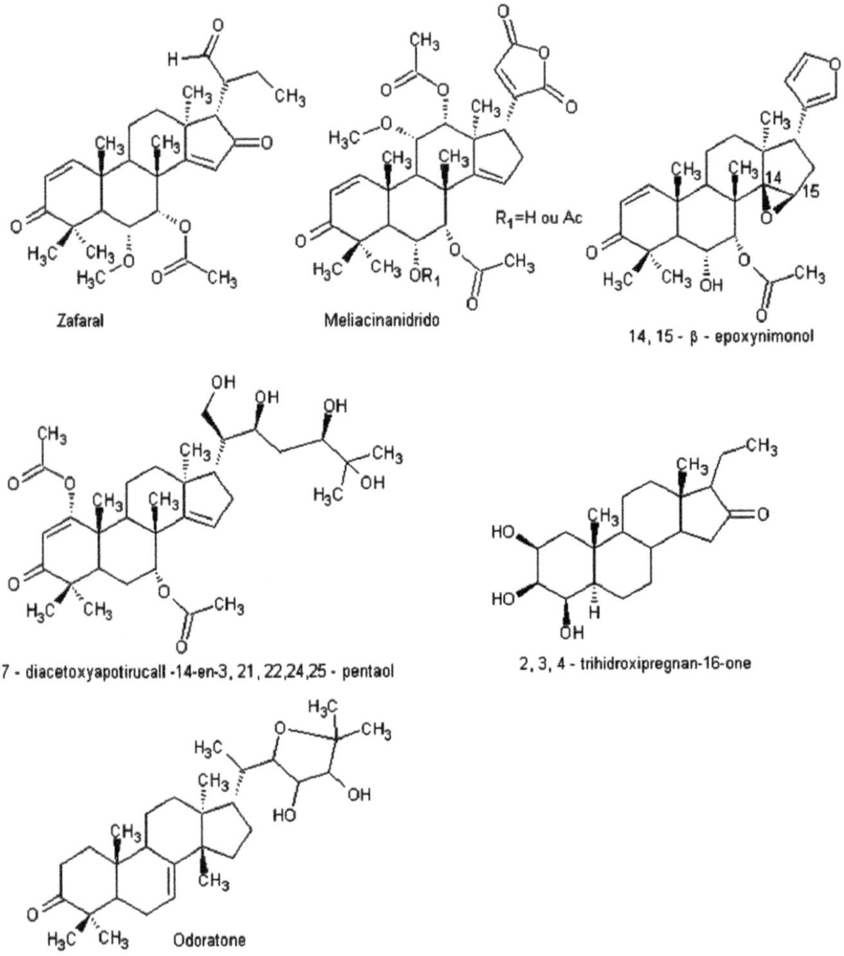

Zafaral

Meliacinanidrido

R_1=H ou Ac

14, 15 - β - epoxynimonol

1, 7 - diacetoxyapotirucall -14-en-3, 21, 22,24,25 - pentaol

2, 3, 4 - trihidroxipregnan-16-one

Odoratone

FIGURE 12.2   Limonoids present in *Azadirachta indica* A. Juss.

bark. Each has been used in the Indian Ayurvedic, Siddha, folk and Unani medicine, and is now being used in pharmaceutical and cosmetics industries (Brototi and Kaplay, 2011).

The neem tree is listed in various sources as having many uses, from medicinal to agricultural. Neem's pharmacological properties are detailed in "properties and uses of neem." In this chapter, it states that neem seed oil has been used for antimalarial, febrifuge, antihelminthic, vermifuge, and anti-septic and antimicrobial purposes, for bronchitis control, and as a healing

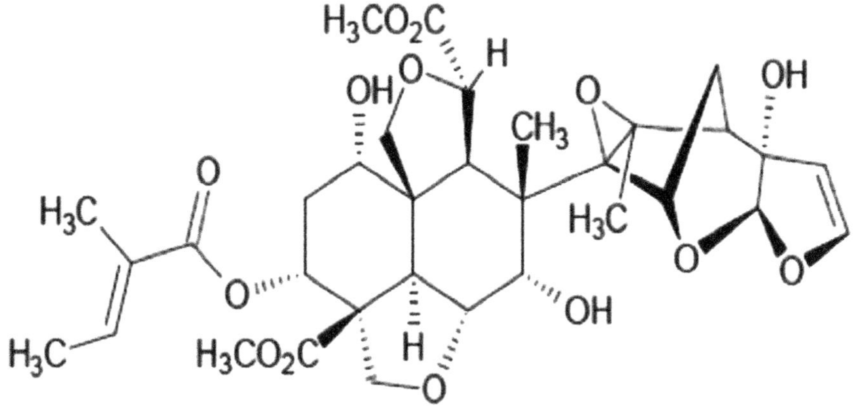

FIGURE 12.3   Chemical structure of 3-tigloilazadiractol.

agent for various skin disorders (Koul et al., 1990). Neem oil is known to control mycobacteria and pathogens, including *Staphylococcus typhosa* and *Klebsiella pneumoniae* (Koul et al., 1990). Analgesic and antipyretic effects have been shown, as well as antinflammatory and antihistaminic properties (Koul et al., 1990).

Hot water extract of the bark is taken orally by the adult female as a tonic and emmenagogue. Hot water extract of the flower and leaf is taken orally as an anti-hysteric remedy, and used externally to treat wound. The dried flower is taken orally for diabetes. Hot water extract of dried fruit is used for piles and externally for skin disease and ulcers. Hot water extract of the entire plant is used as anthelmintic, an insecticide and purgative. Juices of bark of *Andrographis puniculata, Azardiracta indica, Tinospora cardifolia,* are taken orally as a treatment for filariasis. The hot water extract is also taken for fever, diabetes, and as a tonic, refrigerant, anthelmintic. Fruit leaf and root, ground and mixed with dried ginger and 'Triphala" is taken orally with lukewarm water to treat common fever. Leaf juice is given in gonorrhea and leucorrhoea. Leaves applied as poultice to relieve boils, their infusion is used as antiseptic wash to promote the healing of wound and ulcers. A paste of leaves is used to treat wounds, ring worms, eczema and ulcers. Bathing with *Neem* leaves is beneficial for itching and other skin diseases. Leaf juice is used as nasal drop to treat worm infestation in nose. Steam inhalation of bark is useful in inflammation of throat. Decoction can cure intermittent fever, general debility convalescent, and loss of appetite after

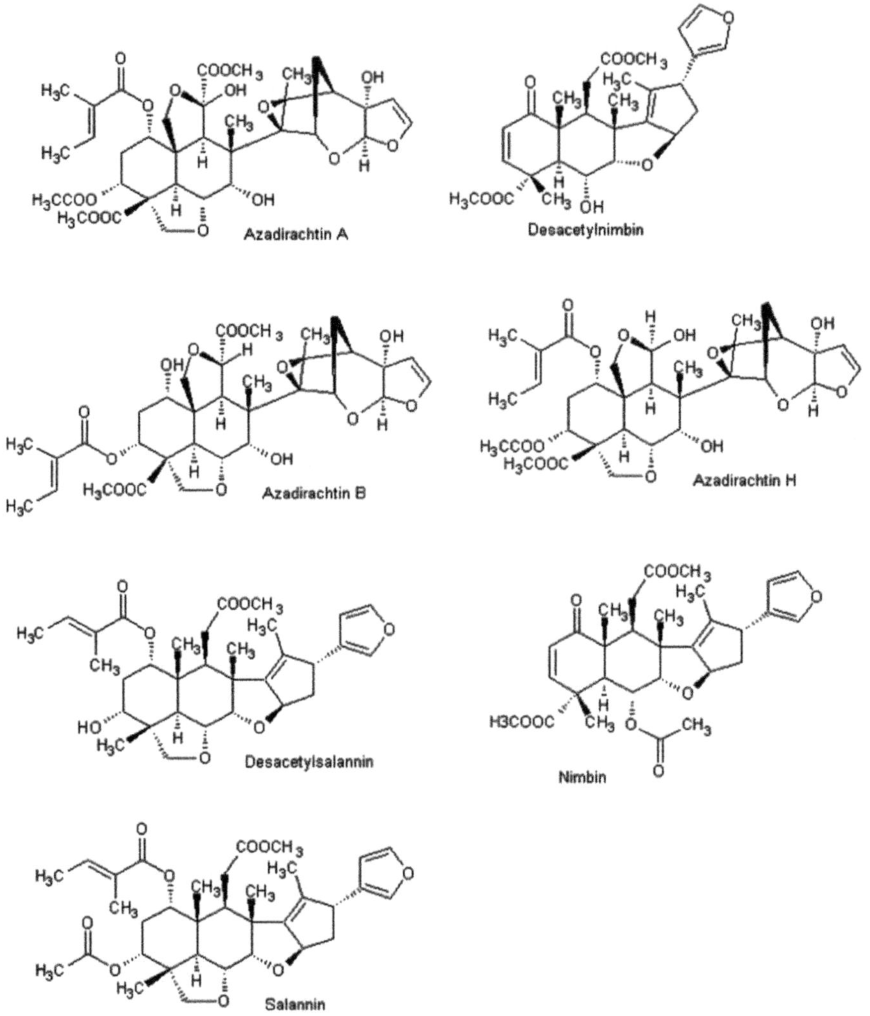

FIGURE 12.4 Chemical structure of some bioactive components isolated from seeds of *Azadirachta indica* (Silva et al., 2007).

fever (Nadkarni, 1994). Infusion of flower is given in dyspepsia and general debility (Chatterjee and Pakrashi, 2010). The tender twigs of the tree are used as toothbrush which is believed to keep the body system healthy, the breath and mouth clean and sweet (Kabeeruddin and Makhzanul, 2007; Tandon and Sirohi, 2010). Seed oil is used in leprosy, syphilis, eczema, chronic ulcer (Ghani and Khazainul, 2004; Kabeeruddin and Makhzanul, 2010). One of the main uses of neem is in malarial fever, both as a preventive

Neem Tree habit   Flowering branch

Vegetative branches   Neem unripe mature fruits

FIGURE 12.5   Various parts of *Azadirachta indica.*

and a curative. Tribals of Odisha use this regularly to control malarial fever. This was proved scientifically (see details in Van der Natetal, 1991).

Irulas of Kodiakkarai reserve forest in Tamil Nadu drink bark extract of *Azadirachta indica* to eliminate stomach worms (Ragupathy and Mahadevan, 1991). Verma et al. (1995) in their study on traditional phytotherapy among the Baiga tribe of Shahdol district of Madhya Pradesh reported that chicken pox and measles are controlled when leaf paste of *Azadirachta indica* is applied on the infected sites. Vihari (1995) while reporting the ethnobotany of cosmetics of Indo-Nepal border reported the following uses for neem: Used as hair shampoo for dandruff and for killing lice; plant exudates is

TABLE 12.1 Some Medicinal Uses of Neem As Mentioned in Indian Traditional Systems of Medicine

Parts	Use	Author/Source
All parts	Fumigation	Sharma, 1996
Bark	Fever, Analgesis, alterative and curative, Leprosy, Vaginal problems	Shodini, 1997; Subramanian and Lakshmanan, 1993
Flowers	Bile suppression, elimination of intestinal worms and phlegm	Varma, 1976
Fruit juice	Eye diseases	Sharma, 1996
Fruits	Piles, intestinal worms, urinary disorder, epistaxis, phlegm, eye problem, diabetes, wounds and leprosy	Ketkar and Ketkar, 1995; Varma, 1976; Sharma, 1996
Gum	Scabies, wounds, ulcers, skin diseases	Varma, 1976
Leaves	Leprosy, eye problem, epistaxis, Phlegen, intestinal worms, anorexia, biliousness, toxicity removal, skin ulcer, vaginal problems, body heat, infections, painful periods, worms, fever, hemorrhage, wounds, jaundice poisoning	Varma, 1976; Shodini, 1997
Neem decoction	Heart diseases, vaginal problems, Gray hairs.	Sharma, 1996
Oil	Leprosy and intestinal worms	Varma, 1976
Root bark	Tooth problems	Sharma, 1996
Seeds	Leprosy and intestinal worms, piles, poisoning	Varma, 1976; Sharma, 1996
Twigs	Cough, asthma, piles, phantom tumor, intestinal worms, spermatorrhoea, obstinate urinary disorder, diabetes, dental problems	Varma, 1976
Total plant	Blood morbidity, biliary afflictions, itching, skin ulcer, burning sensation and leprosy	Varma, 1976

used as gum for pasting 'Bindi' on the forehead; oil is used in making soap; leaf juice is widely used for treating skin diseases, lice infection, dandruff and wounds; twig is very popular as toothbrush. Henry et al. (1996) reported that Palliyan tribes in the Southern Western Ghats are using Neem for treatment of Epilepsy. Leaves of *Acalypha indica* with *Cardiospermum halicacabum* boiled in neem oil and the extract is given internally to cure epilepsy (Henry et al., 1996). Epileptic attacks are cured by Santhal and Paharia tribes of Santhal Paragana, Bihar in the following manner. Fresh stem bark of *Azadirachta indica* is crushed with leaves and roots of *Cissampelos pareira* and *Aristolochia indica* to make extract, 2–3 drops of this extract is applied in nostrils during epileptic attacks (Kumar and Goel, 1998). Mohanty and Padhy (1996) while giving the traditional phytotherapy for diarrheal diseases in Ganjam and Phulbani districts of South Orissa has given the following: 2 to 3 g gum of neem dissolved in rice water and administered twice a day to children to check diarrhea. Kumar and Jain (1998) while giving the ethnomedicinal uses of the ethnic tribes of Surguja district in Madhya Pradesh gave the following uses of Neem. One teaspoon full of juice of flowers given 3 times a day to control vomiting. 5 g of leaves pounded and applied on wounds and pimples at bed time for 7 days. Leaf juice given to treat boils and fever, one cup in the morning for 8–15 days (Kumar and Jain, 1998). Fresh tender twig of Neem is used for brushing the teeth (commonly called Dutoon) by most of the rural people and tribal people all over India even now. The stick or twig is crushed at one end to make it brush-like. Flexible fibers of the crushed end of the stick are used for cleaning teeth surfaces and teeth crevices acting like a brush. The toothbrush is used to clear the decaying teeth, to stop bleeding of gums, to treat severe toothache, infected gums and pyorrhea, to strengthen teeth and gums, to remove foul smell from the mouth, to arrest swellings on the gums and to remove deposits of scaly yellowish or brownish hard chalk-like substance from the surfaces of teeth (Punjani, 1998). Bhatt and Mitaliya (1999) reported that tribals of Vicoria Park reserve forest give the juice of leaves of *Azadirachta indicia* internally in piles, jaundice, fever, etc. and paste is applied on wounds, ringworm, eczema, scorpion sting, etc. Subramani and Goraya (2003) reported that the leaves of *Azadirachta indica* are ground with ginger and black pepper applied externally for poisonous insect bites. Ravikumar and Vijaya Sankar (2003) reported that the Malayali tribes of Javvadhu hills follow the following method for abortion: Few pieces of stem bark crushed with that of *Carica papaya* and *Holoptelia integrifolia* made into a soup; 200 ml of the soup orally administered to women for abortion. Paul (2003) while reviewing ethnobotany of *Azadirachta indica* reported the following uses. *Stem bark:* The bark is very useful in skin diseases. It contains a resinous bitter principle and is used in malarial fever. Fresh tender twigs are used as toothbrush in pyorrheal diseases. *Stemming gum:* Trees grow near water courses exude a sap commonly from the stem tip. This soap is considered by different ethnic groups as refrigerant, nutrient and tonic. It is useful in skin diseases, consumption, a tonic for dyspepsia and general debility. *Leaves:* The leaf paste is put on boils as poultice for suppuration. Leaf decoction is used for washing septic ulcer and is much valuable in pneumonia, typhoid and other infective fevers. It is also applied on eczema, prurigo and sycosis with addition of *Curcuma longa*. Rural folk use tender leaves along with *Piper nigrum* against intestinal worms and blood sugar. The leaf paste is used on cow-pox. The fresh leaf paste along with seed paste of *Psoralea corylifolia* and *Cicer arietinum* tender leaf powder is applied on leucoderma. Fresh tender leaf powder is taken in West Bengal as anti-pox agent. *Flower:* The dried flowers are eaten by the tribal either raw or in curries and in soups; it is used as a fried dish in South India. The dried flower powder is used as tonic in dyspepsia, in general debility and as stomachic. *Fruits:* Tribals use the pulp of the berries as purgative, emollient and anthelmintic.

Upadhyay and Chauhan (2003) reported the following uses of gum of *Azadirachta indica* used by the Gond and Baiga tribes: A pinch of gum dissolved in cold water is used to treat inflammatory eyes; it is applied to treat cracked soles and decoction of the gum is used to treat inflammatory gums. Pattanaik et al. (2007) investigated the traditional medicinal practices of the tribal people of Malkangiri district of Orissa and reported that crushed dried leaves of *Azadirachta indica* in water are applied locally till cure for skin diseases. Bapuji and Ratnam (2009) reported that tribals of Gangaraju Madugula mandal in Visakhapatnam district of Andhra Pradesh are using stem bark of *Azadirachta indica* for treating skin troubles. Babu et al. (2010) reported that tribals of Kotia hills of Vizianagaram district applied leaf paste of *Azadirachta indica* mixed with turmeric on the affected areas twice a day to treat chicken pox. Dahare and Jain (2010) reported that crushed leaves are used to cure many skin diseases by Korku and Gond tribes of Tahsil Multai in Betul district of Madhya Pradesh. Alagesboopathi (2011a) reported that ethnic tribes of Kanjamalai hills in Salem district of Tamil Nadu are using decoction of bark of *Azadirachta indica* as liver tonic, while leaf paste is applied on affected parts of skin diseases. He also reported that seed oil is used for curing leprosy and for wound healing. Leaf ground with castor oil is used to cure small pox by the Kurumba tribals in Pennagaram region of Dharmapuri district in Tamil Nadu (Alagesboopathi, 2011b). Alwa and Ray (2012) reported that tribals in Dhar district in Madhya Pradesh believed that brushing the teeth daily with a stick of neem the body becomes resistant against snake bite and bathing cure skin affections. Das and Choudhury (2012) reported that tribes of Manipuri tribes of Tripura use leaves of neem boiled in water to bathe patient with malaria and chicken pox. They also reported that smoke produced by burning leaves is used as mosquito repellant and bark paste made to tablets is administered in severe jaundice. Senthilkumar et al. (2013) reported that leaves paste of *Azadirachta indica* is used by the Malayali tribals in Yercaud hills in Tamil Nadu for curing skin diseases. Tribals of Alirajpur, Madhya Pradesh mixed 40 gms of bark of neem with 40 gms of bark of *Acacia nilotica*, boiled and filtered it and 50 ml was taken in empty stomach in the early morning for 7 days to treat white discharge. Irula tribes of Tamil Nadu are using bark of *Azadirachta indica* for treating snake bite (Gnanavel and Jose, 2014). Sarkel (2014) made an ethnobotanical survey of folk lore plants used in treatment of snake bite in Paschim Medinipur district in West Bengal reported that the leaf ash or crushed leaves rubbed into scarification around the snake bite as antidote and leaf juice is given as decoction. Rajeswari et al. (2016) reported that Malayali tribes in Jarugumalai in Salem district, Tamil Nadu take raw leaf extracts of *Azadirachta indica* mixed with little water is taken at a dose of 2–3 teaspoons daily in empty stomach to cure diabetes.

To date many reviews have been published on pharmacological properties of Neem. Bhowmik et al. (2010) reviewed the medicinal properties, chemical constituent and commercial uses of *Azadirachta indica.* Dubey and Kashyap (2014) reviewed the pharmacological properties of *Azadiraachta indica* investigated by various researchers. They recorded that Neem has antibacterial, antifungal, antiviral, anti-plasmodial, antiparasitic, anti-inflammatory, antioxidant, neuroprotectivee, immunomodulatory, anti-anxiety, liver protection, antipyretic, contraceptive, anti-gastric ulcer, wound healing anticancer, antidiabetic, anti HIV AIDS, anthelmintic, pesticidal properties and is useful in curing anemia, urticaria, asthma, dysmenorrhea, post delivery care, dental care and regulates hormonal levels.

12.3.3   ETHNOVETERINARY USES

In India, Neem has been used for centuries to provide health cover to live stock in various forms. It has also been very widely used as animal feed. The epic of Mahabharata (3000 B.C.) refers to two Pandava brothers Nakul and Sahadeva, who used to treat wounded horses and elephants with neem oil and leaves, preparations. Neem extracts having antiulcer, antibacterial, antiviral properties are used successfully to treat cases of stomach worms, ulcers, coetaneous diseases, intestinal helminthiasis (Girish and Bhat, 2008).

Pande et al. (2007) while reviewing ethnoveterinary plants of Uttaranchal reported that *Azadirachta indica* is useful in treating broken horns, burns, mange, tympany, indigestion, snake bite, foot and mouth disease, lock jaw (tetanus) and retention of urine.

In Telangana and Andhra Pradesh and other states the leaves of the plant is regularly fed to cattle and goats to get more milk, immediately after parturition (Paul, 2003). Some of the present uses of neem for animals are dog soap and shampoo, cattle feed supplement which kills worms, neem cream, fly and mosquito repellent, and wound dressings (Koul et al., 1990). Bathing soap, toothpaste, tooth powder, and mouthwash are made from neem products (Koul et al., 1990). Reddy et al. (1998) reported that smoke of 500 g of leaves of neem is inhaled daily twice for 10 days to cure ephemeral fevers in cattle.

12.3.4 OTHER USES

In Africa and Caribbean, users of this plant, especially children, eat ripe fruits of Neem. In India, since ancient times the tender leaves of Neem are consumed as food and for tea preparations. Domestic animals are also fed with Neem leaves (Hedge, 1993). Despite *A. indica* being known for its pesticidal properties there are no records of Neem toxicity to humans, probable by avoiding higher doses. In fact, it was observed that, toxic effects of Neem oil in mammals occur only at higher doses (Deng et al., 2013). This toxicity is not lower compared to the natural compound rotenone (largely used as a broad spectrum insecticide, piscicide and pesticide) (Coats, 1994). Woollen and other cloths are stored with dried neem leaves, due to insecticidal properties as also various cereals and other grains for long term storage.

The importance of neem seed cake has been realized by ancient Indian people for a very long time. Not only has the cake been in use as a cattle and poultry feed, but also as important manure and as a soil-amendment agent. The cake not only provides nutrients to the plant, but also controls plant parasitic soil nematodes, fungi and bacteria (Mojumder, 1995).

12.4 HEALTH AND PERSONAL CARE PRODUCTS

Neem personal care products derived from seed, oil and leaf include; Skin care – including eczema cream, antiseptic cream, and nail care; hair care – shampoo, and hair oils; oral hygiene – toothpaste and neem twigs; therapeutic – loose Neem leaves – tea, vegetarian capsules, powders; household products – soaps, insect repellent (spray and lotion), and candles.

12.5   PHYTOCHEMISTRY

Biswas et al. (2002) review deals mainly on the biological activities of some of the neem compounds isolated, pharmacological actions of the neem extracts, clinical studies and plausible medicinal applications of neem along with their safety evaluation. Biologically active principles isolated from different parts of the plant mainly come under two groups of triterpenoids respectively with *euphol* and *tirucallol* skeletons and these includes: *azadirachtins, meliacins, gedunins, nimbidinins, nimbolinins, salannins, nimbins, vilasinins, azadirones, azadiradiones, nimocin, meliantriol, amoorastations, vepinins*, and other triterpenoid derivatives. There are also diterpenoids and non- terpenoidal compounds such as flavonoids. Meliacin form the bitter principles of Neem oil, the seed also contain tignic acid responsible for the distinctive odor of the oil (Table 12.2). All parts of *A. indica* are used for indigenous medicine purposes, especially to combat Viruses, bacteria fungi and wide variety of insect pests (Luo et al., 2000) of the more than 300 compounds identified from *A. indica*, among them azadirachtin was identified as the most toxic metabolite (Soon and Bottrell, 1994). Azadirachtin (AZ) is rapidly biodegradable maintaining the maximum antifeedant effect for two weeks. It consists of closed isomers compounds ranged from AZ-A to AZ-G, the isomer AZ-A the most important component present in the Neem seed extract (Neves et al., 2003). The insecticidal oil is extracted by pressing the seeds obtaining a maximum oil yield of 47% enriched with about 10% of azadirachtin. The seeds residue is very rich in AZ and can be dried and subsequently used for the preparation of insecticides extracts, after mixing with water and filtration. In addition showing nematicide effect and can be used as organic fertilizer (Neves et al., 2003). Neem has been shown to control almost all groups of insects, mites and nematodes that spoil plants or plant products (Latum, 1985). There is no detailed information about specific doses to kill insect species. However, the following doses have shown efficacy in the control of vegetable pests (Neves et al., 2003).

Fresh green leaves are used control ticks attacks on bovines (250 g/100 L of water), and dogs (500 g/3 L of water). For this the leaves should be left

TABLE 12.2   Some Bioactive Compounds from Neem

Neem compound	Source	Biological activity	References
Nimbidin	Seed oil	Anti-inflammatory, Antiarthritic, Antipyretic, Hypoglycemic, Antigastric ulcer, Spermicidal Antifungal, Antibacterial, Diuretic	Bhargava et al., 1970; Pillai and Santhakumari, 1981; David, 1969; Pillai and Santhakumari, 1981; Pillai et al., 1978; Pillaia and Santhakumari, 1984; Sharma and Saksena, 1959; Murthy and Sirsi, 1958; Bhide et al., 1958
Sodium nimdidate	Seed oil	Anti-inflammatory	Bhargava et al., 1970
Nimbin	Seed oil	Spermicidal	Sharma and Saksena, 1959
Nimbolide	Seed oil	Antibacterial, Antimalarial	Rochanakij et al., 1985; Khalid et al., 1989; Rojanapo et al., 1985
Gedunin	Seed oil	Antibacterial, Antimalarial	Rao et al., 1977; Khalid et al., 1989
Azadirachtin	Seed	Antimalarial	Butterworth and Morgan, 1968
Mahmoodin	Seed oil	Antibacterial	Devakumar and Sukhdev, 1996
Gallic acid, (-) epicatechin and catechin	Bark	Anti-inflammatory, immunomodulatory	Van der Nat et al., 1991
Margolone, margolonone and isomargolonone	Bark	Antibacterial	Ara et al., 1989
Cyclic trisulfide and cyclic tetrasulphade	Leaf	Antifungal	Pant et al., 1986
Polysaccharides	Bark	Anti-inflammatory	Kakai, 1984
Polysaccharides Gla, Glb	Bark	Antitumor	Fujiwara et al., 1982
Polysaccharides Gla, Glla	Bark	Anti-inflammatory	Fujiwara et al., 1984
NB-II Peptidoglycan	Bark	Immunomodulatory	Van der Nat et al., 1987, 1989
Nim-76	Seed oil	Antifertility	Jacobson, 1995

to infuse for 24 h, filtered and applied by spraying over the animals. Dried powder leaves for vegetable pests should be dried under shade and triturated (30 g–40 g/L of water standing for 24 h) followed by filtering and then spray on vegetables. Oil seeds residue (5 ml/l of water) is used as nematicide. Among the wide diversity of biological functions from the chemical constituents present in different parts of *A. indica*, anti-ulcerogenic, anti-inflammatory, anticancer, hypolipidemic and hepatoprotective activities were proved (Mossini and Kemmelmeier, 2005).

12.6 USES IN PEST AND DISEASE CONTROL

Neem is a key ingredient in Non-Pesticidal Management (NPM), providing a natural alternative to chemical pesticides. Neem seeds are ground into a powder that is soaked overnight in water and sprayed onto the crop. To be effective, it is necessary to spray at least every 10 days. Neem does not directly kill insects on the crop. It acts as a repellent, protecting the crop from damage. The insects starve and die within a few days. Neem also suppresses the hatching of pest insects from their eggs. Neem is not only much less expensive than chemical insecticides; it also has the advantage of not killing predatory insects that provide natural control of pest insects. Neem leaves can be used to protect stored grain from damage due to insect such as weevils (Kulkarni and Kumbhojkar, 1996), and neem cake can be applied to the soil. Neem cake kills pest insects in the soil while serving as an organic fertilizer high in nitrogen (Schmutterer, 1990). Neem is very effective in the treatment of scabies and is recommended for those who are sensitive to permethrin, a known insecticide which might be an irritant (Swami, 2011). Also, the scabies mite has yet to become resistant to neem, so in persistent cases neem has been shown to be very effective. There is also evidence of its effectiveness in treating infestations of head lice in human. The oil is also used in sprays against fleas for cats and dogs. Numerous studies describe the insecticidal, antifeedant, growth inhibitory, oviposition deterring, antihormonal, and antifertility activities of neem against a broad spectrum of insects (Koul et al., 1990).

12.7 NEEM OIL

Neem oil is unique, comprising non-lipid associates, commonly known as bitters, and sulfur compounds that impart peculiar odor to the oil. The phospholipid in the oil mainly consists of phosphatidylcholine (3.93%) phosphatidylethanol-amine (39.4%), cardiolipin (10.3%) and phosphatidylinositol (36.4%), oleic acid (46%), palmitic acid (19%), stearic acid (18%) and linoleic acid (14%). Large variability in terms of individual fatty acids and fatty acid composition has been observed. High oleic and low linoleic acid contents are desirable for the stability of the oil. The high-oleic acid (>50%) trees can be utilized for improving the fatty acid profile of those having comparatively lower oleic acid.

12.8 TOXICOLOGICAL PROPERTIES AND SIDE EFFECTS

India stands first in neem seed production and about 4,42,300 tons of seeds are produced annually yielding 88,400 tons of neem oil and 3,53,800 tons of neem cake. With beneficial effect sometimes it has also bad effect on living organism. With banning of broad spectrum, toxic insecticides, such as DDT, the use of neem in crop protection has been increasing considerably year after year (Raj, 2014). Despite extensive of work on pharmacological activity of neem extracts, toxicological evaluation work is limited. It is reported that leaves of neem cause toxic effects on sheep (Ali and Salih, 1982) goats and guinea pigs (Ali, 1987). A higher dose is lethal to guinea pigs. However, 200 mg/kg in the same route was found to be non-toxic to rabbits (Thompson and Anderson, 1978).

It is generally believed that medicines or pesticides of plant origin are safe and can be used without any precaution. This is untrue as this has serious side effects. Hence, medicines of plant origin should be treated with the some caution as medicines of synthetic origin. Neem oil seems to be particular concern. Its consumption, although widely practiced in different parts of Asia, is not recommended. The toxicological nature of neem is also harmful for the pregnant women. A higher dose can cause mortality. The leaves or leaf extracts also should not be consumed by people or fed to animals over a long period. There are reports of renal failure in Ghanaians who were drinking leaf teas as malarial treatment. Each preparation needs detailed toxicological evaluation before its commercial use.

12.9 IPR – PATENTS OF NEEM

Recently Singh et al. (2011) reviewed the Intellectual Property Rights (IPR) of Neem. Since the 1980s, many neem related process and products have been patented in Japan, USA and European countries. The first US patent was obtained by Terumo Corporation in 1983 for its therapeutic preparation from neem bark. In 1985, Reobert Larson from USDA obtained a patent for his preparation of neem seed extract and the Environmental Protection Agency approved this product for use in US market. In 1988 Robert Larson sold the patent on an extraction process to the US Company W. R. Grace (presently Certis). In 1990 patent was given for a method of producing neem extract that can be stored well. The abstract says: Storage stable pesticide compositions comprising neem and extracts which contain azadirachtin as the active pesticidal ingredient wherein the compositions are characterized by their non-degrading solvent systems. In 1994 patent was given for a specific method of extracting and treating active substances from neem seeds so that the resulting solution is stable enough to store. The abstract says the patent is for a process for the production of stable azadirachtin solutions comprising extracting ground neem seeds with a solvent having azadirachtin extract solution and then adding an effective amount of 34 Angstrom molecular sieves to selectively remove water from the extract to yield a storage-stable azadirachtin solution having less than 5% water by volume. In 1995, WR Grace patented neem-based bio pesticides, including Neemix, for use on food crops. Neemix suppresses insect feeding behavior and growth in more than 200 species of insects. Having gathered their patents and clearance from the *Environmental Protection Agency*, four years later, Grace commercialized its product by setting up manufacturing plant in collaboration with P. J. Margo Pvt. Ltd. in India and continued to file patents from their own research in USA and other parts of world. Aside from Grace, neem based pesticide were also marketed by another company, Agri Dyne Technologies Inc., USA, the market competition between the two companies was intense. In 1994, Grace accused Agri Dyne a non-exclusive royalty-bearing license. European Patent Office initially granted the patent to the US Department of Agriculture and multinational WR Grace in 1995. In 1992, Grace secured its rights to the formula that used the emulsion from the neem tree's seeds to make a powerful pesticide. It also began suing Indian companies for making the emulsion. But the Indian government successfully argued that the medicinal neem tree is part of traditional Indian knowledge. The backbone of the challenge was that the fungicide qualities of the neem tree and its use had been known in India for over 2,000 years. The winning challenge comes in the year 2000 after years of campaigning and legal efforts against so-called "bio-piracy".

During this period in India large number of companies also developed stabilized neem products and made them available commercially. The number of patents filed in this period were limited and geographically confined to few countries. According to Rekhi (2006), 171 products of neem have been patented till now while United States 54, Japan 59, Germany 05, EPO 05, Great Britain 02, India 36, others 10 (Austria, Belgium, Denmark, Ireland, France, Greece, etc.).

It is only these specific newly invented processes that are covered by the patents. Farmers always have and will continue to be free to use neem in any traditional way they desire. The use of neem extract, or its seeds or leaves, cannot be patented, since they have been used for thousands of years in ancient India. Its properties can only be patented if they are considerably modified. For instance, any synthetic variation of a naturally occurring product is patentable, as it does not occur in nature in that form.

12.10 CONCLUSION

To conclude the authors have made an effort to update the Ethnobotanical use of *Azadirachta indica* on global basis with particular attention to India. It may be recalled that in recent years, ethno-botanical and traditional uses of natural compounds, especially of plant origin received much needed attention as they are well tested for their efficacy and generally believed to be safe for human use. It is best classical approach in the search of new molecules for management of various diseases. Thorough screening of literature available on *Azadirachta indica* depicted the fact that it is a fairly popular remedy among the various ethnic groups, Ayurvedic, Unani, and traditional practitioners for treatment of ailments. In general, the toxicity of leaf and bark extracts and isolated limonoides is very low. However, the seed oil is toxic and hence its use in large quantity may prove hazardous (Nat Vander et al., 1991). Researchers are exploring the therapeutic potential of this plant as it has more therapeutic properties which are presently not known.

ACKNOWLEDGEMENT

T. Pullaiah is thankful to the authorities of Missouri Botanical Garden for permission to consult the Library.

KEYWORDS

- *Azadirachta indica*
- bio pesticides
- ethnobotany
- ethnoveterinary medicine
- neem

REFERENCES

Alagesaboopathi, C. (2011a). Ethnobotanical studies of useful plants of Kanjamalai Hills of Salem District of Tamil Nadu, Southern India. *Arch. Appl. Sci. Res. 3(5)*, 532–539.

Alagesaboopathi, C. (2011b). Ethnomedicinal plants used as medicine by the Kurumba tribals in Pennagaram region, Dharmapuri of Tamil Nadu, India. *Asian J. Exp. Biol. Sci. 2(1)*, 140–142.

Alawa, K. S., & Ray, S. (2012). Ethnomedicinal plants used by tribals of Dhar district, Madhya Pradesh, India. *CIBTech J. Pharmaceut. Sci. 1*, 7–15.

Ali, B. H. (1987). The toxicity of *Azadirachta indica* leaves in goats and guinea pigs. *Veterinary Human Toxicol, 29*, 16–19.

Ali, B. H., & Salih, A. M. M. (1982). Suspected *Azadirachta* toxicity in sheep (Letter). *Veterinary Record, 111*, 494

Babu, N. C., Naidu, M. T., & Venkaiah, M. (2010). Ethnomedicinal plants of Kotia hills of Vizianagaram district, Andhra Pradsh, India. *J. Phytology 2*(6), 76–82.

Bapuji, J. L., & Ratnam, S. V. (2009). Traditional uses of some medicinal plants by tribals of Gangaraju Madugula Mandal of Visakhapatnam District, Andhra Pradesh. *Ethnobotanical Leaflets 3(2)*, 388–398.

Bhargava, K. P., Gupta M. B., Gupta G. P., & Mitra, C. R. (1970). Antiinflammatory activity of saponins. *Indian J. Med. Res. 58*, 724–730.

Bhatt, D. C., & Mitaliya, K. D. (1999). Ethnomedicinal plants of Victoria Park (Reserved Forest) of Bhavanagar, Gujarat, India. *Ethnobotany 11*, 81–84.

Bhide, N. K., Mehta, D. J., & Lewis, R. A. (1958). Diuretic action of sodium nimbidinate. *Indian J. Med. Sci. 12, 141–145.*

Bhowmik, D, Chiranjib, J. Y. Tripathi, K. K., & Sampath Kumar, K. P. (2010). Herbal remedies of *Azadirachta indica* and its medicinal application. *J. Chem. Pharmaceut. Res., 2*, 62–72.

Biswas, K., Chattopadhyay, I., Banerjee, R. K., & Bandhyopadhyay, U. (2002). Biological activities and medicinal properties of Neem (*Azadirachta indica*). *Curr. Sci, 82 (11)*, 1336–1345.

Brototi, B., & Kaplay R. D. (2011). *Azadirachta indica* (Neem): its Economic utility and chances for commercial planned plantation in Nanded District, *Int. J. Pharma, 1(2)*, 100–104.

Butterworth, J. H., & Morgan, E. D. (1968). Isolation of a substance that suppresses feeding in locusts. J. Chem. Soc. Chem. Commun. 1, 23–24.

Chatterjee, A., & Pakrashi S. C. (2010). The Treatise on Indian Medicinal Plants, New Delhi: National Institute of Science Communication (CSIR) 3, 75–78.

Coats, J. R. (1994). Risks from natural versus synthetic insecticides. *Annual Review of Entomology, 39,* 489–515.

Conrick, J. (2001). Neem: The Ultimate Herb. Lotus Press, Wisconsin, USA.

Dahare, D. K., & Jain, A. (2010). Ethnobotanical studies on plant resources of Tahsil Multai, District Betul, Madhya Pradesh, India. *Ethnobotanical Leaflets 14,* 694–705.

Das, S., & Choudhury, M. D. (2012). Ethnomedicinal uses of some traditional medicinal plants found in Tripura, India. *J. Med. Plants Res. 6(35),* 4908–4914.

David, S. N. (1969). Anti-pyretic of neem oil and its constituents. *Mediscope. 12, 25–27.*

Deng, Y., Cao, M., Shi, D., Yin, Z., Jia, R., Xu, J., Wang, C., Lv, C., Liang, X., He, C., Yang, Z., & Zhao, J. (2013). Toxicological evaluation of neem (*Azadirachta indica*) oil: acute and subacute Toxicity. *Environmental Toxicology and Pharmacology, 35*(2), 240–246.

Dubey, S., & Kashyap, (2014). *Azadirachta indica*: A plant with versatile potential. *J. Pharm. Sci. 4,* 39–46.

Fujiwara, T., Sugishita, E. Y., Takeda, T., Ogihara, Y., Shimizu, M. et al. (1984). Further studies on the structure of polysaccharides from the bark of *Melia azadirachta. Chem. Pharm. Bull., 32,* 1385–1391.

Fujiwara, T., Takeda, T., Okihara, Y., Shimzu, M., Nomura, T., & Tomita, Y. (1982). Studies on the structure of polysaccharides from the bark of *Melia azadriachta. Chem. Pharm. Bull., 30,* 4025–4030.

Girish, K & Shankara Bhat, S. (2008). Neem – A Green Treasure. *Electronic J. Biol. 4(3),* 102–111.

Gnanavel, R., & Jose, F. C. (2014). Medicinal plant based antidote against snake bite by Irula tribes of Tamil Nadu, India. *World. J. Pharm. Sci. 2(9),* 1029–1033.

Govindachari, T. R., Malathi, R., Gopalakrishnan, G., Suresh, G., & Rajan, S. S. (1999). Isolation of a new tetranortriterpenoid from the uncrushed green leaves of *Azadirachta indica. Phytochemistry 52,* 1117–1119.

Hedge, N. G. (1993). Improving the productivity of Neem trees. World Neem Conference. *Indian J. Entomol., 50,* 147–150.

Henry, A. N., Hosagoudar, V. B., & Ravikumar, K. (1996). Ethno-medico-botany of the southern Western Ghats of India. In: Jain, S. K. (ed.) Ethnobiology in Human Welfare. Deep Publications, New Delhi, India, pp. 173–180.

http://www.neemfoundation.org/

Kabeeruddin, H., & Makhzanul. (2007). Mufradat. New Delhi, India: Aijaz publishing house, 400–411.

Kakai, T. K. (1984). Anti-inflammatory polysaccharide from *Melia azadirachta. Chem. Abstr., 100,* 913.

Ketkar, A. Y., & Ketkar, C. M. (1995). Medicinal uses including pharmacology in Asia. In: Schumutterer, H. (Ed.). The Neem Tree. VCH Publishers Inc., New York. pp. 518–525.

Khalid, S. A., Duddect, H., & Gonzalez-sierra, M. J. (1989). Neem seed and leaf extracts effective against malarial parasite. *J. Nat. Prod. 52,* 922–927.

Koul, O., Isman, M. B., & Ketkar, C. M. (1990). Properties and uses of Neem *Azadirachta indica. Canadian J. Bot. 68,* 1–11.

Kulkarni, D. K., & Kumbhojkar, M. S. (1996). Pest control in tribal areas of Western Maharashtra – An ethnobotanical approach. *Ethnobotany 8,* 56–59.

Kumar, K., & Goel, A. K. (1998). Lees known ethnomedicinal plants of Santhal and Paharia tribes in Santhal Paragana, Bihar, India. *Ethnobotany 10*, 66–69.

Kumar, V., & Jain, S. K. (1998). A contribution to Ethnobotany of Surguja district in Madhya Pradesh, India. *Ethnobotany 10*, 89–96.

Latum, E. V. (1985). Neem tree in agriculture, its uses in low-input pest management. Ecoscript No: 31. Free University of Amsterdam, Netherlands.

Luo X. D., Wu, S. H., Ma Y. B., & Wu, D. G. (2000). A new triterpenoid from *Azadirachta indica.*, *Fitoterapia, 71*, 668–672.

Mohanty, R. B., & Padhy, S. N. (1996). Traditional phytotherapy for diarrheal diseases in Ganjam and Phulbani districts of south Orissa, India. *Ethnobotany 8*, 60–65.

Mojumder, V. (1995). Nematoda, Nematodes. In: Schmutterer, H. (Ed.). The Neem Tree. VCH Publishers Inc., New York. pp. 129–150.

Morgan E. D. (2009). Azadirachtin, a scientific gold mine. *Bioorganic & Medicinal Chem.*, *17*, 4096–4105.

Mossini, S. A. G., & Kemmelmeier, C. (2005). A árvore Nim (*Azadirachta indica* A, Juss): Múltiplos uses. *Acta Farmaceutica Bonaerense, 24*(1), 139–148.

Murthy, S. P., & Sirsi, M. (1958). Pharmacological studies on *Melia azadirachta. Indian J. Physiol. & Pharmacol. 2, 387–396.*

Nadkarni, K. M. (1994). Indian Materia medica, Popular Prakashan, Bombay, India.

Neves, B. P., Oliveira, I. P., & Nogueira, J. C. M. (2003). Cultivo e Utilização do Nim Indiano., EMBRAPA Circular Técnica 62, Santo Antônio de Goiás, Goiás, Brazil.

Ogbuewu, I. P., Odoemenam V. U., Obikaonu H. O., Opara M. N., Emenalom O. O., Uchegbu M. C. et al., (2011). The Growing Importance of Neem (*Azadirachta indica* A. Juss) In Agriculture, Industry, Medicine and Environment: A Review. *Res. J. Med. Plant., 5*(3), 230–245.

Pande, P. C., Tiwari, L., & Pande, H. C. (2007). Ethnoveterinary plants of Uttaranchal-A review. *Indian J. Trad. Knowl., 6(3),* 444–458.

Pant, N., Garg, H. S., Madhusudanan, K. P., & Bhakuni, D. S. (1986). Sulfurous compounds from *Azadirachta indica* leaves. *Fitoterapia, 57,* 302–304.

Paritala V., Chiruvella, K. K., Thammineni, C., Ghanta, R. G., & Mohammed, A. (2015). Phytochemicals and antimicrobial potentials of mahogany family. *Revista Brasileira de Farmacognosia, 25*(1), 61–83.

Parotta, J. A. (2001). Healing plants of Peninsular India. , New York, CABI Publishing, p.p. 495–496.

Pattanaik, C., Sudhakar Reddy, C., Das, R., & Manikya Reddy, P. (2007). Traditional medicinal practices among the tribal people of Malkangiri district, Orissa, India. *Nat. Prod. Radiance 6*(5), 430–435.

Paul, C. R. (2003). Botany and ethnobotany of *Azadirachta* A. Juss. (Meliaceae) in India. In: Singh, V., & Jain, A. P. (eds.). Ethnobotany and Medicinal plants of India and Nepal. Vol. 1. Scientific Publishers, Jodhpur, India, pp. 17–19.

Pillai, N. R., & Santha Kumari, G. (1981). Hypoglycemic activity of *Melia azadirachta. Indian J. Med. Res. 74, 931–933.*

Pillai, N. R., & Santhakumari, G. (1984). Toxicity studies on nimbidin. *Planta Med., 50,* 143– 146.

Pillai, N. R., Suganthan, D., Seshadri. C., & Santhakumari, G. (1978). Anti-gastric ulcer activity of nimbidin. *Indian J Med Res. 68, 169–175.*

Pillai, N. R., & Santhakumari, G. (1981). Anti-arthritic and anti-inflammatory actions of nimbidin. *Planta Medica 43*, 59–63.

Punjani, B. L. (1998). Plants used as toothbrush by tribes of district Sabarkantha (North Gujarat). *Ethnobotany 10,* 133–135.

Ragasa, C. Y., Nacpil, Z. D., Natividad, G. M., Tada, M., Coll, J. C., & Rideout, J. A. (1997). Tetranortriterpenoids from *Azadirachta indica. Phytochemistry, 46*(3), 555–558.

Raj, A. (2014). Toxicological effect of *Azadirachta indica. Asian J. Multidisciplinary Studies 2*(9), 29–36.

Rajeswwari, R., Selvi, P., & Murugesh, S. (2016). Ethnobotanical survey of anti-diabetic medicinal plants used by the Malayali tribes in Jarugu Malai, Salem district, Tamil Nadu. *Species 17,* 40–47.

Randhawa, G (1997). Cyber India Foundation http://www.neemfoundation.org

Ragupathy, S., & Mahadevan, A. (1991). Ethnobotany of Kodaikkarai reserve forest, Tamil Nadu, India. *Ethnobotany 3,* 79–82.

Ravikumar, K., & Vijay Sankar, R. (2003). Ethnobotany of Malayali tribes in Melpattu village, Javvadhu hills of Eastern Ghats, Tiruvannamalai district, Tamil Nadu. *J. Econ. Taxon. Bot. 27,* 715–726.

Reddy, R. V., Lakshmi, N. V. N., & Venkata Raju, R. R. (1998). Ethnomedicine for ephemeral fevers and anthrax in cattle from the hills of Cuddapah district, Andhra Pradesh, India. *Ethnobotany 10,* 94–96.

Rekhi, J. S. (2006). The patent system in India. Office of DC (SSI), Ministry of Industry, New Delhi, India.

Rochanakij, S., Thebtaranonth, Y., Yenjal, C. H., & Yuthavong, Y. (1985). Nimbolide, a constituent of *Azadirachta indica* inhibits *Plasmodium falciparum* in culture. *Southeast Asian. J. Trop. Med. Public Health, 16,* 66–72.

Rojanapo, W., Suwanno, S., Somjaree, R., Glinsukon, T., & Thebtaranonth, Y. (1985). Mutagenic and antibacterial activity testing of nimbolide and nimbic acid. *J. Sci. Thailand, 11,* 117–188.

Sarkhel S (2014). Ethnobotanical survey of folklore plants used in treatment of snake bite in Paschim Medinipur district, West Bengal. *Asian Pac J Trop Biomed. 4(5),* 416–420.

Schmutterer, H. (1990). Properties and potential of natural pesticides from the neem tree, *Azadirachta indica. Annual Review of Entomology. 35,* 271–297.

Schmuttere, H. (Ed.) (1995). The Neem Tree. VCH Publishers Inc. New York.

Senthilkumar, K., Aravindhan, V., & Rajendran, A. (2013). Ethnobotanical survey of medicinal plants used by Malayali tribes in Yercaud hills of Eastern Ghats, India. *J. Natural Remedies 13*(2), 118–132.

Sharma P., Tomar L., Bachwani M &, Bansal V., (2011). Review on Neem (*Azadirechta indica*): Thousand problem one solution, *Int. Res. J. Pharmacy 2,* 97–102.

Sharma, P. V. (1996). Classical Uses of Medicinal Plants. Chaukbambha Visvabharati. Varanasi 1, India.

Sharma, V. N., & Saksena, K. P. (1959). Sodium nimbidinate. *In vitro* study of its spermicidal action. *Indian J. Med. Res. 13,* 1038.

Sharma, V. N., & Saksena, K. P. (1959). Spermicidal action of sodium nimbinate. *Indian J. Med. Res., 47,* 322–324.

Shodini. (1997). Touch-Me, Touch-me-not. Women, plants and Healing. Kali for women, New Delhi, India.

Siddiqui B., Afshan, S. F., Gulzar, T., & Hanif, M. (2004). Tetracyclic triterpenoids from the leaves of *Azadirachta indica. Phytochemistry, 65,* 2363–2367.

Silva, J. C. T., Jham, G. N., Oliveira, R. D. L., & Brown, L. (2007). Purification of the seven tetranortriterpenoids in Neem (*Azadirachta indica*) seed by counter-current chromatography sequentially followed by isocratic preparative reversed-phase high-performance liquid chromatography. *J. Chromatography A, 151*, 203–210.

Singh, O., Khanam, Z., & Ahmad, J. (2011). Neem (*Azadirachta indica*) in context of Intellectual Property Rights (IPR). *Recent Res. Sci. Tech. 3*(6), 80–84.

Soon, L. G., & Bottrell, D. G. (1994). Neem pesticides in rice: potential and limitations. Manila, IRRI – International Rice Research Institute, pp. 1–64.

Subramani, S. P., & Goraya, G. S. (2003). Some folklore medicinal plants of Kolli hills: Record of a Natti Vaidyas Sammelan. *J. Econ. Taxon. Bot. 27*, 665–678.

Subramanian, M. S., & Lakshmanan, K. K. (1993). *Azadirachta indica* Juss. stem bark as an antileprosy source. World Neem Conference (Bangalore, India). Abstract Page No: 83.

Tandon, P., & Sirohi, A. (2010). Assessment of larvicidal properties of aqueous extracts of four plants against *Culex quinquefasciatus* Larvae. *Jordan J. Biol. Sci. 3*(1), 1–6.

Tariq, A., Mussarat, S., & Adnan, M. (2015). Review on ethnomedicinal, phytochemical and pharmacological evidence of Himalayan anticancer plants. *J. Ethnopharmacol., 64(22),* 96–119.

Thakur, A., Naquivi, S. M. A., Aske, D. K., & Sainkhedia, J. (2014). Study of some ethnomedicinal plants used by Tribals of Alirajpur, Madhya Pradesh, India. *Res. J. Agric. For. Sci. 2*(4), 9–12.

Upadhyay, R., & Chauhan, S. V. S. (2003). Ethnobotanical uses of plant gums by the tribals. *J. Econ. Taxon. Bot. 27,* 601–602.

Van der Nat, J. M., Kierx, J. P. A. M., Van Dijk, H., De Silva, K. T. D., & Labadie, R. P. (1987). Immunomodulatory activity of aqueous extract of *Azadirachta indica* stem bark. *J. Ethnopharmacol., 19,* 125–131.

Van der Nat, J. M., Hart, L. A. T., Van der Sluis, W. G., Van Dijk, H., Van der Berg, A. J. J. et al., (1989). Characterization of anti complement compounds from *Azadirachta indica*. *J. Ethnopharmacol., 27,* 15–24.

Van der Nat, J. M., Van der Sluis, W. G., Hart, L. A., Van Disk, H., De Silva, K. T. D., & Labadie, R. P. (1991). Activity of guided isolation and identification of *Azadirachta indica* A. Juss. (Meliaceae) bark extract constituents, which specifically inhibit human polymorph nuclear leucocytes. *Planta Med., 57,* 65–68.

Van der Nat, J. M., Van der Sluis W. G., De Silva, K. T. D., & Labadie, R. P. (1991). Ethnopharmacolognostical survey of *Azadirachta indica* A. Juss. (Meliaceae). *J. Ethnopharmacol., 35,* 1–24.

Varma, G. S. (1976). Miracles of Neem Tree. Rasayan Pharmacy, New Delhi, India.

Vartak, V. D., & Ghate, V. (1990). Ethnobotany of neem. *Biol. Ind. 1*, 55–59.

Verma, P., Khan, A. A., & Singh, K. K. (1995). Traditional phytotherapy among the Baiga tribe of Shadol district of Madhya Pradesh, India. *Ethnobotany 7*, 69–73.

Vihari, V. (1995). Ethnobotany of cosmetics of Indo-Nepal border. *Ethnobotany 7*, 89–94.

CHAPTER 13

ETHNOGENOMICS OF SOME TRADITIONALLY USED PLANTS: AN EMERGING DISCIPLINE OF BIOLOGY

BIR BAHADUR,[1] GORTI BALA PRATYUSHA,[2] and E. CHAMUNDESWARI[1]

[1]Department of Botany, Kakatiya University, Warangal – 560009, Telangana, India, E-mail: birbahadur5april@gmail.com

[2]Department of Genetics, Shadan P. G. Institute of Biosciences for Women, Osmania University, Hyderabad – 500004, Telangana, India, E-mail: bala.pratyusha@yahoo.com

CONTENTS

ABSTRACT

Ethnobotany Genomics is a new emerging discipline which is synthesis of ethnobotany and genomics and based on the ancient knowledge of biodiversity variation among different cultures around the world and combines with modern genomic tools such as DNA barcoding has been applied to various organisms including plants in exploring the natural genetic variability and biodiversity found among various plant groups followed by high-throughput Automated Identification Technology (AIT) system. These are novel approaches that of late have revolutionized especially botanical research and technological innovations and are founded on the concept of 'assemblage' of biodiversity knowledge, traditional knowledge (TK) and scientific knowledge (SK) and employs modern genomics technology, as an important tool for identifying cryptic species, which are routinely recognized as ethnotaxa using the TK classification systems of local ethnic cultures in India and elsewhere. This paper reviews some well studied plant species belonging to *Acacia, Biophytum, Cardiospermum, Tripogon* and minor millets of Tamil Nadu, south India, India, that have been studied recently ethnogenomically and the variations among the cryptic taxa have been identified.

13.1   INTRODUCTION

The term genomics was coined by Tom Roderick to describe an approach to the study of DNA at the level of entire genomes, chromosomes, or large clusters of genes and has now emerged as a vital tool for the researchers engaged in plant biodiversity, but also deals with the inventory and management of earth's immense biodiversity. Identification at the species level is a pre-requisite for quality assurance, identifying the crude plant product and also evaluating its pharmaceutical quality (*see* Wagner et al., 2011). Genomics in recent times has become a powerful tool for the identification and authentication of biodiversity species (Pereira et al., 2008). As on today, DNA bar coding (Paul D. N. Hebert, founder of DNA barcoding technology) has emerged as an important scientific area that provides a unique forum for exchange of information in biological studies and serves as a rapid and cost-effective method for identifying biodiversity. Genomics is a discipline in genetics that applies recombinant DNA, DNA sequencing methods and bioinformatics to sequence, assemble and analyze the function and structure of genomes, i.e., the complete set of DNA within a single cell of an organism. This study draws on the ancient body of knowledge concerning the variation in biological diversity that is present in different cultures combined with modern genomic tools such as DNA barcoding or metabarcodes from next-generation sequencers (Wilson et al., 2016) as a modern molecular tool and system for species to identify closely related or newly evolved species. For their short gene sequencing from a standardized region of the genome (Hebert et al., 2003; Kress and Erickson, 2007, 2008, 2009; Newmaster et al., 2006, 2009a; Kun Luo et al., 2010) and enables one to explore the natural genetic variations present among plant species and other organisms. Ethnobotany genomics is a novel approach that has already impacted botanical/zoological research and is bound to create wave in the years to come. The application of DNA barcoding is a new and novel approach to ethnobotany and the term was ethnobotany genomics proposed by Newmaster and Ragupathy (2010). This new branch of science is founded on the concept of 'assemblage' of biodiversity knowledge, traditional knowledge (TK) and scientific knowledge (SK) and employs modern genomic technology, DNA barcoding, as an important tool for identifying cryptic species, which were already recognized as ethnotaxa using the TK classification systems of local cultures in the Velliangiri Hills of Tamil Nadu, India (Ragupathy et al., 2009). Ethnobotany genomics engages modern tools that can overcome taxonomic impediments to exploring biodiversity. Biodiversity genomics study as on today involves intensive sampling of organisms at taxonomic levels for the same genomic region and provides a link between variation in taxa, sequence evolution and genomic structure, function, and good estimate of evolutionary process. This approach helps integrate genomic thinking with natural occurring variation in ecosystems to explore biological diversity. DNA barcoding indeed is a critical technique for ethnobotany genomics research. The TK classifies three broad categories of traits: (1) morphological (plant height, seed shape, size, etc.), (2) agricultural (grain yield, drought tolerance, etc.), and (3) cultural value traits (gastronomic and medicinal) to the farmers (Newmaster et al., 2013).

The investigation of plants and their uses has been one of the most primary human concerns and has been practiced by all cultures for tens, if not hundreds, of thousands of years, though it wasn't called 'Ethnobotany' then. Ethnobotany is the scientific study of plant lore and agricultural customs of a people since ancient times. Given their extensive range of knowledge of medicinal plants, indigenous people remain the ultimate resource for retrieving this information for the purpose of application, particularly in modern medicine. Ethnobotany is a rapidly growing science and is now multidisciplinary in nature, attracting people with widely varying academic background and interests and still predominantly linked to Economic Botany, and hence pursued to determine the potential economic value of various plants as potential sources for life saving drugs that have proven important in the treatment of various serious diseases such as AIDS and cancer and improve healthcare. There is revival of ethnobotany during the last few decades and the subject has become a hot topic of research and new foci have been developed by critical research raising thereby the credibility of Traditional Knowledge (TK) in modern scientific studies which have gained credibility and have evolved considerably in recent times (Schultes, 1962; MacDonald, 2009).

Ethnobotanists describe, document and explain the complex relationships between various cultures and the utility of plants which includes how plants are used, managed and perceived across human societies around the world for food, medicines, textiles, building materials and cosmetics, etc.; within cultural deviation, rituals and religions. On the other hand ethnobotanic genomics is a novel approach that is poised to create botanical discoveries and innovations in a new era of exploratory research which is founded on the concept of assemblage of biodiversity knowledge including species variation and valorizing value to both tradition knowledge and scientific ethnobotany.

The collision, influence and to some extent convergence of eastern knowledge and advance western scientific technology has resulted in a unique synthesis of medical belief and practice, along with the development and processing of innovative and effective drugs both of Traditional Chinese Medicine and Indian Ayurveda, other systems but also the knowledge and practices that have been orally transmitted in various cultures over the centuries (Macdonald, 2009).

Genomics draws on the ancient body of knowledge concerning the variation in biological diversity that is found in different cultures combined with modern genomic tools such as DNA barcoding also explores the natural genetic variations found among higher plants and other organisms. The application of DNA barcoding in a new and novel approach to ethnobotany and Newmaster and Ragupathy (2010) proposed the term "ethnobotany genomics" 'which is founded on the concept of 'assemblage' of biodiversity knowledge, traditional knowledge (TK) and Scientific Knowledge (SK). They employed modern genomic technology, DNA barcoding, as an important tool for identifying cryptic species, which were already recognized ethnotaxa using the TK classification systems of local cultures in the Velliangiri Hills of Tamil Nadu, India. Ethnobotany genomics engages modern tools that can overcome taxonomic impediments to exploring biodiversity. Contemporary biodiversity genomics include intensive sampling of organisms at taxonomic levels for the same genomic region. This study provides a link between variation in taxa, sequence evolution and genomic structure and function, providing thereby reasonably good estimate of evolutionary process. The approach integrates genomic thinking with natural variation encountered in ecosystems to explore biological diversity. DNA barcoding is an important technique for studying ethnobotany genomics. Bar coding systems in land plants is much more challenging as the plant genome substitution rates are considerably lower than those observed in animal mitochondria, suggesting that a much greater amount of sequence data from multiple loci is needed to barcode plants using a tiered approach wherein highly variable loci are nested under a core barcoding gene. Analysis of over 10,000 *rbcL* sequences from GenBank demonstrated that this locus could serve well as the core region, with sufficient variation to discriminate among species in approximately 85% of congeneric pair-wise comparisons (Newmaster et al., 2006).

Modern Automated Identification Technology (AIT), using DNA barcoding, provides a rapid, repeatable and reliable tool for identifying ethnotaxa and variation in cryptic species. Recent development of this system for plants identification indicates the efficacy of an AIT system in saving of time, resources, and provides quick, reliable automatable identification (Newmaster et al., 2009a). DNA barcoding has been used widely to discriminate the cryptic ethno-taxa for several plants and to mention few cases: *Tripogon* (Newmaster et al., 2008b), *Cardiospermum halicacabum* (Ragupathy et al., 2008a), *Biophytum* (Newmaster et al., 2010) and Rutaceae (72 genera, 192 species) of China (Kun Luo et al., 2010). These authors proposed that a DNA bar code to be a quick and reliable tool to identify ethnotaxa, which also legitimizes the validity of (TK), rendering it stable, testable, meaningful and globally acceptable.

The impact of ethnobotany genomics has been phenomenal and now has been extended beyond biodiversity science and is being used effectively even for lower and higher animal species. Explorations of the genomic properties across the expanse of life are now possible using DNA barcoding to assemble sequence information for a standard portion of the genome from large assemblages of species. This is in contrast to the usual focus of large-scale genomics projects which acquire sequence information for all genes in single taxon. The barcode region is a genomics entity in which nucleotide composition of the plant barcode region closely mirror those in the rest of the genome. As the library of species expands it will be possible to flag species whose genomes show unusual nucleotide composition, allowing them to be probed in more detail. Shifts in sequence composition may also reveal idiosyncrasies of sequence and amino acid change. The most important contribution of barcode projects will leave/impact an important legacy; a comprehensive repository of high-quality DNA extracts that will facilitate future genomic investigations. This will also help in correct authentication of genuine ethnobotanicals. The paper reviews some best studied plants from Tamil Nadu, South India.

13.2   CASE STUDIES

13.2.1   ACACIA

The genus *Acacia* family Mimosaceae is an economically important and comprises of about 1350 species and sub-divided into three subgenera: subg. *Acacia* (c. 161 species), subg. *Aculiferum* (c. 235 species), and subg. *Phyllodineae* (c. 960 species), with many cryptic sister species showing pantropical distributions (Maslin et al., 2003; Newmaster and Ragupathy, 2009b). *Acacia* species are well adapted to arid conditions are of great utility in the forest industry; timber, fuel wood, fiber, medicine, food, handicrafts, domestic utensils, environmental amelioration, soil fertility, livestock fodder, ornamental/horticultural planning, gum, and tannins, etc. (Wickens et al., 1995; McDonald et al., 2001: Midgley and Turnbull, 2003). However, taxonomic ambiguity exits because many *Acacia* species are difficult to identify on morphological and micromorphological and other related characters (Bentham, 1842; Wardill et al., 2005). Therefore, correct identification is essential to distinguish the rare species (Byrne et al., 2001) and economically useful species (Midgley and Turnbull, 2003). Prickly *Acacia* (*A. nilotica* subsp. *indica*) is an invasive species in northern Australia and believed to have been introduced from India into Australia but the present distribution pattern appears variable throughout India. This variability perhaps includes new species, which could be due to invasive weedy nature and new food/ medicinal value. The aboriginal cultures recognize several ethnotaxa of this species, which have been commonly used as timber, tools, furniture, fodder for sheep and used for personal hygiene; the young twigs are used as toothbrush by tribals to cure infected gums while *Acacia leucophloea* is used for making liquor.

In view of the importance of *Acacia* described above, Newmaster and Ragupathy (2009b) developed a reliable identification method to differentiate *Acacia* species using only leaf samples. A classification of tree based on DNA barcoding sequence data (*rbcl*, ribulose-1,5-bisphospate carboxylase *matK* – Maturase K and trnH-*psbA*) clearly resolved 12 *Acacia* species and identified considerable intraspecific variation. In this study the authors chose sister species of *Acacia* that are difficult to distinguish. The defining characters of many acacias are found in the small flowers that appear during short periods of time during the year. Since the vegetative characters are variable and hence less reliable for species identification. Their study revealed wrongly identified herbarium specimens that only had vegetative material, which hampered identification of the species. In contrast, molecular studies utilizing DNA bar to classify previously/unidentified species/undetermined specimens due to lack of morphological characters and as a classification tool where specimens have proven difficult to classify (Wardill et al., 2005). Many of these studies employed fragments of DNA from various regions such as *ITS1* and *trnL* which are useful for sub- species identification (Fagg and Greaves, 1990; Wardill et al., 2005) and created an *ITS1* genotype library that was used as an identification tool to match exactly to the genotypes of other herbarium specimens identified by taxonomists. Although this *ITS1* genotype library is a useful tool for Acacias, this has not been found to be a suitable region for DNA barcoding because it is not possible to sequence this region for many different groups of plants (*see* Erickson et al., 2008). Their findings indeed confirmed a recent taxonomic split in the genus *Acacia*. In the classification, DNA barcode using *rbcl, matK or trnH-psbA* and enables distinguish a new taxon *Vachellia* from *Acacia* species. Variation in *rbcl* alone could be used to differentiate *Vachellia* species from that of the *Acacia* species. These results are also supported by previous phonetic analysis (*see* Newmaster and Raghupathy, 2009a). These results are also supported by other phylogenetic studies in which *Vachellia* species are placed in a separate clad (100% bootstrap support); all *Acacia* species other than *Vachellia* species be replaced in a different clad (66% boot strap support), suggesting that *Vachellia* is distantly related to *Acacia* (Luckow et al., 2003; Miller and Bayer, 2001; Seigler et al., 2006), *Vachellia* (Acacieae, *Acacia* subg. *Acacia*) and thus rightly recognized as a distinct taxon from the 'true' *Acacia* as described in the earlier taxonomic literature (*see* Wight and Arnott, 1834).

13.2.2 MYRISTICA (THE NUTMEG)

The family Myrsticaceae is comprised of about 500 species of canopy to sub-canopy woody trees and native to tropical rainforest including some recently evolved species (Janovec and Harrison, 2002). Although information on its history and cultivation have been studied earlier, ethnobotanical studies of several *Myristica* species was unavailable till recently. *Myristica fragrans* Houtt., is a species endemic to the Maluku Province of Indonesia and has long been used both as a spice was studied in the Indonesian provinces of Maluku and Central and East Java. Historical and current indigenous uses of the fruit and seed and information on medicinal aspects is well known. It is well known that *M. fragrans* is still commonly used for culinary and medicinal purposes not only in its area of origin but in south India in particular. Identification of *Myristica* species in the past was difficult because many species share similar leaf morphology, hence difficult to identify and identification of species relies mainly on small flowers that bloom for just few weeks during the flowering season. Incorrect or misidentification of Myristicaceae species is c. 25% and is considered as ecological. Plastid DNA barcode and multilocus gene marker have been used for diagnostic method of all the genes *mat k* (maturase k) appears to have evolved rapidly and shows high level of variation making it a perfect marker for Nutmeg species. Using *mat k*, the genus *Myristica* can be separated from the genus *Virola* (up to 99.25%). Other related taxa viz., *Virola* and *Compsoneura* are used in several South American countries, as wood for veneer and timber while in some Neotropical countries, Brazil and Coloumbia, exports of *Virola* sp. are comparable in economic importance to big leaf mahogany (Macedo and Anderson, 1993). Thus, the genera *Virola* and *Compsoneura* show considerable intraspecific genomic variation (Newmaster et al., 2008a).

The genus *Compsoneura,* is considered as an ideal group for testing barcoding in plants as the species present a taxonomic hinderance since the family has low levels of molecular variation compared to other closely related families of Magnoliales (*see* Sauquet et al., 2003) *Compsoneura* contains some recently described taxa (Janovec and Neill, 2002) and a new species split (Janovec and Harrison, 2002). A recent ethnobotany genomic study by Newmaster et al. (2008b) showed the utility of six coding (Universal Plastid Amplicon UPA, *rpoB, rpoc1, accD, rbcl, matK*) and one non-coding (*trnH-psbA*) chloroplast loci for barcoding in the genus *Compsoneura* using both single and multi region approaches. Five of the regions tested by them were predominantly invariant across species (UPA, *rpoB, rpoC1, accD, rbcl*). Two of the regions (*matK* and *trnH-psbA*) showed significant variation and hence considered ideal for barcoding in Myristicas. This study clearly demonstrated that a two-gene approach utilizing a moderately variable region (*matK*) and a more variable region (*trnH-psbA*) provides resolution among all the *Compsoneura* species sampled including *C. sprucei* and *C. mexicana*. A classification analyzes based on non metric multi dimensional scaling ordination concluded that the use of two regions showed a decreased range of intraspecific variation relative to the distribution of interspecific divergence with 95% of the samples being correctly identified in a sequence identification analysis (see Newsmaster et al., 2008a; Newmaster and Ragupathy, 2010). Further research by them revealed cryptic diversity within the current species concepts, which has been recognized earlier by various aboriginal cultures. The classification tree from recent DNA barcoding sequence data (*rbcl, matK* and *trnH-p sbA*) reveals considerable intraspecific variation.

13.2.3   *BIOPHYTUM*

Species of the genus *Biophytum*, family Oxalidaceae are predominantly pantropical to sub-tropical in distribution. *Biophytum* species in India are distributed mostly in the south India and show a considerable diversity. There are 17–19 species of which 4 of them are said to be endangered and many are heterostylous including *B. sensitivum* (Mayura Devi, 1964). The various aboriginals/tribes the "Malasara and Irulas" living in the Velliangiri hills of south India, reportedly used 177 different plant species for various purposes including some *Biophytum* species [Thottal sinungi in Tamil] meaning "Touch me not" (Murugesan et al., 2009; Ragupathy et al., 2008b). Taxonomists identified taxa with 97% accuracy, the various species based on traditional and scientific knowledge. DNA barcoding has validated the presence of cryptic species including 'Vishamuruchi' (meaning detoxification of the poison); *Biophytum coimbatorense,* a new species), which is used as an antidote for poisonous scorpion bite, 'Thear chedi' (translation from Tamil – Chariot umbrella; *Biophytum tamilnadense*, a new species) is used as a bait plant for fish and crab and 'Idduki poondu' (translation from Tamil – between the rock; *Biophytum velliangirianum*; yet another new species) is used for curing ear ache. A classification tree from DNA barcoding sequence data (*rbcL, matK and trnH-psbA* 41 quantitative variables) resolved 19

Biophytum species and varieties including the new species stated above. DNA barcoding clearly discriminated the cryptic new ethnotaxa *Biophytum coimbatorense* from the morphologically similar species *B. longipedunculatum* (Thottal sinungi). DNA amplification data were found to be highly specific with a clear background in the agarose gel. Although there were no differences in the *rbcL* or *atpF* sequences for these two cryptic species, the *matK* and more variable non-coding spacer regions such *astrnH-psbA* sequences were found to be consistently distinct and different. Several segregating sites in the *matK* sequences were recorded consistently among the five distant populations. Studies by Newmaster and Raghupathy (2010) have also shown that closely related species are not distinguished by several plastid regions like *rbcL* or *atpF.*

13.2.4   *TRIPOGON*

The genus *Tripogon* Roem. & Schult. family Poaceaeae comprise of about 40 species of tropics and sub-tropics (Peterson et al., 1997; Clayton et al., 2006).The diversity of this taxon has been studied by Peterson et al. (1997) while Ruguolo-Agrasar and Vega (2004) reported that Indo-Asian region constitutes the center of diversity for this genus, with 23 species of which 16 species are native to China and 21 species including eight species endemics are native to India (Newmaster and Raghupaty, 2009a). A new species of *Tripogon cope* Newm. has recently been discovered during an ethnobotanical and genomics study in the Nilgiri Biosphere Reserve, Western Ghats, India by Newmaster and Raghupaty (2009b). Taxonomic identification of seven taxa from the 40 specimens with 96% (RF) accuracy among individuals. Aboriginal informants local tribes identified eight taxa from the same 40 specimens with 98% RF among the informants. DNA barcoding revealed the new species. Classification tree from DNA barcoding sequence data (*rbcL*, *matK* and *trnH-psbA*) clearly distinguished the 12 *Tripogon* known species from *T. cope*. The DNA amplifications were found to be highly specific with a clear background in the agarose gel. The *matK* and *trnH-psbA* sequences showed several segregating sites in sequences that were found consistently among the distant populations. However, TK classification of *Tripogon* is hierarchical, employing a series of characters, that is, morphological, nutritional, medicinal and ritual. For more informatiom the reader may refer Newmaster and Ragupathy (2010).

13.2.5 CARDIOSPERMUM

The genus *Cardiospermum,* family Sapindaceae has about 14 species world-wide (Willis, 1951) The species *C. halicacabum* is a dioecious climber with dissected leaves, small white flowers and balloon like heart shaped fruits hence the common name "balloon vine"or Heart seed Vine. The traditional classification of *Cardiospermum* is complex and many aboriginal communities such as the Irulas of Tamil Nadu, India and elsewhere in south India use for various purposes. Traditionally, the Irulas tribals are mainly gatherers and depend on the forest produce for food and medicine. A study of the Irulas and Malasras tribals of Tamil Nadu, West coast of India, provides evidence for the use of an ancient traditional remedy that has been used for centuries to treat rheumatoid arthritis (Newmaster et al., 2006; Newmaster and Ragupathy, 2007; Ragupathy et al., 2008a,b). This medicinal recipe is made from a plant (*Cardiospermum halicacabum*) "*Modakathon*"(Tamil translation) *modaku* = crippling joint pain; *thon* = remedy) and is still being used by some locals of Tamil Nadu to treat rheumatoid arthritis as well as by other communities of Asia and Africa. Support for this claim was substantiated by Kumaran and Karunakaran (2006), Venkatesh Babu and Krishnakumari (2005) and Naik et al. (2014) reported that the tender, young shoots of *C. halicacabum* have been traditionally used to treat stiffness of limbs and several other ailments, in addition used as a vegetable. Their phytochemical, ethnopharmacological activity of crude ethanolic extract has anti-inflammatory, antioxidant, analgesic, antipyretic and antidiabetic activities that are yet to be commercially formulated as modern herbal medicines, even though they have been acclaimed for their therapeutic properties in the traditional systems of medicine by the Irula tribals. Unfortunately, *C. halicacabum* is said to be poisonous as it contains cyanide. Hence, it is necessary that one should search for the non-cyanide accessions among the haplotypes which can be commercialized with no side effects.

Ragupathy and Newmaster (2009) in his ethnobotanical survey of the Irulas, which is comprised of small Dravidian tribal community of Negroid race in Thanjavur district of Tamil Nadu recorded the food and medicinal uses of several ethnotaxa especially *C. halicacabum*. They noted that most tribal informants were familiar with this species as a food and could identify other ethnotaxa used for various purposes. The stem and leaves of this plant are routinely used to make soup/curry, while the seeds of some ethnotaxa are used as oral pain relievers/applied to aching joints as a paste. This clearly indicates that the Irulas classify several ethnotaxa with a specific utility as these tribals believed in the concept *'Neenda aauil'*, in Tamil meaning "living a long healthy life". Surveys from non-traditional communities indicate that around 20% people are familiar even now with some of the basic traditional knowledge concerning the utility of balloon vine. Interestingly it was also noted that over 75% people in modern urban centers still use balloon vine to treat rheumatoid arthritis, but less than 5% are familiar with the traditional knowledge of balloon vine. It is presently not known if the balloon vine remedies of urban centers are made from balloon vine and therefore recognizing a complex traditional classification system. Team of researchers at the Biodiversity Institute of Ontario Herbarium, University of Guelph in association with University of Madras and Bharatihiar University, Tamil Nadu, India have investigated the ethnobiological classification of balloon vine using a unique approach that bridges the aboriginal multi-mechanistic approach (Newmaster et al., 2006, 2007, 2010) with modern molecular tools such as DNA barcoding and biochemical analyzes to evaluate the classification at genomic and other levels (Newmaster and Ragupathy, 2009a).

13.2.6   MINOR MILLETS

Minor millets are important for local food security and genetic diversity in the arid and semi-arid regions of southern India and are widely cultivated because of their short duration and drought tolerance, constitute a group characterized by shorter, slender culms and small coarse seeded nutritionally rich cereals viz., *Eleusine coracana, Setaria italica, Panicum miliaceum, Echinochloa frumentacea, Paspalum scorbiculatum,* etc.

Maloles et al. (2011) investigated variation in minor millets of Kolli Hills, southern India in the context of traditional and scientific knowledge including ethnobotany genomics to understand and examine the biodiversity, and to detect natural variation in plastid regions *rbcL, trnH-psbA* and *matK* among 19 TK landraces, but noted that these regions were invariant among species within the context of existing classifications among 19 TK landraces, Elaborating on this study, Newmaster et al. (2013) using the nuclear regions *ITS* (*ITS, ITS1* and *ITS2*) to examine variation between 15 landraces of 174 millet samples for both TK and SK Malayali informants and recorded 96 morphological characters and even studied DNA barcoding. Quantitative multivariate classification analysis of these plants revealed that the Malayali millet classification to be hierarchical and recognized considerable fine scale variation with high consensus which was analyzed using morphometric and DNA barcoding methods that revealed existence of fewer taxa. Furthermore, they also found that the plastid region *trnH-psbA* allowed differentiation for eight out of 15 landraces.

Recently, Ragupathy et al. (2016) based on their DNA bar coding studies utilized a tiered approach using *ITS2* DNA barcode to make 100% accurate landrace (32 landraces) and six species assignments for all 160 blind samples used by indigenous farmers located in the rain fed areas of rural India and noted considerable variation in various traits and DNA sequences. They provided details of the various millet species studied by them. They also recorded precious TK of nutritional value, ecological and agricultural traits used by the local farmers for each of these traditional landraces. This work clearly demonstrates the immense potential of DNA barcoding as a reliable identification tool for evaluating and conserving genetic diversity of small millets, documenting protecting farmers rights and Traditional Knowledge.

Genomic diversity of landraces of millet was investigated by Newmaster et al. (2013) a key uncertainty that will provide a framework for a DNA barcode method that could be used for fast, sensitive, and accurate identification of millet landraces, and millet landrace conservation including biocultural diversity. They found considerable intraspecific variation among 15 landraces representing six species of small millets using nuclear regions (ITS, ITS1, and ITS2); without any variation in plastid regions (*rbcL, matK, and trnH-psbA*). An efficient ITS2 DNA barcode enabled them make 100% accurate landrace assignments for 150 blind samples representing 15 landraces revealing that genomic variation is aligned with a fine-scale classification of landraces using traditional knowledge (TK) of local farmers. Significantly the landrace classification was found to be highly correlated with traits morphological, agricultural, and cultural traits of utility associated with factors such as yield, drought tolerance, growing season, medicinal properties, and nutrition. This could provide a DNA-based model for conservation of genetic diversity and the associated bicultural diversity (TK) of millet landraces, which has sustained marginal farming communities in harsh environments for many generations. It may be recalled that Kun Luo et al. (2010) proposed that ITS2 is a promising candidate barcode for plant species identification region exhibited the highest inter-specific divergence, and that this is significantly higher than the intraspecific variation in the "DNA barcoding gap" assessment and Wilcoxon two-sample tests. The ITS2 locus had the highest identification efficiency among all tested regions which works as internal species tags.

13.2.7 *ELECTRONIC SUPPLEMENTARY MATERIAL*

Despite the extensive use of small millet landraces as an important source of nutrition for people living in semi-arid regions, they are presently marginalized and their diversity and distribution are threatened at a global scale. Local farmers have developed ancient breeding programs entrenched in TK that has sustained rural cultures for thousands of years. The convention on biological diversity seeks fair and equitable sharing of genetic resources arising from local knowledge and requires signatory nations to provide appropriate policy and legal framework to farmers' rights over plant genetic resources and associated TK. DNA barcoding employed in this study is proposed as a model for conservation of genetic diversity and an essential step towards documenting and protecting farmers' rights and TK. Study by Ragupathy et al. (2016) focuses on 32 landraces of small millets that are still used by indigenous farmers located in the rain fed areas of rural India and Nepal. Traditional knowledge of traits and utility was gathered using participatory methods and semi-structured interviews with key informants. DNA was extracted and sequenced (*rbcL*, *trnH-psbA* and *ITS2*) from 160 samples. Both multivariate analysis of traits and phylogenetic analyzes were used to assess diversity among small millet landraces. Their research revealed considerable variation in traits and DNA sequences among the 32 small millet landraces. They utilized a tiered approach using *ITS2* DNA barcode to make 100% accurate landrace (32 landraces) and species (6 species) assignments for all 160 blind samples in our study. They also recorded precious TK of nutritional value, ecological and agricultural traits used by local farmers for each of these traditional landraces. This research demonstrates the potential of DNA barcoding as a reliable identification tool and for use in evaluating and conserving genetic diversity of small millets. They suggest ways in which DNA barcodes could be used in the Protection of Plant Varieties and Farmers' Rights in India and Nepal.

13.3 DISCUSSION

From the foregoing discussion of various case studies it is obvious that DNA barcoding and ethnogenomics is a rapidly developing frontier technology and science that is gaining not only global attention but has proved beyond doubt its utility in solving problems by providing reliable results in distinguishing species and other cryptic species. Similar work in India is being carried out at some Indian universities but elsewhere abroad in advanced countries considerable work has been carried out in this direction. It may be mentioned here that ethnobotany is different from agri-genomics which deals with the study of the make up of and interaction between genes in crops and combinatorial chemistry and should not be confused with ethnobotany. Varah and Desai (2015) have recently made the first ever scientometric global analysis of the number of publications on genomics in relation to ethnobotany genomics programs for their biodiversity and observed noticeable increase of research in the subject globally, as seen by increasing number of publications originating mostly from USA, Canada, UK, and China. According to the Scopus database, the ethnobotany genomics researches are being published in 23 different subject like Agricultural and Biological Sciences and Biochemistry, Genetics, Molecular Biology, Medicine, Pharmacology, Toxicology and Pharmaceutics, etc., the latter are concerned with medicinal properties of plants. Presently, 72 countries are participating in the ethnobotany genomics research. United States is much ahead with 171, followed by China (158), Canada (78), UK (75), France (60), and India (43). In terms of average citation per article, the highest (35.21) is achieved by UK, followed by Canada (24.17), France (19.55), USA (18.42), followed by others. On the other hand, developing countries including China and India have shown an increase in their publications share in ethnobotany genomics (*see* Varah and Desai, 2015) since China and India share old history and tradition of advanced development in the field of medicine, especially Ayurvedic medicines in India are one such ancient tradition which are still in existence. However, in the case of genomics research, western countries dominate advancement in modern science and technology, largely because of better resources in terms trained manpower and sound finance. In Chapter 14, the authors have briefly discussed how the various modern Omic technologies, and System biology would likely to impact in the coming years better understanding of identification of various active bio-components, clinical testing and development of quality standards for safety and efficacy of ancient medical systems. Thus, the ethnogenomic approach is gaining considerable edge in validating indigenous drugs *vis-a-vis*, traditional knowledge much needed by the present day society for sustained and cheaper health care practices. It may be relevant to point out that Saslis-Lagoudakis et al. (2011) have shown the use of phylogeny to interpret cross cultural patterns in plant use and guide medicinal plant discovery using *Pterocarpus*. Traditional knowledge of medicinal plants has led to important discoveries that have helped combat several diseases and has improved healthcare. However, the development of quantitative measures that can assist in our quest for new medicinal plants has not greatly advanced in recent years. Phylogenetic tools have entered many scientific fields in the last few decades to provide explanatory power, but have been overlooked in ethnomedicinal studies. Several studies show that medicinal properties are not randomly distributed in plant phylogenies, suggesting that phylogeny shapes ethnobotanical use. Nevertheless, empirical studies that explicitly combine ethnobotanical and phylogenetic information are rare and needs more intensive investigations. Furthermore, ethnobotany genomics can also be used profitably to determine the distribution of rare species, their ecological requirements, including traditional ecological knowledge so that conservation strategies can be implemented. This aspect incidentally is aligned with the Convention on Biological Diversity (CBD) which was signed by over 150 nations, and thus the *"world's complex array of human-natural-technological relationships has effectively been reorganized"* as rightly stated by Ragupathy et al. (2016).

ACKNOWLEDGEMENT

We are thankful to National Centre for Biotechnology Information at the U.S. National Library of Medicine for providing some of the literature cited in this paper.

KEYWORDS

- *Acacia*
- *Biophytum*
- *Cardiospermum*
- ethnobotany
- ethnogenomics
- minor millets
- *Tripogon*

REFERENCES

Bentham, G. (1842). Notes on Mimoseae with a short synopsis of species. *J. Botany (Hooker). 4,* 323–418

Byrne, M., Tischler, G., Macdonald, B., Coates, D. J., & McComb, J. (2001). Phylogenetic relationships between two rare Acacias and their common, wide spread relatives in south-western Australia. *Conservation Genetics. 2,* 157–166.

Clayton, W. D., Harman, K. T., & Williamson, H. (2006). Grassbase—the Online World Grass Flora. Available from URL: http://www.kew.org/data/grasses-db/sppindex. htm#T (accessed 8 November 2006).

Erickson, D. L., Spouge, J., Resch, A., Weigt, L. E., & Kress, W. J. (2008). DNA barcoding in land plants: developing standards to quantify and maximize success. *Taxon 57,* 1304–1316.

Fagg, C. W., & Greaves, A. (1990). *Acacia nilotica* 1869–1988, Annotated Bibliography No. F42 (ed. Langdon K), CAB International, published in collaboration with the Oxford Forestry Institute, Wallingford, UK.

Hebert, P. D. N., Cywinska, A., Ball, S. L., & De Waard, J. R. (2003). Biological identification through DNA barcodes. *Proc. Roy. Soc. B: Biol Sci. 270,* 313–321.

Janovec, J. P., & Harrison, J. S. (2002). A morphological analysis of the *Compsoneura sprucei* complex (Myristicaceae), with a new combination for the Central American species *Compsoneura mexicana. Sys. Bot. 27,* 662–673.

Janovec, J. P., & Neill, A. K. (2002). Studies of the Myristicaceae: an overview of the *Compsoneura atopa* complex, with descriptions of new species from Columbia. *Brittonia 54,* 251–261.

Kress, W. J., & Erickson, D. L. (2007). A two-locus global DNA barcode for land plants: the coding *rbcl* gene complements the non-coding *trnH-psbA* spacer region. *PLoS. 2,* e508. doi: 10.1371/journal.pone.0000508.

Kress, W. J., & Erickson, D. L. (2008). DNA barcodes: genes, genomics, and bioinformatics. *Proc. Natl. Acad. Sci., USA. 105,* 2761–2762.

Kress W. J., & Erickson D. L. (2009). Plant DNA barcodes and a community phylogeny of a tropical forest dynamics plot in Panama. *Proc. Nat. Acad. Sci. USA, 106,* 18621–18626.

Kumaran, A., & Karunakaran, R. J. (2006). Antioxidant activities of the methanol extract of *Cardiospermum halicacabum. Pharm. Biol. 44,* 146–151.

Kun Luo, ShiLin Chen, Ke Li Chen, Jing Yuan Song, Hui Yao, Xinye Ma, Ying Jie Zhu, Xiao Hui Pang & Hua Yu. (2010). Assessment of candidate plant DNA barcodes using the Rutaceae family. *China Life Sciences, 53*(6), 701–708.

Luckow, M., Miller, J. T., Murphy, D. J., & Livshultz, T. (2003). A phylogenetic analysis of the Mimosoideae (Leguminosae) based on chloroplast DNA sequence data. In: Advances in Legume Systematics, Part 10, Higher Level Systematics (eds. Klitgaard, B. B., & Bruneau, A.), Royl Bot Gard, Kew, UK. pp. 197–220.

MacDonald, I. (2009). Current trends in Ethnobotany: Editorial. *Trop. J. Pharmaceut. Res, 8*(4), 295–296.

Macedo, D. S., & Anderson, A. B. (1993). Early ecological changes associated with logging in an Amazon Floodplain. *Biotropica. 25(2),* 151–163.

Maloles, J. R., Berg, K., Ragupathy, K., Balasubramaniam, C., Nirmala, K., Althaf, A., Vadaman Palanisamy, C, & Newmaster, S. G. (2011). The fine scale ethnotaxa classification of millets in Southern India. *J. Ethnobiol., 31*(2), 262–287.

Maslin, B. R., Miller, J. T., & Seiger, D. S. (2003). Overview of the generic status of *Acacia* (Leguminosae: Mimosoideae). *Austral. Sys. Bot. 16*, 1–18.

Mayura Devi, P. (1964). Heterostyly in *Biophytum sensitivum* DC. *J. Genet. 59*, 41–48.

McDonald, M. W., Maslin, B. R., & Butcher, P. A. (2001). Utilisation of Acacias. In: Flora of Australia, Volume 11A, Mimosaceae, *Acacia* Part 1 (eds. Orchard, A. E., & Wilson, A. J. G.), ABRS/CSIRO Publishing, Melbourne, Australia. pp. 30–40.

Midgley, S. J. &Turnbull, J. W. (2003). Domestication and use of Australian Acacias: an overview. *Austral. Sys. Bot. 16*(1), 89–102.

Miller, J. T., & Bayer, R. J. (2001). Molecular phylogenetics of *Acacia* (Fabaceae: Mimosoideae) based on the chloroplast matK coding sequence and flanking trnK intron spacer region. *Am. J. Bot. 88*(4), 697–705.

Murugesan, M., Ragupathy, S., Balasubramaniam, V., Nagarajan, N., & Newmaster, S. G. (2009). Three new species of the genus *Biophytum* DC. (Oxalidaceae-Geraniales) from Velliangiri hills in the Nilgiri Biosphere Reserve, Western Ghats, India. *J. Econ. Taxon. Bot. 33*, 10–26.

Naik, V. K. M., Babu, S. K., Latha, J., & Prabhakar, V. (2014). A review on its ethnobotany, phytochemical and pharmacological profile of *Cardiospermum halicacabum* (L). *Intl. J. Pharmaceut. Biosci. 3*(6), 392–401.

Newmaster, S. G., Fazekas, A. J., & Ragupathy, S. (2006). DNA barcoding in the land plants: evaluation of *rbcL* in a multigene tiered approach. *Canad. J. Bot. 84*, 335–341.

Newmaster, S. G., Fazekas, A. J., Steeves, R., & Janovec, J. (2008a). Testing candidate plant barcode regions in the Myristicaceae. *Mole. Ecol. Resour. 8*, 480–490.

Newmaster, S. G., Murugesan, M., Ragupathy, S., & Balasubramaniam, V. (2009a). Ethnobotany Genomics study reveals three new species from the Velliangiri Holy Hills in the Nilgiri Biosphere Reserve, Western Ghats, India. *Ethnobotany 21*, 2–24.

Newmaster, S. G., & Ragupathy, S. (2007). Exploring ethnobiological classifications for novel alternative medicine: A case study of *Cardiospermum halicacabum* L (Modakathon, Balloon Vine) as a traditional herb for treating arthritis, *Ethnobot. 19*, 1–20.

Newmaster, S. G., & Ragupathy, S. (2009a). Ethnobotany genomics – use of DNA barcoding to explore cryptic diversity in economically important plants. *Indian J. Sci. Technol. 2*(5), 1–8.

Newmaster, S. G., & Ragupathy, S. (2009b). Testing plant barcoding in a sister species complex of pantropical Acacias (Mimosoideae, Fabaceae). *Mol. Ecol. Resour. 9(Suppl. 1)*, 172–180.

Newmaster, S. G., & Ragupathy, S. (2010). Ethnobotany Genomics – Discovery and innovation in a new era of exploratory research. *Bio. Med. J. Ethnobiol. Ethnomed., 6*, 1–11.

Newmaster, S. G., Ragupathy, S., Balasubramaniam, N. C., & Ivanoff, R. F. (2007). The multi-mechanistic taxonomy of the Irulas in Tamil Nadu, South India. *J. Ethnobiol. 27*, 31–44.

Newmaster, S. G., Ragupathy, S., & Janovec, J. (2009b). A botanical renaissance: State-of-the-art DNA Barcoding Facilitates an Automated Identification Technology (AIT) System for Plants. *Intl. J. Comput. Appl. Technol. 35*(1), 51–60.

Newmaster, S. G., Velusamy, B., Murugesan, M., & Ragupathy, S. (2008b). *Tripogon cope*, a new species of *Tripogon* (Poaceae: Chloridoideae) in India with a morphometric analysis and synopsis of *Tripogon* in India. *Syst. Bot. 33*(4), 695–701.

Newmaster, S. G., Ragupathy. S., Dhivya, S., Jijo, C.J, Sathishkumar, R., & Patel, K. (2013). Genomic valorization of the fine scale classification of small millet landraces in southern India. *Genome 56*(2), 123–127.

Pereira. F, Carneiro, J., & Amorim, A. (2008). Identification of species with DNA based technology: Current Progress and Challenges. *Recent Pat DNA Gene Seq*, 2, 187 99

Peterson, P. M., Webster, R. D. &Valdes Reyna, J. (1997). Genera of new world Eragrostideae (Poaceae: Chloridoideae). *Smithsonian Contrib. Bot, 87,* 1–50.

Ragupathy, S., & Newmaster, S. G. (2009). Valorizing the 'Irulas' traditional knowledge of medicinal plants in the Kodiakarai reserve forest, India. *J. Ethnobiol. Ethnomed. 5,* 10 doi: 10.1186/1746-4269-5-10.

Ragupathy, S., Newmaster, S. G., Gopinadhan, P., & Newmaster, C. (2008a). Exploring ethnobiological classifications for novel alternative medicine: a case study of *Cardiospermum halicacabum* L. (Modakathon, Balloon Vine) as a traditional herb for treating rheumatoid arthritis. *Ethnobotany. 19,* 1–20.

Ragupathy, S., Newmaster, S. G., Murugesan, M., Velusamy, B., & Huda M. (2008b). Consensus of the 'Malasars' traditional aboriginal knowledge of medicinal plants in the Velliangiri holy hills, Indian. *J. Ethnobiol. Ethnomed. 4* (8), 1–14.

Ragupathy, S., Newmaster, S. G., Velusamy, B., & Murugesan, M. (2009). DNA barcoding discriminates a new cryptic grass species revealed in an ethnobotany study by the hill tribes of the Western Ghats in southern *India. Mol. Ecol. Res. 9*(Suppl), 1172–1180.

Ragupathy, S., Dhivya, S., Pate, K., Sritharan, A., Sambandan, K., Gartaula, H., Sathishkumar, R., Balasubramanian, K., Nirmala, C., Kumari, A. N., & Newmaster, S. G. (2016). DNA record of some traditional small millet landraces in India and Nepal. *Biotech. 6*(2), 133.

Ruguolo de Agrasar Z. E., & Vega A. S. (2004). *Tripogon nicorae* a new and synopsis of *Tripogon* (Poaceae: Chorloideae) in America. *Systematic Botany, 29*(4), 874–882.

Saslis-Lagoudakis, C. H., Klitgaard, B. B., Forest, F., Francis, L., Savolainen, V., Williamson, E. M., & Hawkins J. A. (2011). The use of phylogeny to interpret cross-cultural patterns in plant use and guide medicinal plant discovery: an example from *Pterocarpus* (Leguminosae). *PLoS One. 6*(7), e22275.

Sauquet, H., Doyle, J. A., Scharaschkin, T., Borsch, T., Hilu, K. W., Chatrou, L. W., & Le Thomas, A. (2003). Phylogenetic analysis of Magnoliales and Myristicaceae based on multiple data sets: implications for character evolution. *Bot. J. Linn. Soc. 142,* 125–186.

Schultes, R. E. (1962). The role of the ethnobotanists in the search for new medicinal plants. *Lloydia 25,* 257–266.

Seigler, D. S., Ebinger J. E. &. Miller. J. T. (2006). New combinations in the genus *Senegalia* (Fabaceae: Minosoideae) from the New World. *Phytologia 88,* 38–93.

Varah, F., & Desai, P. N. (2015). Ethnobotany genomics research: Status and future prospects. *J. Scientometric Res. 4,* 29–39.

Venkatesh Babu, K. C., & Krishnakumari, S. (2005). Anti-inflammatory and antioxidant compound, rutin in *Cardiospermum halicacabum* leaves. *Ancient Sci. Life. 25*(2), 47–49.

Wagner, H., Bauer, R., Melchart, D., Xiao, P. G., & Staudinger, A. (2011). Chromatographic Fingerprint Analysis of Herbal Medicines: Thin layer and High Performance Liquid Chromatography of Chinese Drugs. 2nd ed., Vol. I, II, Springer, Wien.

Wardill, T. J., Graham, G. C., Zalucki, M., Palmer, M, William A., Playford & Julia, P. (2005). The importance of species identity in the biocontrol process: identifying the subspecies of *Acacia nilotica* (Leguminosae: Mimosoideae) by genetic distance and the implications for biological control. *J. Biogeography. 32,* 2145–2159.

Wickens, G. E., Seif-El-Din, A. G., Sita, G., & Nahal, I. (1995). Role of *Acacia* species in the rural economy of dry Africa and the Near East. FAO Conservation Guide No. 27, Food and Agriculture Organization, Rome, Italy.

Wight, R., & Arnott, G. A. W. (1834). Prodromus Florae Peninsulae Indiae Orientalis. Parbury, Allen & Amp. Co., London.

Wilson, J. J., Kong-Wah Sing, Ping-Shin Lee, & Wee A. K. S. (2016). Application of DNA barcodes in wild life conservation in Tropical East Asia. *Conservation Biol.*, *30*(5), 982–989.

Willis, J. C. (1951). A Dictionary of the Flowering Plants and Ferns. Cambridge University Press.

CHAPTER 14

ETHNOBOTANY POST-GENOMIC HORIZONS AND MULTIDISCIPLINARY APPROACHES FOR HERBAL MEDICINE EXPLORATION: AN OVERVIEW

MANICKAM TAMIL SELVI[1] and ANKANAGARI SRINIVAS[2]

[1]Value Added Corporate Services Pvt. Ltd, Chennai – 600090, Tamil Nadu, India

[2]Department of Genetics, Osmania University, Hyderabad – 500007, Telangana, India, E-mail: srinivasmessage@gmail.com

CONTENTS

ABSTRACT

Plant based herbal medicines are getting consideration all over the world. The importance of herbal medicine and medicinal plants is realized globally as a source for drug discovery and development. Post genomic era offers the great opportunity to screen lot of bioactive components from the herbal medicinal plants with its various advancing technologies. Environmental changes occurring globally by the various activities of human beings due to which biodiversity of important medicinal plants are endangered and their decline from natural habitat is gaining significance for improvement, identification, analysis and conservation in the wake of sustainable development. The process of herbal drug manufacturing, standardization prescribes standards that impart safety, efficacy and consistency of herbal medicines. Hence, regulatory submission requirements for scientific data on quality, safety, and efficacy are indispensable to evaluate and permit to market an herbal drug on similar lines as synthetic, chemical moieties. In addition, global harmonization of herbal medicines combined with modern 'omic' technologies can facilitate enhanced support of public health and treatment of disease.

14.1   INTRODUCTION

Plant derived substances which are not having any industrial process and used to treat illness are called herbal medicines. Herbal medicines are considered one of the main forms of complimentary and alternative therapy to modern medicine. There are more than 11,000 species of herbal medicinal plants in use, out of which about 500 species are commonly used in Asian and other countries. Large number of people relies on traditional medicines to meet healthcare needs. Although modern medicine exists, but traditional medicine being popular for its history and culture is getting attention all over the world.

Herbal medicines are available commercially in developed countries Germany, USA, UK and Australia. In Germany, herbal products are sold as 'phytomedicines,' and are subjected to the same criteria for efficacy, safety and quality as are other drug products. In USA, by contrast, most herbal products in the market are sold as dietary supplements, and the same does not require pre-approval of products. Herbal drugs, annual sales have gone up considerably in the recent times in Germany and in the USA. Use of herbal drugs increased 25% annually. In international market, the total European market for homeopathic medicines was £590 m in 1991, in recent times sale of herbal medicine is gone up to £1.45bn. Over 60% of the population in Germany, and 80 90% of the population in China uses herbs regularly (Li, 2002). In Europe, North America and Asia, herbal medicines are popular and always there is a demand (8–15% annually) for efficient herbal medicines (Grünwald and Büttel, 1996).

In recent years, the importance of herbal medicinal plants is realized globally as a source for drug discovery and development. In addition to low cost along with no side effects creates them ideal targets for newer drugs. Recent advances in 'omic' technologies, hyphenated technologies, biotechnology, nanotechnology, etc. as applied to clinical cases, evaluation of phytochemical constituents and their effective formulation have enriched herbal medicine. Validating the traditional herbal medicine has set forth a new theory of reverse pharmacology. In this chapter we have focused on multidisciplinary approaches of herbal drug discovery and development in the perspective of post-genomic era.

14.2 MULTIDISCIPLINARY APPROACHES FOR HERBAL DRUG DISCOVERY AND DEVELOPMENT

14.2.1 ETHNOBOTANY

Ethnobotany, in general defined as the "science of people's interaction with plants" deals with therapeutic applications of plants (Turner, 1996). The objective of ethnobotany is to develop new chemical moieties which illuminate pharmacological activities. For example, Kava, a social beverage, is one of the examples of ethnobotanical lead with new pharmacological knowledge. Today the goal of ethnobotany is to understand relationship between human population and plants associated ecosystems. Rich biodiversity regions, called hotspots, are the focal point of ethnobotanical guided research to find on medicinal active plants. Most of the traditional medicines are coming from the wild for domestic supply, while harvesting the medicinal plants care was not taken for the habitat, thus leading to deterioration of its density The 41st World Health Assembly (1988) resolution WHA41.19 along with Chiang Mai Declaration endorsed the call for international cooperation and coordination to protect and conserve medicinal plants (Guidelines on the Conservation of Medicinal Plants, 1993). Ethnobotany is a valuable tool in exploring the plant genetic resources for utilization in industries as well (Ogunkunle and Ladejobi, 2011).

14.2.1.1   Biodiversity

Biodiversity is universality of life forms living in various ecosystems. Biodiversity of important medicinal plants are endangered due to the environmental changes occurred globally from the various activities of human beings, i.e., land degradation, shortage of freshwater. This reduction in the biodiversity may have fewer discoveries of natural chemical substances, thus leading to reduction in vital medical and biological applications, which will have high impact in medical discoveries in future. Many ethno-medicinal plants found to grow in northeast India, examples are *Swertia angustifolia* Buch. (Gentianaceae), used against fever and malaria; *Stemona tuberosa* Lour. (Stemonaceae), used for the treatment of asthma and tuberculosis and *Dillenia indica* L. (Dilleniaceae), used against diarrhea and dysentery (Yacoub et al., 2014). Currently with the advancing technology one can use smaller quantity of plant material for example cell and tissue culture and these can be subjected to various stresses like chemicals or hot or cold climates to synthesize phytochemicals.

14.2.1.2   Phytochemistry

Phytochemicals are secondary metabolites produced as by products from the primary metabolism. Secondary metabolites are alkaloids, phenolics, glucosinolates, amino acids, terpenoids, oils and waxes. Research findings showed that these compounds can be used for medical purposes. The phytochemical constituents present in the plant are bioactive molecules which have been shown to treat for various physiological conditions of people. Screening of phytochemicals serves to identify various compounds which can be used for treating various diseases (Ganesh and Vennila, 2011).

14.2.1.3   Traditional Knowledge and IPR

The herbal traditional medicine practices vary between regions to region. The traditional medicine has strong indigenous culture and supporting evidence. WHO also agrees with herbal traditional medicine and its practitioners playing an important role in treating the disease and improving the quality of life (Abbot, 2014). Traditional medicinal knowledge (TK) is generally defined as "consisting of the medicinal and remedial properties of plants in indigenous culture," including genetic resources. TK is often "defined by its general characteristics: creation through a long period of time which has been passed down from generation to generation; new knowledge is integrated to the existing, as knowledge is improved; improvement and creation of knowledge is a group effort; and ownership of indigenous knowledge varies between indigenous peoples." Having access to use herbal medicine is a human right but it has a difference when it comes in terms of property rights for indigenous community. For example, vinblastine and vincristine from the Madagascar rosy periwinkle (*Catharanthus roseus*); the drugs have earned the firm around $100 million per year, either the government or shamans of Madagascar have not received any monetary benefit for their contribution. Commodification is a breach of human rights. IPR is not suited for protection of indigenous herbal drugs, since IPR recognize only individual contribution not a community (Arihan and Ozkan, 2007).

The preservation of traditional knowledge of herbal medicine practice from the practitioners should be globally recognized and the knowledge source should be easily identifiable (Batugal et al., 2004). Globally TK has gained special significance and demand for herbal medicines. In such trend there are issues regarding patenting and sharing benefits of traditional medicine. For example patent involved in the Jeevani case shows the partnership between TK and herbal and pharmaceutical industries. Knowledge focus in establishing an equitable IP agreement needs shift from patenting alone to benefit sharing with or without patents. In addition to community-based initiatives to protect local TK with respect to indigenous medicine, two important database initiatives, designed against patenting of indigenous medical products and applications of plants were established. One in India operates internationally for the defense of national TK, and the other, established by the Science and Human Rights Program of the American Association for the Advancement of Science (AAAS), operates internationally for the defense of TK globally (Chatterjee, 2002).

14.2.2   ETHNOPHARMACOLOGY

Ethnopharmacology deals with the study of indigenous medical systems, which connects the region where treating the disease is done and its importance in medical practices and continue to provide new drugs and lead molecules for the pharmaceutical industry. The recent introduction of artemisinin as an effective antimalarial is a good example of this as the source of this compound, *Artemisia annua,* was used to treat fevers and malaria-like symptoms in traditional Chinese medicine. Currently modern medicines used for treating major disease are developed from medicinal plants, for example, reserpine, withanolide and curcumin, etc. Herbal medicines are offering vast range of structural diversity for pharmacological treatment of various diseases. Herbal medicines are having synergic effects that deactivate the side effects produced by modern medicines (Mukherjee et al., 2014). Secondary metabolites such as alkaloids, diterpenes, flavones, phenolics, and triterpenes have been isolated and some of these have been shown corresponding biological activities like antibacterial and antiparasitic (Koay et al., 2013) and reported to be anti-pyretic, antidiabetic, anti-inflammatory, anti-malarial and health maintaining properties (e.g., *Tinospora crispa*). The hydroethanolic extracts obtained from *Funtumia elastica, Raphyostylis beninensis, Butyrospermum paradoxum, Serataria caudula, Parkia biglobosa* and *Curculigo pilosa* plant species showed significant antimicrobial activities and are used to treat skin diseases in Southwest Nigeria (Adebayo-Tayo et al., 2010).

14.2.2.1 Indian Herbal Systems

Indian medicinal systems of treating the disease are about several thousand years old. Indians believes in traditional medical care and treatment, which is based on the concepts and practices of three ancient codified Indian Systems of Medicine (ISMs): Ayurveda, Unani and Siddha. Indian traditional medicine is the dictionary of herbal formulations. These traditional herbal medicines are capable of treating potential chronic diseases. For example in Indian Ayurvedic monographs depicted that reserpine (*Rauwolfia serpentina*) was used for treatment of high blood pressure (Vickers and Zollman, 1999). India has traditional knowledge for anti-malarial activities of various medicinal plants from the times of Charaka and Susruta (Gupta et al., 2015). In India traditional practice of treating disease is well practiced by Vaidyas and Hakims. Presently, there is slow deterioration of traditional knowledge of herbal plants in many countries.

India has enormous facilities to support herbal drug research which includes the Central Drug Research Institute (CDRI), Council of Scientific and Industrial Research (CSIR), Central Institute of Medicinal and Aromatic Plants (CMAP), National Botanical Research Institute (NBRI), Regional Research Laboratories (RRL), and National Chemical Laboratory (NCL) are playing pivotal role (Sen and Chakraborty, 2015). Herbal drug information, standardization of drug, quality control, and strict monitoring are the few complexities involved in the promotion of traditional Indian herbal products. In recent years several regulatory guidelines are released to overcome such problems. Quality control is required to establish the effectiveness and safety of herbal products. In order to provide improved health care facilities to mankind around the world, scientific integration of Indian traditional herbal medicine along with the international system of herbal medicine into evidence-based clinical management of diseases is essential.

14.2.2.2 International Herbal Systems

Globally traditional herbal medicines are getting significant attention in relation to health. Traditional Chinese medicines have been a major player in treating chronic respiratory syndrome. In African countries 80% of population uses traditional herbal medicines. In Kenya most of people choose combination of herbal and modern medicine, especially for treatments of HIV/AIDS, hypertension, infertility, cancer and diabetes (Nagata et al., 2011). Sri Lanka has well reserve of herbal plants compared to any other country in Asia.

14.2.3 GENETICS

The herbal medicines have long term healing effects in the body. Genes or genetic material are modified when the herbal medicines are used. Among the existing herbal medicines one third of medicines are having the property to alter genes. Research showed that 36% of these medicines interact with the enzymes responsible for altering histone which is present in the human chromosome. These changes in the enzymes promote condensation of chromatin. Almost most of the herbal medicines are found to alter the histones. Moreover, these medicines make changes in the miRNA of cells and modify genetic epigenome. The curative effect of herbal medicine differs from the chemical medicine. Each herbal plant have tens of minimum of active components each will have effect on the body. The combination of all bioactive components will effect metabolism of the human body by interacting with the immune system, organ functions, different tissue systems and nerve cells (Hsieh et al., 2011, 2013).

14.2.3.1 Bioprospecting

Looking for genetic and biochemical properties of biological material which has commercial value is called Bioprospecting (Reid et al., 1993). Mainly there are two types of search or developing new drugs from herbal plants, one is from the lead component from existing traditional use, and another one is random screening for highly diversified chemical molecule from the medicinal plants. Assays are done to detect the biological activity after random screening. Diverse collection of natural products from genetic sources is the key for success in the development. For example, notable drug cyclosporine A from a fungus (*Tolypocladium inflatum*), rapamycin from a microbe (*Streptomyces hygroscopicus*) and anti-cancer agent paclitaxel are identified as a result of large-scale screening of plant extracts (Frisvold and Day-Rubenstein, 2008). India's estimated biodiversity is 8% of the world with 3500 species of plants of medicinal value and 500 species being used in Ayurveda. Indian Government has constituted National Biodiversity Authority (NBA) to oversee utilization of plants of medicinal value for sustainability and equity (Singh, 2006).

14.2.3.2 Genetic Engineering

Genetic engineering tools are important in genetic advancements and genetic transformation. Plant genetic engineering technologies combined with the culture of hairy root and crown gall paves a way for research and development to produce active pharmaceutical ingredients (Huangi, 2012). New transformation techniques allow silencing specific genes or stacks in the same region of chromosome. Due to the advances in DNA technology, obtaining genetic background of chemically important molecules aid to identify whether genes are able to produce bioactive molecules in plants (Huangi, 2012).

14.2.3.3 Molecular Markers

According to general guidelines for methodologies on research and evaluation of traditional medicines by WHO, first step is ensuring quality, safety, and efficacy of traditional medicines and correct identification and this can be done successfully by molecular markers. Molecular markers are genetic markers to identify plants at the gene level and develop new standards in standardizing herbal quality control. Molecular markers are highly polymorphic nature, and reproducibility makes data transfer facile between laboratories, for example, RFLP, SNP and AFLP (Srivastava and Mishra, 2009).

14.2.3.4   Combinatorial Approach

A method called combinatorial biosynthesis in which genes from different organisms are put together for synthesis of a novel plant product of pharmaceutical value. This method can be used for synthesis of new drugs as well as for enhancement of efficacy of existing drugs. Present development of modern experimentation envisaged on herbal drugs helps to fight against various diseases (Biotechnology Forum, 2013).

14.2.3.5   Metabolic Engineering

In plants increased production of important metabolites is accomplished through genetic engineering. Metabolic engineering improves the metabolite composition at the cellular levels thereby eliminating the undesired effects and enhances the production of existing secondary metabolites in plants (Ramesh Kumar, 2016). Recent advances in metabolic engineering have allowed increase in the concentration of lead components and identifying non characterized pathways. For example, metabolic engineering was used to produce paclitaxel under *in vitro* conditions (Engel et al., 2008).

14.2.3.6   Conservation Genetics

Two approaches for conserving plant genetic resources are (*in situ*) farming and gene bank (*ex situ*). Over the period of time there is a genetic change that has occurred based on locations. By quantifying genetic change between these locations allows the genetic resource personnel to validate the gene bank protocols and recollect with appropriate intervals. This will help to address in case of species that can prevent further decline in vulnerable *in situ* populations. For example *Trifolium pratense* is used

in traditional medicine of India as deobstruent, antispasmodic, expecto-
rant, sedative, anti-inflammatory and antidermatosis agent (Stephanie et
al., 2014).

14.2.4   HERBAL BIOTECHNOLOGY

The outcome from synthetic compounds didn't fulfill the expectation from
high throughput screening (HTS) platforms of pharmaceutical industry.
This situation lead to discovery of plant based drug in herbal medicines.
The Drugs from Nature Targeting Inflammation (DNTI) program aimed
at identifying and characterizing natural products with anti-inflammatory activity by the combined and synergistic use of computational techniques, ethnopharmacological knowledge, phytochemical analysis and
isolation, organic synthesis, plant biotechnology, and a broad range of *in
vitro*, cell-based, and *in vivo* bioactivity models (Fakhrudin et al., 2014).
Mostly plant-derived substances are identified based on forward pharmacology approach. Reverse pharmacology uses *in vitro* screening of large
number of plant-derived compounds against pre-characterized disease-
relevant protein targets to identify the bioactive molecules (Zheng et al.,
2013).

14.2.4.1   Plant Cell and Tissue Culture

Biotechnology tools are equally important in multiplication, plant cell
culture-selective metabolite production, and tissue culture-vegetative
propagation (Siahsar et al., 2011). Plants tissue culture helps vegetative propagation in a short duration, which has the capacity to reproduce
millions of genetically and physiologically identical medicinal plants.
Another application of plant tissue culture is to isolate the drug from suspension culture, in which uniformity in production of the desired products
and reduce considerable time and manpower with increase in production
of the drug (Hussain et al., 2012). Various tissue culture studies were done
exhaustively to enhance the production of chemicals in medicinal plants
for, for example, *Swertia chirayita, Cathranthus roseus, Panax ginseng,
Stevia rebaudiana, Artemisia annua, Elettaria cardamomum, Allium chi-
nense, Camellia sinensis* (Pradhan et al., 2013; Pant and Thapa, 2012;
Nongdam and Chongtham, 2011).

14.2.4.2   Bioreactor Technology

Bioreactor is designed for mass cultivation of plant shoots and plantlet cultures which can result in high number of plant secondary products/bioactive compounds compared to the whole plants. Bioreactor has several advantages, not only in giving nutrition in a controlled environment for the growth of mass cell culture, but also helps plant cells to carry out biochemical transformation which leads to synthesis of bioactive components. Bioreactors provide better control, constant regulation of plant cell growth, simple and trouble free harvest and uptake of nutrients, thus leading to high multiplication rate and high yield of bioactive compounds. It gives great hope to the pharmaceutical industry. Furthermore, bioreactor prevents deterioration of natural herbs and helps mass cultivation. This method can be used for most of the lifesaving drugs (Popovic and Portner, 2012).

14.2.4.3   Conservation Biotechnology

Tools of biotechnology are increasingly applied to conserve plant genetic resources. Several *in vitro* techniques have been developed for storage of vegetative propagation and recalcitrant seed producing species, which includes: (i) slow growth procedures, (ii) cryopreservation (Kasagna and Karmuri, 2011). Cryopreservation helps to store germplasm at −196° C to ensure storage efficiently of plant cells for a longer time. Maintaining genetic integrity is one of essential criteria while preserving plant cells under cryopreservation. Flow cytometry technique helps to check genetic stability in *in vitro* regenerated plants as well as cryopreserved regenerated plants (e.g., *Oncidium flexuosum* Sims for its wound healing property) (Galdiano et al., 2013).

14.2.5   HERBAL GENOMICS

To study the structure and function of genomes, recombinant DNA technology, sequencing, and bioinformatics are applied and is known as genomics (Freyhult et al., 2008). Draft of human genome published in the year of 2003 lead post genomic period received more attention and progress on personalized health care with the advent of genomic profiling. In addition it made way to every single disease risk and the approach towards the healthcare management (Mendoza and Maria, 2010). Advances in high throughput sequencing technologies help us to analyze a complex chemical mixture which have more than one lead component and is termed as herbal genomics. It helps to understand the quality and identify of herbs at molecular level. Modification of chromatin structure without changing primary DNA sequences makes alteration of expression in genes called epigenetic regulation (Boonsanay et al., 2012). High advances in technology leads to integrating epigenomics, spatial organization and genomic evolution into the total cell profiles, the data can be useful if it is clearly interpreted by the researcher. Post genomic era offers the great opportunity to screen lot of bioactive components from the herbal medicinal plants with its various advancing technology especially with epigenomics and bioinformatics technologies (Falcone, 2014).

14.2.5.1 Next-Generation Sequencing and DNA Barcoding

Next-Generation Sequencing (NGS) produces millions of short DNA sequence in short time with less cost. It includes template preparation, sequencing and imaging, and data analysis in which protocol distinguishes one technology from another and by the amount of the data produced from each platform. In the NGS techniques, DNA templates are randomly read along the entire genome in a massively parallel sequencing by splitting the entire genome into small pieces followed by adapter ligation to the fragmented DNA (Zhang et al., 2011). DNA barcoding is meant to identify small, standardized gene sequences in a rapid, accurate, and cost-effective manner in the plant materials. Chloroplast/ nuclear regions are used by researchers as universal barcodes for the authentication/adulteration of phyto-medicines. Plastome sequencing/ superbarcode is an enhanced progress in DNA barcoding through Next Generation Sequencing. It identifies the presence of various plant species in herbal mixtures. By using DNA barcoding technologies, integrity and authenticity of herbal medicines can easily be identified and it also protects the health of mankind from adulteration of herbal products. Barcode databases available for plants are as follows: GenBank – USA, BOLD – Canada, Medicinal Materials DNA Barcode database – China (Balachandran et al., 2015).

14.2.5.2   Transcriptomics

Plant tissue used to generate cDNAs from mRNA populations and sequenced to generate Expressed Sequence Tags (ESTs) represent the transcriptome. The transcriptome sequences are annotated for putative function using a suite of bioinformatics approaches such as sequence searches of protein databases, motif/domain identification, biochemical pathway mapping, and sub cellular localization predictions. Transcript abundance data can also be used to provide in-depth expression profiles of individual genes on a per tissue/treatment basis. The deduced function, coupled with expression frequency, can facilitate identification of candidate genes pertinent to the pathway of interest as well as non-pathway targets (e.g., primary/intermediary metabolism) whose expression is consistent with synthesis of compounds.

DNA microarray is an orderly arranged sequence of genes on impermeable sheet like microchip and nylon. It allows a detailed analysis and investigation of gene expression. In herbal drug discovery it helps to identify the genes responsible for the lead components, identification of specific herbal components and its validation in pharmaceutical industry (Khan et al., 2009). Genetic variants involved in the gene expression in the toxicity profile and difference in effect between patients can be analyzed through Micro array technologies (Hanna, 2012).

14.2.5.3   Metabolomics

Metabolomics is the complete quantitative and qualitative analysis of all metabolites present in a specific cell, tissue, or organism (Shyur and Yang, 2008). Metabolomics is one of the key approaches of systems biology that consists of studying biochemical networks having a set of metabolites, enzymes, reactions and their interactions (Tagore and Chowdhury, 2014). The challenge is the subsequent compound identification and handling the noise or false-positive peaks in a spectrum of complex metabolic profiles. Metabolomics is an important emerging technology in field of phytomedicine, drug development, and toxicology, combined with hyphenated analytical methods (e.g., gas chromatography–mass spectrometry (GC–MS), liquid chromatography–mass spectrometry (LC–MS), and nuclear magnetic resonance (NMR) spectroscopy) and data mining tools to generate comprehensive metabolic profiles of an organism. Advancement in metabolomics provides an opportunity to segregate and measure complex matrices of plant tissues which helps to assess any modifications of the metabolic pathways.

Some metabolomics studies have characterized plant metabolites of nutritional importance and significance (Hall et al., 2008) and shown intricate relationships association between the intake and metabolism of dietary phytochemicals with human health (Manach et al., 2009). Metabolome analysis with LC–MS is a unique method for profiling of pharmacologically bioactive secondary metabolites (e.g., carotenoids, flavonoids, saponins, alkamides, alkaloids, and glycosidic derivatives) (Matsuda et al., 2009; Moco et al., 2009; Hou et al., 2010). Metabolomics aid to recognize biomarkers for drug action.

Post genomic era advances helps in targeted genome modification, specific gene expression of tissue, cell and organelle and controlled expression pathways of multigenes (Wilson and Roberts, 2014). The experimental system and tools are readily available in omics era to approach multiple molecular gene targets. Evolution of metabolic engineering with high throughput methods for gene discovery and functional analysis is a new platform for testing heterologous pathways in plants. Providing the metabolome for medicinal plant to know and have a deeper understanding of the metabolic potential of plants gives an opportunity to the investigators to identify the sites of synthesis and accumulation of structurally diverse compounds. Determining the metabolome for a plant involves the profiling of the small molecules varying in size from 100 to 2,500 atomic mass units found throughout the plant in all the tissues and organs, and in response to various growth conditions. Single analytic method is not sufficient to study all the chemical diversity within a plant, but by using LC-TOF (liquid chromatography-time of flight mass spectrometry) method one can document the non-targeted (documenting the metabolites without any bias) profile of small molecules (metnetdb.org).

14.2.5.4 Bioinformatics

Bioinformatics act as an essential tool for the identification of genes and its pathways that correlate with important bioactive secondary metabolites in medicinal plants. The International Ethnobotany Database (ebDB), NAPALERT and USDA maintain a database for medicinal plants. The majority of genomics resources for plants have come from ESTs. Transcript-level information could be valuable to molecular biology-based research relative to medicinal plants. Bioinformatics approaches can be used to create coexpression networks from transcriptome data, providing possible leads to gene discovery in related plant species. In particular, the use of comparative genomics provides basis for exchange of information among the different species (Sharma and Sarkar, 2013).

14.2.5.5 Proteomics

Proteomics used to interpret and investigate physiological conditions, mutations, changes in response to external factors, and adaptation. Proteomics is an analytical tool, two-dimensional gels coupled with tandem mass spectrometry-based isobaric tags, allows the systematic quantitative and qualitative mapping of the whole proteome in case of any diseased condition (Hussain and Huygens, 2012). Protein alterations are a potential target to understand drug's mechanism of action by using proteomics tools. Proteomics can also involve in the identification of post-translational protein modifications (Mann and Jensen, 2003; Zhang and Ge, 2011). To study the changes in the protein and protein interaction *in vivo* and *in vitro* before and after herbal medicine treatment proteomics is used to study the mechanism of action of remedies (Cho, 2007). Proteomics can also be used to understand the mechanism of action of herbal remedies for diseases including neurological diseases, cancer, and diabetes. For, for example, *Podophyllum hexandrum* in anticancer treatment (Bhattacharyya et al., 2012).

14.2.5.6 Systems Biology

Systems biology includes all omics (genomics, epigenomics, proteomics, and metabolomics, lipidomics, etc.) that explains the interactions between biological system components and to understand the biological system behavior (Oberg et al., 2011). To answer how each component assembles and form a structure in the biological system and how its interaction produce complex system behaviors, and how changes in conditions may alter these behaviors systems biology has come out as one of the important area of research in drug discovery that leads to the interpretation of genome data at systems level and understand molecular basis of disease and mechanism of drug action. It is visible that cancer disease is more complex, due to combination of multiple molecular abnormalities, which supports a novel network perspective of complex diseases. Systems biology makes possible for translating preclinical discoveries into clinical benefits for, for example, Biomarkers (Ziu et al., 2013).

Recent technologies that are available to interpret different information on complex biological phenomena at systems level have increased. The development of high-throughput technologies, data analysis methods and omics technologies have made it possible to understand biological phenomena because of the huge information obtained. Along with it bioinformatics, data mining, machine learning, made possible to understand and predict interactions and patterns of biological systems. Very recently developed database of medicinal materials and chemical compounds in northeast Asian traditional medicine (TMMC), has tried to overcome the problems faced by prior databases, such as redundancy of herbs in a system, and contains medicinal materials from Korean, Chinese and Japanese pharmacopeias. Another database is the anticancer herbs database of systems pharmacology (Cancer HSP) is a specialized database on anticancer herbs. It has anticancer activities from 492 cancer cell lines, helps to define the molecular mechanisms of anticancer activities. Shortcomings of origin, synonyms of the medicinal plant and language are solved and the researchers have made many studies based on databases (Lee, 2015).

14.2.6 HERBAL NANOTECHNOLOGY

Nanotechnology in herbal medicine is playing a lead role in drug delivery. The reach of plant molecules at the affected parts in human body is poor due to low permeability, low solubility and low bioavailability, and these properties are masked by using them with encapsulated nanomaterials for safe herbal drug delivery. Nanotechnology helps to develop and commercialize macromolecules to deliver the bioactive molecule at intracellular level precisely. Herbal medicines incorporated into novel drug delivery system, such as nanoparticles, microemulsions, matrix systems, solid dispersions, liposomes, solid lipid nanoparticles are delivered carefully at the site of action without side effect and enhance the bioavailability (Sachan and Gupta, 2013). Nanotechnology is used to reduce the toxicity and side effects in pharmaceutical medicine. Herbal nanocarriers help to treat the dangerous diseases like cancer. The nanoparticles can enhance the therapeutic index and pharmacokinetics of herbal drugs, for example, Liposomes in cancer drug delivery. Nanoparticles introduce the gene and activate it accurately and manage with no side effect after its use (Pandey and Pandey, 2013).

14.3 ANALYTICAL HERBAL MEDICINE

14.3.1 HERBAL TOXICOLOGY

Medicinal plants are widely used for treating illness and assumed to be safe, but these medicines are potentially toxic due to misidentification of the plants or incorrect preparation and administration by inadequately trained personnel (Nasri and Shirzad, 2013). Plants produce primary and secondary metabolites that have various toxic elements. Intake of such plants will give negative effects, thus, it is imperative to study the elemental contents of medicinal plants to highlight safety and efficacy of traditional herbal medicines and it is important to have herbal medicines with no side effects. In practice, herbal plants can be identified from a toxic/safety point of view. Herbal plants have pharmaceutical concentrations of poisonous constituents which should not be taken internally, for example, *Arnica* spp., *Atropa belladonna*, *Aconitum* spp. and *Digitalis* spp. Herbs with potential actions should be used only under appropriate conditions. There is a distinct group of herbs which has specific kind of toxicity, for example, hepatotoxicity of pyrrolizidine-alkaloid, Comfrey, *Dryopteris, Viscum,* and Corynanthe (Nasri and Shirzad, 2013). In Tanzania 25% of the corneal ulcer and in Nigeria and Malawi 26% of the childhood blindness were associated with the use of traditional eye medicine as per the studies conducted (Sushma et al., 2011).

14.3.2 HYPHENATED TECHNOLOGIES

Harmonized standards of toxicity testing methods for herbal medicine toxicological characterization was done globally. Next generation sequencing and computer-based modeling and simulation tools are used to predict the potential toxicity of herbal medicine which may arise from herbs administered alone or concomitantly with other herbs and/or drugs. To ensure chemical uniformity and detect chemical adulterants in herbal products, hyphenated technologies are used (Ifeoma and Oluwakanyinsola, 2013). Hyphenated technology is the combination of separation technique along with spectroscopic detection technique. Using various hyphenated technologies which includes for example, GC-MS, LC-MS, LC-FTIR, LC-NMR, CE-MS, etc. helps to isolate from crude extracts or fraction from various natural sources and on-line detection of natural products, using LC as the separation tool. Hyphenated techniques are used for the qualitative and quantitative determination of compounds in natural product extracts. Since these techniques are coupled with the separation and detection techniques, it helps to solve complex structure of natural products. The physical connection of HPLC and MS or NMR has increased the capability of solving structural problems of complex natural products. Considering trace analysis is vital, LC-MS has been considered one of the potential tools (Patel et al., 2010).

14.4 HERBAL PRACTICES AND QUALITY STANDARDS

14.4.1 HERBAL PRACTICES

The process of herbal drug manufacturing, standardization prescribes a set of standards that imparts safety, efficacy and consistency and it is meant for quality control (Kunle et al., 2012). Standardization includes all quality measures taken at the time of manufacturing process (Shinde et al., 2009). Quality control ensures safe, consistent and predictable performance and is dependent on before and after the harvesting processes. This covers Good Agricultural Practices (GAPs) and Good Manufacturing Practices (GMPs). The World Health Organization (WHO) has encouraged the development of national standards and guidelines to evaluate the quality of traditional drugs. It has also emphasized on the need to develop national pharmacopeia and monographs of medicinal plants as well as the protection of biodiversity (WHO, 2005; Shinde, et al., 2009).

14.4.2 QUALITY STANDARDS

The current development of herbal medicine is focused on the manufacture process, quality control standards, material basis and clinical research. Herbal medicine should be safe, consistent and predictable performance. Quality control should be followed during the manufacturing process, which is based on collection and extraction processes (Chege et al., 2015). Quality control is a foundation step for the manufacturing herbal product and which should suit for biological and clinical studies. A typical drug preparation includes mixture of many plants and having vast chemical component that function synergistically to get a therapeutic effect. A combination of analytical finger printing and quality control is useful to monitor consistency in each batch and to assure authenticity and quality through multi-component assay that has been developed and applied for several Chinese herbal medicine preparations, for example, *Salvia miltiorrhiza* (Guo et al., 2015).

Good Clinical Practice (GCP) provides assurance that a study's results are credible and accurate. It includes conduct, performance, monitoring, auditing, recording, analysis and reporting and that the rights and confidentiality of the study subjects are protected. It is necessary to have a system management of safety evaluation, for toxicology research of herbal drugs based on standard of Good Laboratory Practices (GLP) (Gao et al., 2012). GLP inspections are to review the non-clinical safety, toxicological and pharmacological studies proposed in human applications for herbal marketing authorizations and various post-authorization applications submitted to any regulatory agencies (EMA, 2015). GLP ensures the quality and reliable data by complying qualified Study Director for each study, a quality assurance unit, adequate test system care facilities, characterized test samples, equipment that has been calibrated to perform its function and procedures for the activities. The GLP compliances need to have documentation of all laboratory activities. It should include the result of original observations and activities of a non-clinical laboratory study, since it is essential for reconstruction and evaluation of specified study. All the study related information should be archived properly to retrieve raw data, documents, protocols/plans, and specimens generated as a result of a non-clinical laboratory study.

14.5 HERBAL TRANSLATIONAL RESEARCH

Herbal medicines are alternatives to modern medicine for treating complex disease. To validate the effects of herbal medicine, recent technology is used. To transfer the knowledge from laboratory to clinical setup the barrier between clinical and basic medical sciences is overcome with Translational Research. For bronchial asthma, herbal formulation (UNIM-352) is used. Clinical and experimental studies were conducted to validate their observed effects. The results indicate that this poly-herbal agent could be used as an alternative/adjunct in the treatment of bronchial asthma (Gulati, 2015). Ethnopharmacology is contributing the evidence-based data. Traditional medicines are taken only when the availability of experimental data and clinical data are established. The standard studies are ensuring the safety and efficacy of medicine. Clinical studies should focus on effectiveness by including various populations in the usual healthcare setting (Leonti and Casu, 2013).

14.5.1   REVERSE PHARMACOLOGY

Ethnopharmacology not only applies forward-translational research but also back-translational research, starting with traditional treatment observations of patients undergoing trial and the findings into para-clinical trials and laboratory studies. This strategy, referred to as "reverse pharmacology", has the advantage that it is more efficient and much faster than with traditional approach. Reverse pharmacology comprises three domains. The first one is the experiential phase that includes documentation of clinical observations/formulation effect from traditional medicine or developed drug which has all the details of drug formulation. Second, is the exploratory study to evaluate target activity by para-clinical study using relevant *in vitro* and *in vivo* models. Third phase includes experimental studies, to identify and validate reverse pharmacological correlates of safety and efficacy of drug candidates. Finally, reverse pharmacology can improve efficiency with minimal toxicity (Surh, 2011). For example, the decoction of *Argemone mexicana,* used as an antimalarial traditional medicine in Mali used the method of reverse pharmacology.

14.5.2   COMPARATIVE EFFECTIVENESS RESEARCH

Comparative effectiveness research (CER), a current trend in clinical medicine research supports the real world evidence based criteria. CER aims to provide evidence from the real life that helps clinicians and patients to select the treatment options which is best suited for individual preference and needs. It uses a combination of research technologies i.e., omics technologies. It is useful in identifying, connecting and interdependence of each component with various level of organization (Witt et al., 2015).

14.5.3   CLINICAL TRIALS

Herbal medicine consumption is 80% around the world, according to statistics given by WHO. There is a challenge in the safety and efficacy of herbal drug and there is no scientific evidence to support the issue. It is known that most of the population is consuming herbal drugs, thus it is the need of the hour to prove the efficacy of herbal drugs through clinical trials and is advised to use single and consistent batches of formulations (FDA, 2004). This issue can be solved by using modern technologies in clinical trials. Several clinical trials were performed, but there is a wide gap to comply international medicine policy. In general clinical trial should be carefully designed protocol and prove the safety and efficacy of the drug when it is used first in human. In herbal clinical trials inclusion criteria includes both modern diagnosis and traditional diagnosis, thus treatment is given by using traditional medicine and outcomes are analyzed by using modern and traditional systems (Leung, 2004; Jonas and Linde, 2002). Not having standard methods in preclinical trials and too many clinical trials on herbal products is poor quality of clinical trials. Regulatory authorities need to assess the purity of medicinal plant substances used and contraindication of the substances with other medicine (Jäger, 2015). In India DCGI released guidelines for safety and efficacy of herbal drugs in the year of 1993.

Herbal clinical trials have several challenges such as quality, finance and ethics. Quality and consistency of the traditional herbal products can be made possible through advanced scientific methodologies and carefully planned clinical trials. Safety and efficacy of clinical trials start with randomized clinical trials. Current herbal medicine clinical studies outcome show deficiency in the quality trial reports such as design, execution and analysis and the information of drug. Though traditional use does not ensure the safety and effectiveness of herbal medicines but it is useful guide for identification of new pharmacologically active substances in plants. A reverse pharmacology/toxicology or "bedside-to-bench" approach starting with a rigorous collection of clinical data in field surveys, as suggested by Graz (2013), may be a fruitful strategy to improve knowledge on the safety of traditionally used herbal medicines (Moreira et al., 2014).

China, Japan, and Germany have their national policy and laws on regulations of traditional herbal medicines (Parveen et al., 2015). To prove the efficacy of clinical trials, it is advised to use single and consistent batches of formulations (FDA, 2004). World Health Organization (WHO) released guidelines and regulatory requirements needed to support clinical trials of herbal products (WHO, 2005). In herbal clinical trials, inclusion criteria can be based either on modern medicine or herbal medicine diagnosis to understand the nature of the disease (Leung, 2004). Therefore, it becomes difficult to define inclusion and exclusion criteria and hence Jonas and Linde have devised a "double classification method" where subjects are primarily diagnosed using modern diagnostic criteria and then are classified according to the traditional system. Treatments are given according to traditional classification and outcomes are evaluated by criteria for both the systems (Jonas and Linde, 2002). In randomized clinical trials (RCT), blinding is a gold standard that eliminates bias and isolates placebo effects. Treatment allotment is not known by the investigator and the subject.

Consistency in high quality, effectiveness of treatment, safety and patient affordability are the major player in the drug development. The principles underlying in translational research is the standard bottom up bench to bed side. Clinical studies should be conducted in parallel. A well-defined methodology for standardized assessment of the quality, efficacy, and safety is must before starting of any basic project (Yang et al., 2014). For topical treatment of external genital and perianal warts Veregen (sinecatechins) is the first FDA approved herbal drug in the year of 2006 and the second is Fulyzaq (crofelemer) for symptomatic relief of noninfectious diarrhea in patients with HIV/AIDS on antiretroviral therapy in 2012. It shows that herbal drugs can be developed to comply the FDA standards of quality, safety, and efficacy (Lee, 2015).

14.5.4 REGULATORY AGENCIES

Around the world various indigenous herbal medicines have evolved from diverse ethnic, cultural and geographical backgrounds. Ensuring safety and efficacy of these products is not well standardized and the regulatory requirements of these products, varying from country to country, present an important challenge. Consumer interest in the herbal products globally leads pharmaceutical industry to focus on herbal medicines for economical reason. Most of the countries are allowing legalized trade of herbal products. But complexities in quality, safety and efficacy data, differences in the status of ingredients and excipients are delaying growth of herbal drug industry (EMA, 2004). In Europe, for the marketing approval is based on Traditional medicinal use provisions accepted on the basis of continuous use for medical treatment at least 10 years with in Europe and safety and efficacy data from the literature or developed from own treatment.

14.5.4.1 India

In India, herbal drugs are regulated under the Drug and Cosmetic Act (D and C) 1940. Department of AYUSH (Ayurveda, Unani, Siddha medicine) is the regulatory authority in India. According to AYUSH norms it is mandatory for any herbal drug manufacturing company should get the license (Malik, 2013). The D and C Act extends the control over licensing, formulation composition, manufacture, labeling, packing, quality, and export under the category of ASU drugs and Patent. Schedule "T" meant for Good Manufacturing Practice (GMP) to be followed. The quality standards of the medicines are found in official pharmacopeias and formularies. Ayurvedic medicines are in increasing demand from India and (Chaudhary and Singh, 2011) several drugs from ISMs have undergone clinical trials to verify and prove their efficacy.

Ayurvedic products have been successfully evaluated in clinical trials for the treatment of bronchial asthma, rheumatoid arthritis, ischemic heart disease, and cancer, among other illnesses (Patwardhan et al., 2005; Gupta et al., 1998, 2000; Chopra et al., 2000; Kumar et al., 1999). To establish the effects of drug and to compare the potency of traditional medicines with allopathic medicines clinical trials are done in India. In Indian regulations, the major class of drugs includes Ayurveda, Siddha, or Unani (ASU) drugs. Classical ASU drugs, issue of license to manufacture are based on citation in authoritative books and published literature, unless the drug is meant for a new indication when proof of effectiveness is required. Safety and efficacy studies are not required for marketing approval, as per the Drugs and Cosmetics Act of 1940 (DCA). Patent or proprietary medicine makes use of ingredients from authoritative texts but with intellectual intervention and invention to manufacture products vary from the existing classical medicine. For this category issue of a license to manufacture requires proof of effectiveness, based on the pilot study protocol relevant to ASU drugs. In India, ASU drugs have been under the scrutiny of Department of AYUSH. In 2015 regulatory requirements for phytopharmaceuticals are under the purview of the Central Drugs Standards Control Organization (CDSCO).

Regulatory submission requirements for scientific data on quality, safety, and efficacy are essential to evaluate and permit marketing for an herbal drug on similar lines to synthetic, chemical moieties. In Schedule Y, the newly added Appendix I-B describes data to be submitted along with the application to conduct clinical trial or import or manufacture of a phyto-pharmaceutical drug in the country. The regulatory requirements for NDA for the phytopharmaceutical drug include standard requirements for a new drug-safety and pharmacological information, human studies, and confir-matory clinical trials. For phytopharmaceutical drug, there is a lot of stress on available information on the plant, formulation and route of administra-tion, dosages, and therapeutic class for which it is indicated and the claims to be made for the phytopharmaceutical, and supportive information from published literature on safety and efficacy and human or clinical pharmacol-ogy information and data generated on manufacturing process and quality control.

The new regulation for phytopharmaceutical is in line with regulations in USA, China, and other countries involving scientific evaluation and data generation. In India, USA and China regulation is expected to promote inno-vations and development of new drugs from botanicals in a scientific way and would help in the acceptance of the use of herbal products by modern medical profession. A report of a global survey on national policy on tradi-tional medicine and regulation of herbal medicines indicated that most of the countries including China, Japan, and Germany have their national policy and laws on regulations of traditional herbal medicines and efforts should be made for the integration of traditional medicine into national healthcare systems.

14.5.4.2   Malaysia

In Malaysia herbal products need to be registered with the Malaysian Registrar of Business. Herbal drugs are classified into Traditional products and Health supplements and labeling is mandatory (Shak and Mohamed, 2011).

14.5.4.3   Australia

Herbal products are categorized under complementary medicine by Therapeutic Goods Administration (TGA) regulatory agency of Australia.

Herbal medicine has to be registered before marketing in case of high risk whereas the low risk medicines are listed under complementary medicine and should support for the claim (TGA, 2014).

14.5.4.4   United States of America

Any herbal drug should be marketed under Approved New Drug Application as per the regulatory norms of USFDA (NDA), Dietary Supplement Health and Education Act of 1994, FDA, manufacturer has to ensure safety of the herbal product, though pre market approval is not necessary and labeling as per the FDA compliance and the drugs should be manufactured as per the present GMP for dietary supplements (FDA 2004, 2014). FDA Botanical Drug Development Guidance recommends that IND should have data to support for the drug is safe if it is given to the humans. In addition pharmacology/toxicology studies, to prove/ensure safety and efficacy of the drug by product documentation which has studies on animals and the raw materials used, i.e., evidence to support quality control and evidence to ensure therapeutic consistency are needed.

14.5.4.5   Canada

In Canada from 2004, before initiation of any activity related to the herbal drugs, manufacture or marketer has to register with Health Canada. The process involves registration of the manufacturing site(s) along with the products. Complete data on product composition, standardization, stability, microbial and chemical contaminant testing methods and tolerance limits, safety and efficacy along with ingredient characterization, quantification by assay and comply with GMP norms (Health Canada, 2013).

14.5.4.6   European Union

The European Medicine Agency (EMA) has two types of registration of herbal medicinal products: (1) A full marketing authorization by submission of a dossier includes the information on quality, safety and efficacy of the medicinal products including the physicochemical, biological or microbial tests and pharmacological, toxicological and clinical trials data; under directive 2001/83/EC. (2) Traditional herbal medicine which has, long traditional use and not having scientific evidence can submit a simplified procedure under directive 2004/24/EC. The product should comply with the quality standards in European Pharmacopoeia. The evidence should support that the medicine was used for at least 30 years out of which 15 years the medicine should be practiced in the Europe.

14.5.4.7 *China*

Herbal medicines in China are normally considered as medicinal products with special requirements for marketing, for example a quality dossier, safety and efficacy evaluation, and special labeling. New drugs have to be examined and approved according to the Drug Administration Law. After approval, a New Drug certificate is granted an approval number. The factory is then permitted to put the product on the market. This procedure reflects the respect in which traditional experiences are held, while modern scientific and technical knowledge is used in appraising the therapeutic effects and the quality of the modified traditional medicines, and contributes administratively to the exploitation of traditional Chinese medicine (Wang, 1991).

14.6 CONCLUSIONS

Medicinal plants are serving as rich resources of traditional medicine and curing the diseases for long. Currently allopathic medicines have become very costly thus leading to the development of herbal drugs and being considered equal to modern therapy in developing countries. Therefore, it is a crucial time to develop plant-based drugs involving ethanobotanist, researchers from diverse fields and pharmaceutical companies. Post genomic era which offers latest technologies could help us to derive the chemicals/plant based medicine with the traditional knowledge of our very ancient system of medicinal practices as well the plant based medicinal script which was written in ancient times. In terms of development of herbal drug, when compared to modern medicine, it has high success rate and enhance the drug development in near future in order to have a safe drug to boost national health.

KEYWORDS

- **expressed sequence tags**
- **gas chromatography–mass spectrometry**
- **good agricultural practices**
- **good clinical practice**
- **good laboratory practices**
- **good manufacturing practices**
- **liquid chromatography–mass spectrometry**
- **nuclear magnetic resonance**

REFERENCES

Abbott, R. (2014). Documenting traditional medical knowledge. World Intellectual Property Organization.

Adebayo-Tayo, B. C., Adegoke, A. A., Okoh, A. I., & Ajibesin, K. K. (2010). Rationalizing some medicinal plants used in treatment of skin diseases. *African J. Microbiol. Res. 4*(10), 958–963.

Arihan, O., & Ozkan, A. M. G. (2007). Traditional medicine and intellectual property rights. *Ankara Ecz. Fak. Derg. 36*(2), 135–151.

Balachandran, K. R. S., Mohanasundaram, S., & Ramalingam, S. (2015). DNA barcoding: a genomic-based tool for authentication of phytomedicinals and its products. *Dove Press 5*, 77–84.

Batugal, P. A., Kanniah, J., Lee, Sy. & Oliver, J. T. (2004). Medicinal plants research in Asia, Vol. 1. The framework and project work plans. International Plant Genetic Resources Institute.

Bhattacharyya. D., Sinha. R., Ghanta. S., Chakraborty. A., Hazra. S., Chattopadhyay. S. (2012). Proteins differentially expressed in elicited cell suspension culture of *Podophyllum hexandrum* with enhanced Podophyllotoxin content. *BMC Proteome Sci.*, *10*, 34.

Biotechnology Forum (2013): http://www.biotechnologyforums.com/thread-2575.html.

Boonsanay, V., Kim, J., Braun, T., & Zhou, Y. (2012). The emerging role of epigenetic modifiers linking cellular metabolism and gene activity in cardiac progenitor cells. *Trends in Cardiovascular Medicine 22*, 77–81.

Chatterjee, A. (2002). Traditional knowledge herbal medicine and intellectual property: A debate over rights. http://www.smpborissa.org.in/pub.html.

Chaudhary, A., & Singh, N. (2011). Contribution of World Health Organization in the global acceptance of Ayurveda. *J. Ayurveda Integr. Med. 2*(4), 179–186.

Chege, I. N., Okalebo, F. A., Guantai, A. N., Karanja, S., & Derese, S. (2015). Herbal product processing practices of traditional medicine – practitioners in Kenya – Key informant interviews. *J. Health, Medicine and Nursing, 16,* 11–23.

Cho, W. C. S. (2007). Application of proteomics in Chinese medicine research. *American J. Chinese Med. 35*(6), 911–922.

Chopra, A., Lavin, P., Patwardhan, B., & Chitre, D. (2000). Randomized double blind trial of an ayurvedic plant derived formulation for treatment of rheumatoid arthritis. *J Rheumatol. 27* (6), 1365–1372.

Engels, B., Dahm, P., & Jennewein, S. (2008). Metabolic engineering of taxadiene biosynthesis in yeast as a first step towards Taxol (Paclitaxel) production. *Metab Eng. 10*, 201–206.

European Medicines Agency (2004). Science, Medicines, Health, European Union Herbal monograph.

European Medicines Agency (2015). Science, Medicine, Health. Good Laboratory Practice Compliance.

Fakhrudin, N., Waltenberger, B., Cabaravdic, M., Atanasov, A. G., Malainer, C., Schachner, D., et al., (2014). Identification of plumericin as a potent new inhibitor of the NF-κB pathway with anti-inflammatory activity in vitro and in vivo. *Br. J. Pharmacol. 171*, 1676–1686.

Falcone, E. (2014). Looking beyond the post-genomic era. *Genome Biol. 14*, 313.

Food and Drug Administration (2004). Department of Health and Human Services, Centre for Drug Evaluation and Research (CDER): Guidance for Industry, Botanical Drug Products Online Resource.

Food and Drug Administration (2014). Department of Health and Human Services, Dietary Supplements Guidance Documents and Regulatory Information Online Resource.

Freyhult, E., Edvardsson, S., Tamas, I., Moulton, V., & Poole Fisher, A. M. (2008). A program for the detection of H/ACA snoRNAs using MFE secondary structure prediction and comparative genomics—assessment and update. *BMC Research Notes. 1*, 49.

Frisvold, G., & Day-Rubenstein, K. (2008). Bioprospecting and Biodiversity Conservation: What happens when discoveries are made? *Arizona Law Review 50*, 545.

Galdiano Jr. R. F, Lemos E G., & Vendrame, W. A. (2013). Cryopreservation, early seedling development, and genetic stability of *Oncidium flexuosum* Sims. *Plant Cell, Tissue and Organ Cult., 114*(1), 139–148.

Ganesh, S., & Vennila, J. J. (2011). Phytochemical analysis of *Acanthus ilicifolius* and *Avicennia officinalis* by GC-MS Res. *J. Phytochem. 4*, 109–111.

Gao, Y., Ma, Z., & Zhang, B. (2012). Significance of re-evaluation and development of Chinese herbal drugs. *Zhongguo Zhong Yao Za Zhi. 37*(1), 1–4.

Graz, B. (2013). What is "clinical data"? Why and how can they be collected during field surveys on medicinal plants? *J. Ethnopharmacol. 150(2)*, 775–779.

Grünwald, J., & Büttel. K. (1996). The European phytotherapeutics market. *Drugs Made In Germany 39*, 6–11.

Guidelines on the Conservation of Medicinal Plants (1993). The World Health Organization (WHO) IUCN -The World Conservation Union WWF – World Wide Fund for Nature.

Gulati, K. (2015). Translational research and herbal drug development: An experience with bronchial asthma. Global Summit on Herbals & Natural Remedies. October 26–27, 2015, Chicago, USA.

Guo, D. A., Wu, W. Y., Ye, M., Xuan Liu, X., & Cordel, G. A. (2015). A holistic approach to the quality control of traditional Chinese medicines. *Science 347 (6219 Suppl)*, S27–S50.

Gupta, I, Gupta, V, Parihar A, et al. (1998). Effects of *Boswellia serrata* gum resin in patients with bronchial asthma: results of a double-blind, placebo-controlled, 6-week clinical study. *Eur J Med Res. 3*(11), 511–514.

Gupta, S., Parvez, N., Bhandari, A., & Sharma, P. K. (2015). Microspore based on herbal activities: the less explored ways of disease treatment. *Egypt Pharmaceut J. 14*, 148–157.

Hall, R. D., Brouwer, I. D., & Fitzgerald, M. A. (2008). Plant metabolomics and its potential application for human nutrition. *Physiol. Plant. 132*, 162–175.

Hanna, N. W. (2012). Pharmacogenomics: The significance of genetics in the metabolism of natural medicines. *J. Biomaterials and Nanobiotechnol. 3*, 452–461.

Health Canada (2013). Ottawa: Natural Health Products – Drugs and Health Products.

Hou, C. C., Chen, C. H., Yang, N. S., Chen, Y. P., Lo, C. P., Wang, S. Y., Tien, Y. J., Tsai, P. W., & Shyur, L. F. (2010). Comparative metabolomics approach coupled with cell- and gene-based assays for species classification and anti-inflammatory bioactivity validation of Echinacea plants. *J. Nutr. Biochem. 21*, 1045–1105.

Hsieh, H. Y., Chiu, P. H., & Wang, S. C. (2011). Epigenetics in traditional Chinese pharmacy: a bioinformatic study at pharmacopeia scale. *Evid. Based Complement. Alternat. Med.* 816714.

Huangi, Lu-Qi. (2012). Molecular Pharmacognosy. Scientific and Technical Publishers. Shangai.

Hussain, M. A., & Huygens, F. (2012). Proteomic and bioinformatics tools to understand virulence mechanisms in *Staphylococcus aureus. Current Proteomics 9*, 2–8.

Ifeoma & Oluwakanyinsola. (2013). Screening of herbal medicines for potential toxicities. InTech Science, Technology and Medicine Open Access Publisher.

Jäger, A. K. (2015). Medicinal plant research: A reflection on translational tasks. John Willey & Sons Ltd, UK.

Jonas, W. B., & Linde, K. (2002). Conducting and evaluation clinical research on complementary and alternative medicines. In: Gallin, J. J. (ed.). Principle and Practice of clinical research. San Diego (CA), Academic Press, pp. 401–426.

Kasagana, V. N., & Karumuri, S. S. (2011). Conservation of medicinal plants (Past, Present & Future Trends) *J. Pharm. Sci. & Res. 8*, 1378–1386.

Khan, M. H., Aliabbas, S., Kumar, V., & Rajkumar, S. (2009). Recent advances in medicinal plant biotechnology. *Indian J. Plant Biotech. 8*, 9–22.

Koay, Y. C., & Faheem Amir, F. (2013). A review of the secondary metabolites and biological activities of *Tinospora crispa* (Menispermaceae) *Trop. J. Pharmaceut. Res. 12(4)*, 641–649.

Kumar, P. U., Adhikari, P., Pereira, P., & Bhat, P. (1999). Safety and efficacy of Hartone in stable angina pectoris – an open comparative trial. *J. Assoc. Physicians India. 47*(7), 685–689.

Kunle, O. F., Egharevba, Omoregie, H., Ahmadu, & Peter, O. (2012). Standardization of herbal medicines – A review. *Intern. J. Biodivers. Conserv. 4(3)*, 101–112.

Lee, S. (2015). Systems biology – A pivotal research methodology for understanding the mechanisms of traditional medicine. *J. Pharmacopuncture. 18*(3), 11–18.

Leonti, M., & Casu, L. (2013). Traditional medicines and globalization: current and future perspectives in ethnopharmacology. *Front Pharmacol. 4*, 92.

Leung, P. C. (2004). Textbook of Clinical Trials. In: Machin, D., Day, S., & Green, S. (eds.) John Wiley and Sons: Chi Chester, pp. 63–84.

Li, W. (2002). Botanical Drugs: The Next New Thing? http://nrs.harvard.edu/urn-3:HUL. InstRepos: 8965577.

Malik, V. (ed.) (2013) Law Relating to Drugs and Cosmetics. 23rd edn., Lucknow: Eastern Book Company.

Manach, C., Hubert, J., Llorach, R., & Scalbert, A. (2009). The complex links between dietary phytochemicals and human health deciphered by metabolomics. *Mol. Nutr. Food Res.* *53*, 1303–1315.

Mann, M., & Jensen, O. N. (2003). Proteomic analysis of post-translational modifications. *Nature Biotech. 21*, 255–261.

Matsuda, F., Yonekura-Sakakibara, K., Niida, R., Kuromori, T., Shinozaki, K., & Saito, K. (2009). MS/MS spectral tag-based annotation of non-targeted profile of plant secondary metabolites. *Plant J., 57,* 555–577.

Mendoza & Maria, C. (2010). "HIM and the path to personalized medicine: Opportunities in the Post-Genomic Era." *J. AHIMA 81* (11), 38–42.

Moco, S., Schneider, B., & Vervoort, J. (2009). Plant micrometabolomics: the analysis of endogenous metabolites present in a plant cell or tissue. *J. Proteome Res. 8,* 1694–1703.

Moreira, D. L., Teixeira, S. S., Monteiro, M. H. D., Ana Cecilia A. X., De-Oliveira., A. C. A. X., & Paumgartten, F. J. R. (2014). Traditional use and safety of herbal medicines. *Rev. Bras. Farmacogn. Curitiba. 24(2)*, 248–257.

Mukherjee, P. K., Nema, N. K., Bhadra, S., Mukherjee, D., Braga, F. C., & Matsabisa, M. G. (2014). Immunomodulatory leads from medicinal plants. *Indian J. Tradit. Knowl. 13,* 235–256.

Nagata, J. M., Jew, A. R., Kimeu, J. M., Salmen, C. R., Bukusi, E. A., & Cohen, C. R. (2011). Medical pluralism on Mfangano Island: use of medicinal plants among persons living with HIV/AIDS in Suba District, Kenya. *J. Ethnopharmacol. 135,* 501–509.

Nasri, H., & Shirzad, H. (2013). Toxicity and safety of medicinal plants. *J Herb Med Plarmacol. 2(2),* 21–22.

Nongdam, P., & Chongtham, N. (2011). *In vitro* rapid propagation of *Cymbidium aloifolium* (L.) Sw.: a medicinally important orchid via seed culture. *J. Biol. Sci. 11(3),* 254–260.

Oberg, A. L., Kennedy, R. B., Li, P., I. G., & Ovsyannikova, G. A. (2011). Poland systems biology approaches to new vaccine development. *Current Opinion in Immunology 23,* 436–443.

Ogunkunle, A. T. J., & Ladejobi, A. T. (2011). Ethnobotanical and phytochemical of five [5] species of *Senna* in Nigeria. *Afr. J. Biotechnol. 5*, 2020–2023.

Pandey & Pandey (2013). Usefulness of nanotechnology for Herbal medicines. *Plant Archives 13(2),* 617–621.

Pant, B., & Thapa, D. (2012). *In vitro* mass propagation of an epiphytic orchid *Dendrobium primulinum* Lindl. through shoot tip culture. *Afr. J. Biotechnol. 11(42),* 9970–9974.

Parveen, A., Parveen, B., Parveen, R., & Ahmad, S. (2015). Challenges and guidelines for clinical trial of herbal drugs. *J. Pharm. Bioallied Sci. 7(4),* 329–333.

Patel, K. N., Patel, J. K., Patel, M. P., Rajput, G. C., & Patel, H. A. (2010). Introduction to hyphenated techniques and their applications in pharmacy. *Pharm. Methods 1(1),* 2–13.

Patwardhan, B., Warude, D., Pushpangadan, P., & Bhatt, N. (2005). Ayurveda and traditional Chinese medicine: a comparative overview. *Evid. Based Complement. Alternat. Med. 2(4),* 465–473.

Popović, M. K., & Pörtner, R. (2012). Bioreactors and cultivation systems for cell and tissue culture. Encyclopedia of Life Support Systems (EOLLS). Eolss Publishers, Oxford.

Pradhan, S., Paudel, Y. P., & Pant, B. (2013). Efficient regeneration of plants from shoot tip explants of *Dendrobium densiflorum* Lindl., a medicinal orchid. *Afr. J. Biotechnol. 12(12)*, 1378–1383.

Ramesh Kumar, B. (2016). A review on metabolic engineering approaches for enrichment and production of new secondary metabolites in basella species. *World J. Pharmacy and Pharmaceutical Systems. 5(4)*, 652–671.

Reid, W. V., Laird, S. A., Gámez, R., Sittenfeld, A., Janzen, D. H., Gollin, M. A., & Juma, C. (1993). A new lease on life. In: Biodiversity Prospecting: Using Genetic Resources for Sustainable Development. World Resources Institute. pp. 1–52.

Sachan, A. K., & Gupta, A. (2015). A review on nanotized herbal drugs. *Int. J. Pharm. Sci. Res. 6(3)*, 961–70.

Sen, S., & Chakraborty, R. (2015). Toward the integration and advancement of herbal medicine: a focus on traditional Indian medicine. *Botanics: Targets and Therapy. 5*, 33–44.

Shak, R., & Mohamad, J. (2011). Guidelines on registration of traditional and health supplement products, Revised ed. Version 1.0. Kuala Lumpur. Malaysian Biotechnology Corporation SDN BH.

Sharma, V., & Sarkar, I. N. (2013). Bioinformatics opportunities for identification and study of medicinal plants. *Brief Bioinform. 14(2)*, 238–250.

Shinde V. M., Kamlesh Dhalwal, K., Manohar Potdar, M., & Mahadik, K. R. (2009). Application of quality control principles to herbal drugs. *Intern. J. Phytomed. 1*, 4–8.

Shyur, L. F., & Yang, N. S. (2008). Metabolomics for phytomedicine research and drug development. *Curr. Opin. Chem. Biol. 12*, 66–71.

Siahsar, B., Rahimi, M., Tavassoli, A., & Raissi, A. S. (2011). Application of biotechnology in production of medicinal plants. *American-Eurasian J. Agric. & Environ. Sci., 11 (3)*, 439–444.

Srivastava, S., & Mishra, N. (2009). Genetic markers – A cutting-edge technology in herbal drug research. *J. Chem. Pharmaceut. Res. 1(1)*, 1–18.

Stephanie L. Greene, S.l., Kisha, T. J., Yu, L. X., & Parra-Quijana, M. (2014). Conserving plants in gene banks and nature: Investigating Complementarity with *Trifolium thompsonii* Morton. *PLosOne, 9*(8), e105145.

Surh, Y. J. (2011). Reverse pharmacology applicable for botanical drug development – Inspiration from the legacy of traditional wisdom. *J. Tradit. Complement. Med. 1(1)*, 5–7.

Sushma, G., Debnath, S. C. S. K., & Chandu, A. N. (2011). Quality and regulatory affairs of herbal drugs: A world-wide Review. *IAJPR. 1(12)*, 389–396.

Tagore, S & Chowdhury, R. K. (2014). De Analyzing methods for path mining with applications in metabolomics. *Gene 534*, 125–138.

Therapeutic Goods Administration (2014). Australian Government Department of Health. North Sydney.

Turner, D. M. (1996). Natural product source material use in the pharmaceutical industry: the Glaxo experience. *J. Ethnopharmacol. 51*, 39–43.

Vickers, A., & Zollman, C. (1999). ABC of complementary medicine: Herbal medicine. *British Medical Journal* 319, 1050–1053.

Wang, X. (1991). Traditional herbal medicines around the globe: Modern perspectives. China: Philosophical basis and combining old and new. In: *Proceedings of the 10th General Assembly of WFPMM, Seoul, Korea* (pp. 16–18).

WHO (2005). *National Policy on Traditional Medicine and Regulation of Herbal Medicines.* Report of a World Health Organization Global Survey. Geneva, Switzerland.

Wilson, S. A., & Roberts, S. C. (2014). Metabolic engineering approaches for production of biochemicals in food and medicinal plants. *Current Opinion in Biotechnol. 26*, 174–182.

Witt, C. M., Liu, J. &Robinson N (2015). Combining omics and comparative effectiveness research decision making for Chinese medicine. *Science 347(6219)*, S50–S51.

Yacoub, K., Cibis, K., Risch, C., Surmann, E. M., Efferth, T., Kersten, C., Lenz, S., Wich, J., Antunes, A., Dreis, T., & Gartner, C. (2014). Biodiversity of Medicinal Plants. In *Biodiversity, Natural Products and Cancer Treatment* (pp. 1–32).

Yang, B., Liang, B. C. W., Pan, Y., Cai, S. F., Du, J. X., & Peng, D. X. (2014). Clinical observation of Yanshu injection combined with chemotherapy of advanced pancreatic carcinoma, *Liaoning. J. Traditional Chinese Medicine 41,* 1926–1927.

Zhang J., Chiodini R., Badr, A., & Zhang, G. (2011). The impact of next-generation sequencing on genomics. *J. Genet. Genomics 38,* 95–109.

Zhange, H., & Ge, Y. (2011). Comprehensive analysis of protein modifications by top-down mass spectrometry. *Circulation-Cardiovascular Genetics 4,* 711.

Zheng, W., Thorne, N., & McKew, J. C., (2013). Phenotypic screens as a renewed approach for drug discovery. *Drug Discov. Today* 18, 1067–1073.

Ziu, M., Zheng, M. W., Li, G., & Su, Z. G. (2013). Advanced systems biology methods in drug discovery and translational biomedicine. *Biomed Res Int.* Article ID 742835, 8 pages. http://dx.doi.org/10.1155/2013/742835.

CHAPTER 15

INDIAN ETHNOBOTANY: PRESENT STATUS AND FUTURE PROSPECT

CHOWDHURY HABIBUR RAHAMAN

Department of Botany, Visva-Bharati University, Santiniketan – 731235, West Bengal, India, E-mail: habibur_cr@rediffmail.com or habibur_cr@yahoo.co.in

CONTENTS

ABSTRACT

This chapter deals with discussion on present status and future prospect of Indian ethnobotanical research carried out over the last six decades or more. After reviewing a vast literature on ethnobotany in general and Indian ethnobotany in particular, it has been observed that the past ethnobotanical work in India is of descriptive and utilitarian type, simply documenting the use of plants. This classical type of documentation of Indian ethnobotanical works is continuing till date without statistical analysis of the recorded data by quantitative tools. A positive indication is noticed in present-day Indian ethnobotanical research where ethnobotanists are employing quantitative indices for analysis of their research works. There is a giant database on Indian ethnobotanical knowledge which needs a digitized presentation after analyzing its data with the help of suitable statistical indices. In reviewing the present status of Indian ethnobotanical research some lacunae have been identified there in which have been discussed in detail. Finally, a discussion on future prospects of Indian ethnobotany has been made along with some suggestions for future plan of research.

15.1   INTRODUCTION

From the very beginning of human civilization, a very close relationship between human beings and its immediate environment has been established and in multiple ways they have interacted with the environment for their livelihood, subsistence and survival. Over the time, human culture and tradition associated with its surrounding biotic components lay the foundation of ethnobiology which embodies the knowledge developed by any human society regarding the plant and animal world. Ethnobiology deals with the knowledge of ethnic communities regarding grouping of the plants and animals on the basis of their utility, and various uses of this biological resource for food, shelter, health care and recreation.

Studies in such traditional knowledge and related biological resources are currently being conducted not only by the anthropologists but also the researchers of other fields like botany, zoology, ecology, agronomy, social sciences, etc. The involvement of these researchers of various disciplines reflects the scholastic growth in the field of ethnobiology and its multidisciplinary characteristics that allow a broad spectrum of approaches leading to the emergence of various branches of ethnobiology, such as ethnobotany, ethnozoology, ethnoecology, ethnomedicine, and ethnopharmacology.

Like any other system of knowledge, the traditional practices of plant usages had been shaped into a specific system of knowledge through trial and error method and it is the basis of Ethnobotany, a much highlighted and important branch of ethnobiology. The term "Ethnobotany" was not familiar to us before Harshberger who first coined the term in 1895 (Harshberger, 1896). Ethnobotany can be defined as the total natural and traditional relationship and interactions between men, domesticated animals and their surrounding plant wealth (Jain, 1987). It is now considered as a multi-disciplinary subject which deals with the studies among traditional people for recording their unique knowledge about plant wealth and for search of new resources of herbal drugs, edible plants and other aspects of plants (Mudgal and Jain, 1983). Ethnobotany explores the relationship between people in a culture and plants in an environment.

Now-a-days ethnobotanical works are not only documenting the traditional herbal knowledge but also providing more scientific information about important plant species used by the people of a particular community or culture. Such ethnobotanical information does have immense potential in the field of bioprospecting to develop commercially valuable natural products as well as novel genes, and also for management and conservation of plant resources. For this reason, there is a worldwide resurgence of interest in ethnobotanical as well as ethnomedicinal studies now, to meet the growing needs of agro-industries, herbal drug industries, conservation and development of plant genetic resources. Documentation of this valuable herbal knowledge through ethnobotanical studies is being carried out from almost all regions of the world.

Ethnobotanical study was initiated by the European scientists long ago and flourishes gradually with the interest of a lot of field botanist, taxonomist, herbarium curator, botanist and many more. Subsequently the documentation of ethnobotanical knowledge has steadily been carried forward by the researchers of the other parts of the world. In India, ethnobotanical work had been initiated more than 60 years ago and in later years, the work has been stream lined by organized efforts of the Indian workers covering a wide range of areas of ethnobotanical research. In the early phase of ethnobotanical research, it was only restricted in documentation of various usages of plant resources, but in recent past ethnobotanists throughout the world have focused their efforts on management and conservation of regional phytoresources along with various utilitarian aspects of the local resources.

Starting from its very beginning ethnobotanical research has been gradually going through metamorphosis from its subjectivity to objectivity. It was noticed that the earlier ethnobotanical works mostly embodied the documentation of the plant knowledge, were largely subjective type and data compilations there in such works were very traditional one, without employing the suitable statistical indices for more objectivity in research. To get more objectivity in their ethnobiological research, scientists are now employing various suitable statistical indices which help to identify more important plant species used by the people in the traditional communities. The quantitative analyses done in ethnobotanical research also highlight the exact pattern of utilization of resources which gives clear indication towards management planning of local phytoresources and thus plays a very important role in conservation prioritization of plant species in a region which are identified as rare and most acceptable species in a culture.

Since the 1990s, a number of quantitative techniques have been proposed for ethnobotanical data analysis, and till date many authors have adopted them in their research. Medeiros et al. (2011) reviewed 87 quantitative techniques and explained them under three major groups previously cited by Phillips (1996), i.e., (1) informant consensus, (2) subjective allocation, and (3) aggregation of uses. In a recent study, Mathur and Sundaramoorthy (2013) have mentioned 120 different indices and categorized them into several groups like consensus methods, use value methods, ethno-medicine methods, relative importance methods, equitability methods, methods related to food, and ecological methods for analyzing the ethnobotanical data.

Realizing merit of the quantitative techniques, ethnobotanists all over the world are now employing quantitative indices in their ethnobotanical research specially in the field of bioprospecting and conservation of local phytoresources. Some of the quantitative indices which are most commonly used in ethnobotanical research are informant consensus factor or Fic (Heinrich et al., 1998), fidelity level or FL (Friedman et al., 1986), use value (Rossato et al., 1999), relative frequency of citation or RFC (Shah et al., 2015), relative importance index or RI (Albuquerque et al., 2007a,b), cultural agreement index or CAI (Bruschi et al., 2011), conservation priority index or CPI (Martinez et al., 2006) and local conservation priority index or LCPI (Oliveira et al., 2007).

In this global scenario of ethnobotanical research, an attempt has been made here in present chapter to review the panorama of Indian ethnobotanical research by analyzing its past trends, present status and future prospects.

15.2   INDIAN ETHNOBOTANY: ITS PAST AND PRESENT

Through perusal of relevant literature from India it has been found that a giant volume of ethnobotanical data has been documented from different parts of the country covering various aspects in ethnobotanical research. In recent years, it has also been observed that there is a regular addition of a number of publications to the existing voluminous database on ethnobotany in India. The publication on Indian ethnobotany includes the research papers, short reports, books, popular articles, technical reports, etc. Here, the research work on Indian ethnobotany has been discussed briefly under two broad headings – past work and present status of the work.

15.2.1 PAST WORK

In India, though the usage of plants and also the records of such usage are very ancient, but under the title of 'Ethnobotany' such sort of work had been initiated by Dr. E. K. Janaki Ammal in 1954. She studied subsistence food plants of certain tribes of southern India (Janaki Ammal, 1955–56). With this initiation, Dr. S. K. Jain and other contemporary eminent ethnobotanists stream lined the ethnobotanical studies in India. In subsequent years, organized studies in different disciplines of ethnobotany have been carried forward by a number of Indian ethnobotanists documenting the traditional plant knowledge from different parts of the country. Research on Indian ethnobotany in its early stage was very much of descriptive and utilitarian type, documenting the use of plants through lists of the species. Till date, Indian research activity is mostly restricted within the boundary of classical phase of ethnobotanical research – simply documenting the research findings with traditional type of data compilation. S. K. Jain and his associates carried out organized ethnobotanical studies among the tribes of central India (Jain, 1963, 1965, 1981). A large number of publications have been made from India covering various lines of research such as ethnobotany of specific tribes, of certain regions, of particular plant groups or diseases and on many other miscellaneous sub- or interdisciplinary approaches without employing suitable quantitative indices (Jain and Borthakur, 1980; Jain, 1987, 2004; Jain and Patole, 2001; Singh and Pandey, 1998; Rao and Pullaiah, 2007; Katewa et al., 2001; Patil and Patil, 2006; Sajem and Gosai, 2006; Rahaman and Pradhan, 2011; Shrivastava, 2015).

India is a land of culture, tradition and ethnic diversity. The ethnic communities in India have their own unique customs and knowledge systems. In India, studies in ethnomedicine of a particular tribe or a group of tribes had been initiated long ago, even much earlier than the date of initiation of Indian ethnobotanical studies. Bodding (1925, 1927) is known as pioneer in systematic study of tribal medicine in India. He carried out an extensive monographic work on Santhal medicine and connected folklore and published it in two volumes. Later on, Bressers (1951) documented the ethnomedicinal knowledge of various ethnic groups from Ranchi district of then Bihar state. In recent past, a detailed scientific study on tribal medicine of four major ethnic groups of India (Lodha, Munda, Oraon and Santhal) has been carried out by two eminent ethnobotanists of India, Dr. S. K. Jain and Dr. D. C. Pal, documenting 2000 prescriptions of tribal medicine including introductory notes on basic characteristic features of tribal medicine, its pharmacology, tribal concept of disease diagnosis, treatment, magico-religious belief, taboos, etc. (Pal and Jain, 1998). Another gigantic database on ethnomedicine of 550 ethnic communities all over India has been developed in recent time with financial assistance from Ministry of Environment, Forests and Climate Change, New Delhi which embodies the data of 1,75,000 drug preparations made from about 7500 plant species (Anonymous, 1992–1998). Indian ethnobotanists have successfully documented the ethnobotanical as well as ethnomedicinal knowledge of different tribal communities inhabited in India. Extensive works on ethnobotany of various tribes of India have been done by many workers (Jain, 1991, 1997; Maheshwari, 2000; Trivedi and Sharma, 2004). Maheshwari and Singh (1984) have carried out their investigation on *Bhoxa* tribe of Bijnor and Pauri Garhwal districts of the then Uttar Pradesh. Ethnobotanical knowledge of *Shompens*, a primitive tribe of Great Nicobar Island has been studied by Elanchezhian et al. (2007). Mairh et al. (2010) studied the ethnomedicinal uses of local phytoresources by the *Birhore* tribe of Jharkhand. Phytotherapeutic knowledge of *Dimasa* tribe of Barak valley, Assam has recently been documented by Nath et al. in 2011. Bosco and Arumugam (2012) have reported 35 plant species used as ethnomedicine by the *Irular* tribes live in Red Hills of Tamil Nadu.

On the basis of richness of phyto-resources and concentration of tribal community, ethnobotanical studies have been carried out targeting a specific region of India (Jain and Dam, 1979; Ignacimuthu et al., 2006; Singh and Singh, 2009; Sen et al., 2011). North-eastern region of India can be considered as an ethnobotanical hotspot in respect of its rich biodiversity and varied diversity in ethnic cultures. From this region nearly 1400 ethnomedicinal uses have been recorded a decade ago (Dutta and Dutta, 2005) and still search for new ethnomedicinal uses is being pursued (Purkayastha et al., 2005; Mao et al., 2009; Sen et al., 2011; Ningthoujam et al., 2013; Kichu et al., 2015).

The northern part of India harbors a great diversity of medicinal plants because certain regions of it fall under the splendid Himalayan range. More than 8000 species of angiosperms, 44 species of gymnosperms and 600 species of pteridophytes have been reported from the north Indian part of Himalaya so far studied, of these 1748 species are known to have medicinal value (Kala et al., 2006). A large number of research articles on ethnobotany have so far been published from the states of Uttar Pradesh and Uttarakhand situated in northern India (Alam and Anis. 1987; Bhatt and Negi, 2006; Singh and Narain, 2009; Mathur and Joshi, 2013)

In the northwestern part of India, a number of research works have been carried out and many articles have been published in the recognized journals documenting the ethnobotanical as well as ethnomedicinal uses of local plant resources (Bhattacharyya, 1991; Prakash and Aggarwal, 2010; Sharma and Devi, 2013).

Research activities in various areas of ethnobotany have extensively been carried out in the states of Rajasthan, Gujarat and Maharashtra of western India (Bedi, 1978; Sharma and Kumar, 2007; Punjani, 2002). Books on ethnobotany of particular districts from Maharashtra state have been published documenting the importance of ethno-flora of those districts (Patil and Patil, 2006; Pawar and Patil, 2008). A voluminous documentation on ethnobotany of Rajasthan has been made by Singh and Pandey (1998) covering all the major areas of ethnobotanical studies. From the Union Territories of India like Dadra, Nagar Haveli and Daman, ethnobotanical information on 305 plant species have been documented along with their field collection number, chemical constituents, distribution and phenology (Sharma and Singh, 2001).

In the earlier phase of ethnobotanical research in India, S. K. Jain and his associates have done extensive survey in the tribal areas of Central India. Later on, plenty of ethnobotanical studies have been conducted in different districts of Madhya Pradesh (Sahu, 1982; Saxena, 1988; Jain and Sahu, 1993; Rai et al., 2004; Mahajan, 2007; Shukla et al., 2010; Dahare and Jain, 2010; Jain et al., 2010; Rai, 2012; Singh and Upadhyay, 2012; Wagh and Jain, 2014). On the other hand quite a lesser number of research articles have been published from the state of Chhattisgarh as it has been splitted from Madhya Pradesh in November, 2000 (Tirkey. 2006; Kala, 2009; Ekka, 2011; Pandey et al., 2015; Sinha et al., 2016).

In eastern India, ethnobotanical studies were initiated by the researchers like Saxena and Dutta (1975), Mudgal and Pal (1980) and many others.

From all the states of this region of India research articles are regularly being published by the researchers documenting various traditional uses of plant resources (Sadangi et al., 2005; Pradhan and Badola, 2008; Mallik et al., 2012; Pattanaik and Mohapatra, 2010; Prusti and Behera, 2007; Rout and Panda, 2010; Panda et al., 2011, Mondal and Rahaman, 2012).

In southern India, the Eastern Ghats and Western Ghats regions are very rich in plant resources and many oldest ethnic communities like *Kani, Irular, Malasar*, etc. have their traditional relation with plant resources of these regions. To explore the valuable botanical knowledge organized studies have been undertaken in different locations of this region. As for example, Samya et al. (2008) have documented 72 medicinal plants from southern part of Tamil Nadu, which are used for the treatment of snakebite. Ayyanar and Ignacimuthu (2009) have identified 46 ethnomedicinal species from Kani traditional healers which are used to treat wounds and related injuries. In 2015, Mathew et al. (2015) published a book on the ethnobotanical knowledge of *Paliya* tribe of Kerala.

Bhargava (1983) documented ethnomedicinal knowledge of the native people of Andaman and Nicobar Islands, one of the world's biodiversity hotspots. Recently from this region scientists from ICMR have identified a total of 77 traditional knowledge practitioners who together have been using 132 medicinal plant species to treat 43 ailments (Punnam Chander et al., 2015).

In ethnobotany, a specific group of plants or a particular plant species has also a view point to highlight its importance. Many research works have already been done on various plant groups. As for example, ethnobotanical importance of Zingiberaceae from northeastern India has been evaluated by Basak (2010). Whereas, ethnobotanical uses of ginger in Manipur has been documented by Daimei and Kumar (2014). Ethnomedicinal importance of the genus *Momordica* L. reported from southern Western Ghats (Joseph and Antony, 2008). Ethnomedicinal species belonging to the families Euphorbiaceae (Beg, 2015) and Rubiaceae (Singh and Ali, 2012) have also been studied. Shah (2004) made a detail ethnobotanical survey regarding the uses of *Cannabis sativa* L. (Hemp) in Uttaranchal.

Ethnomedicinal knowledge attached to a specific disease or ailment or particular disease categories has also been targeted by a lot of scientists to identify the important ethnomedicinal plants which can further be exploited for development of effective drugs (Sarkhel, 2014; Devi et al., 2016; Thirumalai et al., 2012; Siddiqui et al., 1989; Dey and De, 2012; Sharma et al., 2012, 2013; Chinsembu, 2016).

Ethnobotanists from different states of India have also shown interest in identifying the important plant parts used by different ethnic community for preparation of ethnomedicine. Swarnkar and Katewa (2008) reported the uses of tuberous part of the plants from different tribal areas of Rajasthan. Ethnobotany of other plant parts like seeds and barks also documented by the Indian scientists (Patel and Parekh, 2013; Kumar et al., 2013).

15.2.2 PRESENT STATUS

Starting from the beginning ethnobotanical research in India was very much of subjective type and till date same trend is continuing in Indian ethnobotanical research. Now, a huge volume of data on ethnobotanical knowledge has been accumulated from different parts of India and the volume of such vast database is gradually being increased by addition of a large number of ethnobotanical documentations, each year. These vast data on Indian ethnobotany are embodied in some technical reports of different projects, in many books, as chapters or research articles in the edited books, etc., very scattered. Many of such literature mentioned at the beginning of this chapter. As for example, a few of such literature are mentioned here. The technical report of the All India Co-ordinated Research Project on Ethnobiology (AICRPE) highlighted ethnobotanical knowledge on 10,000 plant species. Out of which, nearly 7500 medicinal species are used in preparation of 1,75,000 traditional formulations which are used among different ethnic communities of India (AICRPE, Technical Report, 1992–1998). Similarly, in a compiled work entitled *Dictionary of Indian Folk Medicine and Ethnobotany,* about 2500 plant species have been enumerated with 15,000 references of their folk uses (Jain, 1991). Through critical checking of the ethnobotanical literature it has been noticed that the Indian ethnobotanical works mostly embodied the simple documentations of plant knowledge which are largely of subjective type with very traditional type of data compilation, and they are devoid of statistical analysis employing suitable quantitative indices. So, objectives of such ethnobotanical works in India are not more focused and there is very faint or no indication of identifying culturally important or most dependable plant species to the ethnic communities studied. According to the objective of research, ethnobotanists throughout the world are now employing various suitable statistical indices which help to quantify the reliability or importance of the plant species used by the people in the traditional communities. Here in this point of quantitative analysis, most of the Indian works are lagged behind the contemporary global ethnobotanical research. But very recently, this scenario of Indian research has been changed. Some of the Indian ethnobotanists have been documenting their works with quantitative analysis employing statistical indices since last five years. But it is very lesser in percentage of the total work on ethnobotany documented so far from India. Several research articles have been published from different parts of the country analyzing the data with the help of suitable quantitative indices to add more objectivity to the research (Ayyanar and Ignacimuthu, 2011; Namsa et al., 2011; Kumar et al., 2012; Bhat et al., 2014; Mandal and Rahaman, 2014; Tarafdar et al., 2014; Shil et al., 2014; Rahaman and Karmakar, 2015; Shah et al., 2015; Francis et al., 2015). Quantitative indices like use value (UV), informant consensus factor (Fic), fidelity level (FL), relative frequency of citation (RFC) and few others have been used by the Indian ethnobotanists for analyzing their ethnobotanical data. It is a very good indication that Indian ethnobotany is putting its steps down on the global platform of contemporary ethnobotanical research.

After 1980s, ethnobiological research mainly focused and promoted the interaction of ethnobiology with bioprospecting and conservation of bioresources (Albuquerque et al., 2013). In India, there is a vast scope of finding out various promising bioresources through ethnobotanical research employing quantitative indices to measure the authenticity of traditional knowledge, to define the exact pattern of utilization of plant resources by the ethnic communities and to determine the conservation status of local flora. The quantitative approaches in Indian ethnobotanical research will change the dimension of its kind of research which further help in bioprospecting and conservation of its phytoresources. Except the case of bioprospection of *Trichopus zeylanicus* to a herbal formulation 'Jeevani' by the end of 1994 (Pushpangadan, 1994), from India, no other case of bioprospecting from the ethnomedicinal information has so far been noticed. But many examples are there in the field of bioprospecting worldwide where ethno-directed leads played an important role in discovery and development of novel products of medicinal and nutritional importance. For example, "Kolaviron", a natural product, has been patented for commercial exploitation of the fruits of the plant *Garcinia kola* which are traditionally chewed by the African tribes for liver protection (Iwu and Igboko, 1986). India is endowed with a huge volume of database on tribal or ethnomedicine which is an asset for bioprospecting of commercially viable natural products. The reasons behind such sort of poor attention of the scientists to the Indian eth-nomedicinal claims for their bioprospecting may be: the vast ethnomedici-nal data had not been statistically analyzed to identify the most important or reliable ethnomedicinal remedies or claims where success rate in drug discovery is greater than the claims or information taken from database other than statistically analyzed data source. The second probable reason is that Indian scientists are interested more in scientific validation of rem-edies of the Ayurveda, Unani and Siddha medicine systems, the codified and well-established Systems of Medicine in India. They have taken it more authentic than the information embodied in ethnomedicine of India which is not a codified system and has not been proved itself as an established system of medicine, so far.

Similar type of situation is also found here in the field of ethnoecology where no collaborative work has been noticed between ethnobotanists and ecologists in India. In ethnoecology, research findings are analyzed by specific statistical tools like Local Conservation Priority Index (LCPI) and Conservation Priority Index (CPI) which are uniquely devised by the scientists taking selective indices from ethnobotany and ecology. Throughout the globe a number of ethnoecological works have been car-ried out to prioritize the species for conservation by measuring their utili-tarian importance to a particular ethnic culture and distributional status of the species in a specific locality (Martinez et al., 2006; Lucena et al., 2013; Oliveira et al., 2007). But no such type of work from India has been executed employing the suitable statistical tools like LCPI and CPI to study the conservation status of the ethnobotanical phytoresources. Many ecological studies have been carried out employing the ecological tools only to investigate the phytosociology of plant communities in different forest types of India, but no ethnoecological quantitative approaches have been employed there in (Kala, 2000, 2005; Laloo et al., 2006; Sahu et al., 2007; Joshi, 2012; Pilania et al., 2015; Pradhan and Rahaman, 2015). In India, a lacuna in collaborative research between the ecologists and ethnobotanists is predominantly evident and this lack of collaboration among the Indian scientists emerged as a setback to carry out the research on assessing conservation priorities of ethnospecies at local as well as regional level. Researchers, scientists and ethnobiologists engaged in this field of research should adopt quantitative approaches for data analysis to add more objectivity in research and for proper scientific interpretation of the research findings.

15.3   INDIAN ETHNOBOTANY: FUTURE PROSPECTS

Analyzing the present scenario of Indian ethnobotany some lacunae have been identified which need immediate attention of the ethnobotanists so the standard of ethnobotanical research in India will be increased up to the extent of present day global ethnobotanical research. Some of such areas of attention in ethnobotanical research of the country have briefly been discussed here.

India does have a giant volume of database on ethnobotany and it remains scattered in different published and gray literature in the form of books, scientific articles, popular articles and technical reports. Such vast data should be taken together under a digitalized database after compilation and comparison of the data recorded from different states of India. During compilation, a comparison should be made between the information recorded for different ethnic communities, specific plant groups, and various use and disease categories. The statistical analysis of that documented data has to be done using standard software dedicated for such type of work. Similar sort of digitized inventories on ethnobotanical knowledge has to be prepared for each state of the country. Really it's a huge task and time taking also, but has to be done by any means. Government of India should take a serious initiative to carry out such a pilot project on digitalization and scientific inventorization of the huge database on ethnobotany documented from different parts of the country. For this, a 'Task Force' is to be formed taking the experts from the Ministry of Health & Family Welfare, Ministry of Culture, Ministry of Tribal Affairs and experts from the Universities including other academic institutions.

Another important area which is left unattended, is that, preparation of Digital Library (DL) on Indian ethnobotanical knowledge similar to that of Traditional Knowledge Digital Library (TKDL), a digitalized database on traditional knowledge embodied the information from Indian Traditional Systems of Medicine and Yoga, prepared by the Government of India. TKDL is a proprietary database made available to the patent offices for protection of Intellectual Property Rights (IPR) of the traditional knowledge of India and to avoid bio piracy and inappropriate patenting. A digitized database on Indian ethnobotanical knowledge named Ethnobotanical Knowledge Digital Library (EBKDL) has to be prepared which will involve translations of the formulations or use procedures involving ethnomedicinal as well as ethnobotanical plants recorded in various ethnobotanical literatures, in several international languages (English, Spanish, German, French, Japanese and Hindi). Once it is done properly then the IPR of the ethnic people of India will be protected and sharing of benefits come out of exploitation of their knowledge will be ensured.

Henceforth, documentation of all the ethnobotanical findings from India has to be made with statistical analysis employing suitable quantitative ethnobotanical indices. It has to be ensured by the Indian ethnobotanists that all their ethnobotanical documentation will now be made after proper statistical analysis to uplift the Indian ethnobotany at per the global standard of contemporary ethnobiological research. Once it is done meticulously then Indian work will be focused more to its objectives and then be interpreted with utmost scientific rationale. Thus the ethnobotanical work of India can gain confidence of the scientists those who are working in the areas of bioprospecting and conservation of plant resources. To ensure the success or hit rate higher in the field of bioprospecting, the utmost need is to compile and compare the statistically analyzed data and to make a very concise inventory of ethnobotanical knowledge on the basis of use categories, important plant species used by the ethnic community. Scientific validation of such important ethnobotanical leads or claims will then establish the ethnobotanical knowledge system on a solid foundation of authenticity. A sincere and consorted effort between ethnobotanists, folklorists, anthropologists, chemists and physicians has to be needed to make this huge task materialized.

During documentation and subsequently scientific validation of the ethnobotanical knowledge, scientists should be aware of the IPR of the knowledge holders and its protection. It has been observed that very often IPR of the traditional people is neglected in India as most of the Indian works are documented without taking any written PIC (Prior Informed Consent) of the knowledge holders or it is taken in oral form. This unscientific practice is very common among the Indian ethnobotanists and it goes against the Article 8(j) of CBD (Convention on Biological Diversity). PIC is one of the important safe-guards devised for protection of IPR of the knowledge providers. In the global scenario of Bioprospecting it has also been noticed that scientists or multinational companies under which projects they are working, are often ignoring the IPR of the knowledge holders violating the rules of ABS (Access and Benefit Sharing) of CBD. In this regard ethnobotanists all over the world including India should be sensitive enough in safe-guarding the IPR of the traditional knowledge providers and it ultimately accelerates the scientific growth of the ethnobiology and helps to avoid the biopiracy.

So far, two bibliographic works on Indian ethnobotany have been published which cover a narrow range of publication of its kind (Jain et al., 1984; Jain, 2002). An exhaustive bibliographic account covering a wide range of Indian ethnobotanical works has to be prepared highlighting the panorama of Indian ethnobotanical research. It will be helpful to the researchers in understanding the untouched areas of Indian ethnobotanical research and will show them proper direction towards the right goal of such type of research.

15.4 CONCLUSION

In concluding this discussion, it can be said that the future prospect of Indian Ethnobotany is very promising as a good number of enthusiast and sincere researchers are rightly carrying forward the legacy of research on ethnobotany of India. Many of those researchers have timely opted the quantitative ethnobotanical tools and they have been employing these tools regularly in analyzing and interpreting their respective research works. It is a very positive approach noticed recently among the Indian ethnobotanists which ultimately boosted them up in achieving the world standard in their research works. The scope of ethnomedico-botanical research in India is enormous as the country is very rich in its traditional knowledge and associated biological diversity. It is now a priority issue to the Indian scientists to document such knowledge resources after statistical analysis with the help of quantitative indices, for its proper utilization and bioprospecting.

The huge ethnobotanical database which has documented earlier from different parts of India may be a vital source of ethnobotanical leads or information for their bioprospecting only after its proper statistical analysis. Initiative from the Government of India and concern scientists should be taken for digitization and statistical analysis of such vast data, otherwise this valuable database would remain useless. No further scientific exploitation can be done with such statistically not-analyzed data and it will be a great loss for Indian ethnobotany, ethnic people and ultimately for Indian Government.

Finally, it is understood that in doing so mammoth task for the sake of Indian ethnobotany, a multidisciplinary research effort has to be developed keeping active collaboration between botanists, ethnobotanists, anthropologists, folklorists, phytochemists, pharmacologists, medical doctors, policy maker and development officers.

KEYWORDS

- **Indian ethnobotany**
- **present status**
- **future prospects**
- **multidisciplinary research**

REFERENCES

Alam, M. M., & Anis, M. (1987). Ethno-medicinal uses of plants growing in the Bulandshahr district of northern India. *J. Ethnopharmacol., 19*(1), 85–88.

Albuquerque, U. P., Medeiros, P. M., Almeida, A. L. S., Monteiro, J. M., Lins Neto, E. M. F., Melo, J. G., & Santos, J. P. (2007). Medicinal plants of the caatinga (semi-arid) vegetation of NE Brazil: a quantitative approach. *J. Ethnopharmacol., 114*(3), 325–354.

Albuquerque, U. P., Silva, J. S., Campos, J. L. A., Sousa, R. S., Silva, T. C., & Alves, R. R. N. (2013). The current status of ethnobiological research in Latin America: gaps and perspectives. *J. Ethnobiol. Ethnomed., 9*, 72. doi: 10.1186/1746–4269–9-72.

Anonymous, 1992–1998. AICRPE (All India Coordinated Project on Ethnobiology). Final Technical Report, Ministry of Environment and Forests, Government of India, New Delhi.

Ayyanar, M., & Ignacimuthu, S. (2009). Herbal medicines for wound healing among tribal people in southern India: Ethnobotanical and scientific evidences. *Intern. J. Appl. Res. Nat. Prod., 2*(3), 29–42.

Ayyanar, M., & Ignacimuthu, S. (2011). Ethnobotanical survey of medicinal plants commonly used by Kani tribals in Tirunelveli hills of Western Ghats, India. *J. Ethnopharmacol., 134*, 851–864.

Basak, T. S., Sarma, G. C., & Rangan, L. (2010). Ethnomedical uses of Zingiberaceous plants of Northeastern India. *J. Ethnopharmacol., 132*, 286–296

Bedi, S. J. (1978). Ethnobotany of the Ratan Mahal Hills, Gujarat, India. *Economic Bot., 32*(3), 278–284.

Beg, M. J. (2015). Medicinal plants wealth of the family Euphorbiaceae in Azamgarh district (U. P.). *Indian J. Scientific Res., 11*(1), 149–152.

Bhargava, N. (1983). Ethnobotanical studies of the tribes of Andaman and Nicobar Islands, India. I. Onge. *Economic Bot., 37*(1), 110–119.

Bhat, P., Hegde, G. R., Hegde, G., & Mulgund, G. S. (2014). Ethnomedicinal plants to cure skin diseases – an account of the traditional knowledge in the coastal parts of central Western Ghats, Karnataka, India. *J. Ethnopharmacol. 151*(1), 493–502.

Bhatt, V. P., & Negi, G. C. S. (2006). Ethnomedicinal plant resources of Jaunsari tribe of Garhwal Himalaya, Uttaranchal. *Indian J. Trad. Knowl., 5*(3), 331–335.

Bhattacharyya, A. (1991). Ethnobotanical observations in the Ladakh region of northern Jammu and Kashmir state, India. *Economic Bot., 45*(3), 305–308.

Bodding, P. O. (1925). Studies in Santhal medicines and connected folklore: Part I, Santhals and disease. *Mem. Asiatic Soc. Bengal, 10*(1), 1–132.

Bodding, P. O. (1927). Santal medicine. The Book Trust, Calcutta.

Bosco, F. G., & Arumugam, A. (2012). Ethnobotany of Irular tribes in Red Hills, Tamil Nadu, India. *Asian Pacific J. Trop. Dis., 2*(2), S874–S877.

Bressers, J. (1951). The Botany of Ranchi district, Bihar. Catholic Press, Ranchi.

Bruschi, P., Morganti, M., Mancini, M., & Signorini, M. A. (2011). Traditional healers and laypeople: A qualitative and quantitative approach to local knowledge on medicinal plants in Muda (Mozambique). *J. Ethnopharmacol., 138,* 543–563.

Chinsembu, K. C. (2016). Ethnobotanical study of medicinal flora utilized by traditional healers in the management of sexually transmitted infections in Sesheke district, Western Province, Zambia. *Revista Brasileira de Farmacognosia, 26,* 268–274.

Dahare, D. K., & Jain, A. (2010). Ethnobotanical studies of plant resources of tahsil Multai, district Betul, Madhya Pradesh, India. *Ethnobotanical Leaflets, 14,* 694–705.

Daimei, P., & Kumar, Y. (2014). Ethnobotanical uses of gingers in Tamenglong district, Manipur, northeast India. *Genetic Resources and Crop Evolution, 61,* 273–285.

Devi, S., Kumar, D., & Kumar, M. (2016). Ethnobotanical values of antidiabetic plants of M. P. region, India. *J. Med. Plants Studies, 4*(3), 26–28.

Dey, A., & De, J. N. (2012). Ethnobotanical survey of Purulia district, West Bengal, India for medicinal plants used against gastrointestinal disorders. *J. Ethnopharmacol., 143*(1), 68–80.

Dutta, B. K., & Dutta, P. K. (2005). Potential of ethnobotanical studies in North East India: An overview. *Indian J. Trad. Knowl., 4*(1), 7–14.

Ekka, A. (2011). Folklore claims of some medicinal plants used by tribal community of Chhattisgarh, India. *Res. J. Biol., 1*(1), 16–20.

Elanchezhian, R., Kumar, R. S., Beena, S. J., & Suryanarayana, M. A. (2007). Ethnobotany of *Shompens* – a primitive tribe of Great Nicobar Island. *Indian J. Trad. Knowl., 6*(2), 342–345.

Francis, X. T., Kannan, M., & Auxilia, A. (2015). Observation on the traditional phytotherapy among the Malayali tribes in Eastern Ghats of Tamil Nadu, South India. *J. Ethnopharmacol., 165,* 198–214.

Friedman, J., Yaniv, Z., Dafni, A., & Palewitch, D. (1986). A preliminary classification of the healing potential of medicinal plants, based on a rational analysis of an ethnopharmacological field survey among Bedouins in the Negev desert, Israel. *J. Ethnopharmacol., 16,* 275–287.

Harshberger, J. W. (1896). The purpose of ethnobotany. *Bot. Gaz., 21,* 146–158.

Heinrich, M., Ankli, A., Frei, B., Weimann, C., & Sticher, O. (1998). Medicinal plants in Mexico: healers' consensus and cultural importance. *Social Science and Medicine, 47*(11), 1859–1871.

http://www.csir.res.in retrieved on 28.07.2016.

Ignacimuthu, S., Ayyanar, M., & Sivaraman, S. K. (2006). Ethnobotanical investigations among tribes in Madurai District of Tamil Nadu (India). *J. Ethnobiol. Ethnomed., 2,* 25. doi:10.1186/1746-4269-2-25

Iwu, M. M., & Igboko, A. O. (1986). The flavonoids of *Garcinia kola. J. Nat. Prod., 45,* 650–651.

Jain, A. K., & Patole, S. N. (2001). Less known medicinal values of plants among some tribal and rural communities of Pachmarhi forest (M.P). *Ethnobotany, 13,* 96–100.

Jain, A. K., Vairale, M., & Singh, R. (2010). Folklore claims on some medicinal plants used by Bheel tribe of Guna district, Madhya Pradesh. *Indian J. Trad. Knowl. 9(1),* 105–107.

Jain, P., & Sahu, T. R. (1993). An ethnobotanical study of Noradehi Sanctuary Park of Madhya Pradesh, India: Native plant remedies for scorpion sting and snake bite. *J. Econ. Taxon. Bot., 17*(2), 315–328.

Jain, S. K. (1963). Studies in Indian ethnobotany – less known uses of fifty common plants from the tribal areas of Madhya Pradesh. *Bull. Bot. Surv. India, 5,* 223–226.

Jain, S. K. (1965). Medicinal plant lore of the tribals of Bastar. *Economic Bot., 19*(3), 236–250.

Jain, S. K. (1981). Glimpses of Indian Ethnobotany. Oxford & IBH Publishing Co., New Delhi.

Jain, S. K. (1987). A Manual of Ethnobotany. Scientific Publishers, Jodhpur, India.

Jain, S. K. (1991). Dictionary of Indian Folk Medicine and Ethnobotany. Deep Publications, New Delhi.

Jain, S. K. (1997). Contribution to Indian Ethnobotany (3rd edn.). Scientific Publishers, Jodhpur, India.

Jain, S. K. (2002). Bibliography of Indian Ethnobotany. Scientific Publishers, Jodhpur, India.

Jain, S. K. (2004). Credibility of traditional knowledge- the criterion of multilocational and multiethnic use. *Indian J. Trad. Knowl., 3*(2), 137–153.

Jain, S. K., & Borthakur, S. K. (1980). Ethnobotany of the Mikirs of India. *Economic Bot., 34*(3), 264–272.

Jain, S. K., & Dam, N. (1979). Some ethnobotanical notes from northeastern India. *Economic Bot., 33*(1), 52–56.

Jain, S. K., Mudgal, V., Banerjee, D. K., Guha, A., & Pal, D. C. (1984). Bibliography of Ethnobotany. Botanical Survey of India, Calcutta.

Janaki Ammal, E. K. (1955–56). An introduction to the subsistence economy of India. In: William, L. T. (ed.), Man's role in changing the face of the earth. University of Chicago Press, Chicago, pp. 324–335.

Joseph, J. K., & Antony, V. T. (2008). Ethnobotanical investigations in the genus *Momordica* L. in the Southern Western Ghats of India. *Genetic Resource and Crop Evolution; 55,* 713–721.

Joshi, H. G. (2012). Vegetation structure, floristic composition and soil nutrient status in three sites of Tropical dry deciduous forest of West Bengal, India. *Indian J. Fundamental and Applied Life Sci., 2*(2), 355–364.

Kala, C. P. (2000). Status and conservation of rare and endangered medicinal plants in the Indian trans-Himalaya. *Biol. Conserv. 93*(3), 371–379.

Kala, C. P. (2005). Indigenous uses, population density, and conservation of threatened medicinal plants in protected areas of the Indian Himalayas. *Conservation Biol., 19*(2), 368–378.

Kala, C. P. (2009). Aboriginal uses and management of ethnobotanical species in deciduous forests of Chhattisgarh state in India. *J. Ethnobiol. Ethnomed., 5,* 20. doi: 10.1186/1746–4269-5-20

Kala, C. P., Dhyani, P. P., & Sajwan, B. S. (2006). Developing the medicinal plants sector in northern India: Challenges and opportunities. *J. Ethnobiol. Ethnomed., 2,* 32. doi: 10.1186/1746–4269-2-32

Katewa, S. S., Guria, B. D., & Jain, A. (2001). Ethnomedicinal and obnoxious grasses of Rajasthan, India. *J. Ethnopharmacol., 76*(3), 293–297.

Kichu, M., Malewska, T., Akter, K., Imchen, I., Harrington, D., Kohen, J., Vemulpad, S. R., & Jamie, J. F. (2015). An ethnobotanical study of medicinal plants of Chungtia village, Nagaland, India. *J. Ethnopharmacol., 166,* 5–17.

Kumar, A., Pandey, V. C., & Tewari, D. D. (2012). Documentation and determination of consensus about phytotherapeutic veterinary practices among the Tharu tribal community of Uttar Pradesh, India. *Tropical Animal Health and Production, 44,* 863–872.

Kumar, R. S., Venkateshwar, C., Samuel, G., & Rao, S. G. (2013). Ethnobotanical uses of some plant barks used by Gondu tribes of Seethagondi grampanchayath, Adilabad district, Andhra Pradesh, India. *J. Nat. Prod. Plant Res., 3*(5), 13–17.

Laloo, R. C., Kharlukhi, L., Jeeva, S., & Mishra, B. P. (2006). Status of medicinal plants in the disturbed and undisturbed sacred forest of Meghalaya. *Current Sci., 90*(2), 225–232.

Lucena, R. F. P., Lucena, C. M., Araújo, E. L., Alves, Â. G. C., & Albuquerque, U. P. (2013). Conservation priorities of useful plants from different techniques of collection and analysis of ethnobotanical data. *Ann. Brazilian Acad. Sci., 85*(1), 169–186.

Mahajan, S. K. (2007). Traditional herbal remedies among the tribes of Bijagarh of West Nimar district, Madhya Pradesh. *Indian J. Trad. Knowl., 6*(2), 375–377.

Maheshwari, J. K. (2000). Ethnobotany and Medicinal Plants of Indian Sub-continent. Scientific Publishers, Jodhpur, India.

Maheshwari, J. K., & Singh, J. P. (1984). Contribution to the ethnobotany of Bhoxa tribe of Bijnor and Pauri Garhwal districts, U. P. *J. Econ. Taxon. Bot. 5*, 2251–2259.

Mairh, A. K., Mishra, P. K., Kumar, J., & Mairh, A. (2010). Traditional botanical wisdom of Birhore tribes of Jharkhand. *Indian J. Trad. Knowl., 9*(3), 467–470.

Mallik, B. K., Panda, T., & Padhy, R. N. (2012). Traditional herbal practices by the ethnic people of Kalahandi district of Odisha, India. *Asian Pacific J. Trop. Biomed., 2*(2), S988–S994.

Mandal, S. K., & Rahaman, C. H. (2014). Determination of informants' consensus and documentation of ethnoveterinary practices from Birbhum district of West Bengal, India. *Indian J. Trad. Knowl., 13*(4), 742–751.

Mao, A.A, Hynniewta, T. M., & Sanjappa, M. (2009). Plant wealth of Northeast India with reference to ethnobotany. *Indian J. Trad. Knowl., 8*(1), 96–103.

Martinez, G. J., Planchuelo, A. M., Fuentes, E., & Ojeda, M. (2006). A numeric index to establish conservation priorities for medicinal plants in the Paravachasca valley, Cordoba, Argentina. *Biodiversity Conservation, 15*, 2457–2475.

Mathew, A., Philip, A. T., & Mathew, B. (2015). Ethnobotany of Paliya tribe in Idukki district of Kerala. LAP Lambert Academic Publishing, Germany.

Mathur, A., & Joshi, H. (2013). Ethnobotanical studies of the Tarai region of Kumaun, Uttarakhand, India. *Ethnobot. Res. & Appl., 11*, 175–203.

Mathur, M., & Sundaramoorthy, S. (2013). Census of approaches used in quantitative ethnobotany. *Studies on Ethno-Medicine, 7*(1), 31–58.

Medeiros, M. F. T., Silva, P. S., & Albuquerque, U. P. (2011). Quantification in ethnobotanical research: an overview of indices used from 1995 to 2009. *Sientibussérie Ciencias Biológicas, 11*(2), 211–230.

Mondal, S., & Rahaman, C. H. (2012). Medicinal plants used by the tribal people of Birbhum district of West Bengal and Dumka district of Jharkhand in India. *Indian J. Trad. Knowl., 11*(4), 674–679.

Mudgal, V., & Jain, S. K. (1983). Ethnobotany in India. Botanical Survey of India, Culcutta.

Mudgal, V., & Pal, D. C. (1980). Medicinal Plants used by tribals of Mayurbhanj (Orissa). *Bull. Bot. Surv. India, 22*, 59–62.

Namsa, N. D., Mandal, M., Tangjang, S., & Mandal, S. C. (2011). Ethnobotany of the Monpa ethnic group at Arunachal Pradesh, India. *J. Ethnobiol. Ethnomed., 7*, 1–14.

Nath, M., Dutta, B. K., & Hajra, P. K. (2011). Medicinal plants used in major diseases by Dimasa tribe of Barak valley. *Assam University J. Sci. & Tech.: Biol. Environ. Sci., 7*(1), 18–26.

Ningthoujam, S. S., Das Talukdar, A., Potsangbam, K. S., & Choudhury, M. D. (2013). Traditional uses of herbal vapor therapy in Manipur, North East India: an ethnobotanical survey. *J. Ethnopharmacol., 147*(1), 136–147.

Oliveira, R. L. C., Lins Neto, E. M. F., Araújo, E. L., & Albuquerque, U. P. (2007). Conservation priorities and population structure of woody medicinal plants in an area of Caatinga vegetation (Pernambuco state, NE Brazil). *Environmental Monitoring and Assessment, 132,* 189–206.

Pal, D. C., & Jain, S. K. (1989). Notes on Lodha medicine in Midnapur district, West Bengal, India. *Economic Bot. 43*(4), 464–470.

Panda, S. K., Rout, S. D., Mishra, N., & Panda, T. (2011). Phytotherapy and traditional knowledge of tribal communities of Mayurbhanj district, Orissa, India. *J. Pharmacogn. Phytother., 3*(7), 101–113.

Pandey, B., Pandey, P., & Paikara, D. (2015). Some important medicinal plants used by tribal people of Chhattisgarh. *Indian J. Life Sci., 5*(1), 67–69.

Patel, P. K., & Parekh, P. P. (2013). Therapeutic uses of some seeds among the tribals of Banaskantha district, Gujarat, India. *Romanian J. Biol.-Plant Biol., 58*(1), 79–82.

Patil, D. A., & Patil, M. V. (2006). Ethnobotany of Nasik District, Maharashtra. Daya Publishing House, Delhi.

Pattanaik, D. K., & Mohapatra, P. (2010). Ethnomedicinal plants used by the Paroja tribe of Koraput. *Ancient Sci. Life, 30*(2), 42–46.

Pawar, S., & Patil, D. A. (2008). Ethnobotany of Jalgaon District, Maharashtra. Daya Publishing House, Delhi.

Phillips, O. (1996). Some quantitative methods for analyzing ethnobotanical knowledge. In: Alexiades, M. (ed.), Selected Guidelines for Ethnobotanical Research: A field manual (Advances in Economic Botany. Vol. 10), The New York Botanical Garden, New York, pp. 171–197.

Pilania, P. K., Gujar, R. V. Joshi, P. M., Shrivastav, S. C., & Panchal, N. S. (2015). Phytosociological and ethanobotanical study of trees in a tropical dry deciduous forest in Panchmahal district of Gujarat, western India. *Indian Forester 141*(4), 422–427.

Pradhan, B., & Rahaman, C. H. (2015). Phytosociological study of plant species in three tropical dry deciduous forests of Birbhum district, West Bengal, India. *J. Biodiv. Environ. Sci., 7*(2), 22–31.

Pradhan, B. K., & Badola, H. K. (2008). Ethnomedicinal plant use by Lepcha tribe of Dzongu valley, bordering Khangchendzonga Biosphere Reserve, in north Sikkim, India. *J. Ethnobiol. Ethnomed., 4,* 22. doi: 10.1186/1746-4269-4-22.

Prakash, V., & Aggarwal, A. (2010). Traditional uses of ethnomedicinal plants of lower foothills of Himachal Pradesh-I. *Indian J. Trad. Knowl., 9*(3), 519–521.

Prusti, A. B., & Behera, K. K. (2007). Ethno-Medico botanical study of Sundargarh district, Orissa, India. *Ethnobotanical Leaflets, 11,* 148–163.

Punjani, B. L. (2002). Ethnobotanical aspects of some plants of Aravalli hills in north Gujarat. *Ancient Sci. Life, 21*(4), 268–280.

Punnam Chander, M., Kartick, C., & Vijayachari, P. (2015). Herbal medicine and healthcare practices among Nicobarese of Nancowry group of Islands – an indigenous tribe of Andaman and Nicobar Islands. *Indian J. Med. Res., 141*(5), 720–744.

Purkayastha, J., Nath, S. C., & Islam, M. (2005). Ethnobotany of medicinal plants from Dibru-Saikhowa biosphere reserve of northeast India. Fitoterapia, 76(1):121–127.

Pushpangadan, P. (1994). Ethnobiology in India (A status report). Ministry of Enviroment and Forest, Govt. of India, New Delhi.

Rahaman, C. H., & Karmakar, S. (2015). Ethnomedicine of Santal tribe living around Susunia hill of Bankura district, West Bengal, India: The quantitative approach. *J. Applied & Pharmaceut. Sci., 5*(2), 127–136.

Rahaman, C. H., & Pradhan, B. (2011). A study on the ethnomedicinal uses of plants by the tribal people of Birbhum district, West Bengal. *J. Econ. Taxon. Bot.*, 35(3), 529–534.

Rai, R. (2012). Ethnobotanical studies on Korku tribes of Madhya Pradesh. *ENVIS Forestry Bulletin, 12*(2), 86–93.

Rai, R., Nath, V., & Shukla, P. K. (2004). Ethnobotanical studies in Patalkot valley in Chhindwara district of Madhya Pradesh. *J. Trop. Forestry, 20*, 38–50.

Rao, D. M., & Pullaiah, T. (2007). Ethnobotanical studies on some rare and endemic floristic elements of Eastern Ghats-hill ranges of south-east Asia, India. *Ethnobotanical Leaflets, 11*, 52–70.

Rossato, S. C., Leitão Filho, H., & Begossi, A. (1999). Ethnobotany of Caiçaras of the Atlantic forest coast (Brazil). *Economic Bot., 53*, 387–395.

Rout, S. D., & Panda, S. K. (2010). Ethnomedicinal plant resources of Mayurbhanj district, Orissa. *Indian J. Trad. Knowl., 9*(1), 68–72.

Sadangi, N., Padhy, R. N., & Sahu, R. K. (2005). A contribution to medico- ethnobotany of Kalahandi district, Orissa on ear and mouth diseases. *Ancient Sci. Life, 24*(3), 160–163.

Sahu, S. C., Dhal, N. K., Sudhakar Reddy, C. Pattanaik, C., & Brahmam, M. (2007). Phytosociological study of tropical dry deciduous forest of Boudh district, Orissa, India. *Res. J. Forestry, 1*, 66–72.

Sahu, T. R. (1982). An ethnobotanical study of Madhya Pradesh 1: plants used against various disorders among tribal women. *Ancient Sci. Life, 1*(3), 178–181.

Sajem, A. L., & Gosai, K. (2006). Traditional use of medicinal plants by the Jaintia tribes in North Cachar Hills district of Assam, northeast India. *J. Ethnobiol. & Ethnomed., 2*, 33. doi: 10.1186/1746-4269-2-33

Samy, R. P., Thwin, M. M., Gopalakrishnakone, P., & Ignacimuthu, S. (2008). Ethnobotanical survey of folk plants for the treatment of snakebites in southern part of Tamil Nadu, India. *J. Ethnopharmacol., 115*(2), 302–312.

Sarkhel, S. (2014). Ethnobotanical survey of folklore plants used in treatment of snakebite in Paschim Medinipur district, West Bengal. *Asian Pacific J. Trop. Biomed., 4*(5), 416–420.

Saxena, H. O. (1988). Observations on the ethnobotany of Madhya Pradesh. *Bull. Bot. Surv. India, 28*, 149–156.

Saxena, H. O., & Dutta, P. K. (1975). Studies on the ethnobotany of Orissa. *Bull. Bot. Surv. India, 17*, 124–131.

Sen, S., Chakraborty, R., De, B., & Devanna, N. (2011). An ethnobotanical survey of medicinal plants used by ethnic people in West and South districts of Tripura, India. *J. Forestry Res., 22*(3), 417–426.

Shah, A., Bharati, K. A., Ahmad, J., & Sharma, M. P. (2015). New ethnomedicinal claims from Gujjar and Bakerwals tribes of Rajouri and Poonch districts of Jammu and Kashmir, India. *J. Ethnopharmacol., 166*, 119–28.

Shah, N. C. (2004). Indigenous uses and ethnobotany of *Cannabis sativa* L. (Hemp) in Uttaranchal (India). *J. Industrial Hemp, 9*(1), 69–77.

Sharma, P., & Devi, U. (2013). Ethnobotanical uses of biofencing plants in Himachal Pradesh, northwest Himalaya. *Pakistan J. Biol. Sci., 16*, 1957–1963.

Sharma, L. K., & Kumar, A. (2007). Traditional medicinal practices of Rajasthan. *Indian J. Trad. Knowl., 6*(3), 531–533.

Sharma, P. P., & Singh, N. P. (2001). Ethnobotany of Dadra, Nagar Haveli and Daman: Union Territories. Botanical Survey of India, Calcutta.

Sharma, J., Gairola, S., Gaura, R. D., & Painuli, R. M. (2012). The treatment of jaundice with medicinal plants in indigenous communities of the sub-Himalayan region of Uttarakhand, India. *J. Ethnopharmacol., 143*(1), 262–291.

Sharma, J., Gairola, S., Gaura, R. D., Painuli, R. M., & Siddiqi, T. O. (2013). Ethnomedicinal plants used for treating epilepsy by indigenous communities of sub-Himalayan region of Uttarakhand, India. *J. Ethnopharmacol., 150,* 353–370.

Shil, S., Dutta Choudhury, M., & Das, S. (2014). Indigenous knowledge of medicinal plants used by the *Reang* tribe of Tripura state of India. *J. Ethnopharmacol., 152*(1), 135–141.

Shrivastava, M. (2015). Important ethnomedicinal plants used by the Muria tribes of Bastar for the treatment of snake bite. *Indian J. Appl. Pure Biol., 30*(2), 165–168.

Shukla, A. N., Srivastava, S., & Rawat, A. K. S. (2010). An ethnobotanical study of medicinal plants of Rewa district, Madhya Pradesh. *Indian J. Trad. Knowl., 9*(1), 191–202.

Siddiqui, M. B., Alam, M. M., & Husain, W. (1989). Traditional treatment of skin diseases in Uttar Pradesh, India. *Economic Bot., 43*(4), 480–486.

Singh, A., & Singh, P. K. (2009). An ethnobotanical study of medicinal plants of Chandauli district of Uttar Pradesh, India. *J. Ethnopharmacol., 121,* 324–329.

Singh, P., & Ali, S. J. (2012). Ethnomedicinal plants of family Rubiaceae of Eastern UP. *Indian J. Life Sci., 1*(2), 83–86.

Singh, U., & Narain, S. (2009). Ethno-botanical wealth of Mirzapur district, UP. *Indian Forester, 135*(2), 185–197.

Singh, V., & Pandey, R. P. (1998). Ethnobotany of Rajasthan, India. Scientific Publishers, Jodhpur, India.

Singh, B. P., & Upadhyay, R. (2012). Ethnobotanical importance of Pteridophytes used by the tribe of Pachmarhi, central India. *J. Med. Plant Res., 6,* 14–18.

Sinha, M. K., Kanungo, V. K., & Naik, M. L. (2016). Ethnobotany in relation to livelihood security in district Bastar of Chhattisgarh state with special reference to non-timber forest produces. *Current Bot., 7,* 27–33.

Swarnkar, S., & Katewa, S. S. (2008). Ethnobotanical observation on tuberous plants from tribal area of Rajasthan (India). *Ethnobotanical Leaflets, 12,* 647–666.

Tarafdar, R. G., Nath, S., Das Talukdar, A., & Dutta Choudhury, M. (2014). Antidiabetic plants used among the ethnic communities of Unakoti district of Tripura, India. *J. Ethnopharmacol., 160,* 219–226.

Thirumalai, T., David, B. C., Sathiyaraj, K., Senthilkumar, B., & David, E. (2012). Ethnobotanical study of anti-diabetic medicinal plants used by the local people in Javadhu hills, Tamil Nadu, India. *Asian Pacific J. Trop. Biomed., 2*(2), S910–S913.

Tirkey, A. (2006). Some ethnomedicinal plants of family-Fabaceae of Chhattisgarh state. *Indian J. Trad. Knowl., 5*(4), 551–553.

Trivedi, P. C., & Sharma, N. K. (2004). Ethnomedicinal Plants. Pointer Publishers, Jaipur, India.

Wagh, V. V., & Jain, A. K. (2014). Herbal remedies used by the tribal people of Jhabua district, Madhya Pradesh for the treatment

INDEX

E

Printed and bound by CPI Group (UK) Ltd, Croydon, CR0 4YY

23/10/2024

01777704-0020